Intelligent Automation and Computer Engineering

T0191730

Lecture Notes in Electrical Engineering

Sio-Iong Ao • Oscar Castillo • Xu Huang
Editors

Intelligent Automation and Computer Engineering

 Springer

Editors
Sio-Iong Ao
International Association of Engineers
37–39 Hung To Road
Unit 1, 1/F
Hong Kong SAR
publication@iaeng.org

Xu Huang
University of Canberra
School of Information
Sciences & Engineering
Canberra ACT 2601
Australia
xu.huang@canberra.edu.au
publication@iaeng.org

Oscar Castillo
Tijuana Institute of Technology
Dept. Computer Science
P.O. Box 4207
Chula Vista CA 91909
USA
ocastillo@hafsamx.org

ISSN 1876-1100   e-ISSN 1876-1119
ISBN 978-94-007-3224-7   ISBN 978-90-481-3517-2(eBook)
DOI 10.1007/978-90-481-3517-2
Springer Dordrecht Heidelberg London New York

Cover design: eStudio Calamar S.L.

Printed on acid-free paper

Springer is part of Springer Science+Business Media (www.springer.com)

Preface

A large international conference in Intelligent Automation and Computer Engineering was held in Hong Kong, March 18–20, 2009, under the auspices of the International MultiConference of Engineers and Computer Scientists (IMECS 2009). The IMECS is organized by the International Association of Engineers (IAENG). IAENG is a non-profit international association for the engineers and the computer scientists, which was founded in 1968 and has been undergoing rapid expansions in recent years. The IMECS conferences have served as excellent venues for the engineering community to meet with each other and to exchange ideas. Moreover, IMECS continues to strike a balance between theoretical and application development. The conference committees have been formed with over 200 and 50 members who are mainly research center heads, deans, department heads (chairs), professors, and research scientists from over thirty countries. The conference participants are also truly international with a high level of representation from many countries. The responses for the conference have been excellent. In 2009, we received more than one thousand and one hundred manuscripts, and after a thorough peer review process 56.19% of the papers were accepted.

This volume contains 37 revised and extended research articles written by prominent researchers participating in the conference. Topics covered include artificial intelligence, decision supporting systems, automated planning, automation systems, control engineering, systems identification, modelling and simulation, communication systems, signal processing, and industrial applications. The book offers the state of the art of tremendous advances in intelligent automation and computer engineering and also serves as an excellent reference text for researchers and graduate students, working on intelligent automation and computer engineering.

Harvard University, USA
Tijuana Institute of Technology, Mexico
University of Canberra, Australia

Sio-Iong Ao
Oscar Castillo
Xu Huang

Contents

Chapter 1
A Linguistic CMAC vs. a Linguistic Decision Tree for Decision Making

Hongmei He and Jonathan Lawry

Abstract Cerebellar Model Articulation Controller (CMAC) belongs to the family of feed-forward networks with a single linear trainable layer. A CMAC has the feature of fast learning, and is suitable for modeling any non-linear relationship. Combining label semantics and an original CMAC, a linguistic CMAC based on Mass Assignment on labels is proposed to map the relationship between the attributes and the goal variable that is often highly nonlinear. Linguistic Decision Trees based on label semantics have been used as a decision maker in many areas. A linguistic decision tree presents information propagation from input attributes to a goal variable based on transparent linguistic rules. The proposed LCMAC model is functionally equivalent to a linguistic decision tree, and takes the advantage of fast local training of the original CMAC and the advantage of transparency of a linguistic decision tree.

Keywords Linguistic CMAC · Fine grain mapping · Coarse grain mapping · Linguistic decision tree · Label semantics

1 Introduction

For multiple attribute decision making, the underlying uncertain and nonlinear relationship between attributes and goal variable requires an integrated treatment of uncertainty and fuzziness when modeling the propagation of information from low-level attributes to high-level goal variables. It is well recognized that the fuzzy measure plays a crucial role in the fusion of multiple attributes. Wang and Chen [1] used the Choquet fuzzy integral and the g-Lamda fuzzy measure to improve significantly the neural network classification accuracy. Yang et al. [2] and Van-nam et al. [3] have proposed to aggregate evidence from different attributes on the basis

H. He (✉) and J. Lawry
Department of Engineering Mathematics, University of Bristol, Bristol, UK
e-mail: H.He@bristol.ac.uk; J.Lawry@bristol.ac.uk

S.-I. Ao et al. (eds.), *Intelligent Automation and Computer Engineering*,
Lecture Notes in Electrical Engineering 52, DOI 10.1007/978-90-481-3517-2_1,

of weighted combination rules in evidence theory, where the underlying idea is to use random set (mass assignments) to provide a unified model of probability and fuzziness.

Label semantics proposed by Lawry [4, 5], different with the paradigm of computing with words proposed by Zadeh [6], is a random set based semantics for modeling imprecise concepts. Based on this semantics, a tree-structured model, Linguistic Decision Tree (LDT) was proposed by Qin and Lawry [7]. In such an LDT, transparent label semantic rules of the LDT present an effective way for information propagation between low-level and high-level.

Neural networks have been well used for decision making or classification. The Cerebellar Model Articulation Controller (CMAC) [8] is an artificial neural network techniques that models the structure and function of the part of the brain known as the cerebellum. CMAC is a special feed-forward neural network, and has the unique property of quickly training areas of memory without affecting the whole memory structure due to local training property of CMAC. In a CMAC, each variable is quantized and the problem space is divided into discrete states. A vector of quantized input values specifies a discrete state and is used to generate addresses for retrieving information from memory at this state. Information is distributively stored. This property benefits the nonlinear multiple attribute decision making or classification. Lee et al. proposed a self-organizing hierarchical CMAC (HCMAC) classifier, which is comprised of Gaussian-based CMACs [11]. In this paper, a linguistic CMAC (LCMAC) based on Label Semantics is proposed to map the relationship between the attributes and the goal variable. We investigate the equivalence between the LCMAC and an LDT. A case study will illustrate how an LCMAC works.

2 Label Semantics

2.1 Overview

Label semantics [5] proposes two fundamental and inter-related measures of the appropriateness of labels as descriptions of an object or value. Given a finite set of labels $\mathcal{L}$ from which a set of expressions LE can be generated through recursive applications of logical connectives, the measure of appropriateness of an expression $\theta \in LE$ as a description of instance x is denoted by $\mu_\theta(x)$ and quantifies the agent's subjective belief that θ can be used to describe x based on the one's (partial) knowledge of the current labeling conventions of the population. From an alternative perspective, when faced with an object to describe, an agent may consider each label in $\mathcal{L}$ and attempt to identify the subset of labels that are appropriate to use. Let this set be denoted by $\mathcal{D}_x$. In the face of their uncertainty regarding labeling conventions the agent will also be uncertain as to the composition of $\mathcal{D}_x$, and in label semantics this is quantified by a probability mass function $m_x : 2^{\mathcal{L}} \to [0, 1]$ on subsets of labels. The relationship between these two measures will be described below.

Unlike linguistic variables [9], which allow for the generation of new label symbols using a syntactic rule, label semantics assumes a finite set of labels $\mathcal{L}$. These are the basic or core labels to describe elements in an underlying domain of discourse Ω. Based on $\mathcal{L}$, the set of label expressions LE is then generated by recursive application of the standard logic connectives as follows:

Definition 1. Label Expressions
The set of label expressions LE of $\mathcal{L}$ is defined recursively as follows:

- If $L \in \mathcal{L}$, then $L \in LE$.
- If $\theta, \varphi \in LE$, then $\neg\theta, \theta \wedge \varphi, \theta \vee \varphi \in LE$.

A mass assignment m_x on sets of labels then quantifies the agent's belief that any particular subset of labels contains all and only the labels with which it is appropriate to describe x.

Definition 2. Mass Assignment on Labels
$\forall x \in \Omega$ a mass assignment on labels is a function $m_x : 2^{\mathcal{L}} \to [0, 1]$ such that $\sum_{S \subseteq \mathcal{L}} m_x(S) = 1$.

Now depending on labeling conventions there may be certain combinations of labels which cannot all be appropriate to describe any object. For example, *small* and *large* cannot both be appropriate. This restricts the possible values of $\mathcal{D}_x$ to the following set of focal elements:

Definition 3. Set of Focal Elements
Given labels $\mathcal{L}$ together with associated mass assignment $m_x : \forall x \in \Omega$, the set of focal elements for $\mathcal{L}$ is given by:

$$\mathcal{F} = \{S \subseteq \mathcal{L} : \exists x \in \Omega, \, m_x(S) > 0\} \tag{1}$$

The appropriateness measure, $\mu_\theta(x)$, and the mass m_x are then related to each other on the basis that asserting 'x is θ' provides direct constraints on $\mathcal{D}_x$. For example, asserting 'x is $L_1 \wedge L_2$', for labels $L_1, L_2 \in \mathcal{L}$ is taken as conveying the information that both L_1 and L_2 are appropriate to describe x so that $\{L_1, L_2\} \subseteq \mathcal{D}_x$. Similarly, '$x$ is $\neg L$' implies that L is not appropriate to describe x so $L \notin \mathcal{D}_x$. In general we can recursively define a mapping $\lambda : LE \to 2^{2^{\mathcal{L}}}$ from expressions to sets of subsets of labels, such that the assertion 'x is θ' directly implies the constraint $\mathcal{D}_x \in \lambda(\theta)$ and where $\lambda(\theta)$ is dependent on the logical structure of θ. For example, if $\mathcal{L} = \{low, medium, high\}$ then $\lambda(medium \wedge \neg high) = \{\{low, medium\}, \{medium\}\}$ corresponding to those sets of labels which include *medium* but do not include *high*. Hence, the description $\mathcal{D}_x$ provides an alternative to Zadeh's linguistic variables in which the imprecise constraint 'x is θ' on x, is represented by the precise constraint $\mathcal{D}_x \in \lambda(\theta)$, on $\mathcal{D}_x$.

Definition 4. λ-mapping $\lambda : LE \to 2^{\mathcal{F}}$ is defined recursively as follows: $\forall \theta, \varphi \in LE$

- $\forall L_i \in \mathcal{L} \ \lambda(L_i) = \{F \in \mathcal{F} : L_i \in F\}$
- $\lambda(\theta \wedge \varphi) = \lambda(\theta) \cap \lambda(\varphi)$
- $\lambda(\theta \vee \varphi) = \lambda(\theta) \cup \lambda(\varphi)$
- $\lambda(\neg\theta) = \lambda(\theta)^c$

Therefore, based on the λ-mapping we define the appropriateness measure as below:

Definition 5 (Appropriateness Measure). Appropriateness measure $\mu_{\theta(x)}$ is evaluated as the sum of mass assignment m_x over those subsets of labels in $\lambda_\theta(x)$, i.e., $\forall \theta \in LE, \forall x \in \Omega, \mu_{\theta(x)} = \sum_{F \in \lambda(\theta)} m_x(F)$.

For example, if $\mathcal{L} = \{low(l), medium(m), high(h)\}$ with focal sets $\{\{l\}, \{l, m\}, \{m\}, \{m, h\}, \{h\}\}$ and $\theta = l \wedge \neg m$, then

$$\mu_{l \wedge \neg m}(x) = \sum_{F:l \in F, m \notin F} m_x(F) = m_x(\{l\}).$$

2.2 Consonance Assumption

Appropriateness measures are not a one-to-one function of mass assignments, since m_x cannot be uniquely determined from $\mu_L(x) : L \in \mathcal{L}$. However, in the presence of additional assumptions the calculus can be functional. One such assumption, based on an idea of ordering, which is often rather natural for labels defined in multi-attribute models, is as follows:

Definition 6 (Consonance in Label Semantics). Given non-zero appropriateness measure on basic labels $\mathcal{L} = \{L_1, L_2, \ldots, L_n\}$ ordered such that $\mu_{L_i}(x) \geq \mu_{L_{i+1}}(x)$ for $i = 1, \ldots, n$ then the consonant mass assignment has the form:

$$m_x(\{L_1, \ldots, L_n\}) = \mu_{L_n}(x),$$

$$m_x(\phi) = 1 - \mu_{L_1}(x),$$

$$m_x(\{L_1, \ldots, L_i\}) = \mu_{L_i}(x) - \mu_{L_{i+1}}(x) \text{ for } i = 1, \ldots, n.$$

In this context the consonant assumption is that each $x \in \Omega$, we first identify a total ordering on the appropriateness of labels. We then evaluate our belief value m_x about which labels are appropriate to describe x in such way so as to be consistent with this ordering. For example, in Fig. 1, $\mu_{vl}(x) = 1$ and $\mu_l(x) = x - 1, x \in [1, 2]$. According to the consonance in Label Semantics, we have: $m_x(\{vl, l\}) = x - 1$ and $m_x(\{vl\}) = 1 - (x - 1) = 2 - x, x \in [1, 2]$. Figure 2 shows the mass assignments generated from the appropriate measure in Fig. 1.

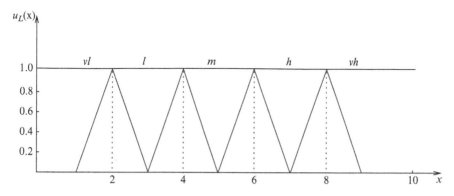

Fig. 1 Appropriateness measures for $L \in \mathcal{L}$

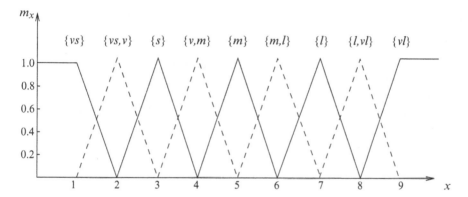

Fig. 2 Mass assignments generated from the appropriateness measures in Fig. 1

3 LCMAC Based on Label Semantics

3.1 Basic CMAC

The basic CMAC is a machine that is analogous to the process of cerebellum's work. In CMAC, the input vectors are viewed as sensory cell firing patterns $\mathcal{X}$, which may be either binary vector or R-ary vector. The appearance of an input vector $\mathcal{X}$ on the sensory cells produces an association cell vector $\mathcal{A}$, which also either binary or R-ary. The association cell vector $\mathcal{A}$ multiplied by the weight matrix W produces a response vector $\mathcal{P}$. There are two mapping in CMAC:

$$f : \mathcal{X} \longrightarrow \mathcal{A}, \; g : \mathcal{A} \longrightarrow \mathcal{P}$$

where, $\mathcal{X}$ is sensory input vectors, $\mathcal{A}$ is association cell vectors, $\mathcal{P}$ is response output vectors. The function f is generally fixed, but the function g depends on the values

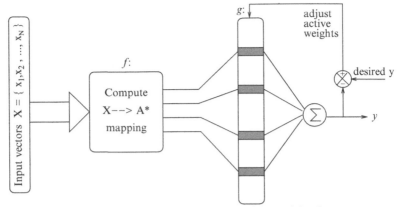

Fig. 3 The structure of basic CMAC

of weights which may be modified during the data storage (or training) process. When an input vector $\mathcal{X} = (x_1, x_2, \ldots, x_N)$ is presented to the sensory cells, it is mapped into an association cell vector $\mathcal{A}$. Define $\mathcal{A}^*$ to be a set of active or nonzero elements of $\mathcal{A}$ shown as in Fig. 3. The response cell sums the values of the weights attached to active association cells to produce the output vector $\mathcal{Y}$. Only the non-zero elements comprising $\mathcal{A}^*$ will affect this sum. The input vector $\mathcal{X}$ can be considered as an address. If for any input $\mathcal{X}$, it is expected to change the contents $\mathcal{Y}$, then we only need to adjust the weights attached to association cells in $\mathcal{A}^*$.

3.2 Mapping with Linguistic Labels of Input Vectors

The new LCMAC is based on the mass assignments on the focal sets for each input attribute. In the LCMAC, the first mapping is the fuzzy discretisation of input attributes. Given appropriateness measure for each attribute, mass assignments on focal elements can be obtained according to the consonance assumption presented in Section 2.

Given input vector $\mathcal{X} = (x_1, x_2, \ldots, x_N)$, for each attribute $x_i, i = 1, \ldots, N$, the label set $\mathcal{L} = \{L_1, L_2, \ldots, L_n\}$ is used to describe the attribute. The focal set for the attribute will be $\mathcal{F} = \{\{L_1\}, \{L_1, L_2\}, \{L_2\}, \ldots, \{L_{n-1}, L_n\}, \{L_n\}\}$. The size of the focal set is $f_n = 2n - 1$. F_{ij} denotes the jth focal element of the ith attribute. For example, Fig. 4 illustrates an LCMAC with 2-dimension input space, where each focal element is associated to one unit of memory. Given a value of the input vector (x_1, x_2), where each attribute can be described with three labels, we can calculate the mass assignments $m_x(F_{1j})$ and $m_x(F_{2j})$, $j = 1, \ldots, 5$. For each attribute, usually there exist two neighbouring focal elements on which the mass assignments are not zero. Thus four units of memory are active. If $m_x(F_{1i}) \neq 0$,

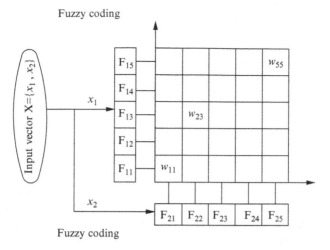

Fig. 4 The structure of LCMAC

$m_x(F_{1(i+1)}) \neq 0$, $m_x(F_{2j}) \neq 0$, $m_x(F_{2(j+1)}) \neq 0$, then units M_{ij}, $M_{(i+1)j}$, $M_{i(j+1)}$, and $M_{(i+1)(j+1)}$ are active.

Theorem 7. *If every focal element of an attribute is associated to a unit of memory, for N-dimension input vector $\mathcal{X} = (x_1, x_2, \ldots, x_N)$, the active space of memory is in an N-dimension hypercube with edge length 2 (i.e., 2^N units of memory).*

Proof. According to labeling conventions and consonance assumption between appropriateness and mass assignment in Section 2, for any pair of neighbouring focal elements F_i and F_{i+1}, $\exists x$, $m_{F_i}(x) \neq 0$ and $m_{F_{i+1}}(x) \neq 0$. In other words, $\forall x$, at most on one pair of neighbouring focal elements F_i and F_{i+1}, $m_{F_i}(x) \neq 0$ and $m_{F_{i+1}}(x) \neq 0$. There are two possible extreme cases: $m_{x_d}(F_{d_1}) \neq 0$, but $m_{x_d}(F_{d_2}) = 0$, and $m_{x_d}(F_{d_{(2n-2)}}) = 0$, but $m_{x_d}(F_{d_{(2n-1)}}) \neq 0$, where n is the number of labels that are used to describe x_d. If $m_{F_i}(x) \neq 0$, for non-neighbouring focal elements $F_{i\pm k}$, $k > 1$ and $2n - 1 \geq i \pm k > 0$, $m_{F_{i\pm k}}(x) = 0$. Therefore, the active space is in a N-dimension hypercube with edge length 2, which holds 2^N units of memory.

3.3   Response Mapping

3.3.1   Fine Grain Mapping

Each unit of memory is used to store a weight, which represents the probability that an input region described by label expressions occurs in the current database. The input region is constrained by mass assignments on focal elements of each attribute in the input vector (see Fig. 4). A unit is addressed with the focal element indices

$(a_1, \ldots, a_N)$ for all input attributes $(x_1, \ldots, x_N)$. For example, in Fig. 4, vector $\mathcal{X} = (x_1, x_2)$, the weight w_{11} responds to the unit whose address is $(1, 1)$, where the first '1' is indicated by the first focal element F_{11} of attribute x_1, and the second '1' is indicated by the first focal element F_{21} of attribute x_2. Given an input vector $\mathcal{X} = (x_1, \ldots, x_N)$, the probability $Pr(M|\mathcal{X})$ that a unit is located is the product of all mass assignments on focal elements of all input attributes, and formalized as below:

$$Pr(M|\mathcal{X}) = \prod_{d=1}^{N} m_{x_d}(F_{d_i}), \tag{2}$$

where, M denotes a unit of memory, and F_{d_i} is a focal element for the dth attribute, and the address of M is given by focal element indices $(d_1, d_2, \ldots, d_N)$. Assuming there are l possible labels $\mathcal{L}_y = \{L_1, \ldots, L_l\}$ to describe the goal variable $\mathcal{Y}$. The output of the neural network is a vector $\mathcal{Y} = \{y_1, \ldots, y_l\}$, where, y_k indicates how appropriate L_k is used to describe the goal based on the neural network given an input vector. The weight is a vector $W_a = \{w_1, w_2, \ldots, w_l\}$, which represents the distributed probability that the goal belongs to a class (label) in the unit of memory. Therefore, according to Jeffery's rule [10], the probability of a label L_k is the sum of probabilities in all active units of memory, and formalized as below:

$$y_k = P(L_{y_k}|\mathcal{X}) = \sum_{a \in \mathcal{A}^*} w_k Pr(M_a|\mathcal{X}) = \sum_{a \in \mathcal{A}^*} w_k \prod_{d=1}^{N} m_{x_d}(F_{d_i}). \tag{3}$$

The probability $Pr(M|\mathcal{X})$, the product of the mass assignments on labels in Eq. 3, is equivalent to the Gaussian basis function with N-dimensions in the conventional CMAC described in [11].

3.3.2 Overlapping Coarse Grain Mapping

Each active area of memory responses to a weight, which suggests the probability of occurrences of the input region described with label expressions in the current database. Obviously, each input region corresponds to an active area (see Fig. 5). According to the formula in Definition 5, given an input vector $\mathcal{X}$, the probability that an active area $\mathcal{A}^*$ is located can be calculated by:

$$Pr(\mathcal{A}^*|\mathcal{X}) = \prod_{d=1}^{N} \mu_\theta = \prod_{d=1}^{N} (m_{x_d}(F_{d_i}) + m_{x_d}(F_{d_{i+1}})), \tag{4}$$

where, $\mathcal{X}$ is N-dimension vector, F_{d_i} and $F_{d_{i+1}}$ are two neighbouring focal elements for attribute x_d, and the corresponding mass assignments of x_d on the two focal elements are not zero. Any pair of neighbouring active areas overlap. The probability that an active area is located is only related to the given input vector. Similar as in fine grain mapping, assuming there are l possible labels $\mathcal{L}_y = \{L_1, \ldots, L_l\}$

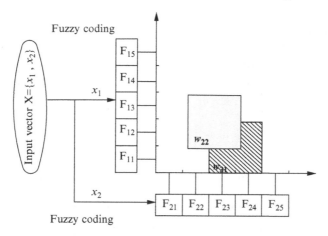

Fig. 5 The overlapping map of LCMAC

to describe the goal variable $\mathcal{Y}$. The output of the neural network is a vector $\mathcal{Y} = \{y_1, \ldots, y_l\}$, where, y_k indicates how appropriate L_k is used to describe the goal based on the neural network given an input vector. Given input vector $\mathcal{X}$, according to Jeffery's rule [10], the probability that a label L_k is appropriate to describe goal variable is the product of the weight in the active area and the probability that the active area is located. So, according to Formula (4), it can be written as:

$$y_k = P(L_k|\mathcal{X}) = w_k Pr(\mathcal{A}^*|\mathcal{X}) = w_k \prod_{d=1}^{N} (m_{x_d}(F_{d_i}) + m_{x_d}(F_{d_{i+1}})). \quad (5)$$

4 The Convergence of the LCMAC

The purpose of training the neural network is to adjust the weights to make the LCMAC approach the desired output. We now investigate the convergence of the LCMAC. Hirsch [12] viewed a neural network as a nonlinear dynamic system called Neurodynamics, which uses differential equations to describe activity patterns. Assuming the dynamic system with N state variables $v_1, v_2, \ldots, v_N$, the network motion equation is $\frac{du_i}{dt} = -\frac{\partial E}{\partial v_i}$, where u_i and v_i are the input and output of the ith neuron. Takefuji and Szu has proved [13]:

$$\frac{dE}{dt} = \sum \frac{dv_i}{dt} \frac{\partial E}{\partial v_i} = \sum \frac{dv_i}{dt} \left(-\frac{du_i}{dt}\right)$$

$$= -\sum \left(\frac{dv_i}{du_i}\right)\left(\frac{du_i}{dt}\right)\left(\frac{du_i}{dt}\right) = -\sum \left(\frac{dv_i}{du_i}\right)\left(\frac{du_i}{dt}\right)^2.$$

Therefore, convergence of a neural network does not depend on the model. As long as the output v_i is the continuous, differentiable and monotonous increasing function of input u_i, namely, there exists the relationship between outputs and inputs of neurons $\frac{dv_i}{du_i} > 0$, the neural network always converges with a negative grade. Finally, the neural network arrives at a stable state with $\frac{dE}{dt} = 0$. Here we define the activation function as below:

$$v_i = \begin{cases} 1 & u_i \geq 1 \\ u_i & 0 \leq u_i < 1 \\ 0 & u_i < 0. \end{cases} \tag{6}$$

In LCMAC, each cell of response mapping represents a neuron. The weight in each cell indicates the state of each neuron (i.e., the output of each neuron). As defined in the activation function 6, the output v_i and input u_i of each neuron have the relationship $v_i = u_i$, if $u_i \in [0, 1]$. Therefore, we have:

$$\Delta u = \Delta W = -\frac{\partial E}{\partial W} \times \Delta t. \tag{7}$$

The Least Mean Square (LMS) algorithm is well-known for neural network training. Miller et al. used LMS to train the CMAC [14]. We can define the Mean Square Error as:

$$E_k(t) = \gamma(D_k - y_k(t))^2/2, \tag{8}$$

where D_k indicates if a desired label L_k is appropriate to describe the goal. If the goal is labeled with L_k, then $D_k = 1$, otherwise $D_k = 0$. Assuming updating time $\Delta t = 1$, then we have:

$$\Delta w_k = \gamma(D_k - y_k(t))\frac{\partial y_k(t)}{\partial w_k}, \tag{9}$$

where γ is the learning factor. Given a training sample $\mathcal{X}$, we can calculate the value of Eq. 9, which will be as a correction to each of the memory cells activated by the input vector. For the fine grain mapping, according to Eqs. 3 and 9, the motion function is:

$$\Delta w_k = \gamma \left(D_k - \sum_a w_k Pr(M_a|\mathcal{X}) \right) w_k Pr(M_a|\mathcal{X}). \tag{10}$$

For the coarse grain mapping, according to Eqs. 5 and 9, the motion function is:

$$\Delta w_k = \gamma(D_k - w_k Pr(\mathcal{A}^*|\mathcal{X}))Pr(\mathcal{A}^*|\mathcal{X}). \tag{11}$$

5   Comparing with an LDT

5.1   LDTs Based on Label Semantics

In an LDT [7], the nodes are attributes, such as $x_1, \ldots, x_N$, and the edges are label expressions describing each attribute. A branch B is a conjunction of expressions $\theta_1 \wedge \cdots \wedge \theta_N$, where θ_k is the label expression of an edge in branch B for $k = 1, \ldots, N$. Each branch also is augmented by a set of conditional mass values $m(F|B)$, which is equivalent to $P(F|B)$, for each output focal element $F \in \mathcal{F}_y$.

5.1.1   A Focal Element Linguistic Decision Tree

Qin and Lawry [7] suggested a Focal Element Linguistic Decision Trees (FELDTs) created from databases. In an FELDT, branches have the form $B = (F_{i1}, \ldots, F_{iN})$ where x_{id} is the attribute node at the depth d of B, and $F_{id} \in \mathcal{F}_{id}$ for $d = 1, \ldots, N$. If we use the LID3 algorithm [7] to learning the FELDT, the probabilities $P(F_y|B)$ for a focal element $F_y \in \mathcal{F}_y$ conditional on a branch B can be evaluated from a database DB as below:

$$
\begin{aligned}
P(F_y|B) &= \frac{\sum_{r \in DB_{F_y}} m_{\langle x_{i_1}(r), \ldots, x_{i_N}(r) \rangle}(F_{i_1}, \ldots, F_{iN})}{\sum_{r \in DB} m_{\langle x_{i_1}(r), \ldots, x_{i_N}(r) \rangle}(F_{i_1}, \ldots, F_{iN})} \\
&= \frac{\sum_{r \in DB_{F_y}} \prod_{v=1}^{N} m_{x_{i_v}}(r)}{\sum_{r \in DB} \prod_{v=1}^{N} m_{x_{i_v}}(r)}.
\end{aligned} \tag{12}
$$

According to Jeffery's rule, the mass assignment of goal variable y on a focal element can be calculated as follows:

$$
M_{F_y}(y) = \sum_{i=1}^{b} \left(\prod_{d=1}^{N} (m_{x_{i_d}}(F)) \right) P(F_y|B_i), \tag{13}
$$

where, b is the number of branches, and N is the number of attributes or the depth of a branch in the FELDT; Here we assume without the limitation of the depth, so the depth of all branches is the same as the number of attributes; x_{i_d} is the attribute incident to the edge at the dth layer of branch B_i.

5.1.2   Dual-Edge LDTs

Another kind of LDT is the one whose edge grain is two neighbouring focal elements. However, two neighbouring edges overlapping on a focal element. From each node there are $l - 1$ edges, where l the size of focal set. For an example, attribute x_1 in an LDT has the focal set $\{F_1, \ldots, F_9\} = \{\{vl\}, \{vl, l\}, \{l\}, \{l, m\}, \{m\}, \{m, h\}, \{h\}, \{h, vh\}, \{vh\}\}$. Then we have edges from node x_1, such as $\{F_1, F_2\}, \{F_2, F_3\}, \ldots, \{F_7, F_8\}, \{F_8, F_9\}$. We call the LDT as dual-edge LDT. The revised conditional probability of a focal element $F_y \in \mathcal{F}_y$ that is appropriate to describe a goal given the branch B can be evaluated from DB according to:

$$P(F_y|B) = \frac{\sum_{r \in DB_{F_y}} \prod_{d=1}^{N} (m_{x_{i_d}}(F_j) + m_{x_{i_d}}(F_{j+1}))}{\sum_{r \in DB} \prod_{d=1}^{N} (m_{x_{i_d}}(F_j) + m_{x_{i_d}}(F_{j+1}))} \qquad (14)$$

This dual-edge LDT needs similar space as an FELDT does, but the calculation is based on a unique branch with Eq. 15.

$$M_{F_y}(y) = \prod_{d=1}^{N} \mu(\theta_d) P(F_y|B) = \prod_{d=1}^{N} (m_{x_{i_d}}(F_j) + m_{x_{i_d}}(F_{j+1})) P(F_y|B)),$$
$$(15)$$

where, x_{i_d} is the attribute incident to the edge at the dth layer of branch B, and $m_{x_{i_d}}(F_j)$ and $m_{x_{i_d}}(F_{j+1})$ are the non-zero mass assignments of attribute x_{i_d} on two neighbouring focal elements F_j and F_{j+1}, corresponding to the edge.

5.2 Functionally Equivalent to an LDT

From Section 3.3, whether the response mapping is fine grain or coarse grain, the final output of the neural network is the distributed probabilities that the goal can be described with each label. From this point of view, an LCMAC has the same effectiveness as an LDT, presenting the mass assignments on labels of a goal variable.

Comparing the fine grain mapping LCMAC with FELDT, from Eqs. 3 and 13, we can see that the difference between the two equations lies in w_k for fine grain mapping LCMAC, which implies the probability that the goal can be described with a label is conditional on a unit in the active area, and $P(F_y|B)$ for FELDT, which is the conditional mass assignment of the goal variable y on focal element F_y given branch B. Therefore, a unit in the active area in an fine grain LCMAC is equivalent to a branch in FELDT.

Similarly, comparing the coarse grain mapping LCMAC with the dual-edge LDT, from Eqs. 5 and 15 there exists the difference as above, but w_k indicates the probability that the goal can be described with a label is conditional on the active area. Therefore, an active area in coarse grain mapping LCMAC is equivalent to a branch in dual-edge LDT.

5.3   *Linguistic Interpretability*

Given an LDT, the rules corresponding to the branch B_i can be: $\theta_{i1} \wedge \cdots \wedge \theta_{id} \rightarrow F : m(F|B_i)$ for each focal element $F \in F_y$. For a given sample, the probability that the goal can be described with L_k can be calculated by formula (13) and (14) for a focal element LDT and a dual-edge LDT, respectively.

Given a fine grain LCMAC, the rules corresponding to the unit a can be : $\theta_{i1} \wedge \cdots \wedge \theta_{id} \rightarrow F_k : w_a^k$ for each focal element $F_k \in F_y$. For a given sample, the probability that the goal can be described with L_k can be calculated by formula (3).

Given a coarse grain LCMAC, the rules corresponding to the active area $\mathcal{A}$ can be : $\theta_{i1} \wedge \cdots \wedge \theta_{id} \rightarrow y_k : w_{\mathcal{A}}^k$ for each $F_k \in F_y$. For a given sample, the probability that the goal can be described with L_k can be calculated by formula (4).

5.4   *Different Training Processes*

The large difference between an LCMAC and an LDT should be in the learning process. For an LCMAC, learning algorithms vary with different strategies based on the LMS algorithm, which uses the feedback of the error of desired output and calculated output to correct the state of a neuron, so that the neural network arrives at a stable state with least square error, and the training process only involves the neurons' state in the active area located by a given sample, while for an LDT, the learning algorithm LID3 proposed by Qin and Lawry [7], is an extension of classic ID3 algorithm based on statistics [15]. The search is guided by a modified measure of information gain in accordance with label semantics. The basic step of LID3 is to calculate the conditional probability that a goal can be described with a label, then to decide which attribute is extended to current node in the tree according to the expected entropy.

6   A Case Study

Given a sample $\mathcal{X} = \{x_1, x_2\} = \{1.8, 4.4\}$, and the goal is a binary variable with positive (+) or negative (−). Using a trained LCMAC, we can estimate the distributed probabilities of the goal on the two classes. According to the sample, we have $m_{x_1}(F_3) = 0.4$, $m_{x_1}(F_4) = 0.6$, $m_{x_2}(F_4) = 0.6$, $m_{x_2}(F_5) = 0.4$.

For a fine grain LCMAC, rules for the four units in the active area:

$(x_1 \text{ is } F_3) \wedge (x_2 \text{ is } F_4): W_{a_1} = \{w_{a_1}^+, w_{a_1}^-\} = \{0.4, 0.6\}$
$(x_1 \text{ is } F_3) \wedge (x_2 \text{ is } F_5): W_{a_2} = \{0.7, 0.3\}$
$(x_1 \text{ is } F_4) \wedge (x_2 \text{ is } F_4): W_{a_3} = \{0.25, 0.75\}$
$(x_1 \text{ is } F_4) \wedge (x_2 \text{ is } F_5): W_{a_4} = \{0.6, 0.4\}.$

The probability that '+' is appreciated to describe the goal can be calculated as below:

$$P(+|\mathcal{X}) = w_{a_1}^+ m_{x_1}(F_3) m_{x_2}(F_4) + w_{a_2}^+ m_{x_1}(F_3) m_{x_2}(F_5)$$
$$+ w_{a_3}^+ m_{x_1}(F_4) m_{x_2}(F_4) + w_{a_4}^+ m_{x_1}(F_4) m_{x_2}(F_5)$$
$$= 0.442$$

The probability that '−' is appreciated to describe the goal can be calculated as below:

$$P(-|\mathcal{X}) = w_{a_1}^- m_{x_1}(F_3) m_{x_2}(F_4) + w_{a_2}^- m_{x_1}(F_3) m_{x_2}(F_5)$$
$$+ w_{a_3}^- m_{x_1}(F_4) m_{x_2}(F_4) + w_{a_4}^- m_{x_1}(F_4) m_{x_2}(F_5)$$
$$= 0.558$$

For a coarse grain LCMAC, rules in the active area $\mathcal{A}$:

$((x_1 \text{ is } F_3) \vee (x_1 \text{ is } F_4)) \wedge ((x_2 \text{ is } F_4) \vee (x_2 \text{ is } F_5)),$
$W_A = \{w_A^+, w_A^-\} = \{0.442, 0.558\}$

The probability that '+' is appreciated to describe the goal can be calculated as below:

$$P(+|\mathcal{X}) = w_A^+ (m_{x_1}(F_3) + m_{x_1}(F_4))(m_{x_2}(F_4) + m_{x_2}(F_5)) = 0.442$$

The probability that '−' is appreciated to describe the goal can be calculated as below:

$$P(-|\mathcal{X}) = w_A^- (m_{x_1}(F_3) + m_{x_1}(F_4))(m_{x_2}(F_4) + m_{x_2}(F_5)) = 0.558$$

According to the definitions of mass assignments and focal elements, for an attribute, the sum of non-zero mass assignments on two neighbouring focal elements is 1. Therefore, the following equations are true:

$$P(+|\mathcal{X}) = w_A^+, P(-|\mathcal{X}) = w_A^-. \tag{16}$$

Hence, the coarse grain mapping is directly a linear mapping. Given a sample, we can find its active area, then the weight vector in the active area directly represents the distributed probabilities of the goal's states. However, the accuracy of the coarse grain mapping may be lower than that of the fine grain mapping.

7   Conclusions

For multiple attribute decision making, we presented an LCMAC by combining the Label Semantics based on mass assignments of attributes, and investigated the convergence of the neural network. It is shown that an LCMAC and an LDT are functionally equivalent. A memory unit of an active area in the fine grain mapping LCMAC is equivalent to a branch in an FELDT, while an active area of memory in coarse grain mapping LCMAC is equivalent to a branch in an dual-edge LDT. But they are different in their training processes. The proposed LCMAC take the advantage of fast local training for a convention CMAC and the advantage of transparency rules for an LDT. Comparing the fine grain mapping and the coarse grain mapping, the coarse grain mapping presents a weight vector in active area directly representing the distributed probabilities of goal's states, while the fine grain mapping could have higher accuracy. In order to validate the performance of an LCMAC, simulation of the model and experiments on some benchmark databases will be the further work. We will examine the performance of an LCMAC and an LDT through the further experiments.

References

1. Wang, X.-Z., & Chen, J.-F. (2004). Multiple neural networks fusion model based on choquet fuzzy integral. In *Proceedings of the third international conference on machine learning and cybernetics*, Shanghai, China, pp. 2024–2027, 2004.
2. Yang, J.B., Wang, Y.M., Xu, D.L., & Chin, K.S. (2006). The evidential reasoning approach to MADA under both probabilistic and fuzzy uncertainty. *European Journal of Operational Research, 171*(4), 309–343.
3. Van-Nam, H., Nakamori, Y., Ho, T., & Murai, T. (2006). Multi-attribute decision making under uncertainty: the evidence reasoning approach revisited. In *IEEE Transactions on Systems, Man and Cybernetics: Part A: System and Humans, 36*(4), 804–822.
4. Lawry, J. (2008). Appropriateness measures: an uncertainty measure for vague concepts. *Synthese, 161*(2), 255–269.
5. Lawry, J. (2006). In J. Kacprzyk (Ed.) *Modeling and reasoning with vague concepts*. Springer, Germany.
6. Zadeh, L.A. (1996). Fuzzy logic=computing with words. *IEEE Transaction on Fuzzy systems, 4*(2), 103–111.
7. Qin, Z., & Lawry, J. (2005). Decision tree learning with fuzzy labels. *Information Sciences, 172*, 91–129.
8. Albus, J.S. (1975). A new approach to manipulator control: the cerebellar model articulation controller(CMAC). *Journal of Dynamic Systems, Measurement and Control Transactions on ASME, 97*, 200–227.
9. Zadeh, L.A. (1975). The concept of linguistic variables and its application to approximate reasoning, Part 1. *Information Science, 8*, 199–249.
10. Jeffrey, R.C. (1965). *The logic of decision*. New York: Gordon and Breach.
11. Lee, H.-M., Chen, C.-M., & Lu, Y.-F. (2003). A self-oganizing HCMAC neural-network classifier. *IEEE Transaction on Neural Networks, 14*(1), 15–27.
12. Hirsch, M.W. (1989). Convergent activation dynamics in continuous time networks. *Proceedings of the National Academy of Sciences, 2*, 331–349.

13. Takefuji, Y., & Szu, H. (1989). Design of parallel distributed Cauchy machines [J]. In *Proceedings of the IJCNN internal joint conference on neural networks* (pp. 529–532), 1989.
14. Miller, W.T., Filson, H.G., & Craft, L.G. (1982). Application of a general learning algorithm to the control of robotic manipulators. *The International Journal of Robotics Research, 6,* 123–147.
15. Quinlan, J.R. (1986). Induction of decision trees. *Machine Learning, 1,* 81–196.

Chapter 2
A Multiple Criteria Group Decision Making Model with Entropy Weight in an Intuitionistic Fuzzy Environment

Chia-Chang Hung and Liang-Hsuan Chen

Abstract The theory of intuitionistic fuzzy sets (IFSs) is well-suited to dealing with vagueness and hesitancy. In this study, we propose a new fuzzy TOPSIS group decision making model using entropy weight for dealing with multiple criteria decision making (MCDM) problems in an intuitionistic fuzzy environment. This model can measure the degrees of satisfaction and dissatisfaction of each alternative evaluated across a set of criteria. To obtain the weighted fuzzy decision matrix, we employ the concept of Shannon's entropy to calculate the criteria weights. An investment example is used to illustrate the application of the proposed model.

Keywords Entropy · Intuitionistic fuzzy sets (IFSs) · Multiple criteria decision making (MCDM) · TOPSIS

1 Introduction

A number of multiple criteria decision making (MCDM) approaches have been developed and applied to diverse fields, such as engineering, management, economics, and so on. Among those approaches, TOPSIS (technique for order performance by similarity to ideal solution), first developed by Hwang and Yoon [1], is widely adopted by practitioners and researchers. The primary concept of the TOPSIS approach is that the most preferred alternative should not only have the shortest distance from the positive ideal solution (PIS), but also have the farthest distance from the negative ideal solution (NIS) [1, 2]. The applications of TOPSIS have some advantages, including (a) a simple, rationally comprehensible concept, (b) good computational efficiency, and (c) being able to measure the relative performance of each alternative in a simple mathematical form [3].

C.-C. Hung (✉) and L.-H. Chen
Department of Industrial and Information Management, National Cheng Kung University, Tainan 701, Taiwan, ROC
e-mail: kennyhred@gmail.com; lhchen@mail.ncku.edu.tw

S.-I. Ao et al. (eds.), *Intelligent Automation and Computer Engineering*,
Lecture Notes in Electrical Engineering 52, DOI 10.1007/978-90-481-3517-2_2,
© Springer Science+Business Media B.V. 2010

In 1965, Zadeh [4] first introduced the theory of fuzzy sets. Later on, many researchers have been working on the process of dealing with fuzzy decision making problems by applying fuzzy sets theory. Zadeh's fuzzy sets only assign a single membership value between zero and one to each element. However, the non-membership degree of an element to a fuzzy set may not always be just equal to one minus the membership degree. In 1993, Gau and Buehrer [5] pointed out that this single value could not attest to its accuracy and proposed the concept of vague sets. Bustince and Burillo [6], however, pointed out that the notion of vague sets coincides with that of intuitionistic fuzzy sets (IFSs) proposed by Atanassov [7] almost 10 years earlier. IFSs are represented by two characteristic functions expressing the degrees of membership and non-membership of elements of the universal set to the IFS. IFSs can cope with the presence of vagueness and hesitancy originating from imprecise knowledge or information. In the last two decades, there have been many studies on the theory and application of IFSs, including logic programming, medical diagnosis, fuzzy topology, decision making, pattern recognition, and so on. Different from other studies, in this work, the criteria weights are obtained by conducting Shannon's entropy concept; after that, a fuzzy TOPSIS method is employed to order the alternatives. The proposed model can deal with uncertain problems and its calculation is not difficult, so that it can provide an efficient way to help the decision maker (DM) in making decisions.

2 Preliminaries

2.1 Intuitionistic Fuzzy Sets

Definition 1. [7]. An IFS A in the universe of discourse X is defined with the form

$$A = \{\langle x, \mu_A(x), \nu_A(x)\rangle \,|\, x \in X\},$$

where

$$\mu_A : X \to [0, 1], \ \nu_A : X \to [0, 1]$$

with the condition

$$0 \le \mu_A(x) + \nu_A(x) \le 1, \ \forall x \in X.$$

The numbers $\mu_A(x)$ and $\nu_A(x)$ denote the membership and non-membership degrees of x to A, respectively.

Obviously, each ordinary fuzzy set may be written as

$$\{\langle x, \mu_A(x), \ 1 - \mu_A(x)\rangle \,|\, x \in X\}.$$

That is to say, fuzzy sets may be reviewed as the particular cases of IFSs.

Fig. 1 Membership, non-membership, and hesitancy degrees

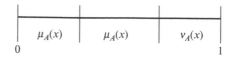

Note that A is a crisp set if and only if for $\forall x \in X$, either $\mu_A(x) = 0$, $\nu_A(x) = 1$ or $\mu_A(x) = 1$, $\nu_A(x) = 0$.

For each IFS, A in X, we will call

$$\pi_A(x) = 1 - \mu_A(x) - \nu_A(x)$$

the intuitionistic index of x in A. It is a measure of hesitancy degree of x to A [7]. It is obvious that $0 \le \pi_A(x) \le 1$ for each $x \in X$. Figure 1 illustrates the three degrees (membership, non-membership, and hesitancy).

For convenience of notation, IFSs(X) is denoted as the set of all IFSs in X.

Definition 2. [8]. For every $A \in$ IFSs(X), the IFS λA for any positive real number λ is defined as follows:

$$\lambda A = \left\{ áx, 1 - (1 - \mu_A(x))^\lambda, (\nu_A(x))^\lambda ñ | x \in X \right\}. \tag{1}$$

2.2 Entropy of IFS

In 1948, Shannon [9] proposed the entropy function, $H(p_1, p_2, \cdots, p_n) = -\sum_{i=1}^{n} p_i \log(p_i)$, as a measure of uncertainty in a discrete distribution based on the Boltzmann entropy of classical statistical mechanics, where p_i ($i = 1, 2, \ldots, n$) are the probabilities of random variable according to a probability mass function P. Later, De Luca and Termini [10] defined a non-probabilistic entropy formula of a fuzzy set based on Shannon's function on a finite universal set $X = \{x_1, x_2, \ldots, x_n\}$. as follows:

$$E_{LT}(A) = -k \sum_{i=1}^{n} [\mu_A(x_i) \ln \mu_A(x_i) + (1 - \mu_A(x_i)) \ln(1 - \mu_A(x_i))], \tag{2}$$

where $k > 0$.

Szmidt and Kacprzyk [11] extended De Luca and Termini's axioms to present four definitions with regard to the entropy measure on IFSs(X) as follows:

EI1: $E(A) = 0$ iff A is a crisp set;

EI2: $E(A) = 1$ iff $\mu_A(x_i) = \nu_A(x_i)$, $\forall x_i \in X$;

EI3: $E(A) \le E(B)$ if A is less fuzzy than B, i.e., $\mu_A(x_i) \le \mu_B(x_i)$ and $\nu_A(x_i) \ge \nu_B(x_i)$ for $\mu_B(x_i) \le \nu_B(x_i)$, $\forall x_i \in X$;

or

$\mu_A(x_i) \geq \mu_B(x_i)$ and $\nu_A(x_i) \leq \nu_B(x_i)$ for $\mu_B(x_i) \geq \nu_B(x_i)$, $\forall x_i \in X$;

EI4: $E(A) = E(A^c)$, where A^c is the complement of A.

Recently, Vlachos et al. [12] presented Eq. 3 as the measure of intuitionistic fuzzy entropy which was proved to satisfy the four axiomatic requirements mentioned above.

$$E_{LT}^{IFS}(A) = -\frac{1}{n \ln 2} \sum_{i=1}^{n} [\mu_A(x_i) \ln \mu_A(x_i) + \nu_A(x_i) \ln \nu_A(x_i)$$

$$-(1 - \pi_A(x_i)) \ln(1 - \pi_A(x_i)) - \pi_A(x_i) \ln 2]. \qquad (3)$$

It is noted that $E_{LT}^{IFS}(A)$ is composed of the hesitancy degree and the fuzziness degree of the IFS A.

3   Proposed Fuzzy TOPSIS Group Decision Making Model

The procedures of calculation for this proposed model can be described as follows:

Step 1   Construct an intuitionistic fuzzy decision matrix based on opinions of DMs. An MCDM problem can be concisely expressed in matrix format as

$$D = \begin{array}{c} \\ A_1 \\ A_2 \\ \vdots \\ A_m \end{array} \begin{array}{c} C_1 \ C_2 \ \cdots \ C_n \\ \left[\begin{array}{cccc} x_{11} & x_{12} & \cdots & x_{1n} \\ x_{21} & x_{22} & \cdots & x_{2n} \\ \vdots & \vdots & \vdots & \vdots \\ x_{m1} & x_{m2} & \cdots & x_{mn} \end{array} \right] \end{array} \qquad (4)$$

$$W = (w_1, w_2, \ldots, w_n)^T$$

Let $A = \{A_1, A_2, \ldots, A_m\}$ be a set of alternatives which consists of m non-inferior decision-making alternatives. Each alternative is assessed on n criteria, and the set of all criteria is denoted $C = \{C_1, C_2, \ldots, C_n\}$. Let $W = (w_1, w_2, \ldots, w_n)^T$ be the weighting vector of criteria, where $w_j \geq 0$ and $\sum_{j=1}^{n} w_j = 1$.

In this study, the characteristics of the alternatives A_i are represented by the IFS as:

$$A_i = \{\langle C_j, \mu_{A_i}(C_j), \nu_{A_i}(C_j) \rangle | C_j \in C\}, \ i = 1, 2, \ldots, m, \qquad (5)$$

where $\mu_{A_i}(C_j)$ and $\nu_{A_i}(C_j)$ indicate the degrees that the alternative A_i satisfies and does not satisfy the criterion C_j, respectively, and $\mu_{A_i}(C_j) \in [0, 1]$, $\nu_{A_i}(C_j) \in [0, 1]$, $\mu_{A_i}(C_j) + \nu_{A_i}(C_j) \in [0, 1]$. The intuitionistic index $\pi_{A_i}(C_j) = 1 - \mu_{A_i}(C_j) - \nu_{A_i}(C_j)$ has the feature that the larger $\pi_{A_i}(C_j)$ the greater the hesitancy of the DM about the alternative A_i with respect to the criterion C_j.

For a group decision making (GDM) problem, let $E = \{e_1, e_2, \ldots, e_l\}$ be the set of DMs, and $\lambda = (\lambda_1, \lambda_2, \ldots, \lambda_l)^T$ be the weighting vector of DMs, where $\lambda_k \geq 0, k = 1, 2, \ldots, l$ and $\sum_k^l \lambda_k = 1$. Let $\tilde{D}^k = [\tilde{x}_{ij}^k]_{m \times n}$ be an intuitionistic fuzzy decision matrix of each DM, where $i = 1, 2, \ldots, m; \quad j = 1, 2, \ldots, n$. In the process of GDM, all the individual decision opinions need to be aggregated into a group opinion to conduct the collective decision matrix $\tilde{D} = [\tilde{x}_{ij}]_{m \times n}$. In order to do this, an IFWA (intuitionistic fuzzy weighted averaging) operator [13] is used. Here,

$$
\tilde{x}_{ij} = IFWA_\lambda(\tilde{x}_{ij}^1, \tilde{x}_{ij}^2, \ldots, \tilde{x}_{ij}^l) = \lambda_1 \tilde{x}_{ij}^1 \oplus \lambda_2 \tilde{x}_{ij}^2 \oplus \cdots \oplus \lambda_1 \tilde{x}_{ij}^l
$$
$$
= \left\langle 1 - \prod_{k=1}^{l}(1 - \mu_{ij}^k)^{\lambda_k}, \prod_{k=1}^{l}(v_{ij}^k)^{\lambda_k} \right\rangle \tag{6}
$$

Step 2 Determine the criteria weights using the entropy-based method.

The well-known entropy method [1, 2] can obtain the objective weights, called entropy weights. The smaller entropy value reveals that the evaluated values of all alternative A_i $(i = 1, 2, \ldots, m)$ with respect to a criterion are less similar. Consequently, for the decision matrix $\tilde{D} = [\tilde{x}_{ij}]_{m \times n}$ in an intuitionistic fuzzy environment, the expected information content emitted from each criterion C_j can be measured by the entropy value, denoted as $E_{LT}^{IFS}(C_j)$, as

$$
E_{LT}^{IFS}(C_j) = - \frac{1}{m \ln 2} \sum_{i=1}^{m} \left[\mu_{ij}(C_j) \ln \mu_{ij}(C_j) + v_{ij}(C_j) \ln v_{ij}(C_j) \right.
$$
$$
\left. - (1 - \pi_{ij}(C_j)) \ln(1 - \pi_{ij}(C_j)) - \pi_{ij}(C_j) \ln 2 \right]. \tag{7}
$$

where $j = 1, 2, \ldots, n$ and $1/(m \ln 2)$ is a constant to ensure $0 \leq E_{LT}^{IFS}(C_j) \leq 1$.

Therefore, the degree of divergence (d_j) of the average intrinsic information provided by the corresponding performance ratings on criterion C_j can be defined as

$$
d_j = 1 - E_{LT}^{IFS}(C_j), \quad j = 1, 2, \ldots, n. \tag{8}
$$

illustrated as Fig. 2. The value of d_j represents the inherent contrast intensity of criterion C_j, and thus the entropy weight of the jth criterion is

$$
w_j = d_j \left/ \sum_{j=1}^{n} d_j. \right. \tag{9}
$$

It should be noted that w_j is a crisp weight.

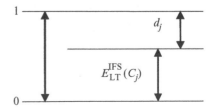

Fig. 2 The divergence degree of information on each criterion

Step 3 Construct the weighted intuitionistic fuzzy decision matrix.

A weighted intuitionistic fuzzy decision matrix $\tilde{Z}$ can be obtained by aggregating the weighting vector W and the intuitionistic fuzzy decision matrix $\tilde{D}$ as:

$$\tilde{Z} = W \otimes \tilde{D} = W \otimes [\tilde{x}_{ij}]_{m \times n} = [\hat{x}_{ij}]_{m \times n}. \tag{10}$$

where

$$W = (w_1, w_2, \ldots, w_j, \ldots, w_n)^T;$$
$$\hat{x}_{ij} = \langle \hat{\mu}_{ij}, \hat{v}_{ij} \rangle = \left\langle 1 - (1 - \mu_{ij})^{w_j}, v_{ij}^{w_j} \right\rangle, \ w_j > 0.$$

Step 4 Determine the intuitionistic fuzzy positive-ideal solution (IFPIS, A^+) and intuitionistic fuzzy negative-ideal solution (IFNIS, A^-).

In general, the evaluation criteria can be categorized into two kinds, benefit and cost. Let G be a collection of benefit criteria and B be a collection of cost criteria. According to IFS theory and the principle of classical TOPSIS method, IFPIS and IFNIS can be defined as:

$$A^+ = \left\{ \left\langle C_j, \left((\max_i \hat{\mu}_{ij}(C_j) | j \in G), (\min_i \hat{\mu}_{ij}(C_j) | j \in B) \right), \right. \right.$$
$$\left. \left. \left((\min_i \hat{v}_{ij}(C_j) | j \in G), (\max_i \hat{v}_{ij}(C_j) | j \in B) \right) \right\rangle \middle| i \in m \right\}. \tag{11a}$$

$$A^- = \left\{ \left\langle C_j, \left((\min_i \hat{\mu}_{ij}(C_j) | j \in G), (\max_i \hat{\mu}_{ij}(C_j) | j \in B) \right), \right. \right.$$
$$\left. \left. \left((\max_i \hat{v}_{ij}(C_j) | j \in G), (\min_i \hat{v}_{ij}(C_j) | j \in B) \right) \right\rangle \middle| i \in m \right\}. \tag{11b}$$

Step 5 Calculate the distance measures of each alternative A_i from IFPIS and IFNIS.

We use the measure of intuitionistic Euclidean distance (refer to Szmidt and Kacprzyk [14]) to help determine the ranking of all alternatives.

$$d_{IFS}(A_i, A^+)$$
$$= \sqrt{\sum_{j=1}^{n} \left[(\hat{\mu}_{A_i}(C_j) - \hat{\mu}_{A^+}(C_j))^2 + (\hat{v}_{A_i}(C_j) - \hat{v}_{A^+}(C_j))^2 + (\hat{\pi}_{A_i}(C_j) - \hat{\pi}_{A^+}(C_j))^2 \right]}$$

$$\tag{12a}$$

$$d_{IFS}(A_i, A^-)$$
$$= \sqrt{\sum_{j=1}^{n} \left[(\hat{\mu}_{A_i}(C_j) - \hat{\mu}_{A^-}(C_j))^2 + (\hat{v}_{A_i}(C_j) + \hat{v}_{A^-}(C_j))^2 + (\hat{\pi}_{A_i}(C_j) - \hat{\pi}_{A^-}(C_j))^2 \right]}$$

$$\tag{12b}$$

Step 6 Calculate the relative closeness coefficient (CC) of each alternative and rank the preference order of all alternatives.

The relative closeness coefficient (CC) of each alternative with respect to the intuitionistic fuzzy ideal solutions is calculated as:

$$CC_i = d_{\text{IFS}}(A_i, A^-) \Big/ \big(d_{\text{IFS}}(A_i, A^+) + d_{\text{IFS}}(A_i, A^-)\big), \qquad (13)$$

where $0 \le CC_i \le 1$, $i = 1, 2, \ldots, m$.

The larger value of CC indicates that an alternative is closer to IFPIS and farther from IFNIS simultaneously. Therefore, the ranking order of all the alternatives can be determined according to the descending order of CC values. The most preferred alternative is the one with the highest CC value.

4   Illustrative Example

An example is provided [15] in this section in order to demonstrate the calculation process of the proposed approach. An investment company wants to invest an amount of money. There are five possible companies A_i ($i = 1, 2, \ldots, 5$) in which to invest: (1) A_1 is a car company; (2) A_2 is a food company; (3) A_3 is a computer company; (4) A_4 is an arms company; and (5) A_5 is a TV company. An expert group is formed of three experts e_k ($k = 1, 2, 3$) with the weighting vector $\lambda = (0.4, 0.3, 0.3)^T$. Each possible company will be evaluated across three criteria with regard to the: (1) economic benefit (C_1); (2) social benefit (C_2); and (3) environmental pollution (C_3), where C_1 and C_2 are benefit criteria, and C_3 is a cost criterion.

The proposed fuzzy TOPSIS GDM model is applied to solve this problem, and the computational procedure is described in a step-by-step way, as below:

Step 1 The ratings for five possible companies with respect to the three criteria are represented by IFSs, and the three experts construct the intuitionistic fuzzy decision matrices $\tilde{D}^k$ ($k = 1, 2, 3$), as listed in Tables 1–3. The three individual decision matrices are then fused into a collective intuitionistic fuzzy decision matrix $\tilde{D}$ in Table 4.

Step 2 Determine the criteria weights. Using Eq. 7, the entropy values for criteria C_1, C_2 and C_3, respectively, are: 0.4477, 0.4985, and 0.9679. The degree of divergence d_j on each criterion 'C_j($j = 1, 2, 3$) may be obtained by Eq. 8 as 0.5523, 0.5015, and 0.0321, respectively. Therefore, the criteria weighting vector can be expressed as $W = (0.509, 0.462, 0.030)^T$ by applying Eq. 9.

Step 3 After determining the criteria weighting vector, using Eq. 10, the weighted intuitionistic fuzzy decision matrix $\tilde{Z}$ is then obtained as Table 5.

Step 4 In this case, criteria C_1 and C_2 are benefit criteria, while C_3 is a cost criterion. Using Eq. 11a and b, each alternative's IFPIS (A^+) and IFNIS (A^-) with respect to the criteria can be determined as

Table 1 Intuitionistic fuzzy decision matrix $\tilde{D}^1$

	C_1	C_2	C_3
A_1	⟨0.70, 0.20⟩	⟨0.85, 0.10⟩	⟨0.30, 0.50⟩
A_2	⟨0.90, 0.05⟩	⟨0.70, 0.25⟩	⟨0.40, 0.50⟩
A_3	⟨0.80, 0.10⟩	⟨0.85, 0.10⟩	⟨0.30, 0.60⟩
A_4	⟨0.90, 0.00⟩	⟨0.80, 0.10⟩	⟨0.20, 0.70⟩
A_5	⟨0.80, 0.15⟩	⟨0.75, 0.20⟩	⟨0.50, 0.40⟩

Table 2 Intuitionistic fuzzy decision matrix $\tilde{D}^2$

	C_1	C_2	C_3
A_1	⟨0.80, 0.15⟩	⟨0.90, 0.05⟩	⟨0.35, 0.55⟩
A_2	⟨0.90, 0.05⟩	⟨0.80, 0.15⟩	⟨0.35, 0.60⟩
A_3	⟨0.80, 0.10⟩	⟨0.70, 0.20⟩	⟨0.40, 0.55⟩
A_4	⟨0.85, 0.05⟩	⟨0.80, 0.15⟩	⟨0.30, 0.50⟩
A_5	⟨0.85, 0.90⟩	⟨0.80, 0.10⟩	⟨0.55, 0.35⟩

Table 3 Intuitionistic fuzzy decision matrix $\tilde{D}^3$

	C_1	C_2	C_3
A_1	⟨0.90, 0.05⟩	⟨0.85, 0.05⟩	⟨0.30, 0.35⟩
A_2	⟨0.85, 0.10⟩	⟨0.90, 0.00⟩	⟨0.30, 0.60⟩
A_3	⟨0.85, 0.05⟩	⟨0.75, 0.15⟩	⟨0.45, 0.50⟩
A_4	⟨0.90, 0.05⟩	⟨0.80, 0.10⟩	⟨0.35, 0.55⟩
A_5	⟨0.80, 0.05⟩	⟨0.80, 0.15⟩	⟨0.45, 0.50⟩

Table 4 Intuitionistic fuzzy decision matrix $\tilde{D}$

	C_1	C_2	C_3
A_1	⟨0.81, 0.12⟩	⟨0.87, 0.07⟩	⟨0.32, 0.46⟩
A_2	⟨0.89, 0.06⟩	⟨0.81, 0.00⟩	⟨0.36, 0.56⟩
A_3	⟨0.82, 0.08⟩	⟨0.78, 0.14⟩	⟨0.38, 0.55⟩
A_4	⟨0.89, 0.00⟩	⟨0.80, 0.11⟩	⟨0.28, 0.59⟩
A_5	⟨0.82, 0.18⟩	⟨0.78, 0.15⟩	⟨0.50, 0.41⟩

Table 5 Weighted intuitionistic fuzzy decision matrix $\tilde{Z}$

	C_1	C_2	C_3
A_1	⟨0.5706, 0.3399⟩	⟨0.6104, 0.2927⟩	⟨0.0115, 0.9770⟩
A_2	⟨0.6749, 0.2388⟩	⟨0.5357, 0.0000⟩	⟨0.0133, 0.9828⟩
A_3	⟨0.5822, 0.2765⟩	⟨0.5032, 0.4032⟩	⟨0.0142, 0.9822⟩
A_4	⟨0.6749, 0.0000⟩	⟨0.5246, 0.3607⟩	⟨0.0098, 0.9843⟩
A_5	⟨0.5822, 0.4178⟩	⟨0.5032, 0.4162⟩	⟨0.0206, 0.9736⟩

$A^+ = (⟨0.6749, 0.0000⟩⟨0.6104, 0.0000⟩⟨0.0098, 0.9843⟩)$

$A^- = (⟨0.5706, 0.4178⟩⟨0.5032, 0.4162⟩⟨0.0206, 0.9736⟩)$

Step 5 Calculate the distance between alternatives and intuitionistic fuzzy ideal solutions (IFPIS and IFNIS) using Eq. 12a and b.

Step 6 Using Eq. 13, the relative closeness coefficient (CC) can be obtained.

The distance, relative closeness coefficient, and corresponding ranking of five possible companies are tabulated in Table 6 Therefore, we can see that the order of

Table 6 The distance measure, relative closeness coefficient and ranking of the five alternatives

Alternatives	$d_{\text{IFS}}(A_i, A^+)$	$d_{\text{IFS}}(A_i, A^-)$	CC_i	Rank
A_1	0.8498	0.2858	0.2517	3
A_2	0.4476	0.7991	0.6409	1
A_3	0.8617	0.2215	0.2045	4
A_4	0.4616	0.6165	0.5718	2
A_5	1.0820	0.0165	0.0150	5

ranking among the five alternatives is $A_2 \succ A_4 \succ A_1 \succ A_3 \succ A_5$, where "$\succ$" indicates the relation "preferred to". Therefore, the best choice would be A_2 (food company). From the above processes, we can conclude that the proposed approach is suitable for dealing with fuzzy MCDM problems in GDM by using IFSs.

5 Conclusion

In this work, we propose an entropy-based multiple criteria GDM model, in which the characteristics of the alternatives are represented by IFSs. In information theory, the entropy is related to the average information quantity of a source. Based on this principle, the optimal criteria weights can be obtained by the proposed entropy-based model. The main difference between this method and the classical TOPSIS is the introduction of objective entropy weight in an intuitionistic fuzzy environment with the former. Although the example provided here is for selecting an optimal investment company, the proposed approach can be applied to many different fields. However, this proposed model considers using only objective criteria weights. To overcome this limitation, future work will examine situations in which the DMs can provide and modify their preferences with regard to the criteria weights incorporated in the proposed model.

References

1. Hwang, C.L., & Yoon, K. (1981). *Multiple attribute decision making–methods and applications: a state-of-the-art survey*. New York: Springer.
2. Zeleny, M. (1982). *Multiple criteria decision making*. New York: McGraw-Hill.
3. Yeh, C.-H. (2002). A problem-based selection of multi-attribute decision-making methods. *International Transactions in Operational Research, 9*, 169–181.
4. Zadeh, L.A. (1965). *Fuzzy sets. Information and Control 8*(3), 338–356.
5. Gau, W.L., & Buehrer, D.J. (1993). Vague sets. *IEEE Transactions on Systems Man and Cybernetics, 23*(2), 610–614.
6. Bustince, H., & Burillo, P. (1996). Vague sets are intuitionistic fuzzy sets. *Fuzzy Sets and Systems, 79*, 403–405.
7. Atanassov, K. (1986). Intuitionistic fuzzy sets. *Fuzzy Sets and Systems, 20*, 87–96.

8. De, S.K., Biswas, R., & Roy, A.R. (2000). Some operations on intuitionistic fuzzy sets, *Fuzzy Sets and Systems, 114*, 477–484.

9. Shannon, C.E. (1948). The mathematical theory of communication. *Bell System Technical Journal, 27*, 379–423; 623–656.

10. De Luca, & Termini, S. (1972). A definition of a non-probabilistic entropy in the setting of fuzzy sets theory. *Information and Control, 20*, 301–312.

11. Szmidt, E., & Kacprzyk, J. (2001). Entropy of intuitionistic fuzzy sets. *Fuzzy Sets and Systems, 118*, 467–477.

12. Vlachos, I.K., & Sergiadis, G.D. (2007). Intuitionistic fuzzy information – Applications to pattern recognition. *Pattern Recognition Letters, 28*, 197–206.

13. Xu, Z. (2007). Intuitionistic fuzzy aggregation operators. *IEEE Transactions on Fuzzy Systems, 15*, 1179-1187.

14. Szmidt, E., & Kacprzyk, J. (2000). Distances between intuitionistic fuzzy sets. *Fuzzy Sets and Systems, 114*, 505–518.

15. Xu, Z. (2008). On multi-period multi-attribute decision making. *Knowledge-Based Systems, 21*, 164–171.

Chapter 3
Emergency HTN Planning

Hisashi Hayashi, Seiji Tokura, Fumio Ozaki, and Tetsuo Hasegawa

Abstract Integration of deliberation and reaction has been an important research topic concerning agents in view of the need for an agent to react tentatively and immediately to the changing world when unexpected events occur while executing a plan. An agent is not supposed to think for a long time before reacting. Also, its reaction is not supposed to change the world greatly. However, there are some cases where deliberation is necessary for achieving an emergency goal or where the emergency plan execution prevents the resumption of the suspended plan execution. This chapter presents a new concept of on-line interruption planning that integrates deliberation and emergency deliberation. When an emergency goal is given while executing a plan, our agents suspend the current plan execution, make and execute an emergency plan, and resume the suspended plan execution. Because our agents continuously modify the suspended plans while executing an emergency plan, they can resume the suspended plans correctly and efficiently even if the world has changed greatly due to the emergency plan execution.

Keywords Agent · Intelligent agent · Planning · Emergency planning · Interruption planning · Deliberation and reaction · Robotics · Intelligent robotics

1 Introduction

Handling asynchronous "emergency" goals is a very important subject of research in planning. Asynchronous goals are inputted to the planning agent even while executing a plan for another goal. The simplest way of handling asynchronous goals

H. Hayashi (✉), S. Tokura, and F. Ozaki
Corporate Research and Development Center, Toshiba Corporation, 1 Komukai-Toshiba-cho, Saiwai-ku, Kawasaki, 212-8582 Japan
e-mail: hisashi3.hayashi@toshiba.co.jp; seiji.tokura@toshiba.co.jp; fumio.ozaki@toshiba.co.jp

T. Hasegawa
Corporate Software Engineering Center, Toshiba Corporation, 1 Komukai-Toshiba-cho, Saiwai-ku, Kawasaki, 212-8582 Japan
e-mail: tetsuo3.hasegawa@toshiba.co.jp

S.-I. Ao et al. (eds.), *Intelligent Automation and Computer Engineering,*
Lecture Notes in Electrical Engineering 52, DOI 10.1007/978-90-481-3517-2_3,
© Springer Science+Business Media B.V. 2010

is to plan and execute goals in a first-come-first-served manner. However, in this approach, even when receiving an emergency goal, it will be served last.

A better approach for handling asynchronous emergency goals is to prioritize multiple goals where the goal with the highest priority is served first. Intention scheduling in BDI (Belief, Desire, Intention) agents is researched in [19]. Here, intentions are committed plans for different goals. Priorities and interferences between goals are taken into consideration. Most current BDI agents [1, 4, 10, 12, 14] are based on PRS [6]. However, the problem to use PRS-like BDI agents is that they can neither plan nor replan. Instead, they reactively select and commit to ready-made plans.

There is a planning agent called ROGUE [7] that can make plans for multiple asynchronous goals with priorities. ROGUE uses an on-line partial-order backward-chaining planner called PRODIGY. When ROGUE receives a new goal while executing a plan, it adds the new goal to the search space of the current plan. Then, PRODIGY incrementally expands the plan for the new goal without invalidating the current plan. The execution order is calculated based on the priorities of goals. It seems that this approach is promising. However, in the case of emergency, the agent should suspend the current plan execution, execute the emergency plan immediately, and modify the suspended plan accordingly. When planning for the emergency goal, it is not necessary to keep the suspended plan valid as in PRODIGY. Although we share a similar motivation with ROGUE and PRODIGY, we would like to execute the best plan (in terms of costs) for the emergency goal rather than an incrementally expanded plan for multiple asynchronous goals. Therefore, the motivation is rather different.

In this chapter, we will present the new concept of "interruption planning" for asynchronous emergency goals. Here, interruption planning means that when the planning agent receives an emergency goal, it suspends the current plan execution, makes emergency plans, executes an emergency plan immediately, and resumes the suspended plan execution. In our new interruption planning, while executing a plan for the emergency goal, the agent keeps and continuously updates the emergency plans and the suspended initial plans. This means that our interruption planning is on-line planning. Unlike ROGUE, in our interruption planning, we make the best plan (in terms of cost) for the emergency goal without trying to reuse any parts of plans for the suspended goals. Also, in most planning systems including PRODIGY, actions are treated as the primitive tasks that cannot be suspended. In our new interruption planning, in order to handle an emergency goal immediately, the planning agent tries to suspend the current action execution if it does not contribute to the achievement of the emergency goal. This is especially important if it takes time to execute an action. We will also show that this action suspension is effective for realizing interruption planning in stratified multi-agent systems. In stratified multi-agent planning, the parent agent makes and executes rough plans, and the child agent makes and executes detailed plans. An action of the parent agent corresponds to a whole plan of the child agent. This means that the action execution time of the parent agent is generally long and changes the world greatly. Therefore, it is very important to suspend unnecessary action execution of the parent agent.

2   Museum Guide Scenario

In this section, we introduce a museum guide scenario as an example to illustrate interruption planning. Subsequently, this scenario will also be used for experimental evaluation. Figure 1 shows the map of a museum where the robot moves. Nodes are places where the robot localizes itself relative to the map with the help of, for example, markers which can be recognized through image processing. Especially, nodes are set at intersections of paths or points of interests. The robot moves from one node to the next node along an arc. When the user specifies the destination (node), the robot takes the person there.

The museum is divided into some areas. Given the destination, the robot first searches a rough route that connects only areas. Then the robot searches a detailed route in the first area that connects nodes. For example, when moving from n1 (area1) to n40 (area8), the robot first makes a rough plan: *area*1 → *area*3 → *area*5 → *area*7 → *area*8. The robot then thinks about how it should move to the next area: *n*1 → *n*6 → *n*10.

While taking a person to a node, suppose that the robot is told to go to a toilet. This is an emergency goal and the user cannot wait until the tour guide ends. Therefore, the robot should suspend the current plan execution, take the person to the toilet node, and resume the tour guide.

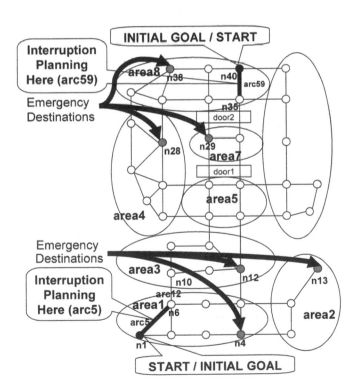

Fig. 1 The map

Note that we use this scenario to evaluate not the efficiency of route planning but the efficiency of on-line interruption planning.

3   On-Line Planning in Dynagent

In order to implement interruption planning, we will use an on-line forward-chaining HTN planning agent called Dynagent [8, 9] in our pilot implementation. Here, an on-line planning agent means an agent that interleaves planning, belief updates, and execution. When new information is found and the belief is updated unexpectedly the on-line planning agent modifies its plans even while executing a plan.

HTN planning [3, 13, 15–17] is different from standard planning which just connect the "preconditions" and "effects" of actions. It makes plans, instead, by decomposing abstract tasks into more concrete subtasks or subplans, which is similar to Prolog that decomposes goals into subgoals. Forward-chaining HTN planning [8, 9, 13] is especially suitable for a dynamically changing world because some task decompositions can be suspended when planning initially and resumed, using the most recent knowledge, just before the abstract tasks are executed. Other merits of HTN planning, such as efficiency, expressiveness of domain knowledge and planning control knowledge are discussed in [5, 13, 18].

Dynagent keeps several alternative plans and incrementally modifies the alternative plans while executing a plan. These alternative plans can contain abstract tasks but only the first task of each alternative plan has to be an executable action. As shown in Fig. 6, in order to implement on-line interruption planning based on Dynagent, the agent has to maintain not only the plans for the emergency goals but also the plans for the suspended initial goal. Dynagent estimates the cost of each plan using A*-like heuristics and searches the best plan in terms of costs.

4   When Suspending an Action

When an emergency goal is given, the agent should suspend the current plan execution. If the time for action execution is short, for the planning agent to wait till the action executor finishes the action execution poses no problem. However, the action execution time is generally long in such areas as robotics, and we would like to suspend the current action execution immediately and start the execution of the emergency plan. Therefore, as Fig. 2 shows, before executing a plan for an emergency goal, the planning agent asks if the action executor can suspend the current action execution. If it is possible, the planning agent tells the action executor to stop the current action execution. After the action suspension, the planning agent updates its belief. For example, if the planning agent suspends the action to go from $n1$ to $n6$

Fig. 2 Action suspension

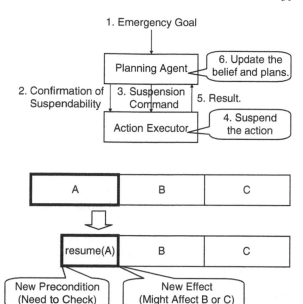

Fig. 3 Replacing with the resuming action

along *arc5*, then the location of the robot will be on *arc5*. We assume the planning agent knows the effect of action suspension.

If the execution of an emergency plan does not affect the suspended plan, then the agent can resume the initial plan execution after resuming the suspended action. If the suspended plan is invalidated by the effect of action suspension, then the plan has to be changed to an alternative plan, in which case we need to rollback the suspended action if necessary.

Figure 3 shows how to replace the suspended action in a plan with the "resuming action." We assume that the planning agent knows the precondition and effect of the resuming action. For example, if the planning agent suspends the action to go to *n6* along *arc5*, then the precondition of the resuming action is that the location of the robot is on *arc5*, and the effect is that the location of the robot becomes *n6*. The resuming action might have the effect that invalidates the rest of the plan. Also the precondition of the resuming action needs to be checked. Therefore, we need to recheck the satisfiability of the preconditions of actions in the modified plan.

On the other hand, the "rollback action" should be added to each alternative plan, as shown in Fig. 4, if the first action is different from the suspended action. We assume that the planning agent knows the precondition and effect of the rollback action. For example, if the agent suspends the action to go to *n6* along *arc5*, then the precondition of the rollback action is that the location of the robot is on *arc5*, and the effect is that the location of the robot becomes *n1*. The rollback action might have an effect that invalidates the rest of the plan. Also the precondition of the rollback action needs to be checked. Therefore, we need to recheck the satisfiability of the preconditions of actions in the modified plan.

Fig. 4 Adding the rollback
action

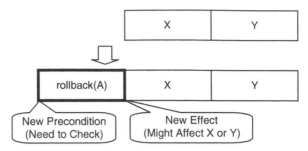

Considering the effects of action suspension, the resuming actions, and rollback
actions, we define the following algorithm for the modifications of the belief and
plans when suspending an action. In the following algorithm, we assume that there
exists the resuming action for each suspendable action. We do not care even if
rollback actions do not exist, in which case we do not add the rollback actions to
alternative plans.

Algorithm 1 *(Modifications of Belief B and Plans PS when Suspending Action A.)*

1. *(Belief Update) Update the current belief B based on the effect of suspension of the
   action A.*
2. *(New Plan Creation) If new valid plans can be created, then add the new valid plans to
   PS. (To find new valid plans, we use the algorithm of Dynagent [8, 9].)*
3. *(Plan Modification) For each plan P in PS, modify P as follows:*
   (a) *(Adding Resuming Actions) If the first action of the plan P is identical to A, then
       replace the occurrence of A in P with the resuming action resume(A) of A.*
   (b) *(Adding rollback Actions) If the first action of the plan P is not identical to the
       suspended action A, and if there exists the rollback action rollback(A) of A, then
       add rollback(A) to the top of the plan.*
   (c) *(Removing Invalid Plans) Based on the current belief B, if a precondition of an
       action in P is not satisfied, then remove P from PS.*

5 Agent Algorithm

This section introduces three algorithms for on-line interruption planning. (See
Fig. 5.) Two types of goals are given to the planning agent: normal goals and emergency goals. Algorithm 4 keeps normal goals in a waiting goal list and handles
them sequentially as usual. On the other hand, when the planning agent receives an
emergency goal, Algorithm 3 changes the status of the current goal to "suspended,"
suspends current action execution (if possible), and handles the emergency goal as
soon as possible. After handling the emergency goal, it resumes the plan execution
of the suspended goal. These algorithms start planning, replanning and plan execution using Algorithm 2.

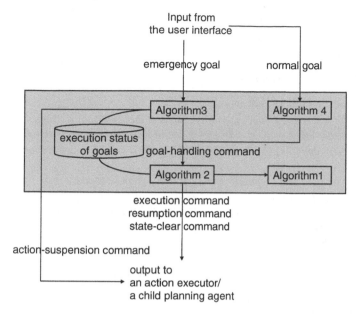

Fig. 5 Agent algorithm

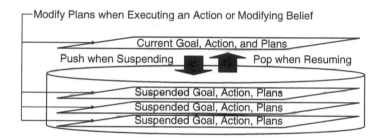

Fig. 6 Stack for suspended goals

As Fig. 6 shows, the agent keeps the suspended goal, the suspended plans, and the suspended action in "the stack for suspended goals." While executing an "emergency plan," the agent continuously updates not only the plans for the emergency goal but also the plans in the stack. To modify these plans, when updating the belief or executing an action, we can use the algorithm of Dynagent [8, 9]. Also, in order to resume the suspended action, the action executor might keep some information while handling an emergency goal. When the action resumption becomes no longer necessary, because of the change of the plan, then the planning agent tells the action executor to clear the recorded state for the action resumption.

Given a goal, the following algorithm starts planning and plan execution. This algorithm is also used when resuming the suspended plan execution. In this case, the suspended plans and the suspended action are recorded in association with the given goal. When another thread (Algorithm 3 or Algorithm 5) makes the status of the goal "suspended," Algorithm 2 updates the belief and plans using Algorithm 1 and finishes the plan execution process.

Algorithm 2 *(Planning and Plan Execution)*

1. *(Goal Input) A goal is given as an input.*
2. *(Input of Suspended Plans/Action) If the status of the given goal is "suspended," then the suspended plans and the suspended action (if it exists) are also given as inputs.*
3. *(Planning) Make the plans for the goal. (Use the HTN planning algorithm of Dynagent [8, 9].)*
4. *(Plan Selection) Select a plan to execute from the alternative plans.*
5. *(Clearance of the Suspended Action) If the status of the goal is "suspended," there exists a suspended action, and the suspended action is different from the next action of the selected plan, then tell the action executor to abandon the recorded state for the action resumption.*
6. *Set the status of the goal to "active."*
7. *(Plan Execution Loop) Repeat the following procedure while the status of the goal is "active:"*
   - (a) *(Action Execution) Following the selected plan, tell the action executor to execute the next action and wait for the result ("success", "failure", or "suspended") that is reported from it.*
   - (b) *(Update of the Belief and Plans) If the result of the action execution is either "success" or "failure," then modify the belief and all the plans, including the plans recorded in the stack for suspended goals, following the plan modification algorithm of Dynagent [8, 9].*
   - (c) *(Update of the Belief and Plans) If the result of the action execution is "suspended," then modify the belief and all the plans, including the plans recorded in the stack for suspended goals, following Algorithm 1.*
   - (d) *(Successful Plan Execution) If one of the plans is successful, then change the status of the goal to "success."*
   - (e) *(Plan Execution Failure) If no alternative plan exists, then change the status of the goal to "failure."*
   - (f) *(Plan Selection) If the status of the goal is "active," then select a plan from alternative plans.*
8. *Output the status of the goal ("success", "failure", or "suspended.")*
9. *If the status of the goal is "suspended," then output the suspended plans and the suspended action (if it exists.)*

The following algorithm shows how to process emergency goals. When an emergency goal is given, the planning agent tries to suspend the current plan execution to execute a plan for the emergency goal as soon as possible. The suspended plan will be resumed after the emergency plan execution. Note that when the following algorithm changes the status of the goal to "suspended," the plan execution process (Algorithm 2) is finished.

Algorithm 3 *(Emergency Goal Handling)*

1. *(Goal Input) An emergency goal A is given as an input.*
2. *(Normal Plan Execution) If there does not exist a goal whose plan is being executed by Algorithm 2, then start the plan execution process (Algorithm 2) for A.*
3. *If there exists a goal B whose plan is being executed, then execute the following procedure:*
   (a) *Change the status of B to "suspended."*
   (b) *(Action Suspension) If an action is being executed, ask the action executor if it is possible to suspend the action that is being executed. If the action can be suspended, then tell the action executor to suspend the action execution.*
   (c) *(Plan Suspension) Wait till the plan execution process (Algorithm 2) for B is finished and receive the suspended plans and the suspended action (if it exists.)*
   (d) *(Pushing to Stack) Push the set of the suspended goal B, the suspended plans, and the suspended action (if it exists) to the stack for suspended goals.*
   (e) *(Emergency Plan Execution) Start the goal handling process (Algorithm 2) for A and wait for the result.*
   (f) *(Popping from Stack) Pop the set of the suspended goal B, the suspended plans, and the suspended action (if it exists) from the stack for suspended goals.*
   (g) *(Plan Resumption) Restart the plan execution algorithm (Algorithm 2) for B, inputting the suspended plans and the suspended action (if it exists) with B.*

The following algorithm shows how to process normal goals. As explained before, normal goals are kept in a waiting goal list and the planning agent handles them sequentially as usual.

Algorithm 4 *(Normal Goal Handling)*

1. *A normal goal A is given as an input.*
2. *Add A to the waiting goal list as the last element.*
3. *Wait till A becomes the first element of the waiting goal list and no goal is being processed by Algorithm 2, and the stack for suspended goals becomes empty.*
4. *Remove A from the waiting goal list.*
5. *Start the plan execution process (Algorithm 2) for A.*

6 Stratified Multi-agent Interruption Planning

In stratified multi-agent planning systems, the parent planning agent executes a rough plan by giving subgoals (=actions of the parent planning agent) to its child planning agents. Given a subgoal, the child planning agents make and execute a detailed plan. For the parent planning agent, the child planning agents are just action executors, and the parent planning agent does not know how its action executors or child planning agents are implemented. Given a subgoal from the parent planning agent, the child planning agent starts planning and plan execution using Algorithm 2. It is the parent planning agent who decides to start, suspend, and resume planning and plan execution for each goal. Therefore, the child planning agent

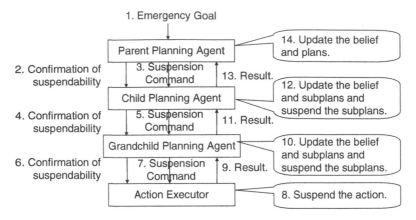

Fig. 7 Subplan suspension

does not use Algorithms 3 and 4. On-line planning in a stratified multi-agent system is researched in [9]. However, interruption planning has not been incorporated into stratified multi-agent systems.

In stratified multi-agent systems, an action of the parent agent corresponds to a whole plan of the child planning agent. Therefore, it is important to suspend meaningless action execution. As shown in Fig. 7 (compare with Fig. 2), a plan suspension instruction from the parent planning agent causes another action suspension of the child planning agent. Similarly, a plan resumption (or clearance) instruction causes another action suspension (respectively, clearance) of the child planning agent. The following algorithm shows how this can be done. In the same way, it is possible to use a grandchild planning agent which is a child of the child planning agent. It is also possible for the parent planning agent to have more than two child planning agents. However, because Dynagent does not execute actions in parallel, two child planning agents do not work at the same time. When the following algorithm changes the status of the goal to "suspended," the plan execution process (Algorithm 2) is finished.

Algorithm 5 *(Subplan Suspension)*

1. *(Instruction Input) A suspension instruction of the current goal G is given as an input (from the parent planning agent.)*

2. *Change the status of G to "suspended."*

3. *(Action Suspension) If an action is being executed, ask the action executor if it is possible to suspend the action that is being executed. If the action can be suspended, then tell the action executor to suspend the action execution.*

4. *(Plan Suspension) Wait till the plan execution process (Algorithm 2) for G is finished and receive the suspended plans and the suspended action (if it exists.)*

5. *(Pushing to Stack) Push the set of the suspended goal G, the suspended plans, and the suspended action (if it exists) to the stack for suspended goals.*

6. *(Instruction Waiting) Wait for the resumption or state clearance instruction of G from the parent planning agent.*

7. *(Popping from Stack) Pop the set of the suspended goal G, the suspended plans, and the suspended action (if it exists) from the stack for suspended goals.*

8. *(State Clearance) When receiving the state clearance instruction of G, abandon the suspended goal, the suspended plans, and the suspended action and tell the action executor to abandon the recorded state for the action resumption.*

9. *(Plan Resumption) When receiving the resumption instruction of G, restart the plan execution algorithm (Algorithm 2) for G, inputting the suspended plans and the suspended action (if it exists) with G.*

7 Experiments

This section evaluates the efficiency of replanning when resuming the suspended plans by means of experiments based on the museum guide scenario explained in Section "Museum Guide Scenario". However, it is not the aim of the experiments to measure the efficiency of route planning. Our planning algorithm can be used for other purposes.

We use two stratified planning agents. The parent planning agent is in charge of the area movement planning and makes a plan to move from one area to another. (See Fig. 1.) The parent planning agent tells the child planning agent to execute the action to move to the next area or a node inside an area. The child planning agent is in charge of the node movement inside an area and makes a plan to move from one node to another. When moving to the next area, the child planning agent also updates the area map in the memory, following the instructions from the parent planning agent. For the planning agent, the child planning agent is an action executor. We assume that the doors are always open. However, these agents can dynamically change the plans, as shown in [9], if a door on the route is closed during the plan execution, but that is not what we wish to show in this chapter.

Initially the robot is at $n40$ in $area8$. We give the goal to go to $n1$ in $area1$ to the parent planning agent. While the robot is moving from $n40$ in $area8$ to $n35$ along $arc59$, we give an emergency goal to go to another place ($n38$ in $area8$ or $n28$ in $area4$ or $n29$ in $area7$). After executing an emergency plan and visiting the node, the parent planning agent resumes the initial plan to go to $n1$ in $area1$. We measure this replanning time for the plan resumption. We compare the naive replanning method to plan from scratch and our on-line replanning method. We measure this replanning time for the plan resumption. Ideally, we would like to compare our on-line interruption planning technique with other on-line interruption planning techniques. However, because the concept of interruption planning is new, our agent is the only on-line interruption planning agent.

We conducted similar experiments. Initially the robot is at $n1$ in $area1$. We give the goal to go to $n40$ in $area8$ to the parent planning agent. While the robot is moving from $n1$ in $area1$ to $n6$ along $arc5$, we give an emergency goal to go to another place ($n13$ in $area2$ or $n4$ in $area1$ or $n12$ in $area3$). After executing an emergency plan and visiting the node, the parent planning agent resumes the initial plan to go to $n40$ in $area8$. We measure this replanning time for the plan resumption.

Table 1 Total replanning time

Starting point	Initial destination	Place of interruption	Emergency destination	On-line replanning	Naive replanning
n40(area8)	n1(area1)	arc59(area8)	n38(area8)	0.06 sec	0.52 sec
n40(area8)	n1(area1)	arc59(area8)	n28(area4)	0.22 sec	0.41 sec
n40(area8)	n1(area1)	arc59(area8)	n29(area7)	0.05 sec	0.35 sec
n1(area1)	n40(area8)	arc5(area1)	n13(area2)	0.29 sec	0.56 sec
n1(area1)	n40(area8)	arc5(area1)	n4(area1)	0.04 sec	0.44 sec
n1(area1)	n40(area8)	arc5(area1)	n12(area3)	0.07 sec	0.38 sec

We compare our on-line replanning method with the naive replanning method to plan from scratch. Each experiment was conducted three times and the average time is shown in Table 1. The agent system is implemented in Java and the planner that the planning agents use is implemented in Prolog and Java. We used a PC (Windows XP) equipped with a Pentium4 2.8 GHz and 512 MB of RAM.

Table 1 shows the total replanning time of the parent planning agent and the child planning agent. Our on-line replanning method is twice as efficient as the naive replanning method when the parent planning agent needs to modify the plan. When the parent planning agent does not need to correct the plan, our on-line replanning method is much faster than the naive replanning approach.

8 Related Work

Integration of deliberation and reaction has been an important subject of research to realize autonomous agents working in dynamic environments. While executing a plan for a goal, even the deliberative agent needs to react to an unexpected event or situation in the case of an emergency. Normally, when combining plan execution and reaction, the agent does not plan to react, and the reaction does not change the world greatly. On the other hand, in our approach, the agent does plan for an emergency goal, and the emergency plan might change the world greatly.

One way [11] of combining plan execution and reaction is to repeat the following cycle: 1. observe; 2. react if necessary; 3. update the plan; 4. act following the plan; 5. go to 1. It seems that the agent can react to the changing world while executing a plan. However, this algorithm does not take into account the time for action execution. If it takes time to act, the agent cannot react quickly as long as we use this algorithm.

Another way of solving this problem is to use the idea of layered architecture [2] where the (higher-level) deliberative planning agent controls the (lower-level) action executor. Here, the planning agent and the action executor are working concurrently. Therefore, the action executor reacts immediately to the world without considering the plan execution. Unless the reaction causes action failure, the action executor does not have to report the reaction to the planning agent. In the case of action failure, the action executor reports it to the planning agent, and the planning agent replans.

9 Conclusions

We have introduced the new concept of interruption planning and shown how interruption planning can be implemented. When receiving an emergency goal, the agent suspends the current plan execution, generates and executes the emergency plan as soon as possible, and modifies the suspended plans accordingly. We have also shown how interruption planning can be implemented in stratified multi-agent systems. As soon as the planning agent receives an emergency goal, it tries to interrupt unnecessary action execution. In the same way, as soon as the planning agent receives an emergency goal, it tries to interrupt unnecessary plan execution of its child planning agent. Our interruption planning is a kind of on-line planning, and the planning agent keeps and continuously updates the suspended plans. Therefore, when resuming the plan execution, the agent replans as quickly as our experiments have shown.

The new concept of interruption planning also combines deliberation and reaction because when the agents quickly make and execute a short emergency plan, the plan can be regarded as a reaction. Unlike previous approaches to combine deliberation and reaction, where reaction is not produced by deliberation, our interruption planning combines deliberation and emergency deliberation.

References

1. Braubach, L., Pokahr, A., & Lamersdorf, W. (2005). Jadex: A BDI-agent system combining middleware and reasoning. In *Software agent-based applications, platforms and development kits* (pp. 143–168). Birkhäuser Book.
2. Brooks, R. (1986). A robust layered control system for a mobile robot. *IEEE Journal of Robotics and Automation, 2*(1), 14–23.
3. Currie, K., & Tate, A. (1991). O-plan: The open planning architecture. *Artificial Intelligence, 52*(1), 49–86.
4. Dastani, M., van Riemsdijk, B., Dignum, F., & Meyer, J-J. (2003). A programming language for cognitive agents: Goal directed 3APL. *Proceedings of international workshop on programming multiagent systemslanguages, frameworks, techniques and tools (ProMAS03)* (pp. 111–130).
5. desJardins, M., Durfee, E., Ortiz, C., & Wolverton, M. (1999). A survey of research in distributed, continual planning. *AI Magazine, 20*(4), 13–22.
6. Georgeff, M., & Ingrand, F. (1989). Decision-making in an embedded reasoning system. In *Proceedings of the international joint conference on artificial intelligence (IJCAI89)* (pp. 972–978).
7. Haigh, K., & Veloso, M. (1998). Interleaving planning and robot execution for asynchronous user requests. *Autonomous Robots, 5*(1), 79–95.
8. Hayashi, H., Tokura, S., Hasegawa, T., & Ozaki, F. (2006). Dynagent: An incremental forward-chaining HTN planning agent in dynamic domains. In *The post-proceedings of the international workshop on declarative agent languages and technologies (DALT05)*, LNAI 3904 (pp. 171–187). Springer, Germany.
9. Hayashi, H., Tokura, S., & Ozaki, F. (2009). *Towards real-world HTN planning agents*, volume 170 of *Studies in Computational Intelligence*, Chapter 2 (pp. 13–41). Springer, Germany.
10. Howden, N., Rönnquist, R., Hodgson, A., & Lucas, A. (2001). JACK: Intelligent agents – summary of an agent infrastructure. In *Proceedings of the international workshop on infrastructure for agents, MAS, and scalable MAS (IAMSMAS01)*.

11. Kowalski, R., & Sadri, F. (1999). From logic programming towards multi-agent systems. *Annals of Mathematics and Artificial Intelligence, 25,* 391–419.
12. Morley, D., & Myers, K. (2004). The SPARK agent framework. In *the international joint conference on autonomous agents and multi-agent systems (AAMAS04)* (pp. 712–719).
13. Nau, D., Cao, Y., Lotem, A., and Mūnoz-Avila, H. (1999). SHOP: simple hierarchical ordered planner. In *the proceedings of the international joint conference on artificial intelligence (IJCAI99)* (pp. 968–975).
14. Rao, A. (1996). AgentSpeak(L): BDI agents speak out in a logical computable language. In *the European workshop on modelling autonomous agents in a multi-agent world (MAAMAW96)* (pp. 42–55).
15. Sacerdoti, E. (1977). *A Structure for plans and behavior.* American Elsevier, USA.
16. Tate, A. (1977). Generating project networks. In *Proceedings of the international joint conference on artificial intelligence (IJCAI77)* (pp. 888–893).
17. Wilkins, D. (1988). *Practical planning.* San Mateo, CA: Morgan Kaufmann.
18. Wilkins, D., & desJardins, M. (2001). A call for knowledge-based planning. *AI Magazine, 22*(1), 99–115.
19. Yan, S.-B., Lin, Z.-N., Hsu, H.-J., & Wang, F.-J. (2005). Intention scheduling for BDI agent systems. In *Proceedings of the annual international computer software and applications conference (COMPSAC05)* (pp. 133–140).

Chapter 4
Adaptive Ant Colony Optimization with Cranky Ants

Masaya Yoshikawa

Abstract Ant Colony Optimization (ACO) is the algorithm inspired by the feeding behavior of ants and its search mechanism is based on the positive feedback reinforcement using pheromone communication. This chapter discusses a new adaptive ACO algorithm and its characteristics are as follows: (1) a novel cranky ant who behaves strangely is introduced to strengthen the pressure of diversification, (2) a new observation technique for the convergence behavior is employed to judge whether it is trapping at local optimal solution. Experiments using benchmark data prove that the proposed algorithm with the cranky ants and the observation technique enables to control the trade-off between intensification and diversification, in comparison with conventional ACO.

Keywords Ant Colony Optimization · Cranky ant · Adaptive optimization · Intensification and diversification · Convergence behavior

1 Introduction

"Swarm intelligence" is the intelligence that emerges when individuals with simple intelligence gather to form a population. Ant Colony Optimization (ACO) is one of the information processing methods developed based on swarm intelligence, a concept which developed from studying the feeding behavior of ants. Ants secrete a volatile chemical substance called pheromone on a route they pass through during feeding activity. Moreover, while following this pheromone, other ants also secrete a pheromone. Although each ant acts based on simple rules of (1) following a pheromone and (2) secreting a pheromone, these pheromones form a complete route to a feed source, and enable efficient feeding behavior. The mechanism of this effective route search is shown in Fig. 1.

M. Yoshikawa (✉)
Department of Information Engineering, Meijo University, 1–501 Shiogamaguchi Tenpaku Nagoya, 468–8502 Japan
e-mail: evolution{_}algorithm@yahoo.co.jp

S.-I. Ao et al. (eds.), *Intelligent Automation and Computer Engineering*,
Lecture Notes in Electrical Engineering 52, DOI 10.1007/978-90-481-3517-2_4,
© Springer Science+Business Media B.V. 2010

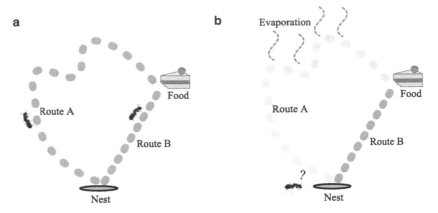

Fig. 1 Example of pheromone communication: (**a**) two routes on initial state and (**b**) positive feedback reinforcement using pheromone information

First, the speed of each ant is presumed to be the same and the secreted pheromone is presumed to evaporate at a fixed rate. As shown in Fig. 1, two ants secrete a pheromone en route from their nest to a feed. Since route A is shorter than route B, the amount of evaporated pheromone on route A is smaller than that on route B; that is, the residual pheromone on route A is larger than that on route B. Therefore, the following ants select the route on which a larger amount of pheromone remains, i.e. they will select route A with its shorter distance. This means a larger amount of pheromone will be added to route A. This is "positive feedback" in terms of ACO.

ACO represents a general name of the algorithm inspired by this feeding behavior of ants. It has been applied to various combinatorial optimization problems such as the travelling salesman problem (TSP) [1,2] the floorplanning problem [3], the quadratic assignment problem [4], and the scheduling problem [5,6]. The basic model of the ACO is the ant system (AS) that was proposed by Dorigo et al. [1]. It originally was introduced to solve the shortest route problems on a graph. Therefore, many ACOs [7,8] applied to TSP are based on the AS. Ant Colony System [2] is one of the expansion algorithm of AS, and it shows better capability than genetic algorithm [9, 10] and simulated annealing [11] when applying to TSP. Therefore, we adopt Ant Colony System as a base algorithm. Hereafter, ACO indicates Ant Colony System.

In this chapter, we propose a new adaptive ACO algorithm. The characteristics of the proposed algorithm are: (1) a novel cranky ant who behaves strangely is introduced to prevent from trapping at the local optima (local optimal solution), (2) a new observation technique for the convergence behavior is adopted to judge whether it is trapping at local optimal solution.

Thus, the proposed algorithm with the cranky ants and the observation technique enables to control the trade-off between intensification (exploitation of the previous solutions) and diversification (exploration of the search space). Experiments using benchmark data prove effectiveness of the proposed algorithm in comparison with the conventional ACO.

Regarding previous work of which ACO is applied to TSP, many works including hybrid approach [12] and pheromone control technique [13] have been reported. G. Shang [12] proposed a hybrid approach which combined ant colony algorithm with genetic algorithm. S.G. Lee [13] introduced a pheromone control technique using a curve-fitting algorithm. However, no studies have ever seen, to our knowledge, adaptive ACO algorithm which introduces the cranky ants.

This chapter is organized as follows: Section 2 describes the search mechanism of ACO. Section 3 indicates the weak points of ACO in terms of the search performance, and explains the proposed algorithm. Section 4 reports the results of computer simulations applied to TSP benchmark data. We summarize and conclude this study in Section 5.

2 Ant Colony Optimization

The search mechanism of ACO utilizes the static evaluation value and the dynamic one. The static evaluation value called heuristic value is peculiar information of the target problem, and usually a reciprocal of the distance between cities is adopted as the heuristic value, when ACO is applied to TSP.

Regarding the selection of ant's move, the concretely procedure is as follows. First, the random number q between from 0 to 1 is generated. Next, q is compared with benchmark (parameter) q_0. When q is smaller than q_0, the city that has the largest value of the product of the static evaluation and the dynamic one is selected as the next destination. This selection rule is called pseudo-random-proportional rule [2]. Otherwise, ant k in city i selects the move to city j according to probability p^k and it is defined as follows.

$$p^k(i, j) = \frac{[\tau(i, j)]\,[\eta(i, j)]^{\beta}}{\sum\limits_{l=n^k} [\tau(l, j)]\,[\eta(l, j)]^{\beta}} \tag{1}$$

Where, $\tau(i, j)$ is a pheromone amount between city i and city j, $\eta(i, j)$ is a reciprocal of the distance between city i and city j, β is a parameter which controls the balance between static evaluation value and dynamic one, and n^k is a set of un-visit cities. Therefore, the selection probability is proportional to the product of the static evaluation and the dynamic one as shown in Fig. 2.

Moreover, a pheromone amount on each route is calculated by using two pheromone update rules. One is local update rule and the other is global update rule. The local update rule is applied to the route which is selected by Eq. 1, and it is defined as follows.

$$\tau(i, j) \leftarrow (i - \psi)\tau(i, j) + \psi\tau_0 \tag{2}$$

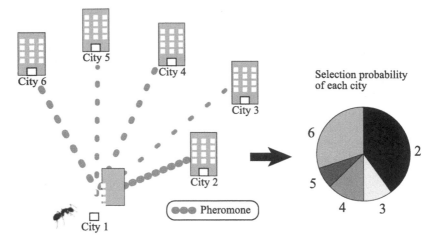

Fig. 2 Example of the selection probability

Where, ψ is a decay parameter in local update rule, τ_0 is the initial value of pheromone. Following the local update rule, whenever an ant selects a route, the amount of pheromone on the route is updated. As shown in Eq. 2, when the amount of pheromone on the selected route is small, it increases. By contrast, when the amount of pheromone on the selected route is large, it decreases. In other words, the local update rule makes the amount of pheromone close to the initial value of τ_0.

The global update rule adds pheromone to the best tour (the completed route) of all tours. The best tour usually indicates the shortest tour. The global update rule is defined as follows.

$$\tau(i, j) \leftarrow (i - \rho)\tau(i, j) + \rho\Delta\tau(i, j) \tag{3}$$

$$\Delta\tau(i, j) = \left\{ \begin{array}{ll} 1 / L^+ & if (i, j) \in T^+ \\ 0 & otherwise \end{array} \right\} \tag{4}$$

Where, T^+ is the best tour, and L^+ is the distance of the best tour.

3 Adaptive Cranky Ant

In a search using meta-heuristics, the balance between diversification and intensification of the search is important to improve the searching ability. If the pressure of intensification is strong, only the near neighbor of a certain solution will be searched for; i.e. this search method is easily led towards the local optimal solution. On the contrary, if the pressure of diversification of the search is strong, this search method will be similar to a random search.

In a search using ACO, the pseudo-random-proportional rule, the global update rule and the positive feedback reinforcement using pheromone information work in favor of intensification, and the local update rule favors diversification. Thus, the escape from the local optimal solution is more difficult than the other meta-heuristics. Therefore, by adding the pressure of diversification, the solution searching success is expected to improve further. Here, the observation technique for the convergence behavior and the escape technique from the local optimal solution are important to add the pressure of diversification.

Regarding the observation technique for the convergence behavior, the proposed algorithm utilizes a transition of the distance of the best tour (the shortest tour). A period of which each ant builds a tour represents one generation. The proposed algorithm judges to be trapped at the local optimal solution if the best tour is not improved across several generations. In contrast, it judges not to be trapped at the local optimal solution while the best tour is improved.

Regarding the escape technique from the local optimal solution, the proposed algorithm newly introduces a cranky ant. Usually, the ant selects the route which is short distance and has a lot of pheromones as shown in Eq. 1. The route which has a lot of pheromones indicates the many-selected route. The cranky ant adopts a reciprocal of the pheromone amount as the dynamic evaluation in the contrary. Thus, the cranky ant chooses the route with few pheromones as shown in Fig. 3.

In other words, the cranky ant selects the route which has not been selected. It enables to change the searching area, and to escape from the local optimal solution as shown in Fig. 4. Using the observation technique and the escape technique, the proposed algorithm achieves the control of the trade-off between intensification and diversification. Specifically, the cranky ants increase when the proposed algorithm judges to be trapped at the local optimal solution. In the proposed algorithm, the normal ants decrease when the cranky ants increase because the total number of ants is constant. Figure 5 shows the relationship between the local optimal solution and the number of the cranky ants.

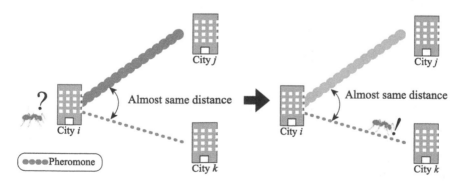

Fig. 3 Example of selection of the cranky ant

Fig. 4 Example of the escape of local optimal solution

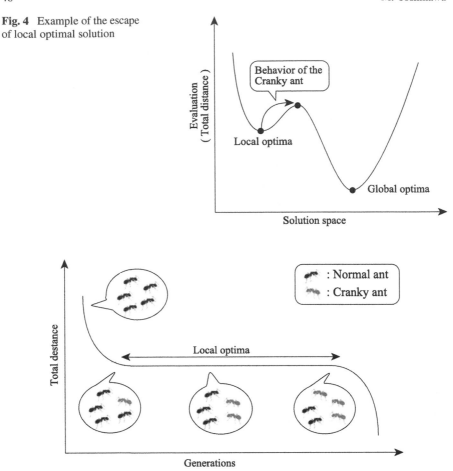

Fig. 5 Example of relationship between the local optimal solution and the number of the cranky ants

4 Experiments and Discussions

To evaluate the proposed algorithm, we conducted several experiments in comparison with the conventional ACO. The experimental platform is Pentium Core 2 Duo with 2G byte memory and the program is described by C language. As experimental data, the travelling salesman library (TSP.LIB) benchmark data of 51 cities (eil51) and that of 100 cities (kroA100) are used. The number of the cranky ants at the initial generation is the several kinds in these experiments.

4.1 Evaluation for Searching Performance

First, we evaluate the searching performance of the proposed algorithm. Experimental results of 10 trials are shown in Tables 1–4. In these tables, #cranky indicates the number of the cranky ants, #normal indicates that of the normal ants, and the value with squares brackets represents the optimal solution in each benchmark data. The parameter q is different in both experiments. The parameters of Tables 1 and 2 are 0.25, that of Tables 3 and 4 are 0.50, respectively.

As shown in both tables, the proposed algorithm found the optimal solution, although the conventional ACO couldn't find it in the case of KroA100. When the ratio of "the number of normal ants" and "the number of cranky ants" was 4:1, the best result was obtained in a small-scale problem, eil51. When the ratio was 1:1, the best result was obtained in a medium-scale problem, KroA100.

These results indicate that since it is more easily led towards the local optimal solution as the scale of a problem increases, strengthening the pressure of diversification (an increase in the number of cranky ants) is an important factor in improving solution searching ability.

Table 1 Results of eil51 using parameter $q = 0.25$

	# Ants	Best	Worst
Conventional ACO	# normal: 51 # cranky: 0	[426]	433
ACO with cranky	# normal: 41 # cranky: 10	427	437
ants (constant)	# normal: 26 # cranky: 25	427	438
	# normal: 10 # cranky: 41	428	459
ACO with Cranky	# normal: 41 # cranky: 10	[426]	432
ants (adaptive)	# normal: 26 # cranky: 25	427	441
	# normal: 10 # cranky: 41	437	459

Table 2 Results of KroA100 using parameter $q = 0.25$

	# Ants	Best	Worst
Conventional ACO	# normal: 100 # cranky: 0	21,792	22,458
ACO with cranky	# normal: 80 # cranky: 20	21,406	22,601
ants (constant)	# normal: 50 # cranky: 50	[21,282]	21,543
	# normal: 40 # cranky: 60	[21,282]	21,706
	# normal: 20 # cranky: 80	21,476	22,547
ACO with cranky	# normal: 80 # cranky: 20	21,370	21,767
ants (adaptive)	# normal: 50 # cranky: 50	[21,282]	21,470
	# normal: 40 # cranky: 60	[21,282]	21,877
	# normal: 20 # cranky: 80	21,597	22,824

Table 3 Results of eil51 using parameter $q = 0.50$

	# Ants	Best	Worst
Conventional ACO	# normal: 51 # cranky: 0	[426]	433
ACO with cranky ants (constant)	# normal: 41 # cranky: 10	[426]	435
	# normal: 26 # cranky: 25	427	440
	# normal: 10 # cranky: 41	430	452
ACO with cranky ants (adaptive)	# normal: 41 # cranky: 10	[426]	432
	# normal: 26 # cranky: 25	428	444
	# normal: 10 # cranky: 41	434	476

Table 4 Results of KroA100 using parameter $q = 0.50$

	# Ants	Best	Worst
Conventional ACO	# normal: 100 # cranky: 0	21,678	22,211
ACO with cranky ants (constant)	# normal: 80 # cranky: 20	21,540	21,821
	# normal: 50 # cranky: 50	[21,282]	21,694
	# normal: 40 # cranky: 60	[21,282]	22,095
	# normal: 20 # cranky: 80	21,505	22,587
ACO with cranky ants (adaptive)	# normal: 80 # cranky: 20	21,340	21,635
	# normal: 50 # cranky: 50	[21,282]	21,694
	# normal: 40 # cranky: 60	[21,282]	21,646
	# normal: 20 # cranky: 80	21,453	23,229

Table 5 Run time of "one iteration" on eil51 using parameter $q = 0.25$

	# Ants	Time (s)
Conventional ACO	# normal: 51 # cranky: 0	0.010527
ACO with cranky ants (constant)	# normal: 41 # cranky: 10	0.010963
	# normal: 26 # cranky: 25	0.010583
	# normal: 10 # cranky: 41	0.011014
ACO with cranky ants (adaptive)	# normal: 41 # cranky: 10	0.010473
	# normal: 26 # cranky: 25	0.010719
	# normal: 10 # cranky: 41	0.010279

4.2 Evaluation for Processing Time

Next, we evaluate the processing speed of the proposed algorithm, using different parameter values. The run time is evaluated in a similar way to the experiments for evaluating performance; ten trials were performed for each case, in which the value of parameter q was set to be either 0.25 or 0.5. Tables 5–8 show the experimental results.

Each table shows the time required for "one iteration". Here, "one iteration" is defined as "a series of processing steps until the global update rule is applied

Table 6 Run time of "one iteration" on KroA100 using parameter $q = 0.25$

	# Ants	Time (s)
Conventional ACO	# normal: 100 # cranky: 0	0.035715
ACO with cranky	# normal: 80 # cranky: 20	0.036565
ants (constant)	# normal: 50 # cranky: 50	0.036155
	# normal: 40 # cranky: 60	0.036313
	# normal: 20 # cranky: 80	0.036456
ACO with cranky	# normal: 80 # cranky: 20	0.037322
ants (adaptive)	# normal: 50 # cranky: 50	0.036017
	# normal: 40 # cranky: 60	0.036924
	# normal: 20 # cranky: 80	0.037424

Table 7 Run time of "one iteration" on eil51 using parameter $q = 0.50$

	# Ants	Time (s)
Conventional ACO	# normal: 51 # cranky: 0	0.010614
ACO with cranky	# normal: 41 # cranky: 10	0.011356
ants (constant)	# normal: 26 # cranky: 25	0.010659
	# normal: 10 # cranky: 41	0.010925
ACO with cranky	# normal: 41 # cranky: 10	0.009825
ants (adaptive)	# normal: 26 # cranky: 25	0.009766
	# normal: 10 # cranky: 41	0.009851

Table 8 Run time of "one iteration" on KroA100 using parameter $q = 0.50$

	# Ants	Time (s)
Conventional ACO	# normal: 100 #cranky: 0	0.035170
ACO with cranky	# normal: 80 #cranky: 20	0.037300
ants (constant)	# normal: 50 #cranky: 50	0.036624
	# normal: 40 #cranky: 60	0.036998
	# normal: 20 #cranky: 80	0.036495
ACO with cranky	# normal: 80 #cranky: 20	0.037480
ants (adaptive)	# normal: 50 #cranky: 50	0.037846
	# normal: 40 #cranky: 60	0.037425
	# normal: 20 #cranky: 80	0.037658

to the best travelling route after each ant has completed one travelling route". As shown in these tables, even when cranky ants and the observation technique for the convergence behavior were introduced, there was no significant difference in the processing time.

4.3   Evaluation for Adaptive Optimization

Finally, we verify the validity of the technique of which the number of cranky ants is adaptively changed. In the verification, KroA100 is used as the objective problem, and 50 normal ants and 50 cranky ants are employed.

Figure 6 shows the changes in the evaluation value, the number of normal ants, and the number of cranky ants.

As shown in Fig. 6, although the evaluation value significantly improved from the initial generation to approximately the 50th generation, the solution did not improve after the 50th generation to approximately the 200th generation. Regarding the number of ants, the number of normal ants decreased and that of cranky ants increased just after the 150th generation.

The reason for this is that the proposed algorithm regarded the search as converged after approximately the 50th generation. In other words, the proposed algorithm judged to be trapped on the local optimal solution.

Thus, by increasing the number of cranky ants, the proposed algorithm escaped from the local solution at approximately the 200th generation. Since the proposed algorithm judged to be trapped on the local optimal solution again after the 400th generation, the number of cranky ants was gradually increased until the algorithm provided an escape from the local optimal solution. This occurred, and the global optimal solution was obtained, near the 900th generation.

In the proposed algorithm, the number of the cranky ants has increased as the generation advances. It means that the cranky ants work to expand the searching

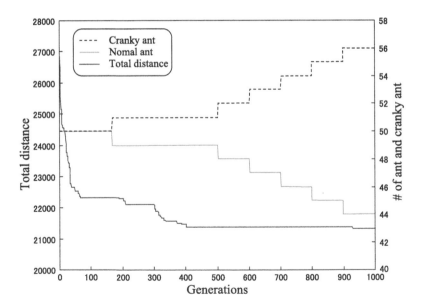

Fig. 6   Relationship between the total distance and the transition of the number of the normal ants

area. Thus, the proposed algorithm enables to improve the searching performance, and it achieves to control the trade-off between intensification and diversification effectively.

5   Conclusion

In this chapter, we proposed a new adaptive ant colony optimization algorithm. It is newly introduced two techniques: the observation technique for the convergence behavior, and the escape technique from the local optimal solution. The observation technique utilized the transition of evaluation, and the escape technique employed the cranky ant which works to strengthen the pressure of diversification. These techniques enabled the search to prevent from trapping at the local optimal solution and achieved the control of the trade-off between intensification and diversification. Experiments using benchmark data proved the effectiveness of the proposed algorithm in comparison with the conventional ACO.

In relation to future work, experiments using large scale data are the most important priority. We will also apply it to other combinatorial optimization problems.

References

1. Dorigo, M., Maniezzo, V., & Colorni, A. (1996). Ant system: Optimization by a colony of cooperating agents. *IEEE Transactions on Systems, Man and Cybernetics, Part B, 26*(1), 29–41.
2. Dorigo, M., & Gambardella, L.M. (1997). Ant colony system: A cooperative learning approach to the traveling salesman problem. *IEEE Transactions on Evolutionary Computation, 1*(1), 53–66.
3. Luo, R., & Sun, R.P. (2007). A novel ant colony optimization based temperature-aware floor-planning algorithm. *Proceedings of Third International Conference on Natural Computation, 4*, pp. 751–755.
4. Ramkumar, A.S., & Ponnambalam, S.G. (2006). Hybrid ant colony system for solving quadratic assignment formulation of machine layout problems. *Proceedings of IEEE Conference on Cybernetics and Intelligent Systems*, pp. 1–5.
5. Sankar, S.S., Ponnambalam, S.G., Rathinavel, V., & Visveshvaren, M.S. (2005). Scheduling in parallel machine shop: An ant colony optimization approach. *Proceedings of IEEE International Conference on Industrial Technology*, pp. 276–280.
6. Yoshikawa, M., & Terai, H. (2006). A hybrid ant colony optimization technique for job-shop scheduling problems. *Proceedings of IEEE/ACIS International Conference on Software Engineering Research, Management & Applications*, pp. 95–100.
7. Yoshikawa, M., & Terai, H. (2007). Architecture for high-speed Ant Colony Optimization. *Proceedings of IEEE International Conference on Information Reuse and Integration*, pp. 1–5.
8. Zhang, H.J., Ning, H.Y., & Hong-yun (2008). New little-window-based self-adaptive ant colony-genetic hybrid algorithm. *Proceedings of International Symposium on Computational Intelligence and Design, 1*, pp. 250–254.
9. Holland, J. (1992). *Adaptation in Natural Artificial Systems*. Ann Arbor, MI: The University of Michigan Press (2nd ed. MIT Press).
10. Goldberg, D.E. (1989). Genetic algorithms in search optimization, and machine learning. Reading, MA: Addison Wesley.

11. Rutenbar, R.A. (1989). Simulated annealing algorithms: An overview. *IEEE Circuits and Devices Magazine, 5*(1), 19–26.
12. Shang, G., Xinzi, J., & Kezong, T. (2007). Hybrid algorithm combining ant colony optimization algorithm with genetic algorithm. *Proceedings of Chinese Control Conference*, pp. 701–704.
13. Lee, S.G., Jung, T.U., & Chung, T.C. (2001). Improved ant agents system by the dynamic parameter decision. *Proceedings of 10th IEEE International Conference on Fuzzy Systems, 2*, 666–669.

Chapter 5
A Kind of Cascade Linguistic Attribute Hierarchies for the Two-Way Information Propagation and Its Optimisation

Hongmei He and Jonathan Lawry

Abstract A hierarchical approach, in which a high-dimensional model is decomposed into series of low-dimensional sub-models connected in cascade, has been shown to be an effective way to overcome the 'curse of dimensionality' problem. We investigate a cascade linguistic attribute hierarchy (CLAH) embedded with linguistic decision trees (LDTs), which can present two-way information propagations. The upwards information propagation forms a process of cascade decision making, and cascade transparent linguistic rules represented by a cascade hierarchy will be useful for analyzing the effect of different attributes on the decision making in a special application. The downwards information propagation presents the constraints to low-level attributes for a given high-level goal threshold. Noisy signals can be thrown out in low level, which could protect from information traffic congestion in wireless sensor networks. A genetic algorithm with linguistic ID3 in wrapper is developed to find optimal CLAHs. Experimental results have shown that an optimal cascade hierarchy of LDTs can not only greatly reduce the number of rules when the relationship between a goal variable and input attributes is highly uncertain and nonlinear, but also achieve better performance in accuracy and ROC curve than a single LDT.

Keywords Cascade linguistic attribute hierarchy · Upwards propagation · Downwards propagation · Cascade decision making · Genetic algorithm in wrapper · Linguistic ID3

1 Introduction

It is required to have an integrated treatment of uncertainty and fuzziness when modeling the information propagation from low-level attributes to high-level goal variables. One of the main drawbacks to fuzzy modeling of systems is known as the

H. He (✉) and J. Lawry
Department of Engineering Mathematics, University of Bristol, Bristol, UK
e-mail: H.He@bristol.ac.uk; Lawry@bristol.ac.uk

S.-I. Ao et al. (eds.), *Intelligent Automation and Computer Engineering*,
Lecture Notes in Electrical Engineering 52, DOI 10.1007/978-90-481-3517-2_5,
© Springer Science+Business Media B.V. 2010

'curse of dimensionality', which is the exponential growth in the number of possible fuzzy rules as a function of the dimension of model input space. A hierarchical approach in which, the original high-dimensional model is decomposed into series of low-dimensional sub-models connected in cascade, has been shown to be an effective way to overcome this problem since it provides a linear growth in the number of rules and parameters as the input dimension increases [12]. Campello and Amaral presented a unilateral transformation that converts the proposed hierarchical model into a mathematically equivalent non-hierarchical one [2].

As a result of the uncertainty and non-linear relationship between different attributes and a goal variable, different cascade hierarchies will have different performance on decision making procedures. In this paper, we propose a cascade hierarchy approach embedded with LDTs representing transparent rules, and describe the process of information propagation through a cascade hierarchy. We then develop a genetic algorithm with the Linguistic ID3 (LID3) [10] algorithm in wrapper to optimise cascade hierarchies. The experiments are performed on benchmark databases from the UCI Machine Learning Repository.

2 Label Semantics

Label semantics [5, 6] proposes two fundamental and inter-related measures of the appropriateness of labels as descriptions of an object or value. Given a finite set of labels $\mathcal{L}$ from which can be generated a set of expressions LE through recursive applications of logical connectives, the measure of appropriateness of an expression $\theta \in LE$ as a description of instance x is denoted by $\mu_\theta(x)$ and quantifies the agent's subjective belief that θ can be used to describe x based on his/her (partial) knowledge of the current labeling conventions of the population. When faced with an object to describe, an agent may consider each label in $\mathcal{L}$ and attempt to identify the subset of labels that are appropriate to use. Let this set be denoted by $\mathcal{D}_x$. In the face of their uncertainty regarding labeling conventions the agent will also be uncertain as to the composition of $\mathcal{D}_x$, and this is quantified by a probability mass function $m_x : 2^{\mathcal{L}} \to [0, 1]$ on subsets of labels.

Unlike linguistic variables [13], which allow for the generation of new label symbols using a syntactic rule, label semantics assumes a finite set of labels $\mathcal{L}$. These are the basic or core labels to describe elements in an underlying domain of discourse Ω. Based on $\mathcal{L}$, the set of label expressions LE is then generated by recursive application of the standard logic connectives as follows:

Definition 1. Label Expressions
The set of label expressions LE of $\mathcal{L}$ is defined recursively as follows:

- If $L \in \mathcal{L}$, then $L \in LE$.
- If $\theta, \varphi \in LE$, then $\neg\theta, \theta \wedge \varphi, \theta \vee \varphi \in LE$.

A mass assignment m_x on sets of labels then quantifies the agent's belief that any particular subset of labels contains all and only the labels with which it is appropriate to describe x.

Definition 2. Mass Assignment on Labels
$\forall x \in \Omega$ a mass assignment on labels is a function $m_x : 2^{\mathcal{L}} \rightarrow [0, 1]$ such that $\sum_{S \subseteq \mathcal{L}} m_x (S) = 1$.

Now depending on labeling conventions there may be certain combinations of labels which cannot all be appropriate to describe any object. For example, *small* and *large* cannot both be appropriate. This restricts the possible values of $\mathcal{D}_x$ to the following set of focal elements:

Definition 3. Set of Focal Elements
Given labels $\mathcal{L}$ together with associated mass assignment $m_x : \forall x \in \Omega$, the set of focal elements for $\mathcal{L}$ is given by:

$$\mathcal{F} = \{S \subseteq \mathcal{L} : \exists x \in \Omega, \, m_x (S) > 0\} \qquad (1)$$

The appropriateness measure, $\mu_\theta (x)$, and the mass m_x are then related to each other on the basis that asserting 'x is θ' provides direct constraints on $\mathcal{D}_x$. For example, asserting 'x is $L_1 \wedge L_2$', for labels $L_1, L_2 \in \mathcal{L}$ is taken as conveying the information that both L_1 and L_2 are appropriate to describe x so that $\{L_1, L_2\} \subseteq \mathcal{D}_x$. Similarly, '$x$ is $\neg L$' implies that L is not appropriate to describe x so $L \notin \mathcal{D}_x$. In general we can recursively define a mapping $\lambda : LE \rightarrow 2^{2^{\mathcal{L}}}$ from expressions to sets of subsets of labels, such that the assertion 'x is θ' directly implies the constraint $\mathcal{D}_x \in \lambda (\theta)$ and where $\lambda (\theta)$ is dependent on the logical structure of θ. For example, if $\mathcal{L} = \{low, medium, high\}$ then $\lambda(medium \wedge \neg high) = \{\{low, medium\}, \{medium\}\}$ corresponding to those sets of labels which include *medium* but do not include *high*. Hence, the description $\mathcal{D}_x$ provides an alternative to Zadeh's linguistic variables in which the imprecise constraint 'x is θ' on x, is represented by the precise constraint $\mathcal{D}_x \in \lambda(\theta)$, on $\mathcal{D}_x$.

Definition 4. λ-mapping $\lambda : LE \rightarrow 2^{\mathcal{F}}$ is defined recursively as follows: $\forall \theta, \, \varphi \in LE$

- $\forall L_i \in \mathcal{L} \; \lambda(L_i) = \{F \in \mathcal{F} : L_i \in F\}$
- $\lambda(\theta \wedge \varphi) = \lambda(\theta) \cap \lambda(\varphi)$
- $\lambda(\theta \vee \varphi) = \lambda(\theta) \cup \lambda(\varphi)$
- $\lambda(\neg \theta) = \lambda(\theta)^c$

Therefore, based on the λ-mapping we define the appropriateness measure as below:

Definition 5 (Appropriateness Measure). Appropriateness measure $\mu_{\theta(x)}$ is evaluated as the sum of mass assignment m_x over those subsets of labels in $\lambda_\theta(x)$, i.e., $\forall \theta \in LE, \forall x \in \Omega, \mu_{\theta(x)} = \sum_{F \in \lambda(\theta)} m_x(F)$.

For example, if $\mathcal{L} = \{low(l), medium(m), high(h)\}$ with focal sets $\{\{l\}, \{l, m\}, \{h\}\}$ and $\theta = l \wedge \neg m$, then
$$\mu_{l \wedge \neg m}(x) = \sum_{F : l \in F, m \notin F} m_x(F) = m_x(\{l\}).$$

3   A Cascade Linguistic Attribute Hierarchy

3.1   Definition of a Cascade Hierarchy

The process of aggregation of evidence in multi-attribute decision problems based on attributes $x_1, \ldots, x_n$ can be viewed as a functional mapping between a high level variable y and input attributes, $y = f(x_1, \ldots, x_n)$, which is often dynamic and nonlinear, and may be imprecisely defined. In some cases, the function f may be approximated by a composition of lower dimensional sub-functions, forming a cascade hierarchy (a binary tree). Each sub-function represents a new intermediate attribute. Figure 1 shows a simple cascade hierarchy. There are $n - 1$ intermediate attributes produced. The last intermediate attribute z_{n-1} corresponds to the goal variable y. The cascade relationship is expressed as following:

$$z_i = \begin{cases} F_1(x_1, x_2) & i = 1, \\ F_i(z_{i-1}, x_{i+1}) & n > i > 1. \end{cases} \tag{2}$$

As proposed in [7], in a linguistic attribute hierarchy, function mappings between parent and child attribute nodes are defined in terms of weighted linguistic rules which explicitly model both the uncertainty and vagueness which often characterises our knowledge of such aggregation functions. These rules will be defined as conditional expressions in the label semantics framework [6] weighted by conditional probabilities. For each attribute, a set of labels and subsequent label expressions is defined. We assume that expressions describing a parent attribute can be (imprecisely) defined in terms of a description of its children. Let $\mathcal{L}_i$, θ_i and $\mathcal{F}_i$ denote the set of labels, a label expression and focal sets respectively, defined for attribute x_i for $i = 1, \ldots, n$. Similarly, let $\mathcal{L}_y$, θ_y and $\mathcal{F}_y$ denote the label set, a label expression and focal set for describing the goal variable y, respectively.

More precisely, the weighted conditional rules can take the form of an LDT. In an LDT, the nodes are attributes, and the edges are label expressions describing each attribute. The depth of an LDT with two input attributes is at most 2. A branch B is a conjunction of expressions $\theta_1 \wedge \theta_2$, where θ_1 and θ_2 are the label expressions of the two edges on the branch B, respectively. Each branch also is augmented by a set of conditional mass values $m(F|B) = P(C_x = F|B)$, for each output focal element $F \in \mathcal{F}_y$. Then the rules corresponding to the branch B would be: $\theta_1 \wedge \theta_2 \rightarrow F : m(F|B)$ for each focal element $F \in \mathcal{F}_y$.

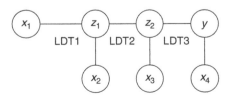

Fig. 1 A cascade hierarchy of LDTs

3.2 Upwards Propagation of Information

The upwards propagation of information through a cascade hierarchy of Linguistic Decision Trees (LDTs) based on label semantics forms a process of cascade decision making. Figure 1 shows the process of bottom-up information propagation through the cascade hierarchy. The only information available regarding the mappings F_1, F_2 and F_3 is in the form of decision trees LDT_1, LDT_2 and LDT_3, which define mapping functions for z_1 in terms of those for x_1 and x_2, for z_2 in terms of those for z_1 and x_3, and for y in terms of those for z_2 and x_4.

However, it is not easy to define the labels for intermediate attributes in terms of their children, as the intermediate attributes are not directly related to basic attributes in the system [2]. Therefore, we suppose all intermediate attributes are approximations of the decision variable y with the same domain and description labels. According to Jeffrey's rule [4], given an LDT, the mass assignment of the decision variable can be calculated by:

$$m_{z_i}(F_y) = \sum_{j=1}^{t_i} \mu_{\theta_1}(z_{i-1}) * \mu_{\theta_2}(x_{i+1}) * m(F_y|B_{i_j}), \qquad (3)$$

where, B_{i_j} is the jth branch in the ith LDT, and $\mu_\theta(x)$ is appropriateness measure, quantifying the degree of our belief that label expression θ is appropriate for x [6]. The appropriateness measure can be calculated with mass assignments of attribute x according to Definition 5.

Information is propagated along the cascade LDTs from low level to high level. For the example in Fig. 1, given mass functions m_{x_1}, m_{x_2}, m_{x_3}, and m_{x_4}, the mass function m_{z_1} is determined by propagating m_{x_1} and m_{x_2} through LDT_1, m_{z_2} is determined by propagating m_{z_1} and m_{x_3} through LDT_2, and finally, m_y is determined by propagating m_{z_2} and m_{x_4} through decision tree LDT_3 (see Fig. 2).

Here we consider only classification problems where the goal variable y belongs to the finite set of classes $\{C_1, \ldots, C_t\}$. In this case, $\mathcal{F} = \{\{C_1\}, \ldots, \{C_t\}\}$, and for input vector $\vec{x}$, $m_y(\{C_i\}) = P(C_i|\vec{x})$.

3.3 Downwards Propagation of Information

Given a high-level goal of the form 'y is θ', for $\theta \in LE_y$, the aim of downward propagation is to identify low-level constraints on $x_1, \ldots, x_n$ by identifying those combinations of attribute values relevant to this goal.

Fig. 2 The cascade upwards information propagation

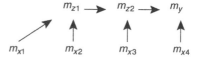

Given a cascade linguistic hierarchy, we recursively trace the regions of all attributes from top to bottom, according to the requirement $\mu_\theta(B) \geq \alpha$, where $\mu_\theta(B) = \sum_{F \in \mathcal{F}_y} m(F|B)$ and α is a threshold value in $[0, 1]$. If the attribute is an intermediate attribute, then further trace the regions of the attributes in the LDT whose output attribute is the intermediate attribute. The task for tracing the regions of attributes in an LDT is to identify the branches in the LDT for which the probability of a high-level goal exceeds the given threshold. This results in a disjunction $B_1 \vee B_2 \vee \cdots \vee B_m$ of branches. Each of these branches is a conjunction of descriptions of attributes at the next lower level of the hierarchy $B_i = \theta_{T_{i-1}} \wedge \theta_k$. A conjunct for an intermediate attribute in each branch is then treated as a new goal and replaced by an equivalent disjunction, and the above process can be iterated down the cascade hierarchy. The top level is goal variable y (i.e., T_{n-1}, for n basic attributes). LE_j is the label expression for branch B_j with two edges. θ_{T_i} is the label expression of the intermediate attribute for the LDT T_i, θ_k is the label expression of a basic attribute in $\vec{x} = \{x_1, \ldots, x_n\}$. Algorithm 1 shows the pseudo code of the downwards algorithm.

Algorithm 1 Downwards propagation(H,θ,α)

1: $i = n - 1, LE = \theta_{T_i} = \theta$;
2: **while** $i > 0$ **do**
3:    **for** (each branch B_j in T_i of H) **do**
4:       $\psi_i = null$;
5:       E=evaluate($\mu_\theta(T_i|B_j) \geq \alpha$)
6:       **if** (E is true) **then**
7:          $LE_j = \theta_{T_{i-1}} \wedge \theta_k$;
8:          $\psi_i = \psi_i \vee LE_j$;
9:       **end if**
10:   **end for**
11:   ψ_i replaces the θ_{T_i} in LE;
12:   $i = i - 1$;
13: **end while**

---

4 GA in Wrapper to Optimise Cascade Hierarchies

4.1 Chromosomes and Reproduction

To learn a linguistic cascade hierarchy, we develop a genetic algorithm as a search agent with the LID3 as an induction algorithm in wrapper. For the optimisation of cascade hierarchies with n attributes, the size of whole search space is $\frac{n!}{2}$. The performance of different hierarchies is judged on the basis of the accuracy for the given classification task.

Fig. 3 Two-point order
crossover

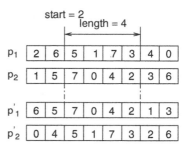

Chromosomes The purpose of a GA is to evolve a population of potential solutions each corresponding to the cascade hierarchies in a multiple-attribute space. Therefore, the GA in wrapper approach conducts a search in the space of possible cascade hierarchies. Different attribute orderings define different cascade hierarchies. So we define any possible permutation of all attributes, $\pi = \{x_1, \ldots x_n\}$, and $\pi \to \mathcal{H}$ as a genome of the genetic algorithm.

Reproduction We use "*roulette-wheel*" selection, according to which, an individual with better fitness has higher probability of being selected. The probability that hierarchy $\mathcal{H}_i$ is selected is given by the nominalised fitness:

$$p_i = \frac{f_i(\mathcal{H}_i)}{\sum_{j=1}^{\Gamma} f_j(\mathcal{H}_j)}. \tag{4}$$

A one-elitism strategy is included since it keeps the current best individual in the next generation, and speeds up the convergence of the evolution process. On the other hand, in order to keep the diversity of solutions, a random hierarchy is generated in each generation.

We use two-point order crossover as follows (Fig. 3): two parental permutations, π_1 and π_2, are chosen randomly depending on the probability chosen in 4. A continuous interval of the permutation π_1 is chosen, and also an interval starting at the same position and of the same length from π_2. The two parameters, '*starting position*' and '*length of interval*', are produced randomly. Two new permutations, π_1' and π_2', are created such that π_1' contains the interval from π_2 with the rest being the other elements of π_1 in the same order as they appeared in π_1. π_2' contains the interval from π_1 with the rest being the other elements of π_2 in the order as they were in π_2 (Fig. 3). Mutation, which is the swapping of two randomly picked elements of a permutation, is carried out with some probability (*m_rate*) on each child in the population.

4.2   Evaluation and Termination Criteria

Here we only consider binary classification problem with two classes '+' and '−'. First, we investigate the ordinary accuracy on a threshold, which is the ratio of the number of correct classifications to the number of testing samples. When the estimated probability $p(C|\vec{x})$ (equivalent to $m_y(\{C\})$) that a sample with measurement vector $\vec{x}$ belongs to class C is larger than a threshold α, then that sample is classified as C. Conventionally we use 0.5 as a threshold. Here we consider two measures of accuracy, integrated accuracy and the area under ROC curve, which measures how well the classifier separates the two classes without reference to a decision threshold. The closer the ROC plot is to the upper left corner, the higher the ordinary accuracy of the test results.

For each possible threshold α for discriminating between the two classes, some positive cases will be correctly classified as positive (TP_α = number of True Positive), but some positive cases will be estimated as negative (FN_α = number of False Negative). On the other hand, some negative cases will be correctly classified as negative (TN_α = number of True Negative), but some negative cases will be classified as positive (FP_α = number of False Positive).

Accuracy For a decision maker, the *Ordinary Accuracy* ($\mathcal{A}_\alpha$) over a threshold α can be calculated as below:

$$\mathcal{A}_\alpha(\mathcal{H}) = \frac{TP_\alpha + TN_\alpha}{\mathcal{M}}, \tag{5}$$

where, $\mathcal{M}$ is the number of test examples. In order to reduce the sensitivity to the threshold α, we define the integrated accuracy to be the integration of accuracies for all $\alpha \in [0.5, 1)$ (Formula (6)):

$$\mathcal{A}_{\tilde{\alpha}}(\mathcal{H}) = \int_{0.5}^{1} \frac{\mathcal{N}_\alpha}{\mathcal{M}} d\alpha \approx \frac{\Delta(\alpha)}{\mathcal{M}} \sum_{i=1}^{m} \mathcal{N}_{\alpha_i}, \tag{6}$$

where, the interval $[0.5, 1)$ is divided into m subintervals with constant step length $\Delta(\alpha)$, and where $\mathcal{N}_{\alpha_i} = TP_{\alpha_i} + TN_{\alpha_i}$.

ROC Curve Receiver Operating Characteristic (ROC) analysis originated from signal detection theory and has been introduced to machine learning in recent years in order to evaluate algorithm performance in an imprecise environment. It is claimed [9] that ROC graphs can offer a more robust framework for evaluating classifier performance than traditional accuracy measure. The true positive rate is calculated with $\eta = \frac{TP}{P}$. The false positive rate is calculated with $\sigma = \frac{FP}{N}$. In a ROC curve, the true positive rate (η) is plotted as a function of the false positive rate (σ) for varying thresholds. Each point on a ROC plot represents a (η, σ) pair corresponding to a particular decision threshold.

Similarly, the integrated accuracy can be defined as the area under ROC curve, which measures how well the decision maker separates the two classes without reference to a decision threshold, as follows:

$$\mathcal{A}_{ROC}(\mathcal{H}) = \int_{0}^{1} \eta\, d\sigma \tag{7}$$

Let $p(+|\vec{x})$ be the estimation of the probability that an instance with measurement vector $\vec{x}$ is positive. If we rank test instances according to increasing positive probabilities, then the area under the ROC curve ($\mathcal{A}_{ROC}$) for a decision making problem with two classes $+,-$ can be calculated [3] by:

$$\mathcal{A}_{ROC} = \frac{\sum_{i=1}^{P} r_i - P(P+1)/2}{PN}, \tag{8}$$

where, P and N are the numbers of positive and negative samples, r_i is the rank of the ith positive instance in the rank list according to the probabilities of the positive class.

Termination Criteria Termination is an important parameter, which affects the running time and quality of solutions. Generally it heavily depends on the size of the chromosome. The maximum generations max_gen is linear function of the number of basic attributes. The evolution procedure will be repeated until the maximum number of generations is reached.

4.3 LID3 Algorithm for the Induction of an LDT

In order to obtain an LAH embedded with LDTs, we need to train in turn all LDTs in the hierarchy. The LID3 algorithm [10] for training cascade LDTs is a black box as part of evaluation in the wrapper of the Genetic Algorithm. LID3, an extension of well-known ID3 algorithm [11], is used to build an LDT based on a given linguistic database. The search is guided by a modified measure of information gain in accordance with label semantics.

Definition 6 (Branch Entropy). The entropy of branch B, for a given goal variable belonging to class set $C = \{C_1, \ldots, C_t\}$, is

$$E(B) = -\sum_{i=1}^{t} P(C_i|B)log_2 P(C_i|B) \tag{9}$$

Given a branch B, suppose x_j is expanded to the branch B, then the Expected Entropy is defined as follows:

Definition 7 (Expected Entropy).

$$EE(B, x_j) = \sum_{F_j \in \mathcal{F}_j} E(B \cup F_j)P(F_j|B). \tag{10}$$

where, $B \cup F_j$ represents the new branch obtained by appending the focal element F_j to the end of branch B. The probability of F_j given B can be calculated as follows:

$$P(F_j|B) = \frac{\sum_{\vec{x} \in \mathcal{D}} P(B \cup F_j|\vec{x})}{\sum_{\vec{x} \in \mathcal{D}} P(B|\vec{x})}, \tag{11}$$

where, $P(B|\vec{x}) = \mu_B(\vec{x}) = \mu_{\theta_1}(x_1) * \mu_{\theta_2}(x_2)$, θ_1 and θ_2 are two label expressions associated with the two edges in the branch B, and x_1 and x_2 are incident to the two edges. Hence, the *Information Gain* can be calculated by:

$$IG(B, x_j) = E(B) - EE(B, x_j). \tag{12}$$

The most informative attribute will form the root of an LDT, and the tree will expand into branches associated with all possible focal elements of this attribute. For each branch, the free attribute with maximal information gain will be next node until the branch reaches the specified maximum depth or the maximum class probability arrives the given threshold. The process forms a level order traversal.

5 Experiments and Evaluation

Experiment Methodologies All attributes are discretised using an entropy-based approach into three labels ($\mathcal{L} = \{small, medium, large\}$), respectively. Each label corresponds to a trapezoidal fuzzy set, which has 50% overlapping with neighbouring label fuzzy sets. Ten-fold cross validation is used for the experiments. Data is split into 10 approximate equal partitions. Each one is used in turn for testing while the remainder is used for training, i.e., 9/10 of data is used for training and 1/10 for testing. The whole procedure is repeated 10 times. A trained hierarchy is evaluated using two types of accuracy measure described in Section 4.2. The ordinary accuracy is evaluated at threshold 0.5. The area under a ROC curve is calculated with Formula (8). We examine the quality of cascade decision making and the cost of a hierarchy, i.e., the total number of branches from all decision trees in a cascade hierarchy, and compare the performance with that of a single LDT providing a direct mapping between input attributes and a classification variable. The experiments are carried out on the Pima Diabetes database from UCI machine learning repository.

5.1 On the Pima Diabetes Database

The Pima Database   The Pima Indian data set is a well-known benchmark problem from the UCI repository [1]. The problem relates to incidents of Diabetes mellitus in the Pima Indian population living near Phoenix Arizona. The target attribute is a binary valued decision variable indicating whether or not the patient shows signs of Diabetes according to World Health Organisation criteria. The database of Diabetes includes 768 samples, in which, 268 positive instances (with Diabetes), 500 instances without Diabetes. There are 8 basic attributes. Table 1 shows the information for all attributes.

Table 1 Attribute information in the database of Pima Diabetes, including Lower Bounds (LB), Upper Bounds (UB)

x_i	Description	LB	UB
x_0	Number of times pregnant	0	17
x_1	Plasma glucose concentration a 2 h in an oral glucose tolerance test	0	199
x_2	Diastolic blood pressure (mm Hg)	0	122
x_3	Triceps skin fold thickness (mm)	0	99
x_4	Two-Hour serum insulin (μ U/ml)	0	846
x_5	Body mass index (weight in kg/(height in m)^2)	0	67.1
x_6	Diabetes pedigree function	0.078	2.42
x_7	Age (years)	21	81
y	+/−. + indicates "tested positive for diabetes"	0	1

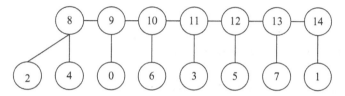

Fig. 4 Optimal cascade hierarchy $\mathcal{H}_2$ for the Pima diabetes database

Table 2 Evaluations of hierarchies obtained by GAW on the Pima database

$\mathcal{H}$	$\mathcal{A}_a$	$\mathcal{A}_{\bar{a}}$	$\mathcal{A}_{ROC}$	β
$\mathcal{H}_1$	0.747396	0.188281	0.783776	115
$\mathcal{H}_2$	0.748698	0.189437	0.790649	115
LDT	0.713542	0.244922	0.769687	1,4845

Solutions and Fitness Values The two orders of attributes corresponding to the optimal cascade hierarchies ($\mathcal{H}_1$ and $\mathcal{H}_2$) obtained by the GAW with fitness values evaluated by $\mathcal{A}_a$ and $\mathcal{A}_{ROC}$ respectively, are: $\mathcal{H}_1$: 3, 4, 2, 5, 6, 7, 0, 1; $\mathcal{H}_2$: 2, 4, 0, 6, 3, 5, 7, 1 (Fig. 4). Table 2 lists the accuracies at threshold 0.5 ($\mathcal{A}_a$), the integrated accuracies ($\mathcal{A}_{\bar{a}}$), the areas under ROC curves ($\mathcal{A}_{ROC}$) and the numbers of branches (β) for $\mathcal{H}_1$, $\mathcal{H}_2$ and the single LDT. It can be seen that $\mathcal{H}_1$ and $\mathcal{H}_2$ achieve similar performance in $\mathcal{A}_a$, $\mathcal{A}_{\bar{a}}$ and $\mathcal{A}_{ROC}$. Their performance in $\mathcal{A}_a$ and $\mathcal{A}_{ROC}$ is better than that of a single LDT, while the single LDT has higher integrated accuracy than $\mathcal{H}_1$ and $\mathcal{H}_2$. The branch numbers for $\mathcal{H}_1$ and $\mathcal{H}_2$ are much less than that for the single LDT.

Accuracy and ROC Curves Figure 5a and b show the accuracy and ROC curves for the two hierarchies and the single LDT, respectively. From the accuracy curves in Fig. 5a, it can be seen that $\mathcal{H}_1$ and $\mathcal{H}_2$ obtain approximately the same accuracy curves, and achieve higher ordinary accuracies at threshold 0.5 than the single LDT does. But the accuracies obtained by $\mathcal{H}_1$ and $\mathcal{H}_2$ decrease as thresholds increase, and become smaller than that for the single LDT when thresholds are over 0.65. Figure 5b shows that the two optimal cascade hierarchies obtain similar ROC curves to the single LDT, although they have different performance in accuracies.

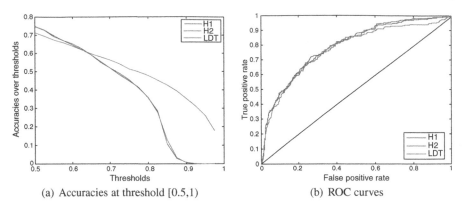

(a) Accuracies at threshold [0.5,1)   (b) ROC curves

Fig. 5 Accuracy and ROC curve for $\mathcal{H}_1$, $\mathcal{H}_2$, and the single LDT on the Pima database

5.2 On the Wisconsin Breast Cancer Database

The Wisconsin Breast Cancer Database The Wisconsin Breast Cancer (WBC) database [1] was created by Dr. William H. Wolberg from the University of Wisconsin Hospitals, Madison [8]. There are 699 samples, in which 458 samples are Benign, and 241 samples are Malignant. There are nine basic attributes, and each attribute is with lower bound 1 and upper bound 10. There are 16 instances that contain a single missing (i.e., unavailable) attribute value. It is claimed that the best result is 93.7% trained on 200 instances and tested on the other 169 in the first group of 369 samples with the 1-nearest neighbor approach in [1].

Solutions and Fitness Values The two permutations of attributes corresponding to the two optimal cascade hierarchies are: $\mathcal{H}_3$:6,2,4,3,8,7,5,1,0; $\mathcal{H}_4$:6,4,3,8,1,7,5,2,0. Table 3 lists the accuracies at threshold 0.5 ($\mathcal{A}_a$), the integrated accuracies ($\mathcal{A}_{\bar{a}}$), and the areas under ROC curves ($\mathcal{A}_{ROC}$) and branch numbers (β) for $\mathcal{H}_3$, $\mathcal{H}_4$ and the single LDT. The experiment results show that $\mathcal{H}_3$ and $\mathcal{H}_4$ have similar performance in ordinary accuracies for different thresholds, and the areas under ROC curves. They have ordinary accuracies at threshold 0.5 better than a single LDT, but they lose performance in the integrated accuracy. The best ordinary accuracy at threshold 0.5 is 96.7% obtained by $\mathcal{H}_3$. Both algorithms for learning a single LDT and a cascade hierarchy have computational complexity $O(n\beta)$, where n is the length of a branch and β is the total number of branches. Table 3 shows that the number of branches for the optimal cascade hierarchies $\mathcal{H}_3$ and $\mathcal{H}_4$ are close to that for the single LDT. However, for each LDT in a cascade hierarchy, there are only two input attributes, thus the length of a branch is at most 2. Therefore, the optimal cascade hierarchies have better computational complexity than the single LDT. Accuracy and ROC Curves Figure 6a and b show the accuracy and ROC curves for the two optimal cascade hierarchies and the single LDT, respectively. From the accuracy curves in Fig. 6a, it can be seen that the ordinary accuracy at threshold 0.5 of $\mathcal{H}_3$

Table 3 Evaluations
of hierarchies obtained by
GAW on the WBC database

$\mathcal{H}$	$\mathcal{A}_a$	$\mathcal{A}_{\bar{a}}$	$\mathcal{A}_{ROC}$	β
$\mathcal{H}_3$	0.967096	0.409156	0.985831	100
$\mathcal{H}_4$	0.962804	0.408530	0.985867	100
LDT	0.934192	0.441863	0.932976	97

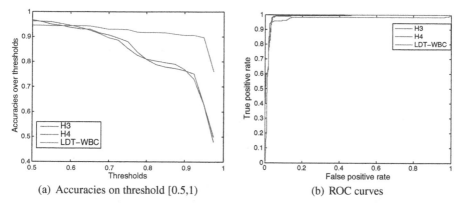

(a) Accuracies on threshold [0.5,1)   (b) ROC curves

Fig. 6 Accuracy and ROC curve for $\mathcal{H}_3$, $\mathcal{H}_4$, and the single LDT on the WBC database

and $\mathcal{H}_4$ is better than that for the single LDT, but their ordinary accuracies when the threshold is larger than 0.6 are worse than for the single LDT. The ROC curves of $\mathcal{H}_3$ and $\mathcal{H}_4$ are slightly better than that for the single LDT.

6   Conclusion

In this paper, we investigated a cascade hierarchy of Linguistic Decision Trees for the two-way information propagation. The upwards propagation forms a process of cascade decision making, through which, we can study the effect of different attributes on the decision making for a special application. Conversely, we can identify and throw out those low-level input vectors that are not located in the regions, which are obtained through the downwards propagation for the given goal over a threshold. A genetic algorithm with the training algorithm LID3 in wrapper was developed to optimise cascade hierarchies for decision making. The optimal cascade hierarchies on the benchmark databases, Pima Diabetes and Wisconsin Breast Cancer, from UCI machine learning repository achieve better performance in the ordinary accuracy at threshold 0.5 and in the area under ROC curves than a single LDT, and the number of rules induced by the optimal cascade hierarchy is much lower than that of a single LDT, when the relationship between a goal and the input attributes is highly uncertain and nonlinear, although accuracy tends to decrease with higher thresholds.

References

1. Asuncion, A., & Newman, D.J. (2007). Uci machine learning repository, irvine, ca: University of california, Department of information and computer science. http://www.ics.uci. edu/ mlearn/MLRepository.html
2. Campello, R.J.G.B., & Amaral, W.C. (2006). Hierarchical fuzzy relational models: Linguistic interpretation and universal approximation. *IEEE Transaction on Fuzzy Systems, 14*(3), 446–453.
3. Hand, D., & Hill, R.J. (2001). A simple generalisation of the area under the roc curve for multiple class classification problems. *Machine Learning, 45*, 171–186.
4. Jeffrey, R.C. (1965). *The logic of decision.* New York: Gordon and Breach.
5. Lawry, J. (2004). A framework for linguistic modeling. *Artificial Intelligence, 155*, 1–39.
6. Lawry, J. (2006). *Modeling and reasoning with vague concepts.* New York: Springer.
7. Lawry, J., & He, H. (2008). Multi-attribute decision making based on label semantics, the international journal of uncertainty, fuzziness and knowledge-based systems. *The International Journal of Uncertainty, Fuzziness and Knowledge-Based Systems, 16*(2), 69–86.
8. Mangasarian, O.L., & Wolberg, W.H. (1990). Cancer diagnosis via linear programming. *SIAM News, 23*(5), 1–18.
9. Qin, Z. (2005). Roc analysis for predictions made by probabilistic classifiers. In proceedings of the international conference on machine learning and cybernetics. In *Proceedingds of the International Conference on Machine Learning and Cybernetics, 5*, 3119–3124.
10. Qin, Z., & Lawry, J. (2005). Decision tree learning with fuzzy labels. *Information Sciences, 172*, 91–129.
11. Quinlan, J.R. (1986). Induction of decision trees. *Machine Learning, 1*, 81–106.
12. Raju, G.U., Zhou, J., & Kiner, R.A. (1991). Hierarchical fuzzy control. *International Journal of Control, 54*(55), 1201–1216.
13. Zadeh, L.A. (1975). The concept of linguistic variable and its application to approximate reasoning part i, part ii. *Information Sciences, 8*(9), 199–249, 301–357, 43–80.

Chapter 6
Simulation Optimization of Practical Concurrent Service Systems

Tad Gonsalves and Kiyoshi Itoh

Abstract Concurrent service systems are modeled using the Generalized Stochastic Petri Nets (GSPN) to account for the multiple asynchronous activities within the system. The simulated operation of the GSPN modeled system is then optimized using the Particle Swarm Optimization (PSO) meta-heuristic algorithm. The objective function consists of the service costs and the waiting costs. Service cost is the cost of hiring service-providing professionals, while waiting cost is the estimate of the loss to business as some customers might not be willing to wait for the service and may decide to go to the competing organizations. The optimization is subject to the management and to the customer satisfaction constraints. The tailor-made PSO is found to converge rapidly yielding optimum results for the operation of a practical concurrent service system.

Keywords Concurrent service systems · Optimization · Meta-heuristics · Swarm Intelligence · Particle Swarm Optimization

1 Introduction

A service system is a configuration of technology and organizational networks designed with the intention of providing service to the end users. Practical service systems include hospitals, banks, ticket-issuing and reservation offices, restaurants, ATM, etc. The managerial authorities are often pressed to drastically reduce the operational costs of active and fully functioning service systems, while the system designers are forced to design (new) service systems operating at minimal costs. Both these situations involve system optimization.

Any optimization problem involves the objective to be optimized and a set of constraints [17]. In this study, we seek to minimize the total cost (tangible

T. Gonsalves (✉) and K. Itoh
Department of Information and Communication Sciences, Sophia University, 7-1 Kioicho, Chiyoda-ku, Tokyo 102-8554, Japan
e-mail: t-gonsal@sophia.ac.jp; itohkiyo@sophia.ac.jp

S.-I. Ao et al. (eds.), *Intelligent Automation and Computer Engineering*,
Lecture Notes in Electrical Engineering 52, DOI 10.1007/978-90-481-3517-2_6,
© Springer Science+Business Media B.V. 2010

and intangible) to the system. The total cost can be divided into two broad categories – cost associated with the incoming customers having to wait for the service (waiting cost) and that associated with the personnel (servers) engaged in providing service (service cost) [1, 6, 13]. Waiting cost is the estimate of the loss to business as some customers might not be willing to wait for the service and may decide to go to the competing organizations, while serving cost is mainly due to the salaries paid to employees.

Business enterprises and companies often mistakenly "throw" capacity at a problem by adding manpower or equipment to reduce the waiting costs. However, too much capacity decreases the profit margin by increasing the production and/or service costs. The managerial staff, therefore, is required to balance the two costs and make a decision about the provision of an optimum level of service.

In recent years, customer satisfaction has become a major issue in marketing research and a number of customer satisfaction measurement techniques have been proposed [2, 5]. Increasing efforts have been made to analyze the causes of customer dissatisfaction and to suggest remedies [4, 20]. In queuing systems, nothing can be as detrimental to customer satisfaction as the experience of waiting for service. For customers, waiting is frustrating, demoralizing, agonizing, aggravating, annoying, time-consuming, and incredibly expensive [10]. Waiting has a negative impact on service quality evaluations [18, 19].

In service systems, customer satisfaction is directly related to the waiting as well as the service experience. In general, the shorter the waiting time and the better the service, the higher is the customer satisfaction. Further, in certain service systems such as restaurants, hospitals and amusement parks, service experience is related to the duration of service (i.e., service time). In such situations, moderately long to sufficiently long service times lead to a higher customer satisfaction. The terms of the type, "moderately long", "sufficiently long" are fuzzy linguistic variables [23, 24] describing the imprecise and subjective experience of the customers. Hence, we define the customer satisfaction constraint as fuzzy sets.

The Particle Swarm Optimization (PSO) is based on the Swarm Intelligence Paradigm of Evolutionary Computation. The algorithm is inspired by the social behavior of birds and fish swarming together to search for food [8, 9]. PSO has been successfully applied to solving optimization problems in diverse disciplines. Compared to other evolutionary computational algorithms, PSO has many desirable characteristics. PSO is easy to implement, can achieve high-quality solutions quickly, and has the flexibility in balancing global and local exploration.

The population-based PSO conducts a search using a population (swarm) of individuals called particles. The performance of each particle is measured according to a predefined fitness function. Particles are assumed to "fly" over the search space in order to find promising regions of the landscape. Each particle is treated as a point in a d-dimensional space which adjusts its own "flying" according to its flying experience as well as the flying experience of the other companion particles. By making adjustments to the flying based on the local best (pbest) and the global best (gbest) found so far, the swarm as a whole converges to the optimum point, or at least to a near-optimal point, in the search space.

In this study, we use the restaurant service system as a practical illustration of the meta-heuristic optimization procedure. Being a concurrent system (independent and asynchronous activities are taking place simultaneously), it is modeled as a Generalized Stochastic Petri Net. The system operation is simulated through a discrete event simulator [3] and the functional aspects of the system are visually verified. The queuing statistics obtained from the simulation are used to compute the waiting costs. The objective function consisting of the service cost and the waiting cost is minimized with the rapidly converging PSO meta-heuristic, subject to the customer satisfaction fuzzy constraints.

2   Petri Net Model of Service Systems

Service systems are inherently concurrent with multiple asynchronous activities. The traditional methods developed for the analysis of sequential systems are found to be inadequate for the analysis of systems exhibiting concurrency and synchronization of independent, asynchronous activities [25]. Petri nets are found to be ideal tools to model distributed and concurrent systems [11, 14]. The original Petri net (PN) is a directed bipartite graph with two types of nodes, called places (represented by circles) and transitions (represented by horizontal or vertical bars). Directed arcs connect places to transitions, and vice versa. Places may contain tokens (represented by black dots). Places represent the conditions to be met before the transitions can fire. A transition is said to be enabled if there is at least one token in each of its input places. An enabled transition can fire by removing a token from each input place and depositing a token in each output place [12]. The transitions fire instantaneously, implying that events do not take any time. Since there is no concept of time duration in the classical PN, it is not complete enough for the study of systems performance. Several concepts of timed Petri nets have been proposed by assigning firing times to the transitions and/or places of Petri nets [7, 15, 16]. Timed PNs in which the firing time is deterministic (constant) are called D-nets, while those in which the firing time is stochastic are called M-nets. Time-nets with transitions containing both kinds of firing times are called Generalized Stochastic Petri Net (GSPN) [25].

In addition, the customer flow and the server roles are made explicit in our Petri net modeled concurrent business systems. The server resides in the serve place (SP), while the customer resides in the customer place (CP) as shown in Fig. 1. The service at a transition T can begin only when there is at least one server in the SP and correspondingly at least one customer in the CP. Making the customer and the server workflows distinct gives a more realistic analysis of the system. We use a GPSN to model concurrent business systems. A restaurant business system modeled as a client server GSPN is shown in Fig. 2.

The performance analyst can easily grasp the workflows of the customers and of the staff in this service system. The customer flow describes the tasks performed by

Fig. 1 Client server GSPN

T0
CP1
SP
T1
CP2
T2
CP3
T3

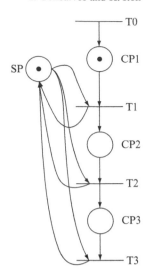

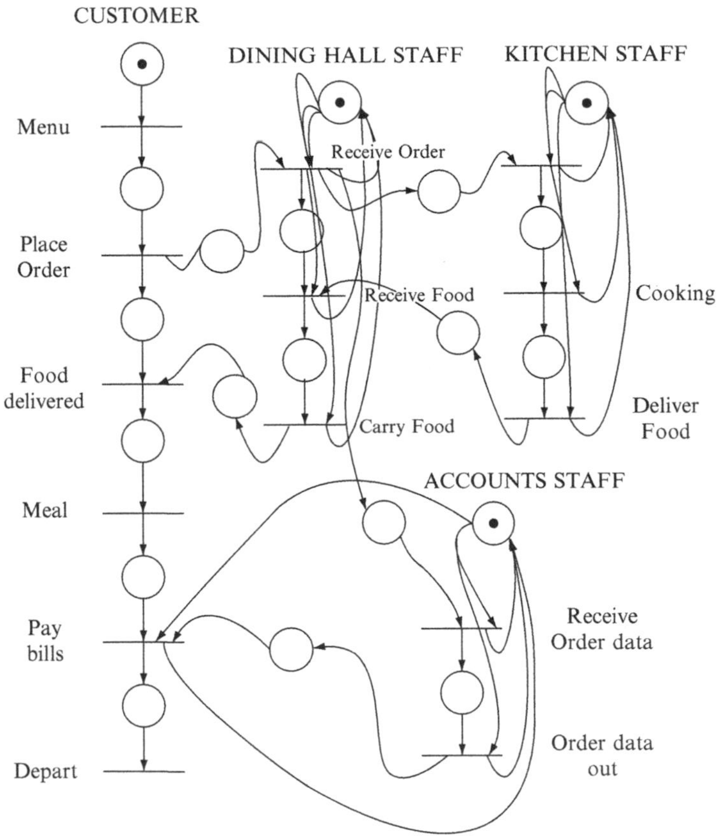

CUSTOMER
DINING HALL STAFF
KITCHEN STAFF
Menu
Receive Order
Place Order
Receive Food
Cooking
Food delivered
Deliver Food
Carry Food
ACCOUNTS STAFF
Meal
Receive Order data
Pay bills
Order data out
Depart

Fig. 2 Concurrent service system

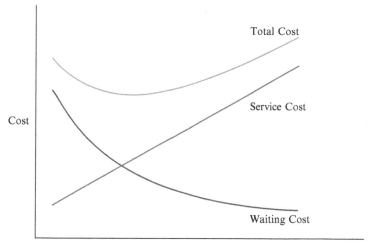

Fig. 3 Service cost and waiting cost

the customer – going through the menu, placing an order, having the meal, paying and departing. The tasks performed by the dining hall staff, the kitchen staff and the accounts staff in providing service to the customers are also described by the PN (Fig. 3).

The authors have also designed a new PN editing and simulating tool. The GSPN model of the concurrent service system is first created by means of the editor. The static model is then executed as a discrete event simulation. The animation facility can also be switched on. Animation shows the firing of transitions and the flow of tokens in the net. This helps the analyst in verifying the functional aspects of the net visually, specially the occurrence of deadlocks. In the GSPN simulation model, the timed-transitions represent the service stations, while the tokens in the server places represent the number of servers assigned to a particular group of service stations or tasks. The customer places act as the queuing locations where the customers queue for service. The transition firings are governed by the average service time allotted for service. The data set associated with the customer places provides the queuing statistics like the average queue length, the average queuing time and the maximum number of customers in the queues. Similarly, the data set associated with the server places provides the average server utilization. The simulation output data is then used to evaluate the objective function to be optimized.

3 Formulation of the Optimization Problem

3.1 Objective Function

If C_W is the average waiting cost per customer per unit time and N_W is the average number of customers waiting for service, then the waiting cost per unit time at a given PN customer place is:

$$W_C = N_W C_W \tag{1}$$

The service cost in the service systems is the sum of the costs required to hire professionals to provide service to customers. If N_S is the number of servers serving at a transition and C_S is the cost per server per unit time, then the service cost at that transition per unit time is:

$$S_C = N_S C_S \tag{2}$$

The objective function (total cost) is given by:

$$f = \sum_{i=1}^{n} N_{Wi} C_{Wi} + \sum_{j=1}^{m} N_{Sj} C_{Sj} \tag{3}$$

where, n is the number of waiting places and m is the number of server groups in the PN.

3.2 Management Constraints

Each service activity has an appropriate service time that is usually drawn from an exponential distribution. The service time constraints can be expressed as:

$$S_{Tmin} < S_T < S_{Tmax} \tag{4}$$

where, S_{Tmin} and S_{Tmax} are respectively the minimum and maximum values of the service time at a given PN transition. The capacity of the server represents the number of servers allotted to a given transition. If N_S is the capacity of a server, serving at a group of transitions, then the constraints are:

$$N_{Smin} < N_S < N_{Smax} \tag{5}$$

Similarly, the priority constraints of the servers with respect to a given transition are:

$$P_{rmin} < P_r < P_{rmax} \tag{6}$$

where, P_{rmin} and P_{rmax} are respectively the minimum and the maximum values of the servers with respect to a given PN transition.

3.3 Customer Satisfaction Constraints

In service systems, customer satisfaction depends on the waiting as well as the service experience. In the restaurant system, if the waiting is too long, the customers are dissatisfied. On the other hand, if they are not allowed to enjoy their meal for a sufficiently long period to time, then they are dissatisfied, too. Consequently, customer satisfaction can be increased by decreasing the waiting time and by increasing the service time (meal time). In this section, we describe the fuzzy membership functions of the waiting and eating experiences.

The membership functions are defined in such a way that they appropriately reflect the changes in the degree of membership in each set, associated with changes in the crisp value [21, 22, 24]. Figure 4 illustrates the membership functions for the fuzzy sets pertaining to the variable waiting. Here, the linguistic variables are Short, Medium and Long. Figure 4 illustrates the membership functions for the variable time spent having meal. The linguistic variables are: Too Short, Short, Medium and Fairly Long (Fig. 5).

The fuzzy rules matrix is presented in Table 1. These rules combine the antecedents of the rules for waiting (time spent in waiting) and those for service (time spent in having meal) to produce a single fuzzy output.

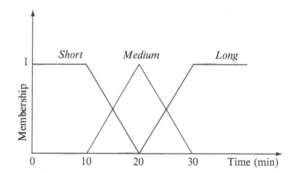

Fig. 4 Membership function of waiting

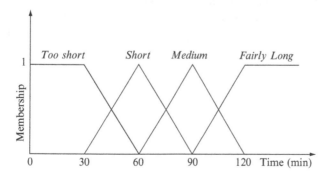

Fig. 5 Membership function of time spent having meal

Table 1 Fuzzy rules matrix

	Waiting		
Dining	Short	Medium	Long
Too short	Poor	Poor	Very poor
Short	Fairly good	Poor	Very poor
Medium	Food	Fairly good	Poor
Fairly long	Very good	Good	Fairly good

Defuzzification is the process by which the output fuzzy variables are converted into a unique (crisp) value. The max method and the centroid methods are well-known methods for obtaining the crisp value from the superposition of the fuzzy membership functions. In our study, the final decision on the waiting and service experience is arrived at by using the centroid method (Eq. 7).

$$FD = \frac{\sum \mu D}{\sum \mu} \qquad (7)$$

4 Particle Swarm Optimization

The Particle Swarm Optimization (PSO) algorithm imitates the information sharing process of a flock of birds searching for food. The population-based PSO conducts a search using a population of individuals. The individual in the population is called the particle and the population is called the swarm. The performance of each particle is measured according to a predefined fitness function. Particles are assumed to "fly" over the search space in order to find promising regions of the landscape. In the minimization case, such regions possess lower functional values than other regions visited previously. Each particle is treated as a point in a d-dimensional space which adjusts its own "flying" according to its flying experience as well as the flying experience of the other companion particles. By making adjustments to the flying based on the local best (*pbest*) and the global best (*gbest*) found so far, the swarm as a whole converges to the optimum point, or at least to a near-optimal point, in the search space. The notations used in PSO are as follows: The (ith) particle of the swarm in iteration t is represented by the d-dimensional vector, $x_i(t) = (x_{i1}, x_{i2}, \ldots, x_{id})$ Each particle also has a position change known as velocity, which for the (ith) particle in iteration t is $v_i(t) = (v_{i1}, v_{i2}, \ldots, v_{id})$ The best previous position (the position with the best fitness value) of the (ith) particle is $p_i(t-1) = (p_{i1}, p_{i2}, \ldots, p_{id})$ The best particle in the swarm, i.e., the particle with the smallest function value found in all the previous iterations, is denoted by the index g In a given iteration t, the velocity and position of each particle is updated using the following equations:

$$v_i(t) = wv_i(t-1) + c_1 r_1 (p_i(t-1) - x_i(t-1)) + c_2 r_2 (p_g(t-1) - x_i(t-1)) \qquad (8)$$

and

$$x_i(t) = x_i(t-1) + v_i(t) \qquad (9)$$

where, $i = 1, 2, \ldots, NP; t = 1, 2, \ldots, T$ NP is the size of the swarm, and T is the iteration limit; c_1 and c_2 are r_2 are random numbers between 0 and 1; w is inertia weight that controls the impact of the previous history of the velocities on the current velocity, influencing the trade-off between the global and local experiences. A large inertia weight facilitates global exploration (searching new areas), while a small one tends to facilitate local exploration (fine-tuning the current search area). Equation 8 is used to compute a particle's new velocity, based on its previous velocity and the distances from its current position to its local best and to the global best positions. The new velocity is then used to compute the particle's new position.

In our application, the decision variables (Table 1) are the particles' "positions" and "velocities". Initially, a group (population) of particles is randomly generated. Their fitness function f (Eq. 3) is evaluated on simulating the system operation. The algorithm is iterated for a fixed number of iterations. The particles' velocities and positions are updated using Eqs. 8 and 9, in every iteration. The lowest value of the fitness function attained by a particle in all the iterations is its pbest, while that of the entire population is the gbest. The latter is the optimal value of the objective function.

5 Results of PSO Optimization

The current optimized values of the decision variables (service time, number of staff members or severs and their priority at each activity) are shown in Tables 2 and 3. These values are bounded between the given minimum and the maximum values.

Table 2 Optimal service times

Transition	Service time (min)		
	Minimum	Current	Maximum
Menu	5	9	10
Order	7	11	12
Food delivery	6	7	15
Meal	30	97	150
Receive Order	5	6	10
Receive food	5	5	10
Carry food	8	12	14
Receive order 2	5	7	10
Cooking	20	21	35
Deliver food	5	9	10
Receive order data	5	10	10
Order data out	4	5	8
Pay bills	4	6	7

Table 3 Optimal service times

Server	Priority			Number		
	Minimum	Current	Maximum	Minimum	Current	Maximum
	1	2	3			
Dining hall staff	1	1	3	1	2	7
	1	3	3			
	1	2	3			
Cooking staff	1	2	3	1	2	5
	1	1	3			
	1	3	3			
Billing staff	1	1	3	1	2	4
	1	1	3			

The restaurant operation is simulated for 6 h for an average inter-arrival time of 15 min. The minimized total cost (sum of the waiting and the serving cost) is found to be 4358,457 yen.

6 Conclusion

In this paper, we have presented the application of the PSO meta-heuristic algorithm in the optimization of the operation of a practical service system, subject to the customer satisfaction constraint. The cost function is expressed as the sum of the service cost and the waiting cost. Service cost is due to hiring professionals or equipment to provide service to end users. Waiting cost emerges when customers are lost owing to unreasonable amount of waiting for service. Waiting can be reduced by increasing the number of personnel. However, increasing the number of personnel, results in a proportional increase in the service cost. The simulation optimization strategy finds the optimum balance between the service cost and the waiting cost. The optimization, however, is subject to the customer satisfaction constraint, which is defined as fuzzy sets quantifying the waiting as well as the service experiences of the customers. The simulation optimization strategy finds the optimum balance between the service cost and the waiting cost without violating the customer satisfaction constraint. PSO obtains the optimum results with rapid convergence even for a very large search space. An extension to this study would be multi-objective optimization.

Acknowledgments This work has been supported by the Open Research Center Project funds from "MEXT" of the Japanese Government (2007–2011).

References

1. Anderson, D.R., Sweeney, D.J., & Williams, T.A. (2003). *An introduction to management science: Quantitative approaches to decision making* (10th ed.). Ohio: Thomson South-Western.
2. Dube-Rioux, L., Schmitt, B.H., & Leclerc, F. (1988). Consumer's reactions to waiting: when delays affect the perception of service quality. In T. Srull (Ed.), *Advances in consumer research*, Association for Consumer Research, *16*, 59–63.
3. Fishman, G.S. (1978). *Principles of discrete event simulation*. New York: Wiley.
4. Folkes, V.S. (1984). Consumer reactions to product failure: an attributional approach. *Journal Consumer Research, 10*(4), 298–409.
5. Folkes, V.S., Koletsky, S., & Graham, J.L. (1987). A field study of causal inferences and consumer reaction: the view from the airport. *Journal of Consumer Behavior, 13*, 534–539.
6. Hillier, F.S. (1963). Economic models for industrial waiting line problems. *Management Science, 10*(1), 119–130.
7. Holliday, M.A., & Vernon, M.K. (1985). A generalized timed Petri net model for performance evaluation. In *Proceedings of the international workshop timed Petri nets*, Torino (pp. 181–190).
8. Kennedy, J., & Eberhart, R.C. (1995). Particle swarm optimization. *Proceedings of the IEEE international conference neural network* (pp. 1942–1948). Piscataway
9. Kennedy, J., Eberhart R.C., & Shi, Y. (2001). *Swarm Intelligence*. San Francisco, CA: Morgan Kaufmann.
10. Maister, D.H. (1985). The psychology of waiting lines. In J. Czepiel, M.R. Solomon, & C.F. Surprenant (Eds.), *The service encounter* (pp. 113–123). Lexington, MA: Lexington Books.
11. Molloy, M.K. (1982). Performance analysis using stochastic Petri nets. *IEEE Transactions on Computers, 31*(9), 913–917.
12. Murata, T. (1989). Petri nets – properties, analysis, and applications. *Proceedings of the IEEE, 77*(4), 541–580.
13. Ozcan, Y.A. (2005). *Quantitative methods in health care management: techniques and applications*. San Francisco, CA: Jossey-Bass/Wiley.
14. Peterson, J.L. (1981). *Petri net theory and the modeling of systems*. New York: Prentice-Hall.
15. Ramamoorthy C.V., & Ho, G.S. (1980). Performance evaluation of asynchronous concurrent systems using Petri nets, *IEEE Transactions on Software Engineering, 6*(5), 440–449.
16. Razouk, R.R. (1984). The derivation of performance expressions for communication protocols from timed Petri nets. *Computer Communications Review, 14*(2), 210–217.
17. Reeves, C. (2003). Genetic algorithms. In F. Glover, & G.A. Kochenberger (Eds.), *Handbook of metaheuristics*. Boston: Kluwer.
18. Scotland, R. (1991). Customer service: a waiting game. *Marketing, 11*, 1–3.
19. Taylor, S. (1994). Waiting for service: the relationship between delays and evaluation of service. *Journal of Marketing, 58*, 56–69.
20. Taylor, S. (1995). The effects of filled waiting time and service provider control over the delay on evaluations of service. *Journal of Academic Marketing Science, 23*(1), 38–48.
21. Terano, T. Asai, K., & Sugeno, M. (1992). *Fuzzy systems theory and its applications*. Boston: Academic.
22. Turksen, I.B. (1991). Measurement of membership functions and their acquisition, *Fuzzy Sets and Systems, 40*, 5–38.
23. Zadeh, L.A. (1965). Fuzzy sets. *Information and Control, 8*, 338–359.
24. Zimmermann, H.-J. (1991). *Fuzzy set theory and its applications* (2nd ed.). Boston: Kluwer.
25. Zuberek, W.M. (1988). D-timed Petri nets and modelling of timeouts and protocols. *Transactions of the Society for Computer Simulation, 4*(4), 331–357.

Chapter 7
A New Improved Fuzzy Possibilistic C-Means Algorithm Based on Weight Degree

Mohamed Fadhel Saad and Mohamed Adel Alimi

Abstract Clustering (or cluster analysis) has been used widely in pattern recognition, image processing, and data analysis. It aims to organize a collection of data items into clusters, such that items within a cluster are more similar to each other than they are items in the other clusters. An improved fuzzy possibilistic clustering algorithm was developed based on the conventional fuzzy possibilistic c-means (FPCM) to obtain better quality clustering results. Numerical simulations show that the clustering algorithm gives more accurate clustering results than the FCM and FPCM methods.

Keywords Fuzzy C-means · Fuzzy possibilistic C-means · Modified fuzzy possibilistic C-means · Possibilistic C-means

1 Introduction

Data analysis is considered as a very important science in the real world. Cluster analysis is a technique for classifying data; it is a method for finding clusters of a data set with most similarity in the same cluster and most dissimilarity between different clusters. Most clustering algorithms do not rely on assumptions common to conventional statistical methods, such as the underlying statistical distribution of data, and therefore they are useful in situations where little prior knowledge exists. The potential of clustering algorithms to reveal the underlying structures in data can be exploited in a wide variety of applications, including classification, image processing, pattern recognition, modeling and identification. The conventional clustering methods put each point of the data set to exactly one cluster. Since 1965, Zadeh proposed fuzzy sets in order to come closer of the physical world [10]. Zadeh introduced the idea of partial memberships described by membership functions. Fuzzy sets could allow membership functions to all clusters in a data set so that

M.F. Saad (✉) and M.A. Alimi
Research Group on Intelligent Machines, University of Sfax, ENIS, BP W 3038, Sfax, Tunisia
e-mail: Fadhel.Saad@isetgf.rnu.tn; mohamed.alimi@ieee.org

S.-I. Ao et al. (eds.), *Intelligent Automation and Computer Engineering*,
Lecture Notes in Electrical Engineering 52, DOI 10.1007/978-90-481-3517-2_7,
© Springer Science+Business Media B.V. 2010

it is very suitable for cluster analysis. Ruspini first proposed fuzzy c-partitions as a fuzzy approach to clustering [4]. Later, the Fuzzy C-Means (FCM) algorithms with a weighting exponent m = 2 proposed by Dunn [8], and then generalized by Bezdek with m > 1 became popular [6]. The FCM uses the probabilistic constraint that the memberships of a data point across classes sum to one. While this is useful in creating partitions, the memberships resulting from FCM and its derivatives, however, do not always correspond to the intuitive concept of degree of belongingness or compatibility. Moreover, the FCM is sensitive to noise. To mitigate such an effect, Krishnapuram and Keller throw away the constraint of memberships in FCM and propose the Possibilistic C-Means (PCM) algorithm [12]. The advantages of PCM are that it overcomes the need to specify the number of clusters and is highly robust in a noisy environment. However, there still exist some weaknesses in the PCM, i.e., it depends highly on a good initialization and has the undesirable tendency to produce coincident clusters [1]. Pal deducted that to classify a data point, cluster centroid has to be closest to the data point, it is the role of membership. Also for estimating the centroids, the typicality is used for alleviating the undesirable effect of outliers. So Pal defines a clustering algorithm called Fuzzy Possibilistic C-Means that combines the characteristics of both fuzzy and possibilistic c-means [11]. The remainder of this paper is organized as follows. In Section 2, preliminary theory algorithms are presented; some drawbacks of them are also mentioned. In Section 3, the improved Fuzzy Possibilistic C-Means is proposed. The proposed IFPCM can solve these drawbacks mentioned in Section 2, and obtain better quality clustering results. In Section 4, we present several examples to assess the performance of IFPCM. The comparisons are made between FCM, FPCM and IFPCM. Finally, conclusions are made in Section 5.

2 Preliminary Theory

2.1 Fuzzy c-Means Clustering Algorithm

The Fuzzy c-means (FCM) can be seen as the fuzzified version of the k-means algorithm. It is a method of clustering which allows one piece of data to belong to two or more clusters. This method (developed by Dunn [8] and improved by Bezdek [6]) is frequently used in pattern recognition. The algorithm is an iterative clustering method that produces an optimal c partition by minimizing the weighted within group sum of squared error objective function J_{FCM}:

$$J_{FCM}(V, U, X) = \sum_{i=1}^{c} \sum_{j=1}^{n} \mu_{ij}^{m} d^2(x_j, v_i) \qquad (1)$$

with $1 < m < +\infty$

Where $X = \{x_1, x_2, \ldots, x_n\} \subseteq R_p$ is the data set in the p-dimensional vector space, p is the number of data items, c is the number of clusters with $2 \leq c \leq n - 1$. $V = \{v_1, v_2, \ldots, v_c\}$ is the c centers or prototypes of the clusters, v_i is the p-dimension center of the cluster i, and $d_2(x_j, v_i)$ is a distance measure between object x_j and cluster centre v_i.

$U = \{\mu_{ij}\}$ represents a fuzzy partition matrix with $\mu_{ij} = \mu_i(x_j)$ is the degree of membership of x_j in the ith cluster; x_j is the jth of p-dimensional measured data. The fuzzy partition matrix satisfies:

$$0 < \sum_{j=1}^{n} \mu_{ij} < n \bigvee i \in \{1, .., c\} \tag{2}$$

and

$$\sum_{i=1}^{c} \mu_{ij} = 1 \bigvee j \in \{1, .., n\} \tag{3}$$

The parameter m is a weighting exponent on each fuzzy membership and determines the amount of fuzziness of the resulting classification; it is a fixed number greater than one.

The objective function J_{FCM} can be minimized under the constraint of U. Specifically, taking of J_{FCM} with respect to μ_{ij} and v_i and zeroing then respectively, tow necessary but not sufficient conditions for J_{FCM} to be at its local extrema will be as the following

$$\mu_{ij} = \left[\sum_{k=1}^{c} \left(\frac{d^2(x_j, v_i)}{d^2(x_j, v_k)} \right)^{\frac{2}{(m-1)}} \right]^{-1} \tag{4}$$

$$v_i = \frac{\sum_{j=1}^{n} u_{ij}^m x_j}{\sum_{j=1}^{n} u_{ij}^m} \tag{5}$$

The algorithm is composed of the following steps:
FCM algorithm

- S1 : Given the data set X, choose the number of clusters $1 < c < n$, the weighting exponent $m > 1$ and the termination tolerance $\varepsilon > 0$.
  Initialize the c cluster centers v_i randomly.

For t = 1,2, Ě,t_{max} do:

- S2: Update $U^t = [\mu_{ij}]$ Eq. 12.

- S3: Update the centers vectors $V^t = [v_i]$ by Eq. 14.

- S4: Compute $Et = \|V^t - V^{t-1}\|$, if $E^t \leq \epsilon$, Stop; Else $t = t + 1$.

End.

2.2 Possibilistic c-Means Clustering Algorithm

The theory of fuzzy logic provides a mathematical environment to capture the uncertainties in much the same human cognition processes. The fuzzy clusters are generated by dividing the training samples in accordance with the membership functions matrix $U = [\mu_{ij}]$. The component μ_{ij} denotes the grade of membership that a training sample belongs to a cluster. Although FCM is a very useful clustering method, its memberships do not always correspond well to the degree of belonging of the data, and may be inaccurate in a noisy environment, because the real data unavoidably involves some noises.

The FCM algorithms use the probabilistic constraint to enable the memberships of a training sample across clusters to sum up to 1, which means the different grades of a training sample are shared by distinct clusters but not as degrees of typicality. To improve this weakness of FCM, and to produce memberships that have a good explanation for the degree of belonging for the data, Krishnapuram and Keller [12] relaxed the constrained condition (10) of the fuzzy c-partition $\{\mu_1, \mu_2, \ldots, \mu_c\}_F$ in FCM to obtain a possibilistic type of membership function with $\{\mu_1, \mu_2, \ldots, \mu_c\}_P$, and propose PCM for unsupervised clustering. The component generated by the PCM corresponds to a dense region in the data set; each cluster is independent of the other clusters in the PCM strategy. The objective function of the PCM can be formulated as:

$$J_{PCM}(V, U, X) = \sum_{i=1}^{c} \sum_{j=1}^{n} \mu_{ij}^m d^2(x_j, v_i) + \sum_{i=1}^{c} \eta_i \sum_{j=1}^{n} (1 - \mu_{ij})^m \qquad (6)$$

Where

$$\eta_i = \frac{\sum_{j=1}^{n} \|x_j - v_i\|^2}{\sum_{j=1}^{n} \mu_{ij}^m} \qquad (7)$$

η_i is the scale parameter at the ith cluster,

$$\mu_{ij} = \frac{1}{1 + \left(\frac{d^2(x_j, v_i)}{\eta_i}\right)^{\frac{1}{m-1}}} \qquad (8)$$

μ_{ij} is the possibilistic typicality value of training sample xj belonging to the cluster i. $m \in [1, \infty)$ is a weighting factor called the possibilistic parameter. Typical of other cluster approaches, the PCM also depends on initialization. In PCM techniques, the clusters do not have a lot of mobility, since each data point is classified as only one cluster at a time rather than all the clusters simultaneously. Therefore, a suitable initialization is required for the algorithms to converge to nearly global minimum.

2.3 *Fuzzy Possibilistic C-Means Clustering Algorithm*

In spite of his good clustering in noisy data samples, PCM has disadvantages: it is very sensitive to initialization and coincident clusters may result, because the columns and rows of the typicality matrix are independent of each other also sometimes this could be advantageous (start with a large value of c and get less distinct clusters) [1, 13]. Pal define a clustering algorithm that combines the characteristics of both fuzzy and possibilistic c-means: Memberships and typicalities are important for the correct feature of data substructure in clustering problem. If a training sample has been classified into a suitable cluster, membership is a better constraint for which the training sample is closest to this cluster. On the other hand, typicality is an important factor for unburdening the undesirable effects of outliers to compute the cluster centers. In accordance with Ref. [11], typicality is related to the mode of the cluster and can be calculated based on all n training samples. Thus, an objective function in the FPCM depending on both memberships and typicalities can be shown as

$$J_{FPCM}(V, U, T, X) = \sum_{i=1}^{c} \sum_{j=1}^{n} (\mu_{ij}^{m} + t_{ij}^{\eta}) d^{2}(x_j, v_i) \tag{9}$$

with the following constraints:

$$\sum_{i=1}^{c} \mu_{ij} = 1 \ \bigvee \ j \in \{1, .., n\} \tag{10}$$

and

$$\sum_{i=1}^{n} t_{ij} = 1 \ \bigvee \ i \in \{1, .., c\} \tag{11}$$

A solution of the objective function can be obtained via an iterative process where the degrees of membership, typicality and the cluster centers are update via:

$$\mu_{ij} = \left[\sum_{k=1}^{c} \left(\frac{d^{2}(x_j, v_i)}{d^{2}(x_j, v_k)} \right)^{\frac{2}{(m-1)}} \right]^{-1} \tag{12}$$

$$t_{ij} = \left[\sum_{k=1}^{n} \left(\frac{d^{2}(x_j, v_i)}{d^{2}(x_j, v_k)} \right)^{\frac{2}{(\eta-1)}} \right]^{-1} \tag{13}$$

$$v_i = \frac{\sum_{j=1}^{n} (u_{ij}^{m} + t_{ij}^{\eta}) x_j}{\sum_{j=1}^{n} (u_{ij}^{m} + t_{ij}^{\eta})} \tag{14}$$

3   A Proposed Improved Fuzzy Possibilistic Clustering Algorithm

The choice of an appropriate objective function is the key to the success of the cluster analysis and to obtain better quality clustering results; so the clustering optimization is based on objective function. To meet a suitable objective function, we started from the following set of requirements: The distance between clusters and the data points assigned to them should be minimized and the distance between clusters should to be maximized [5]. The attraction between data and clusters is modeled by term (9); it is the formula of the objective function. Also Wen-Liang Hung proposed a new algorithm called Modified Suppressed Fuzzy c-means (MS-FCM), which significantly ameliorates the performance of FCM due to a prototype-driven learning of parameter α [14]. The learning process of α is based on an exponential separation strength between clusters and is updated at each iteration. The formula of this parameter is:

$$\alpha = \exp\left(-min_{i \neq k}\frac{\|v_i - v_k\|^2}{\beta}\right) \tag{15}$$

where β is a normalized term so that we choose β as a sample variance. That is, we define β:

$$\beta = \frac{\sum_{j=1}^{n} \|x_j - \overline{x}\|^2}{n} \tag{16}$$

where

$$\overline{x} = \frac{\sum_{j=1}^{n} x_j}{n} \tag{17}$$

But the remark which must be mentioned here is the common value used for this parameter by all the data at each iteration, which may induce in error. We propose a new parameter which suppresses this common value of α and replaces it by a new parameter like a weight to each vector. Or every point of the data set has a weight in relation to every cluster. Therefore this weight permits to have a better classification especially in the case of noise data. So the weight is calculated as follows:

$$w_{ji} = exp\left(\frac{-\|x_j - v_i\|^2}{(\sum_{j=1}^{n} \|x_j - \overline{x}\|^2)} \times \frac{n}{c}\right) \tag{18}$$

where w_{ji} is weight of the point j in relation to the class i. this weight is used to modify the fuzzy and typical partition. All update methods that were discussed in previous sections are iterative in nature, because it is not possible to optimize any of the objective functions reviewed directly. Or to classify a data point, cluster centroid has to be closest to the data point, it is membership; and for estimating the centroids, the typicality is used for alleviating the undesirable effect of outliers. The objective function is composed of two expressions: the first is the fuzzy function and uses a fuzziness weighting exponent, the second is possibililstic function and uses a typical weighting exponent; but the two coefficients in the objective function are only used as exhibitor of membership and typicality.

A new relation, enabling a more rapid decrease in the function and increase in the membership and the typicality when they tend toward 1 and decrease this degree when they tend toward 0. This relation is to add Weighting exponent as exhibitor of distance in the two under objective functions. The objective function of the IFPCM can be formulated as follows:

$$J_{IFPCM}(V, U, T, W, X) = \sum_{i=1}^{c} \sum_{j=1}^{n} (\mu_{ij}^m w_{ji}^m d^{2m}(x_j, v_i) + t_{ij}^\eta w_{ji}^\eta d^{2\eta}(x_j, v_i)) \quad (19)$$

$X = \{x_1, x_2, \ldots, x_n\} \subseteq R_p$ is the data set in the p-dimensional vector space, p is the number of data items, c is the number of clusters with $2 \le c \le n - 1$. $V = \{v_1, v_2, \ldots, v_c\}$ is the c centers or prototypes of the clusters, v_i is the p-dimension center of the cluster i, and $d^2(x_j, v_i)$ is a distance measure between object x_j and cluster centre v_i.

$U = \{\mu_{ij}\}$ represents a fuzzy partition matrix with $\mu_{ij} = \mu_i(x_j)$ is the degree of membership of x_j in the ith cluster and all c centroids; x_j is the jth of p-dimensional measured data. U is defined as:

$$\mu_{ij} = \left[\sum_{k=1}^{c} \left(\frac{d^2(x_j, v_i)}{d^2(x_j, v_k)} \right)^{\frac{2m}{(m-1)}} \right]^{-1} \quad (20)$$

$T = \{t_{ij}\}$ represents a typical partition matrix with $t_{ij} = t_i(x_j)$ is the degree of typicality of x_j in the ith cluster and v_i alone. T is defined as:

$$t_{ij} = \left[\sum_{k=1}^{n} \left(\frac{d^2(x_j, v_i)}{d^2(x_j, v_k)} \right)^{\frac{2\eta}{(\eta-1)}} \right]^{-1} \quad (21)$$

$W = \{w_{ji}\}$ represents a matrix of weight with $w_{ji} = w_i(x_j)$ is the degree of weight of x_j in the ith cluster. W is defined by formula (18).
$V = \{v_i\}$ represents c centers of the clusters, is defined as:

$$v_i = \frac{\sum_{j=1}^{n} (u_{ij}^m w_{ji}^m + t_{ij}^\eta w_{ji}^\eta) x_j}{\sum_{j=1}^{n} (u_{ij}^m w_{ji}^m + t_{ij}^\eta w_{ji}^\eta)} \quad (22)$$

The IFPCM algorithm is summarized as follows:
IFPCM algorithm

- S1 : Given the data set X, choose the number of clusters $1 < c < n$, the weighting exponent $m > 1$, the typical exponent $\eta > 1$ and the termination tolerance $\varepsilon > 0$. Initialize the c cluster centers v_i randomly.

For t $= 1,2,Ě, t_{max}$ do:

- S2 : Compute $W^t = \{w_{ji}\}$ by Eq. 18.
- S3: Compute $U^t = \{\mu_{ij}\}$ Eq. 20.
- S4: Compute $T^t = \{t_{ij}\}$ Eq. 21.
- S5: Modify $\mu_{ij} = w_{ji} * \mu_{ij}$.
- S6: Modify $t_{ij} = w_{ji} * t_{ij}$.
- S7: Update the centers vectors $V^t = [v_i]$ by Eq. 14.
- S8: Compute $Et = \|V^t - V^{t-1}\|$, if $E^t \leq \epsilon$, Stop; Else $t = t + 1$.

End.

4 Experimental Results

In this section, we perform some experiments to compare the performances of these algorithms with some numerical datasets (Tables 2–7). All algorithms are implemented under the same initial values and stopping conditions. The experiments are all performed on an IBM computer with 2.6 GHz Pentium (4) processors using MATLAB (Mathworks, Inc., Natick, MA).

4.1 Example 1 (Data Sets in [7, 11])

In the first experiment, we use a two-cluster data set as presented in [11] shown in Fig. 1. To demonstrate the quality of classification of our approach in relation to the other algorithms (FCM, FPCM) in a case data set without outlier. The clustering results of these algorithms are shown in Fig. 1a–c respectively, where two clusters from the clustering algorithms are with symbols "+" and "o"; also the figure shows that our approach is better than others.

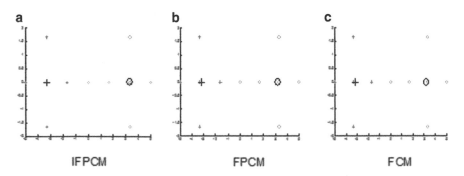

Fig. 1 IFPCM, FPCM, FCM clustering results for the two-cluster data set without an outlier

Table 1 Centers, Performance Index, Mean Square Error and Number of Iterations generated by FCM, FPCM, IFPCM for the experiments in Fig. 1

Centers of clusters					
FCM		FPCM		IFPCM	
X	Y	X	Y	X	Y
−3.1674	0.0000	−3.1980	0.0000	−3.3246	0.0000
3.1674	0.0000	3.1980	0.0000	3.3246	0.0000

Performance index		
FCM	FPCM	IFPCM
−69.66	−130.93	−153.91

Mean square error		
FCM	FPCM	IFPCM
0.24418	0.20083	0.122

Number of iterations		
FCM	FPCM	IFPCM
11	14	14

Table 2 Memberships generated by FCM, FPCM, IFPCM for the experiments in Fig. 1

Points		FCM		FPCM		IFPCM	
X	Y	μ_{i1}	μ_{1j}	μ_{i1}	μ_{1j}	μ_{i1}	μ_{1j}
−5.00	0.00	0.952060	0.047935	0.953910	0.046089	0.998360	0.001638
−3.34	1.67	0.941220	0.058781	0.941890	0.058107	0.996520	0.003479
−3.34	0.00	0.999300	0.000703	0.999530	0.000471	1.000000	0.000000
−3.34	−1.67	0.941220	0.058781	0.941890	0.058107	0.996520	0.003479
−1.67	0.00	0.912560	0.087441	0.910310	0.089691	0.988100	0.011900
0.00	0.00	0.500000	0.500000	0.500000	0.500000	0.500000	0.500000
1.67	0.00	0.087431	0.912570	0.089685	0.910320	0.011900	0.988100
3.34	1.67	0.058780	0.941220	0.058106	0.941890	0.003479	0.996520
3.34	0.00	0.000704	0.999300	0.000472	0.999530	0.000000	1.000000
3.34	−1.67	0.058780	0.941220	0.058106	0.941890	0.003479	0.996520
5.00	0.00	0.047938	0.952060	0.046091	0.953910	0.001638	0.998360

Table 1 shows that the degrees of membership and typicality are better in our approach. The degrees tend toward 1 when the point is near of the center class.

In the second experiment, we use a two-cluster data set with outlier as presented in [7] shown in Fig. 2. The clustering results of these algorithms are shown in Fig. 2a–c respectively, shows that our approach is better than others. The last point (0, 10) is an outlier but it doesn't have an influence on centers although it has the same membership degrees. Table 5 shows that the degrees of membership and typicality are better in our approach. The point (0, 10) has a degree of typicality nearly equal to 0.

The FCM, FPCM and IFPCM are compared in the two previous experiences, using the following criteria for the cluster centers locations: the mean square error

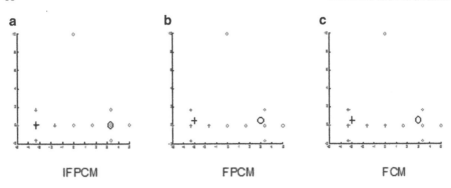

Fig. 2 IFPCM, FPCM, FCM clustering results for the two-cluster data set with an outlier

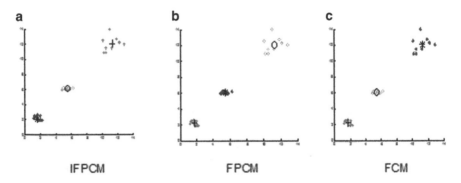

Fig. 3 IFPCM, FPCM, FCM clustering results for the three-cluster data set

(MSE) of the centers ($MSE = \sqrt{\|v_c - v_t\|^2}$), where v_c is the computed center and v_t is the true center) and the number of iterations (NI). The cluster centers found by IFPCM are closer the true centers, than the centers found by FCM and FPCM. The number of iterations tends toward the same value.

In the third experiment, we use a three-cluster data set as presented in [7] shown in Fig. 3. The clustering results of these algorithms are shown in Fig. 3a–c respectively, shows that our approach is better than others.

After a classifier or a cluster model has been constructed, one would like to know how "good" it is. Quality criteria are fairly easy to find for classifiers, or according to Borgelt [2] the quality of a clustering result is calculated while using index of performances or validity index that are used to determine the number of classes. So we can say that IFPCM is better than FCM and FPCM while using criteria index of performances.

Table 3 Typicality generated by FPCM, IFPCM for the experiments in Fig. 1

Points		FPCM		IFPCM	
X	Y	t_{i1}	t_{1j}	t_{i1}	t_{1j}
−5.00	0.00	0.006007300	0.000290310	0.0000000072	0.0000000000
−3.34	1.67	0.006944300	0.000428480	0.0000000073	0.0000000000
−3.34	0.00	0.967420000	0.000456440	1.0000000000	0.0000000000
−3.34	−1.67	0.006944300	0.000428480	0.0000000073	0.0000000000
−1.67	0.00	0.008354900	0.000823330	0.0000000076	0.0000000001
0.00	0.00	0.001907400	0.001907700	0.0000000005	0.0000000005
1.67	0.00	0.000823170	0.008356700	0.0000000001	0.0000000076
3.34	1.67	0.000428400	0.006945600	0.0000000000	0.0000000073
3.34	0.00	0.000456350	0.967410000	0.0000000000	1.0000000000
3.34	−1.67	0.000428400	0.006945600	0.0000000000	0.0000000073
5.00	0.00	0.000290250	0.006008400	0.0000000000	0.0000000072

Table 4 Centers, Performance Index, Mean Square Error and Number of Iterations generated by FCM, FPCM, IFPCM for the experiments in Fig. 2

Centers of clusters					
FCM		FPCM		IFPCM	
X	Y	X	Y	X	Y
−2.98540	0.54351	−3.01160	0.50643	−3.2972	0.0017
2.98540	0.54351	3.01160	0.50643	3.2972	0.0017

Performance index		
FCM	FPCM	IFPCM
−11.064	−48.866	−112.73

Mean square error		
FCM	FPCM	IFPCM
0.76866	0.71622	0.1308

Number of iterations		
FCM	FPCM	IFPCM
11	13	13

Table 5 Memberships generated by FCM, FPCM, IFPCM for the experiments in Fig. 2

Points		FCM		FPCM		IFPCM	
X	Y	μ_{i1}	μ_{1j}	μ_{i1}	μ_{1j}	μ_{i1}	μ_{1j}
−5.00	0.00	0.936360	0.063643	0.938680	0.061323	0.998230	0.001771
−3.34	1.67	0.967310	0.032685	0.966130	0.033870	0.996480	0.003524
−3.34	0.00	0.989660	0.010341	0.991110	0.008892	1.000000	0.000000
−3.34	−1.67	0.899370	0.100630	0.902960	0.097037	0.996450	0.003550
−1.67	0.00	0.915580	0.084418	0.915130	0.084872	0.988610	0.011386
0.00	0.00	0.500000	0.500000	0.500000	0.500000	0.500000	0.500000
1.67	0.00	0.084429	0.915570	0.084869	0.915130	0.011385	0.988610
3.34	1.67	0.032680	0.967320	0.033866	0.966130	0.003524	0.996480
3.34	0.00	0.010342	0.989660	0.008895	0.991110	0.000000	1.000000
3.34	−1.67	0.100640	0.899360	0.097044	0.902960	0.003550	0.996450
5.00	0.00	0.063642	0.936360	0.061326	0.938670	0.001771	0.998230
0.00	10.00	0.500000	0.500000	0.500000	0.500000	0.500000	0.500000

Table 6 Typicality generated by FPCM, IFPCM for the experiments in Fig. 2

Points		FPCM		IFPCM	
X	Y	t_{i1}	t_{1j}	t_{i1}	t_{1j}
−5.00	0.00	0.051546000	0.003367600	0.0000003996	0.0000000007
−3.34	1.67	0.148470000	0.005204700	0.0000004330	0.0000000015
−3.34	0.00	0.595690000	0.005345300	1.0000000000	0.0000000017
−3.34	−1.67	0.044794000	0.004814100	0.0000004296	0.0000000015
−1.67	0.00	0.105530000	0.009787000	0.0000004791	0.0000000055
0.00	0.00	0.023269000	0.023269000	0.0000000284	0.0000000285
1.67	0.00	0.009786700	0.105530000	0.0000000055	0.0000004797
3.34	1.67	0.005204500	0.148470000	0.0000000015	0.0000004335
3.34	0.00	0.005345200	0.595680000	0.0000000017	1.0000000000
3.34	−1.67	0.004813900	0.044794000	0.0000000015	0.0000004301
5.00	0.00	0.003367500	0.051547000	0.0000000007	0.0000004000
0.00	10.00	0.002187700	0.002187700	0.0000000003	0.0000000003

Table 7 Performance Index generated by FCM, FPCM, IFPCM for different datasets

Data set	ND	NC	NDI	PI FCM	PI FPCM	PI IFPCM
Iris	150	3	4	−44527	−46847	−54036
B-c-w-c	683	4	9	−6299	−6402	−16623
Wine	178	3	13	−10751000	−11334000	−21260000
Yeast	528	11	10	117.62	119.28	−1071.6
Auto MPG	398	8	3	−197210000	−202020000	−224870000
Balance scale	625	4	3	1698.20	1711.00	941.51
Buta	345	7	2	81790	77319	20370
Glass	214	9	6	−610970	−668,590	−789,220
Hayes	132	5	3	−132,980	−141,360	−159520
Monk's problem	432	7	2	1006.60	1013.00	821.87
Lettre image	16,000	16	26	57,556	57457	35,644

ND, number of data; NC, number of clusters; NDI, number of data items; PI, performance index.

4.2 Example 2 (Data Sets in [3])

In the experiment, we tested these methods on well-known data sets from the UCI machine learning repository [3] shown in Table performance index for different datasets. The clustering results of these algorithms show that our approach is better than others by using the Performance Index named Fukuyama-Sugeno index, who supposes that the algorithm which has the minimal value of index is the best in relation to others [9].

5 Conclusion

In this paper we presented an improved fuzzy possibilistic clustering algorithm, which is developed to obtain better quality of clustering results. The objective function is based on data attracting cluster centers as well as cluster centers repelling each other and a new weight of data points in relation to every cluster. Comparison of the clustering algorithm and the FCM, FPCM algorithms shows that clustering algorithms will increase the cluster compactness and the separation between clusters. Finally, a numerical example shows that the clustering algorithm gives more accurate clustering results than the FCM and FPCM algorithms for typical problem.

References

1. Barni, M., Cappellini, V., & Mecocci, A. (1996). Comments on "A possibilistic approach to clustering". *IEEE Transactions on Fuzzy Systems, 4*, 393–396.
2. Borgelt, C. (2005). Prototype-based classification and clustering. Habilitation thesis, University of Magdeburg, Germany.
3. Blake, C.L., & Merz, C.J. (1998). UCI Repository of machine learning databases. University of California, Irvine, CA, USA.
4. Ruspini, E.R. (1969). A new approach to clustering. *Information Control 15*(1), 22–32.
5. Timm, H., Borgelt, C., Doring, C., & Kruse, R. (2004). An extension to possibilistic fuzzy cluster analysis. *Fuzzy Sets and Systems, 147*(1), 3–16.
6. Bezdek, J.C. (1981). *Pattern recognition with fuzzy objective function algorithms.* New York: Plenum.
7. Bezdek, J.C., Keller, J., Krishnapuram, R., & Pal, N.R. (1999). *Fuzzy models and algorithms for pattern recognition and image processing.* TA 1650.F89: Kluwer.
8. Dunn, J.C. (1973). A fuzzy relative of the ISODATA process and its use in detecting compact well-separated clusters. *Journal of Cybernetics, 3*(3), 32–57.
9. Lung, K. (2005). A cluster validity index for fuzzy clustering. *Pattern Recognition Letters, 25*, 1275–1291.
10. Zadeh, L. (1965). Fuzzy sets. *Information Control, 8*, 338–353.
11. Pal, N.R., Pal, K., & Bezdek, J.C. (1977). A mixed c-means clustering model. *Proceedings of the Sixth IEEE International Conference on Fuzzy Systems, 1*, 11–21.
12. Krishnapuram, R., & Keller, J. (1993). A possibilistic approch to clustering. *IEEE Transactions on Fuzzy Systems, 1*(2), 88–110.
13. Fayyad, U.M., Piatetsky-Shapiro, G., Smyth, P., & Uthurusamy, R. (1996). *Advances in knowledge discovery and data mining.* Cambridge, MA: MIT.
14. Hung, W.L., Yang, M., & Chen, D. (2005). Parameter selection for suppressed fuzzy c-means with an application to MRI segmentation. *Pattern Recognition Letters, 27*(5), 424–438.

Chapter 8
Low Cost 3D Face Scanning Based on Landmarks and Photogrammetry
A New Tool for a Surface Diagnosis in Orthodontics

Luigi Maria Galantucci, Gianluca Percoco, and Eliana Di Gioia

Abstract Anthropometry is an objective tool serving to evaluate the shape of the face and reveal changes observed in the subject over time, or among different subjects, analyzing quantitative and qualitative differences. It also permits the study of normal and abnormal growth, diagnosis of genetic or acquired malformations, planning and evaluation of surgical and/or orthodontic therapy, and verification of the treatment results by analyzing, measuring and comparing the face shape. Among 3D digitization technologies, photogrammetry shows great promise because it is a low cost, biocompatible, safe and non-invasive methodology, but it still suffers from a need for considerable human intervention. In previous research, the Authors illustrated a new approach based on a 3-Cameras photogrammetric system. After several tests, conducted to verify the validity of this methodology, the present experimental study was carried out using a re-engineered photogrammetric scanning system to obtain landmark-models of human faces, and comparing these results with those achieved with laser scanning, applied to a dummy face (in order to eliminate errors caused by breathing movements in a living subject). Two different, specifically designed experimental 3D photogrammetric setups have been developed and tested to enhance the performance. This research demonstrates the potential of low-cost photogrammetry for medical digitization; further research will be addressed to testing the use of the scanning system on humans to validate its clinical performance.

Keywords 3D Scanning · Biometry · Face digitization · Landmarks · Orthodontics

L.M. Galantucci (✉)
Professor at the Politecnico di Bari, Dipartimento di Ingegneria Meccanica e Gestionale,
Viale Japigia 182, 70126 Bari, Italy
e-mail: galantucci@poliba.it

G. Percoco
Assistant professor at the Politecnico di Bari, Dipartimento di Ingegneria Meccanica e Gestionale,
Viale Japigia 182, 70126 Bari, Italy
e-mail: percoco@poliba.it

E.D. Gioia
Private dentist, Via Dante 97, 70122 Bari, Italy, and Student at the Orthodontic Specialization
School of the Catholic University in Roma, Italy – Head Professor Prof. R. Deli
e-mail: eliana@studiodigioia.com

S.-I. Ao et al. (eds.), *Intelligent Automation and Computer Engineering*,
Lecture Notes in Electrical Engineering 52, DOI 10.1007/978-90-481-3517-2_8,
© Springer Science+Business Media B.V. 2010

1   Introduction

In recent decades, the availability of new digitization and measure systems has prompted the development of anthropometric research using 3D surfaces, aiming to study the 3D geometry and morphology of the main human external tissues.

Anthropometry reveals the face shape and the changes due to time, and allows genetic or acquired malformations to be diagnosed, as well as the planning and evaluation of surgical or orthodontic treatment, the study of normal and abnormal growth and verification of treatments results.

Although the market in this special field is not yet sufficiently exploited [1], many research papers have already been published in literature, describing several approaches for anthropometric measurements that can be subdivided into two-dimensional, three-dimensional and hybrid approaches [2, 3].

Reverse Engineering (RE), a technique that allows reconstruction of the CAD (Computer Aided Design) mathematical model of any existing object, enables the practical implementation of 3D anthropometry. In particular, 3D information can be retrieved about the facial characteristics of a person, allowing construction of a three-dimensional computer model of the human face. Medical applications require a similar 3D precision and accuracy to physical engineering: in this multidisciplinary approach the engineer designs and models physical products, and the doctor diagnoses and treats patients [4, 5] using the physical, engineered products.

2   Background

Two-dimensional anthropometric approaches first emerged more than 30 years ago: essentially they use, as data sources, one or more images of the subject to be identified or measured; data are collected and compared with images of known people for the purposes of identification or recognition. The fast development of data processing technologies, techniques and methods has strongly boosted the implementation of 3D facial recognition. A great many methods have been proposed; the more promising and commonly used in literature are summarized herein.

As regards face matching, in crest lines analysis [6], a set of curves approximates the shape of a 3D object. This increases the real time performance of object recognition algorithms, since one can deal with a specific set of curves as opposed to the entire object. Crest lines are the loci of points on a surface whose largest principal curvature (expressed as an absolute value) is locally maximal in the associated principal direction. In the face recognition context, crest lines correspond to the boundaries between facial features. Since matching of surface shapes can be a complex task, one possible method is to apply fast curve matching algorithms to match crest lines on two surfaces. In fact, curvature profile analysis provides an efficient tool for a rapid comparison of shapes. The information about curvatures could also be exploited by another common methodology, i.e. feature extraction. For this, 3D data are segmented into connected subsets of meshes, called regions or features.

Another existing 3D technique is known as Volume Deformation Energy Cost. This methodology is based on finding out the "energy cost" required to deform one facial surface into another. This is a general method and can either be applied to complete face-to-face deformation or used to compare corresponding features. The energy required to pull or stretch two similar surfaces, one into another, would be low [7].

As regards the acquisition method, studies have been carried out by the Authors, evaluating two different scanning systems, namely photogrammetry and laser scanning, employed for the acquisition and the recognition of human faces [8, 9].

The reconstruction of 3D models of human faces starting from 2D photographs is a complex task, since there are few univocally defined points on the face, such as moles or scars.

It is possible to gain information from 2D images projecting a regular grid of points over the face, and identifying the correspondence between these points;, in this way it is possible to recognize the same point on different photographs and determining its spatial localization.

For each person whose 3D information is available, two point clouds are plotted, coinciding with each side of the face, and each viewpoint of the grid projection. The two point clouds are subsequently merged (the stitching process described by Lane [4]) to obtain a single point cloud for each face. The equipment used for the photogrammetric acquisition is simple and low cost; one or more digital cameras, a projector device and photogrammetric software are enough to obtain the 3D information. The second methodology, used in that study only for comparison purposes, is the laser scanning technique, able to acquire 3D information about the shape of human faces. The authors used a manual Kreon sensor (model KLS51) and an automatic scanning system (Konica Minolta Vivid 910 i); both are based on triangulation. When using a laser scanning system, in order to achieve good acquisition, the person has to collaborate by staying motionless throughout the scanning (15 s). In fact every small movement could produce errors in the resultant point cloud. A special device (cephalostat) is used to hold the head still during 3D acquisition. Results obtained using the two methodologies are analyzed and evaluated to determine whether, although it is a low-cost methodology, a photogrammetric technique is able to provide valid information about facial shape. Data processing is slower with the photogrammetric technique, but the information acquisition, that is the time spent to take the photographs, is very fast (1/5,000 s flashing time). This is an essential aspect, because the "object" acquired is a human being; in fact every movement of the head could make the work useless. Moreover, although less information about facial shape is obtained using photogrammetry, the points whose 3D information are acquired are equally spaced and cover the entire face and after estimation of their spatial localization, need no further processing. A process based on photogrammetry consists of five steps: (a) acquisition of facial images from different directions; (b) determination of the camera positions and calibration parameters; (c) detection of a dense set of corresponding points in the images; (d) computation of 3D coordinates; (e) generation of a surface model. As in an acquisition performed with a 3D scanner, the result of a photogrammetric acquisition is a point cloud, which is then meshed to construct the surface model. Although 3D acquisition and recognition systems have undergone major developments in the past few years, further improvements can still be achieved, such as cost reduction (currently very high), and decreased error in the recognition or authentication process.

Recently, the Authors have proposed a completely automated version of the 3D photogrammetric scanning System, also based on the projection of hybrid and regularly spaced grids [10, 11], to reproduce the whole human face for diagnostic purposes in orthodontics. Using this method it is possible to acquire information about the characteristics of the subject's soft tissues and to make exact measurements.

Nowadays, new methodologies that allow the facial characteristics of orthodontic/surgery patients to be recorded are gaining importance, as a means of comparing affected subjects with controls without using invasive technologies such as radiation.

In the present paper two 3D photogrammetric systems, developed by the authors for measuring soft tissue facial landmarks, are tested and compared with other commercial and research approaches.

The landmarks high relief points can be obtained in three main ways: (i) extraction from a 3D facial model (according to Kovacks [12], Baik [5], Winder [13]); (ii) manual digitization onto the face (according to Sforza [14, 15]); (iii) placement of targets on the face [4].

To obtain surface landmarks (i) starting from a 3D full face surface model, it is necessary to follow a specific measurement protocol.

On the 3D facial models, the protocol illustrated in [5] is used. A consistent Coordinate System is obtained starting from the Nasion N/ as zero point, and establishing the axial referenced-plane by rotating Camper's plane (right nasal ala – both tragus points) 7.5° upward on the axis formed by both tragus points. The sagittal referenced-plane passes through the soft tissue N′ and the midpoint of both tragus points; it is also perpendicular to the axial referenced-plane. Finally, the coronal referenced-plane passes through N′ and is perpendicular to both the axial and sagittal planes (Fig. 1).

Most landmarks used in this study were proposed by Farkas [16]. Simply connecting the landmarks and then calibrating the size (height, width, depth), it is possible to create a simplified 3D facial landmark model; subsequently it is necessary to carry out superimposition, to compare the landmark model to the patient's full face model [5], so as to verify the reliability of the information thus obtained.

The aim is to facilitate a clinical diagnosis for orthodontic/surgery purposes based not only on absolute values (linear-angular distances related to a standard range of measurements), but also on relative values (angles and proportions, to make a 3D soft tissues template that can be related to an average normo-face, that is different for each population) [17].

As regards manual digitization onto the face (ii), in a previous study on soft tissue facial shape based on manual digitations, Ferrario et al. [18] reported that 3D measurements values were not sensitive to head position. Nevertheless, Natural Head Position (NHP) is better exploitable: in fact, if the subject changes his position,

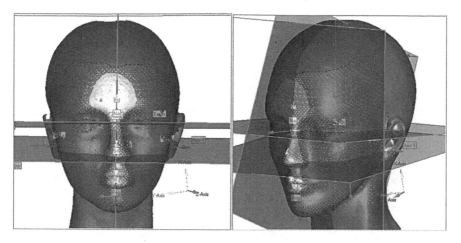

Fig. 1 Landmarks, referenced planes and coordinate system on the 3D model of the acquired mannequin head using the method reported in [5]

the soft-tissue attached to the bone will be stretched or constricted, and soft-tissue drape could also be changed [5].

Moreover, NHP is the most reproducible head position and it is the face's natural orientation for treatment planning [4].

In the recent Ferrario-Sforza Method, described in [15], a 3D computerized electromagnetic contact digitizer (3 Draw Polhemus) is used to collect 50 soft tissue landmarks previously individuated by an expert operator by direct inspection or palpation of the patient's facial soft tissue.

Both for the first [18] and second [15] method the operator needs to be very experienced, and able to identify natural facial landmarks with the greatest precision and accuracy.

Using method (i) it could be difficult to recognize them on the 3D virtual facial model instead of directly on patients, while if method (ii) is adopted the major difficulty lies in the variability of the pressure applied by different operators during direct anatomical landmark digitization, that can modify the registration of the anatomical landmark position due to the elasticity of facial soft tissues. For these reasons, in this study the Authors have decided to adopt method (iii).

3 Proposed Approach

In this study a particular 3D photogrammetric scanning system is presented, that can offer a low-cost solution for craniofacial studies, basing measurements only on few landmarks previously identified and marked on the face using coded targets, and measured without contact.

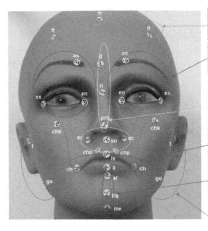

• <u>front,</u> **tr** trichion, **g** glabella, **ft** frontotemporal

• <u>eyes,</u> **ex** hexacanth, **en** endocanth, **os** upper orbital

• <u>nose,</u> **n** nasion, **prn** pronasal, **sn** subnasal, **ac** head of nasal wing, **chp** head of nasal filter

• <u>lips,</u> **ls** upper lip, **li** lower lip, **ch** cheilion

• <u>chin,</u> **sl** sublabial, **pg** pogonion, **me** menton

• <u>lateral surface,</u> **t** tragion, **chk** cheeks, **go** gonion

Fig. 2 Facial landmarks

The landmarks have been described in [18], where a direct manual digitizing technique based on an electromagnetic touch probe instrument is used. In that case the time required for the measurements is very high (some minutes even with an expert operator and collaborating patients). On the other hand, the 3D photogrammetric method does not need an expert operator and collaboration of the patient is not important, because the speed of the takes is such that movements do not pose a problem.

In Fig. 2 thirty coded targets, used to underline the position of the landmarks, are shown; they were chosen to detect the facial characteristics.

Two different 3D photogrammetric setups have been experimented by the Authors. Both have been exploited to scan faces with dense point clouds and compare them with the landmarks.

3.1 First Experimental Setup

In the first setup a specially designed digital 3D photogrammetric system was used, in order to evaluate the performance of this technique when applied to the 3D computer modeling of human faces and face recognition. Three digital cameras were used to capture the images: two "Nikon Coolpix 4500" and a "Nikon Coolpix 990". 3D data were reconstructed utilizing the commercial software Photomodeler.

Normally, the first step of a photogrammetric process is images acquisition from different directions. In this case, the reconstruction of a 3D model of each human face was made starting from the acquisition of three photographs, each one taken from a different direction.

The 3D image acquisition of human faces is more critical as compared to image acquisition of a static object: in fact it is necessary to "freeze motion", that is to avoid breathing and moving effects [4], [17]. If the images are captured at different instants, there could be errors due to major movements (changing the head position) or minor movements (muscle activity, skin or hair surface variation) [12, 15]. Furthermore, any movements could cause errors, due to shifting of the grid projected on to the face. Therefore, a good 3D reconstruction can actually be performed only if the acquisition of photographs is done simultaneously [19].

The three digital cameras were fixed on a single support, designed to accommodate the necessary angulations of the cameras and placed at a suitable distance from the person to be photographed. To cover a wide area of the face [20], the cameras were positioned on a circumference arc with the subject's face in the centre. The central camera was positioned in front of the subject; the lateral ones at an angle of 30° [9–11].

The only hypothesis required during the acquisition sessions was the Natural Head Position and expression of the subjects, which is the most frequent situation represented when the system is employed for personal identification.

For each person three image acquisition sessions were performed [8].

The photographs related to the same person were taken on the same day. Therefore, although time is a factor that affects the results [17], in this case it did not need to be taken into account. The 3D localization of grid points projected over the face was determined using an automatic referencing procedure. Manual referencing helps to improve the referencing process, when automatic referencing is not able to establish the 3D localization of every possible point of the grid projected over the face. After the referencing stage, when 3D points are created, input data are adjusted and errors are minimized.

After a previous calibration and orientation of the cameras, the photogrammetric software Photomodeler 5.0 processes data related to the reference points and calculates their spatial localization.

For each person analyzed, two models related to full face (one for each acquisition session) were created on the basis of the 4,316 points grid [8] (Example in Fig. 3).

Two more models were realized, to identify only the 3D position of the landmarks.

The related acquisition sessions were performed with a different number of points (308 and 4316 points related to grid B and A). Therefore, one point cloud obtained with the 4,316 points grid A was utilized both for the construction of the entire facial model and for the construction of a model constituted only by landmarks.

However, the reconstruction process is not completely automatic. During the orientation stage, the software requires correspondences among a few initial points to be specified, so as to establish the 3D orientation of the cameras. These correspondences must necessarily be obtained with a manual procedure. When the 3D position of the cameras is known, an automatic referencing procedure can be set up for most of the projected points.

Fig. 3 Example of a dense
point cloud

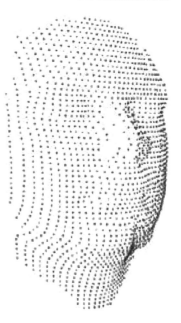

3.2 Second Experimental Setup

Due to the above-mentioned drawbacks and in order to make the system more robust
and industrially valid, the scanning system was re-engineered and firstly applied
only to a dummy head (Fig. 4). The new 3D photogrammetric system consisted
of three 10 megapixel CMOS sensors of commercial cameras and one coded flash
projector synchronized with the cameras by the software. To increase the target
visibility on the investigated surface, the image acquisition was done only by the
light of the projector.

Preliminary automatic orientation of the images is performed exploiting a specifically designed, retro-illuminated pattern. The entire point cloud was compared to
the landmarks to establish the mean error. The images were acquired with the same
procedure used in the preceding cases, but in this case the environment was completely dark, to enhance the contrast.

Instead, the landmarks were acquired using room light conditions, and taking
three photographs. Digitization yielded a point cloud consisting of 2,430 points
(Fig. 5).

Moreover, the dummy head was digitized both with a photogrammetric technique
and with the laser scanner Minolta Vivid 910i. The textured laser scanned point
cloud is shown in Fig. 6.

One of the main problems with projected targets photogrammetry is the low contrast of the targets on the human face surface.

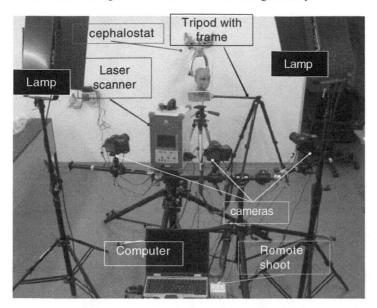

Fig. 4 The re-engineered scanning system

Fig. 5 Photogrammetric
point cloud with the second
setup

For this reason the second experimental procedure was done by directly placing targets on the face, as shown in Fig. 2, and then acquiring the facial images using photogrammetry (Figs. 7 and 8).

The distance values are reported in Table 1.

The maximum linear difference between measured distances was equal to 1.4 mm, while the mean was equal to 0.6 mm. With the previous technique the mean

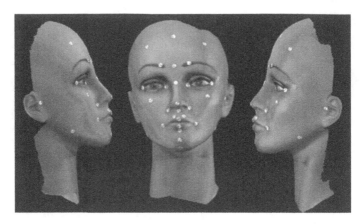

Fig. 6 Textured laser scanned point cloud

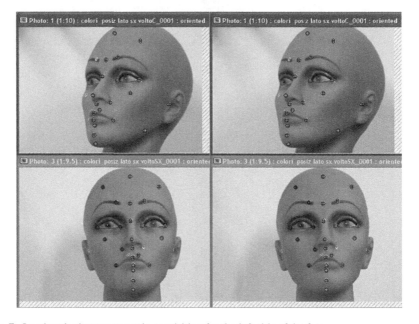

Fig. 7 Landmark photogrammetric acquisition for the left side of the face

between the differences was equal to 2.5 mm: so the presented 3D photogrammetric system resulted very much better.

The comparison among the measurements of some distances done manually with a caliper, or with the photogrammetry and laser scanning measurements, is shown in Table 2.

Figure 9 and Table 3 illustrate an example of angle measurements done on a real subject's face.

Fig. 8 Textured
photogrammetric 3D
landmarks face model

Table 1 Results with the reengineered scanning system

Landmarks distances (mm)	Obtained by photogrammetry	By laser	Difference
Ac(Rt)—ac(Lt)	29.61	30.3	0.7
Ex(Rt)—ex(Lt)	97.15	98.1	1.0
En(Rt)—en(Lt)	30.60	31.0	0.4
Ex(Rt)—en(Rt)	35.06	35.8	0.7
Ex(Lt)—en(Lt)	33.70	33.7	0.0
tr–g	44.91	45.2	0.2
g–n	25.33	25.7	0.3
n–prn	34.84	34.1	0.7
prn–sn	13.01	13.3	0.3
li–sl	7.18	7.4	0.2
sl–pg	13.28	13.8	0.5
pg–me	11.97	12.2	0.2
n–me	93.24	94.3	1.1
tr–n	70.07	71.1	1.0
n–sn	41.33	41.5	0.1
sn–me	43.43	42.0	1.4
T(Rt)—go(Rt)	62.08	62.7	0.6
T(Lt)—go(Lt)	55.50	56.4	0.9
Go(Rt)—me	70.23	70.4	0.1
Go(Lt)–me	74.32	75.2	0.8
T(Rt)—t(Lt)	126.51	128.0	1.4
Chp(Rt)—chp(Lt)	14.77	15.6	0.8
Ch(Rt)—ch(Lt)	46.36	47.1	0.8
Os(Rt)—os(Lt)	40.75	41.2	0.4

When considering the accuracy of the two methods, it should be noted that Konica Minolta declares accuracy measurements for the Vivid 910i laser scanner in the ranges X : ±0.38 mm Y : ±0.31 mm Z : ±0.20 mm, while with the scheme adopted herein it is possible to calculate an accuracy for the photogrammetric

Table 2 Results with some landmarks manually measured

Landmarks distances (mm)	Obtained by photogrammetry	By laser scanning (mm)	Manual measure with caliper	Max difference
n–sn	41.33	41.5	41.51	0.2
T(Rt)–go(t)	62.08	62.7	61.75	0.9
Chp(Rt)–chp(Lt)	14.77	15.6	15.23	0.5

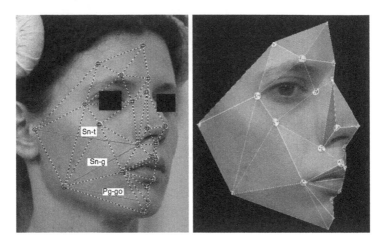

Fig. 9 Angle measurement done on a real face

Table 3 Three dimensional angles

Measured angles (mm)	Using Rhino CAD	Using VIVID 910 Laser scanner
g–n–prn	161.31°	160.5°
n–prn–pg	139.55°	138.5°
Go(Rt)–pg–go(Lt)	83.06°	84.7°

system equal to almost 1/10,000, that means, for the face, $\pm 0.15\,\text{mm}$ on Z and $\pm 0.03\,\text{mm}$ on X and Y with a confidence level of 95%, assuring highly accurate facial measurements.

4 Conclusions

In this paper the validity of photogrammetry for digitization of human faces is demonstrated.

The 3D analysis of facial soft tissue morphology is a precious instrument for clinical purposes, since it is a non-invasive method. Landmark analysis of the human face is a widely used methodology in the medical field in such areas as orthodontics and surgery, being a valid method for making a correct diagnosis and supporting treatment planning, as well as analyzing facial symmetry or asymmetries, and the presence of cranio-facial malformations.

Three-dimensional information regarding each subject can be retrieved, so making subject recognition possible based on the unique features of each face.

Digitization has been performed using 30 landmarks as indicated in literature, and comparing two technologies, namely low-cost photogrammetry and laser scanning, used on the same dummy head.

One of the main drawbacks of photogrammetry is the poor contrast of the targets on the human face surface. Two specifically designed different experimental setups are presented, demonstrating the better performance of the second one applied to a dummy head. This research points out the potential of low-cost photogrammetry for medical digitization. Further research will be addressed to testing the use of the second scanning system on humans to validate its performance.

Acknowledgments This research has been funded by the Italian Ministry of Research and University by the Relevant National Interest Projects Program PRIN 2007 awarded to the Politecnico di Bari University (Coordinator Prof. L.M. Galantucci) and the Università Cattolica del Sacro Cuore di Roma (Coordinator Prof. R. Deli).

References

1. D'Apuzzo, N. (2006). Overview of 3D surface digitization technologies in Europe, three-dimensional image capture and applications VI. In B.D. Corner, P. Li, M. Tocheri (Eds.), *Proceedings of SPIE-IS&T Electronic Imaging, SPIE, vol. 6056*. San Jose, CA, USA.
2. Bowyer, K.W., Chang, K., & Flynn, P. (2005). A survey of approaches and challenges in 3D and multi-modal 3D + 2D face recognition. *Computer Vision and Image Understanding, 101(1)*, 1–15.
3. Zhao, W., Chellappa, R., & Rosenfeld, A. (2003). Face recognition: A literature survey. *ACM Computing Surveys*, pp. 399–458.
4. Lane, C., & Harrell, W. (2008). Completing the 3-dimensional picture. *American Journal of Orthodontics and Dentofacial Orthopedics, 133(4)*, 612–620.
5. Baik, H., Jeon, J., & Leeb, H. (2007). Facial soft-tissue analysis of Korean adults with normal occlusion using a 3-dimensional laser scanner. *American Journal of Orthodontics and Dentofacial Orthopedics, 133(4)*, 612–620.
6. Stylianou, G., & Farin, G. (2004). Crest lines for surface segmentation and flattening. *IEEE Transactions on Visualization and Computer Graphics, 10(5)*, 536–544.
7. Mangan, A., & Whitaker, R. (1999). Partitioning 3D surface meshes using watershed segmentation. *IEEE Transactions on Visualization and Computer Graphics, 5(4)*.
8. Galantucci, L.M., Percoco, G., & Dal Maso, U. (2008). Coded targets and hybrid grids for photogrammetric 3D digitisation of human faces. *Virtual and Physical Prototyping, 3(3 ISSN)*, 1745–2759.
9. Ferrandes, R., Galantucci, L., & Percoco, G. (2004). Optical methods for reverse engineering of human faces. 4th International CIRP 2004 Design Seminar. 16–18 May 2004, Session 6B, pp. 1–12 Cairo, Egypt.
10. Di Gioia, E., Deli, R., Galantucci, L.M., & Percoco, G. (2008). Reverse Engineering and photogrammetry for diagnostics in Orthodontics. *Journal of Dental Research, 87(B)*, 1620.

11. Deli, R., Di Gioia, E., Galantucci, L.M., & Percoco, G. (2008). Non-invasive photogrammetric technique for 3D automatic measurement of faces. 84th EOS Congress of the European Orthodontics Society, #287P, Lisbon (Pt), 10–14 June.

12. Kovacs, L., Zimmermann, A., Brockmann, G., Gu¨hring, M., Baurecht, H., Papadopulos, N.A., Schwenzer-Zimmerer, K., Sader, R., Biemer, E., & Zeilhofer (2006). 3D recording of the human face with a laser scanner. Journal of Plastic, *Reconstructive & Aesthetic Surgery*, *59*, 1193–1202.

13. Winder, R.J., Darvann, T.A., McKnightc, W., Mageed, J.D.M., & Ramsay-Baggs, P. (2008). Technical validation of the Di3D stereophotogrammetry surface imaging system. *British Journal of Oral and Maxillofacial Surgery*, *46*, 33–37.

14. Sforza, C. (2006). Three-dimensional, non invasive analysis of craniofacial growth during deciduous and early mixed dentition. *Ortognatodonzia Italiana*, *13*(1), 53–62.

15. Sforza, C., Peretta, R., Grandi, G., Farronato, G., & Ferrario, V.F. (2007). 3D facial morphometry in skeletal Class III patients: A non-invasive study of soft-tissue changes before and after orthognathic surgery. *British Journal of Oral and Maxillofacial Surgery*, *45*, 138–144.

16. Farkas, L. (1994). *Anthropometry of the head and face*. New York: Raven Press.

17. Kau, C.H., Hunter, L.M., & Hingston, E.J. (2007). A different look: 3D facial imaging of a child with Binder syndrome. *American Journal of Orthodontics and Dentofacial Orthopedics*, *132*(5), 704–709.

18. Ferrario, V.F., Sforza, C., Schmitz, J.H., Miani, A., & Taroni, G. (Dec 1995). Fourier analysis of human soft tissue facial shape: Sex differences in normal adults. *Journal of Anatomy, 187* (Pt 3), 593–602.

19. Ayoub, A.F., Xiao, Y., Khambay, B., Siebert, J.P., & Hadley, D. (2007). Towards building a photo-realistic virtual human face for craniomaxillofacial diagnosis and treatment planning. *International Journal of Oral and Maxillofacial Surgery, 36*, 423–428.

20. Ayoub, A.F., Siebert, P., Moos, K.F., Wray, D., Urquhart, Q.C., & Niblett, T.B. (1998). A vision based 3D capture system for maxillofacial assessment & surgical planning. *British Journal of Oral and Maxillofacial Surgery, 36*, 353–357.

Chapter 9
3D Face Recognition and Compression

Wei Jen Chew, Kah Phooi Seng, Wai Chong Chia, Li-Minn Ang, and Li Wern Chew

Abstract Face recognition using 3D images is an important area of research due to its ability to solve problems faced by 2D images like pose changes. In this chapter, a 3D face range recognition and compression system is proposed and the effect of using compressed 3D range images on the recognition rate is investigated. Compression is used to reduce the file size for faster transmission and is performed using the Set Partitioning in Hierarchical Trees (SPIHT) coding method, which is an improvement of the Embedded Zerotree Wavelet (EZW) coding method. Arithmetic Coding (AC) is also performed after SPIHT to further reduce the amount of bits transmitted. Comparing the uncompressed probe images and probe images compressed using SPIHT coding, simulation results show that the compressed image recognition rate ranges from being lower to being slightly higher than uncompressed probe image recognition rate, depending on bit rate. This proves that a 3D face range recognition system using compressed images is a feasible alternative to a system without using compressed images and should be investigated since the benefits like smaller file storage size, faster image transmission time and better recognition rates are important.

Keywords 3D face recognition · 3D range image compression · SPIHT

1 Introduction

Automatic face recognition is an area that has been researched on for many years. With accurate face recognition systems, industries like the security sector will benefit from them [1]. Besides identifying dangerous people or verifying the identity of

W.J. Chew (✉), K.P. Seng, W.C. Chia, L.-M. Ang, and L.W. Chew
The University of Nottingham, Malaysia Campus, Jalan Broga,
43500 Semenyih, Selangor Darul Ehsan, Malaysia
e-mail: Wei-Jen.Chew@nottingham.edu.my; jasmine.seng@nottingham.edu.my;
keyx7cwc@nottingham.edu.my; kenneth.ang@nottingham.edu.my;
eyx6clw@nottingham.edu.my

S.-I. Ao et al. (eds.), *Intelligent Automation and Computer Engineering*,
Lecture Notes in Electrical Engineering 52, DOI 10.1007/978-90-481-3517-2_9,
© Springer Science+Business Media B.V. 2010

a person, face recognition can also be used for entertainment, like producing more innovative games as well as help lost families reunite by helping them search through a database of millions of faces.

Although two dimensional (2D) images has been use for face recognition for many years, they still suffer from pose changes problem [2]. Therefore, three dimensional (3D) images started to be explored for usage in face recognition. This is because 3D face recognition uses the face surface structure and shape, and can be rotated in any direction to match a face in the database, therefore eliminating the pose changes problem.

An example of a type of 3D image is the range image. This type of image contains the x, y and z coordinate of each face pixel as well as the flag value. Typically, the z-coordinate value is the range value, which is the distance of the face from the camera while the x and y coordinate values are the pixel coordinates. Therefore, when the z-coordinate values are arranged properly in a matrix, then the x and y coordinate values can be ignored. The flag values show whether the pixel has a valid range value or not. However, 3D range images can be large in file size and therefore can pose storage and transmission problems when there is limited space or bandwidth. For example, a 640 × 480 range image is about 13 Mb in size.

A solution to reduce a large image file size is compression. The main purpose of image compression is to reduce the size of the file while ensuring that the quality of the image do not degrade too much. A compression system typically consists of three components, which are source encoder, quantizer and entropy encoder [3]. The source encoder transform the image into coefficients, the quantizer reduces the number of bits of the coefficients and the entropy encoder compresses the quantized coefficients further for better compression [3].

In this chapter, a 3D face recognition and compression system is proposed. This system uses 3D face range images and consists of two parts, which are transmitters and a receiver. A transmitter consists of an image capturing station and at each station, the 3D range images are compressed before being transmitted. The receiver consists of a decompression system and a 3D face matching system. The proposed system consists of several transmitters and one receiver, as shown in Fig. 1.

In the compression system at the transmitter side, the range image will be transformed into coefficients using the Discrete Wavelet Transform (DWT) method. Next, the coefficients will be coded using the Set Partitioning in Hierarchical Trees (SPIHT) [4] and the number of bits further reduced using Arithmetic Coding (AC) [5] before being transmitted over to the receiver side. At the receiver end, the reverse of the compression process will occur to obtain back a range image that will be used for recognition purposes. 3D face matching is performed using a combination of surface matching, Principal Component Analysis (PCA) [6] and Linear Discriminant Analysis (LDA) [7].

The reason to have several small units of 3D camera is to capture 3D range images of unknown people around a building and these images are then sent to the receiver section to perform face recognition to identify the unknown person. The advantage of the proposed setup is to enable the system to have many different

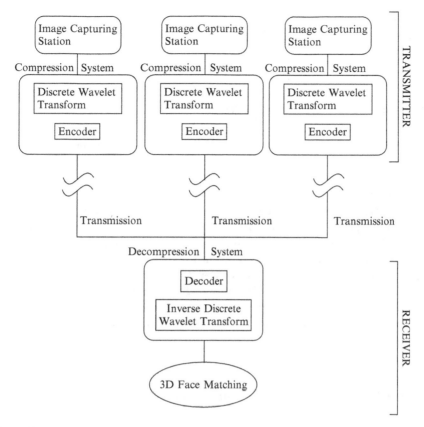

Fig. 1 3D face recognition and compression system

locations to capture images without the need to setup a face recognition system at each location. Instead, all the information just needs to be sent to one central system to perform the face recognition. Hence, compression is a crucial step to reduce transmission time and save bandwidth.

Section 2 discusses about the background and theory of the various methods used while Section 3 discusses about the face recognition and compression method used. The simulation and results are presented in Section 4 and finally the conclusion is in Section 5.

2 Background

The system proposed in this chapter requires compression and recognition methods. After some investigation, it was decided that the compression is performed using Discrete Wavelet Transform (DWT) [3], Set Partitioning in Hierarchical Trees

(SPIHT) [4] and Arithmetic Coding (AC) [5] while recognition will make use of Principal Component Analysis (PCA) [6] and Linear Discriminant Analysis (LDA) [7]. Each of these methods is further discussed below.

2.1 Discrete Wavelet Transform

Discrete Wavelet Transform (DWT) [3] is used to decompose an image into different subbands which contain different frequency components of the original image. Although this is usually performed on a 2D image, for this chapter, DWT will be performed on the range image instead. By performing this transform, the importance of a pixel can be determined. The original image ($x[n]$) will first pass through a series of low pass ($g[z]$) and high pass ($h[z]$) filter, and then subsampled as shown in Fig. 2. The number of subbands created after DWT is affected by the level of decomposition, n. In the example shown in Fig. 2, four subbands which label as LL, LH, HL, and HH are created. Generally, $(3n + 1)$ subbands are created with n level of decomposition. After DWT is applied to the image, most of the important pixels are compacted in the lowest frequency subband (LL subband). The pixel importance reduces when moving from the lowest frequency subband to highest frequency subband.

2.2 Set Partitioning in Hierarchical Trees

After DWT [3] is performed, the image will be encoded using SPIHT [4]. The SPIHT [4] algorithm uses the spatial orientation tree structure to link and encode the wavelet coefficients. Figure 3 illustrates the parent-offspring relationship

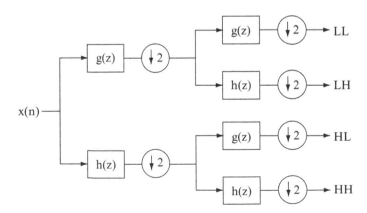

Fig. 2 General filtering process of DWT

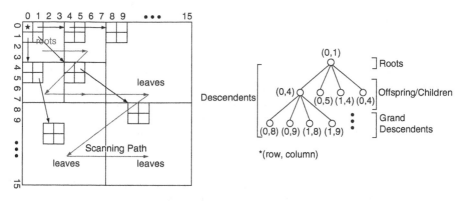

Fig. 3 The parent-offspring relationship of spatial orientation tree structure adopted in SPIHT

of the spatial orientation tree structure. Every wavelet coefficient is treated as a node in the tree structure, except the coefficient with the "*" label. Starting from the lowest frequency subband, the node without the "*" label will serve as the root of a tree. Each of the nodes is connected to either 4 offspring which located in the same spatial region at higher frequency subband, or no offspring (leaves).

The SPIHT [4] algorithm uses three lists - List of Insignificant Pixels (LIP), List of Insignificant Sets (LIS), and List of Significant Pixels (LSP) to control the coding process. Initially, the coordinate of all the coefficients located in the lowest frequency subband is added to the LIP. The entries in the LIS are similar to the LIP, except for those coefficients denoted with "*" in Fig. 3. In addition to this, all the entries in the LIP are marked as Type A entries. Only the LSP is left empty.

Ordinary SPIHT [4] algorithm will start the process by first coding the entries in the LIP. For each entry in the LIP, a '1' bit is sent if the entry is significant. This is followed by the sign bit of the entry. Otherwise, only a '0' bit will be sent. When all the entries in the LIP have been coded, the process is carried on with the entries in the LIS.

First, the process will examine the type of entry. On one hand, if any descendants of the Type A entry are significant, a '1' bit is sent, and the four offspring will be sequentially coded. Otherwise, a '0' bit is sent. The output bits that are used to define the significance of the descendants and offspring will be denoted as DESC bit and LIS SIG bit respectively. On the other hand, if any grand descendants of the Type B entry are significant, a '1' bit is sent, and the four offspring are added to the back of LIS as new Type A entry. Then, the Type B entry will be removed from the LIS. Again, only a '0' bit is sent for the opposite case. The output bit that represents the significance of the grand descendants will be labeled as GDESC bit.

After all the entries in the LIS have been coded, the process continues by sending the refinement bit for each entry in the LSP. Finally, the value of T is divided by two, and the process will start all over again until the desired bit rate is met.

2.3 Arithmetic Coding

In most of the cases, the redundancy in an image cannot be completely removed with image coding. Hence, it is very likely that the redundancy can be further removed with the use of entropy coding. Different from image coding which exploit the redundancy within the image itself, entropy coding will exploit the redundancy among the output bits or symbols, because very likely that certain type of output bits or symbols will have higher probability of occurrence. For example, the output bit stream of SPIHT usually contains more number of bit '0' than number of bit '1'.

Among the various type of entropy coding algorithm, the arithmetic coding (AC) [5] can be considered as the best technique that gives near optimal performance. In AC [5], an interval starting from 0 to 1 is first distributed to all the possible symbols, with respect to their probability of occurrence. Then, the lower and upper limit of the interval is updated to the lower and upper limit of the portion where the symbol to be coded located. After the boundary limit is updated, the interval is redistributed to all the possible symbols. The process will continue until it reaches the last symbol to be coded. In this case, a number that fall within the lower and upper limit of the portion where the last symbol to be coded is located will be chosen. This number is used to represent the entire list of coded symbols.

2.4 Principal Component Analysis

After compressing and decompressing the face range image, the next step in the proposed method is to perform face recognition. The recognition method proposed includes the use of PCA [6] and LDA [7]. PCA [6] is used to feature extraction while LDA [7] is used to further separate the group of data from each other. To use PCA [6] for feature extraction purposes, the first step is to convert each depth values obtained from the database into 1D vector by concatenating the rows or columns into a long vector. Then, the mean is calculated using Eq. 2 by summing the entire database 1D vector together and then dividing by the amount of faces in the database. Each face is then centered by subtracting the mean image from each face image using Eq. 1 [6].

$$\bar{x}^i = x^i - m \tag{1}$$

where

$$m = \frac{1}{p} \sum_{i=1}^{p} x^i \tag{2}$$

Next, the data matrix is created by combining the centered database image side-by-side to create a data matrix. The covariance matrix is then be calculated by multiplying the data matrix with its transpose, as in Eq. 3 [6].

$$\Omega = \overline{XX}^T \tag{3}$$

This is followed by the calculation of the eigenvalues and eigenvectors for the covariance matrix. An eigenspace is created by the sorted eigenvectors matrix. Finally, the centered training images are projected into the eigenspace created. The projection is the dot product of the centered training image with each of the ordered eigenvectors calculated. The more important eigenvectors will have higher eigenvalues.

2.5   *Linear Discriminant Analysis*

For LDA [6], the first step is to calculate the within class scatter matrix which shows the amount of scatter between training images in the same class. The scatter matrix is calculated using Eq. 4 where S_i is the scatter matrix and m_i is the mean of the training images [7].

$$S_i = \sum_{x \in X_i} (x - m_i)(x - m_i)^T \tag{4}$$

The within class scatter matrix, which is the sum of all the scatter matrices, is calculated using Eq. 5 where S_w is the within class scatter matrix and C is the number of classes [7].

$$S_w = \sum_{i=1}^{C} S_i \tag{5}$$

Next, the between class scatter matrix is calculated using Eq. 6 where S_B is the between class scatter matrix, n_i is the number of images in the ith class and m is the total mean of all training images [7].

$$S_B = \sum_{i=1}^{C} n_i (m_i - m)(m_i - m)^T \tag{6}$$

The eigenvectors and eigenvalues of the within class and between class matrices are then calculated. Sorting the non-zero eigenvectors from high to low according to the corresponding eigenvalues, the Fisher basis vector is formed. Calculating the dot product of the training images with each of the Fisher basis vectors, the training images will be projected onto the Fisher basis vectors.

To identify the test image, it is projected onto the Fisher basis vector and the Euclidean distances between the test and training images is calculated. The training image class that the test image belongs to is indicated by the shortest Euclidean distance.

3   3D Face Compression and Recognition

Other works has shown that 3D range face images were successfully used in a 3D face recognition system. However, a drawback of 3D range images is their file size. At about 13 Mb for a single 640×480 image, storing and sending the image becomes a problem when there is limited space or bandwidth. Therefore, it is proposed that the image be compressed.

In this chapter, it was proposed that compression is to be performed using the DWT [3] based image coding technique called SPIHT [4] that was discussed in Section 2. DWT based coding was chosen since it has the advantage over DCT based coding where it decomposes a signal into various subbands, therefore eliminating the need to block the input image [3], hence solving the blocking artifact problem. SPIHT was chosen because it is a progressive compression algorithm that is able to send the most significant bits first, therefore enabling reconstruction of the entire image even at low bit rates. For further bit compression, AC [5] was also performed after SPIHT.

Basically, the 3D face range images would first be decomposed into various subbands using DWT [3] as shown in Section 2.1. The range image would be transformed into coefficients and the more important coefficients would be concentrated in the LL subband, which is the lowest frequency subband. These arranged coefficients would then be encoded using SPIHT [4] which forms parents-offspring links from the DWT coefficients, as shown in Section 2.2. The bit rate at which to encode the coefficients can be controlled using SPIHT. Finally, the encoded coefficients are further compressed before transmission using AC [5] which reduces the amount of bits needed to represent the coefficients.

For the 3D face recognition section, matching was performed using the method used by Chew et al. [8]. This method comprises of 3 parts, which are face feature detection, face alignment and face matching which consist of surface matching, PCA [6] and LDA [7]. Recognition is performed on the decompressed range images.

This method starts by locating key face feature points. These points are then used to translate and rotate the faces so that the unknown probe faces and those in the database are frontal facing, making them easier to be matched to each other more efficiently.

Next, the surface matching method is used. The face row with the nose tip is located and then the horizontal face slice is segmented out slice by slice between the nose row and 100 rows above it, an area which is less susceptible to facial expression. The distance between the database candidates and unknown probe slice is calculated by the vertical distance.

Since the faces are not aligned using Iterative Closest Point (ICP) [9], only surface matching will not provide an accurate face recognition result. Therefore, PCA [6] followed by LDA [7] is further performed on the 20 closest matches from the surface matching result. For the PCA followed by LDA method, instead of using 2D gray images, the proposed method uses 3D range information instead to create the LDA eigenspace. For each range training image, they were each converted into 1D vector and concatenated together to form a matrix where their mean will be calculated using Eq. 2. After that, a centered data matrix is obtained by subtracting the mean from the original data matrix. Next, a covariance matrix is obtained using Eq. 3 and then eigen vectors and eigen values are obtained for the covariance matrix to obtain an eigenspace. The centered training images can be projected into this space to determine their classes. However, sometime the classes between different faces are not clearly defined in the PCA eigenspace. Hence, LDA is further performed to obtain a clearly defined class since it is able to minimize the within class scatter and maximize the between class scatter. After that, each unknown probe face will be projected on the LDA eigenspace and the training face with the shortest Euclidean distance in this eigenspace will be the most likely candidate for the probe.

   This 3D face recognition method is performed on both compressed and uncompressed 3D range images to determine how much degradation the recognition rate will have due to using compressed images.

4   Simulation and Results

The range images are compressed using SPIHT [4] and AC [5] while recognition is performed using the method used by Chew et al. [8]. Compression is performed from 0.03 bits per pixel (bpp) to 0.50 bpp. Figure 4 shows the original image while Fig. 5 shows the decompressed images of the original image at different bit rates using SPIHT.

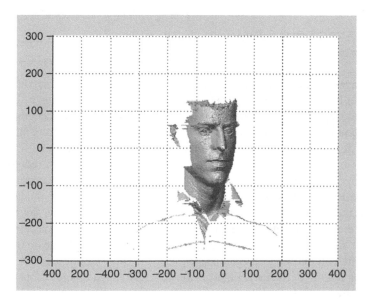

Fig. 4 Original image

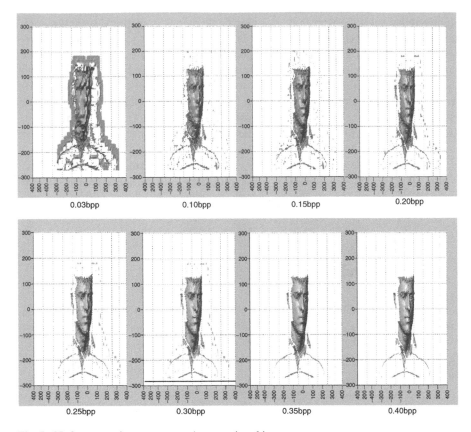

Fig. 5 3D face range image compression at various bit rates

The images in Fig. 5 are the image examples obtained after decompression that are used for recognition. However, before the images in Fig. 5 can be obtained, another vital step needs to be performed. This is because right after decompression, an outline boundary can be observed surrounding the image. Therefore, this outline needs to be removed before the next stage of face recognition can be performed.

To remove the unwanted boundary, the first step is to binarize the reconstructed range image. Then, the image is eroded a few times to obtain a face mask that is slightly smaller that the original face. Finally, the reconstructed range image is compared with the eroded binarized image and only range pixels within the binarized image are retained. Therefore, the unwanted boundary outline was successfully removed. The shrinking of the binarized image to obtain the face range image should not affect the recognition process since most of the information used for recognition is at the middle of the face. Figure 6 shows a face range image before and after removal of the boundary outline.

Once the boundary outline has been removed, the 3D face range image is matched with a 3D face range database using the method discussed in Section 3

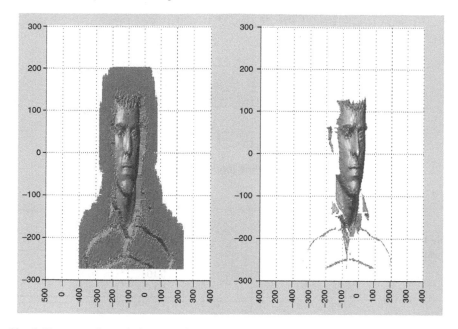

Fig. 6 Face range image before and after boundary outline removal

to determine the identity of the face. The 3D face recognition method consists of performing surface matching followed by PCA and LDA.

For this simulation, the UND database [10, 11] was used. Since training for PCA [6] and LDA [7] requires multiple images per subject, a subset of this database was used. It was decided that the training database consist of three images per subject. This means subjects with at least three images can be selected for training and subjects with at least four images can be selected as the unknown probe image. For this paper, 80 subjects were chosen for training while 61 subjects were chosen as the unknown probe. This means there is a total of 240 training images. All of them were rotated to the front with their nose at position (0,0,0). This is to make it easier for the probe face to align itself to the faces in the database.

Each of the unknown probe faces were compressed and then decompressed before performing face recognition using the method of surface matching followed by PCA [6] and LDA [7]. Figure 7 shows the PSNR obtained for several images compressed with SPIHT [4] coding, Table 1 shows the bit rate difference between using AC and not using AC, Tables 2 and 3 shows the recognition rate obtained for different bit rates after compression without and with AC respectively while Table 4 shows the recognition rate obtained if the images have not been compressed before.

In Fig. 7, it is observed that the PSNR increases as the bit rate increases. This means that the quality of the reconstructed image is better with higher bit rate. The 3D range images used have PSNR values of between 10 to 50 dB.

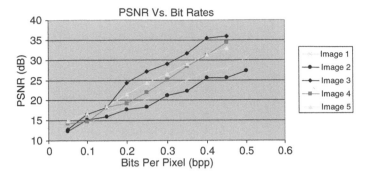

Fig. 7 Graph of PSNR versus bit rates

Table 1 Comparison of bit
rate with and without AC

Bpp with AC	Bpp without AC
0.05	0.08
0.10	0.14
0.15	0.20

Table 2 Recognition rates for compressed images without AC for different bit rates

Recognition rates (%)					
Rank	0.05 bpp	0.10 bpp	0.15 bpp	0.20 bpp	0.25 bpp
1	59.02	65.57	68.85	70.49	72.13
2	65.57	77.05	73.77	73.77	77.05
3	73.77	80.33	75.41	75.41	78.69
Recognition rates (%)					
Rank	0.30 bpp	0.35 bpp	0.40 bpp	0.45 bpp	0.50 bpp
1	75.41	77.05	77.05	75.41	73.77
2	78.69	78.69	78.69	78.69	77.05
3	81.97	81.97	81.97	83.61	81.97

Table 3 Recognition rates for compressed images with AC for different bit rates

Recognition rates (%)					
Rank	0.03 bpp	0.07 bpp	0.11 bpp	0.15 bpp	0.20 bpp
1	59.02	65.57	68.85	70.49	72.13
2	65.57	77.05	73.77	73.77	77.05
3	73.77	80.33	75.41	75.41	78.69
Recognition rates (%)					
Rank	0.24 bpp	0.28 bpp	0.33 bpp	0.37 bpp	0.41 bpp
1	75.41	77.05	77.05	75.41	73.77
2	78.69	78.69	78.69	78.69	77.05
3	81.97	81.97	81.97	83.61	81.97

Table 4 Recognition rates
for uncompressed images

Rank	Recognition rates (%)
1	68.85
2	73.77
3	81.97
4	83.61
5	85.25

In Table 1, each row represents a certain image quality. Therefore, the table shows that to obtain the same image quality, higher bits per pixel is needed when AC is not used. Comparing the results obtained in Table 2 and 3, for the same recognition rate, the bit rate used was lower for a compression system with AC compared with a compression system without AC. Hence, instead of coding till a certain bit rate to get a required recognition rate, the implementation of AC allows the user to code at a lower bit rate to get the same recognition rate.

By comparing Table 3 and 4, it can be observed that the recognition rates for the compressed images increases with higher bit rates till 0.33 bpp before dropping a little. From 0.15 bpp onwards, compressed images have 3 to 5 more Rank 1 images compared to uncompressed images, hence a higher recognition rate. A possible reason is a feature of the probe image that causes it to be an unsuitable match to the correct person in the database became less prominent due to the compression. Therefore, with fewer differences, a better match was achieved. The results show that it is possible to compress the 3D face range image and still be able to use it to perform 3D face recognition. Therefore, it proves that the proposed 3D face recognition system with compression is a feasible system.

5 Conclusion

In this paper, a 3D face recognition system with compression is proposed. This system consists of a transmitter side which contains image capturing and compression systems, and a receiver side which contains a decompression system and a 3D face matching system. First, the captured unknown probe image is compressed and then uncompressed at a determined bit rate using SPIHT and AC. Then, the uncompressed image is processed to remove the boundary outline that formed after compression. Finally, face recognition is performed using the proposed method of surface matching followed by PCA and LDA. From the results obtained, the recognition rate of the proposed 3D face recognition system with compression varies, from lower to higher than a normal 3D face recognition system without compression, depending on bit rate. AC also helps lower the bit rate to obtain a certain recognition rate, when compared with not using AC. Therefore, this shows that adding compression to a 3D face recognition system is a feasible step and warrants further research since recognition rates obtained were good and the benefits obtained from using compressed images, like reduced storage space and faster transmission time, are important when there is limited storage space or bandwidth.

References

1. Identix. (2005). FaceIT surveillance SDK http://www.identix.com/.
2. Zhao, W., Chellappa, R., Rosenfeld, A., & Phillips, P.J. (2000). Face recognition: A literature survey. UMD CfAR Technical Report CAR-TR-948.
3. Saha, S. (2000). Image compression – from DCT to wavelets: A review, http://www.acm. org/crossroads/xrds6–3/sahaimgcoding.html.
4. Said, A., & Pearlman, W.A. (1996). A new, fast, and efficient image codec based on set partitioning in hierarchical trees. *IEEE Transactions on Circuits and Systems for Video Technology, 6*(3), 243–250.
5. Rissanen, J.J., & Langdon, G.G., Jr. (Mar 1979). Arithmetic coding (PDF). *IBM Journal of Research and Development, 23*(2), 149–162.
6. Turk M., & Pentland, A. (Mar 1991). Eigenfaces for recognition. *Journal of Cognitive Neuroscience, 3*(1), 71–86.
7. Belhumeur, P.N., Hespanha, J.P., & Kriegman, D.J. (Jul 1997). Eigenfaces vs. Fisherfaces: Recognition using class specific linear projection. *IEEE Transactions on Pattern Analysis and Machine Intelligence, 19*(7), 711–729.
8. Chew, W.J., Seng, K.P., Liau, H.F., & Ang, L.-M. (2008). New 3D face matching technique for 3D model based face recognition. *Proceeding of International Symposium on Intelligent Signal Processing and Communication Systems (ISPACS2008)*, pp. 236–239.
9. Besl, P., & McKay, N. (1992). A method for registration of 3-D shapes. *IEEE Transaction on Pattern Analysis and Machine Intelligence, 14*(2), 239–256.
10. Flynn, P.J., Bowyer, K.W., & Phillips, P.J. (2003). Assessment of time dependency in face recognition: An initial study. *Audio and Video-Based Biometric Person Authentication*, Springer-Verlag, New York, USA, 44–51.
11. Chang, K., Bowyer, K.W., & Flynn, P.J. (Dec 2003). Face recognition using 2D and 3D facial data. *ACM Workshop on Multimodal User Authentication*, pp. 25–32.

Chapter 10
Head Movement Quantification and Its Role in Facial Expression Study

Fakhrul Hazman Yusoff, Rahmita Wirza O.K. Rahmat, Md. Nasir Sulaiman, Mohamed Hatta Shaharom, and Hariyati Shahrima Abdul Majid

Abstract Temporal modeling of facial expression has been the interest of various fields of studies such as expression recognition, realism in computer animation and behavioral study in psychological field. While various researches are actively being conducted to capture the movement of facial features for its temporal property, works in term of the head movement during the facial expression process is lacking. The absence of head movement description will make expression description to be incomplete especially in expression that involves head movement such as disgust. Therefore, this paper proposes a method to track the movement of the head by using a dual pivot head tracking system (DPHT). In proving its usefulness, the tracking system will then be applied to track the movement of subjects depicting disgust. A simple statistical two-tailed analysis and visual rendering comparison will be made with a system that uses only a single pivot to illustrate the practicality of using DPHT. Results show that better depictions of expression can be implemented if the movement of the head is incorporated in the facial expression study.

Keywords Face expression modeling · Computer graphics · Face tracking · Face animation

1 Introduction

Due to its applicability in various areas such as recognition, robotics and computer animation, facial expression modeling has been the interest of many researches. The computer researches of facial expression study would mainly base their works on Facial Action Coding System (FACS), a face description method proposed by Paul Ekman [8]. Ekman's FACS has been the de-facto reference for face expression

F.H. Yusoff (✉), R.W.O.K. Rahmat, Md. N. Sulaiman, M.H. Shaharom, and H.S.A. Majid
Universiti Teknologi MARA, Universiti Putra Malaysia, Cyberjaya University Collge of Medical Sciences, International Islamic University, Shah Alam, Malaysia
e-mail: fhazman1975@yahoo.com.my; rahmita@fsktm.upm.edu.my; nasir@fsktm.upm.edu.my; hatta@cybermed.edu.my;shahrima@iiu.edu.my

S.-I. Ao et al. (eds.), *Intelligent Automation and Computer Engineering*,
Lecture Notes in Electrical Engineering 52, DOI 10.1007/978-90-481-3517-2_10,
© Springer Science+Business Media B.V. 2010

description where it can be found in works such as [3, 11, 22, 23, 33, 35]. FACS through Action Unit (AU) focuses on describing facial muscle activities during facial expression. While many works utilize the concept of AU in describing facial expression, works in term of head movement, inclusive of the neck muscle is still lacking. Ekman, in his original work includes neck muscle as parts of the FACS. This is obvious from the head turn-left, head turn-right, head-up, head-down, head tilt-left, head tilt right, head forward and head back inclusion in the FACS description [8, 10]. Therefore, in order for facial expression to be properly studied, the head movement should be taken into consideration together with others AUs. The expression of disgust and surprise, for example, will usually be accompanied with some movement of the head which signal the intensity of the action [9].

Currently as most works focus on the non-head AU, the graphical rendering usually is confined to face component movement (mouth, eyes, brows etc.). This reduces the realism factor of the depiction as the head movement plays a vital part in expression production. Systems of facial expression that left out head movement such as [22, 23] will have problems in depicting disgust where the neck may move backward to signal displeasure. We may not be able to graphically depict the displeasure completely without quantifying the head movement. Besides, eventually, if the human head is combined with the body, the head movement needs to be modeled. Without attempt to quantify the movement properly, the head movement will be a form of guessing game at best.

Given that facial expression has been the study of various researches, it is surprising that the head movement has not been given a proper attention. Therefore, this paper intends to highlight the importance of head movement during facial expression study by quantifying and showing the head movement during the facial expression of disgust. Disgust expression is selected because it involves some movements of the head [9]. This paper proposes quantifying the movement of the head using Dual Pivot Head Tracking System (DPHT). A series of experiments on subjects producing facial expression of disgust will then be conducted to show that quantifying the head movement is important for better understanding and depiction of facial expression.

Before going into the details of the proposed method, the following Section 2 will highlight on what constitutes a head movement. This will be followed by Section 3 where some of the works on facial expression will be evaluated on the context of head movement inclusion.

2 Head Motion During Facial Expression

The head motion can be divided into two major motions. The first motion involved the motion of yaw, pitch and roll. The first motion is made possible by movement of the skull with cervical vertebrae known as C2 acts as the pivot point for the motion [19].

The second head motion involved the movement of the cervical vertebrae itself. This cervical vertebrae or commonly known as the neck is responsible in producing the movement of head pulling and pushing which are defined to be part of the head AU in Ekman's FACS. The cervical vertebrae consist of 7 parts which are named as C1, C2, C3, C4, C5, C6 and C7 [19]. Whilst C2 is responsible for the movement of the yaw, pitch and roll, the pivot that causes the movement of the whole neck is not readily apparent. This is because the cervical vertebrae are linked with the thoracic section (human backbone) with no apparent pivot. However, as the cervical vertebrae works in a group as a single component (commonly known as neck) and it is not elastic, it can be generally treated as one single unit which movement is tied on C7, the base of cervical vertebrae. Therefore given the pivot of C2 and the motion of the neck which is based on C7, yaw, pitch, roll, head pulling and head pushing can be generated. Given these useful information on head movements, various researches however did not consider it as part of the facial expression study as can be shown in the following section.

3   Related Work

Facial Action Coding System (FACS) by Paul Ekman is being used by many researches in the field of facial expression. However, the head motion description is not being considered by many of the researches. Ekman, through his Manual of FACS identifies many action units including head motion [10]. Unfortunately, the head motion is not given priority in many studies. Tian, Kanade and Cohn [33] define templates to detect and track specific non-head AU related muscle. The group's work is based on frontal view and focused on the non-head AU. As head muscle AU description is not included, the visualization will be limited only to face. Moreover the group's work in 3D visualization will be limited due to the absence of depth information. Our proposed work on the other hand attempts to quantify movement of the head using both frontal and profile views. This enables 3D visualization as enough information is available to construct 3D coordinate.

Pantic and Patras utilize a rule-based description to establish non-head AU. Their first paper investigates on profile view [22] while the second paper looks at frontal view [23]. Valstar and Pantic [35] is a continuation of rule-based frontal description of non-head AU. While there are frontal and profile elements, these works focus on fiducial points mainly for describing non-head AU. Furthermore, no attempt is being made to implement it in 3D environment. The proposed method attempts to incorporate head motion which will be obtained from frontal and profile view in order for a 3D representation to be built. Using this approach a more encompassing movement can be visualized.

Some researches propose specialized equipment to track movement. Kapoor and Picard [15], for example, work on head nod and shake detector. The group uses a special infra-red device to detect the movement. Meanwhile, Cohn, Schmidt, Gross and Ekman [4] use EMG (Facial Electromyography) device to quantify selected facial muscle during expression. The EMG sensors will be attached to human face to record facial expression. These works on movements admittedly can be used as guidance for Ekman's head AU. However their applicability is rather limited. Usage of devices may be applicable only in controlled environment such as in lab. Moreover, devices such as head gear and EMG wires are intrusive and may hinder subjects from producing good facial expression. It is the goal of this work to avoid intrusion by using image processing approach to detect feature points.

MPEG-4 specification adapted FACS into FAP [21]. The adoption of FACS into FAP can be seen as an acknowledgement on the role of FACS in facial expression description. Unfortunately MPEG-4 FAPs focuses on animating face with the absence of neck. Without the neck, head movement is not truly incorporated in the specification. This will reduce the realistic value of the FAP since some of the facial expression such as surprise and disgust include the neck movement [9]. More importantly, the head AU of Ekman's describe movements that include the neck section. With the absence of the neck, only non-head AU seems to benefit the most in the incorporation of FACS in MPEG-4. Raouzaiou et al. uses MPEG FAPs in their study to describe the archetypal expression [27]. While the effort enables better understanding of facial expression through compartmentalization approach, the work is not going to be complete due to the inadequacies faced by FAP in term of head movement descriptions.

Dornaika and Ahlberg [6] include some discussion on rotational property inferring inclusion of head AU elements during their attempt to track facial expression using template. The work even discuss on the yaw motion that exist during expression. However the implementation is confined to facial component without consideration on the neck motion. Although yaw description is involved, the absence of neck makes discussion on the head to be incomplete.

Reisenzein, Bordgen and Holtbernd [29] attempt to investigate disassociation between emotion and facial display (surprise). However, their work only quantifies the eye area (pupil, brow) movement for sign of surprise. Incorporating description on the head movement may give more information on the surprise response.

The detachment of head movement from facial expression is such that various computer animation books do not touch on head movement while discussing facial expression. The absence of neck from facial expression modeling maybe due to the neck is usually being categorized as part of the body as can be seen in works of [12, 17, 26]. These works infer that neck is part of the body and head is portrayed without the neck.

The concept of neck is part of the body can also be found in books such as [14, 20, 28]. This notion however is not definite as some researchers such as Mao, Qin and Wright model the body without the neck [18]. In situation such as facial expression modeling, neck needs to be part of the expression model. Roberts in [30] points out that neck is actually an integral part of facial expression. Some researchers acknowledge the role of neck in facial expression study by considering it as part of facial expression modeling. Therefore, areas to be considered when it comes to facial expression should be from the base of the neck upwards.

There are actually efforts being done to incorporate head movement during facial expression. Realizing the setback of MPEG-4 FAP Arya and DiPaola proposes Face Modeling and Animation Language for MPEG-4 [2]. Arya and DiPaola aim to improve the face modeling in MPEG so that a more realistic facial expression can be produced. In their proposal neck is included as part of their model. However, Arya and DiPaola's do not suggest any means to quantify the head movement. The proposed method on the other hand will propose the method to quantify this head movement.

Hsu, Kao and Wu touch on the issue of facial appearance for believable facial expression [13]. In their research, proper head shape is mapped to different personality to enable more believable generation of facial expression. The research considers neck to be one of the priorities together with the shape of the eyes, mouth nose and jaw. The research inclusion of neck shows that the dynamics of the neck should also be studied in order to create a proper facial expression model. The proposed research aims to track the dynamic of the neck using DPHT system to further enhance the face expression study done by others.

Perhaps the best statement regarding the lack of study on neck as part of the head movement can be found in Lee and Terzopoulis [16]. Lee and Terzopoulis lament that not many researchers give much attention to the neck. They propose a biomechanical model of the neck. Their neck model is controlled by a number of controllers which simulate actual muscle of human neck. To prove the realistic aspect of their model, the neck is coupled with a facial expression simulation. While Lee and Terzopoulis did not openly state that neck should be part of facial expression, the act of integrating neck in facial expression simulation proves that neck should be part of the facial expression modeling right from the beginning. As the focus of Lee and Terzopoulis is on biomechanical model, the implementation will be too complex if one only needs to look at neck activity in general. The proposed work differs from Lee and Terzopoulis in the context of implementation. The proposed model focuses in the movement of the neck in general where it relies heavily on the observable feature points to be tracked. The biomechanical model of Lee and Terzopoulis on the other hand involves the construction of underlying muscle model which is not visible from naked eye. The advantage of general implementation as proposed by this paper is it can be simply setup and suits studies that suffice itself with the simple existence of neck motion such as in psychological evaluation, believable animation and expression recognition. Therefore considering the acknowledgement that head inclusive of the neck plays important roles, lack of attention on the head movement during the facial expression should no longer be tolerated. In the context of facial expression study, it is apparent that a proper quantification of the head movement needs to be introduced so that the movement can be better described and studied. This paper proposes the usage of DPHT in quantifying the head movement during facial expression. Using DPHT, the paper will show that neck plays a distinct role during facial expression specifically during the expression of disgust. The following sections will elaborate more on the methods used to carry out our proposal.

4 Quantifying Head Motion

Quantifying the head movement requires several steps. The steps include image capturing, feature point tracking, establishment of 3D coordinate, and head motion quantification. Image capturing will capture the subject to be studied. Upon capturing, some feature points will be manually selected and tracked throughout the film frames. These tracked feature points information will be used to form a 3D coordinate of the studied subject. After obtaining the 3D coordinate, the actual motion quantification can be done. In this step, information can be analyzed for significant patterns.

4.1 Image Capturing

Image capturing is done using frontal view and profile view. The two orthogonally arranged views are needed to establish the 3D coordinate. An image distortion correction based on [5] is utilized to correct the distortion that exists during the image capturing process. The proposed method follows the work of [24, 31] in forming the 3D coordinate where basically x coordinate are taken from the frontal view and y, z is taken from the profile view. A slight scaling and rotation similar to [31] is being used to ensure that the views are of the same dimension. The rescaling and rotating is done to correct for difference that exists on the camera focus during recording. Calculating actual coordinate as suggested in [1] is not necessary in the proposed work since relative coordinates retrieved from this orthogonal camera arrangement is sufficient to calculate the motion movements.

4.2 Feature Point Tracking

Tracking the head movement which includes face and neck can be done by tracking some specific feature points. The technique of following feature points can be seen from the work of Pantic and Pantras in [22]. We extend the work of Pantic and Pantras so that the tracking of the neck section can be done. Two pivot points (Point C and D), one secondary point (Point P) and two supporting points (Point E and F) are needed to track the movement of the head that includes the neck. These points skeleton system which will be known as Dual Pivot Head Tracking System (DPHT) are best illustrated using the following Fig. 1.

Based on explanation in Section 2, cervical vertebrae C2 is the first pivot responsible for the yaw, pitch and roll. C2 however is not visible. Yip proposes the location to be estimated using the head ratio [36]. Given its invisibility and variation in human head, this can be a difficult undertaking. Pantic and Pantras [22] on the other hand propose usage of ear hole in lieu of C2. They point out that changes in the ear hole area is minimal during head movement. As such yaw, pitch and roll can still be

a   **b**   **c**

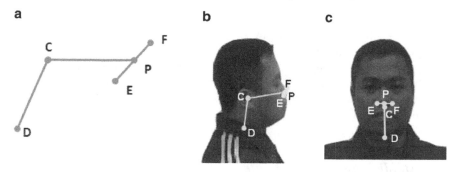

Fig. 1 (**a**) DPHT skeleton – points to be tracked. (**b**) Point placement on the human head – profile. (**c**) Point placement on the human head – frontal

easily quantified if ear hole is used. Yusoff in [37] also shows that usage of ear hole as a pivot point in determining yaw, pitch and roll is practical since only minimal change are recorded at ear hole area while the head is executing some movement.

As such, the proposed method chose the ear hole in lieu of C2 as the first pivot for tracking the standard yaw, pitch and roll of the head. This ear hole will be properly known as Point C. The y and z coordinate of Point C will be defined as the ear hole location seen from profile view. Ear hole is selected as it is visible and can be easily tracked. Moreover, Pantic and Pantras also uses ear hole labeled as P15 in their work in [22]. The x coordinate of Point C will be retrieved from frontal view (see Fig. 1c) where it will be defined to be similar to x coordinate of Point D. This ensures that Point C will be permanently located in the middle of the head. With Point C act as the pivot, introducing Point P, E and F enable for calculation of yaw, pitch and roll.

Yaw for example can be detected by Point P moving away from Point C in horizontal manner. Pitch on the other hand is the up or down movement of Point P with respect to Point C. Finally by fixing Point C, roll is made possible with the existence of slope between Point E and F with respect to Point C.

From the elaboration in Section 2, another point that is responsible for the head movement will be C7. Like C2, C7 is not visible for tracking. As C7 actually refers to the base of the neck, we choose base of the neck as another pivot point responsible for the head movement. The base of the neck will be names as Point D in DPHT system. Point D is proposed to be labeled at the midpoint of the base of the neck (refer to Fig. 1b and c) and the whole body section will be tracked to determine its movement. The proposal is made because of two reasons which will be as follows:

1. As the method focuses on the movement of the head, Point D is used to represent body as one big point. Movement that involves point D will be considered to be the movement of the whole body which is not the focus of this method.
2. Yusoff in [37] shows that marking point arbitrarily near the base of the neck does not make much difference provided that the body is treated as one big point.

All these points (P, C, D, E and F) need to be tracked. By restricting the search area, template matching as detailed in [7] can be used to match the existing points with its previous whereabouts. The proposed method employs optical flow technique with template matching as implemented by Seyedarabi in [32] to search and track the DPHT point movement from one frame to another. Usage of optical flow technique reduces the number of comparison needed during the tracking process as it confines the probable point motion location. Upon confining the probable location using optical flow, the template matching will be utilized to obtain the best solution for the searched feature points.

4.3 Establishing 3D Coordinate

Assuming that scaling between frontal and profile images are equals, translation are equals, human face are symmetrical, and the frontal and profile view are orthogonal, establishing the 3D coordinate of Point P, C, D, E and F can be done using information obtained from the views:

$$P_x = \frac{E_{x:Front} + F_{x:Front}}{2} \tag{1}$$

$$P_y = P_{y:Profile} \tag{2}$$

$$P_z = P_{z:Profile} \tag{3}$$

$$D_x = D_{x:Front} \tag{4}$$

$$D_y = D_{y:Profile} \tag{5}$$

$$D_z = D_{z:Profile} \tag{6}$$

$$C_x = D_x \tag{7}$$

$$C_y = C_{y:Profile} \tag{8}$$

$$C_z = C_{z:Profile} \tag{9}$$

$$E_x = E_{x:Front} \tag{10}$$

$$E_y = E_{y:Front} \tag{11}$$

$$E_z = E_{z:Profile} \tag{12}$$

$$F_x = F_{x:Front} \tag{13}$$

$$F_y = F_{y:Front} \tag{14}$$

$$F_z = F_{z:Profile} \tag{15}$$

By using Eqs. 1–15, 3D coordinates of DPHT can be formed. Using these coordinates head movement can be quantified. The information can be integrated with other facial expression information to create a better temporal rendering of a person expressing an expression.

4.4   Head Motion Quantification

The head motion as elaborated can be sourced from two axis namely the C2 for skull and C7 for neck. The neck movement specifically during head pulling and pushing can be calculated by measuring the angle between Point C and Point D where Point D is treated as the point of origin. Whilst there are a variety of neck movements available, the proposed method will restrict its scope to only the pulling and pushing of head. Therefore the $\angle DC$ of interest will be from the profile view. This is done since the proposed work should be seen as extending the Ekman's based facial expression study where the scope of head movement are confined to only head pulling and pushing (of the neck section) together with the typical yaw, pitch and role (of the skull section) [8, 10]. As such, as far as the Ekman's head AU is concerned, the neck movement of interest will be the one that move forward or backward in near horizontal manner. The $\angle DC$ can be calculated as follows:

$$\angle DC = COS^{-1} \left(\frac{(C_z - D_z)}{\sqrt{(C_z - D_z)^2 + (C_y - D_y)^2}} \right) \tag{16}$$

Meanwhile, to enable tracking of the skull movement (i.e. yaw, pitch and roll), a secondary point (Point P) is introduced. Treating Point C as point of origin, the angle between Point C and Point P can be calculated to quantify the yaw and pitch. The following equations are used to calculate the yaw and pitch respectively:

$$\angle CP_{yaw} = COS^{-1} \left(\frac{(P_x - C_x)}{\sqrt{(P_z - C_z)^2 + (P_x - C_x)^2}} \right) \tag{17}$$

$$\angle CP_{pitch} = COS^{-1} \left(\frac{(P_z - C_z)}{\sqrt{(P_z - C_z)^2 + (P_y - C_y)^2}} \right) \tag{18}$$

To account for roll movement, points are needed to track head tilt. Usage of eyes and noses as in [34] can be unstable as eyes blink. The proposed method as can be seen in Fig. 1b and c, uses the edge of nose hole as nose will not change its state. This approach can also be seen in [22]. While [22] uses a single nose edge via fiducial point P5, the proposed method will be using both sides (left nose and right nose) where the nose edge will be labeled as Point E and F. The roll movement can be calculated as the following Eq. 19:

$$\angle EF = COS^{-1} \left(\frac{(E_x - F_x)}{\sqrt{(E_x - F_x)^2 + (E_y - F_y)^2}} \right) \tag{19}$$

Equations 16 through 19 are sufficient to quantify the movement of pulling/pushing of head, yaw, pitch and roll a typical head movement in facial expression as can be found in Ekman's FACS. Whilst, much has been said about yaw, pitch and roll as can be seen in [2, 6, 15], study on pulling and pushing of head that involves neck motion is rather limited. This pulling and pushing of head can be seen during the facial expression of disgust. In showing the application of DPHT and the existence of neck motion during facial expression of disgust, the following sections will elaborate on the experiments conducted.

5 Experiments

To demonstrate the importance of DPHT in describing facial expression, an experiment that elicits disgust expression has been devised. Eighty subjects participate in the experiment. The subjects are selected from a pool of first semester students majoring in Computer Science. A simple random sampling approach is being used to obtain 80 students. To further minimize variations only students of a single race is being used for this experiment. A narration is read to the subject and followed by display of disgusting images. The session is repeated for three times where during each session a different disgusting image is displayed. The images consists of a person throwing up, a non-flushed toilet filled with faeces and an elephant defecating. Students are specifically requested to react to the narration and images by manifesting disgust. While this approach cannot be said as eliciting natural expression, it is still valid if we assume that human facial expression is a form of social signal [25]. In this particular experiment, it is for the subject to socially signal disgust through manifestation of his facial expression.

Steps in Section 4 are applied to establish the motion data. In studying the motion of head during disgust, the angle of the line between Point D and C will be evaluated ($\angle DC$). Assuming the movement is mainly pull back movement with minimal side ways movement of the neck, the angle $\angle DC$ is calculated as in Eq. 16. By using (16), the angle (in radian) can be tracked and this can be used as the value that quantifies the neck movement. To factor out individual posturing, first frame angle value will be deducted from all the recorded value as the first frame value usually indicate the personal posturing of each individual person. In order to prove that the head movement exists during the facial expression, a simple statistical two-tailed test is being conducted using the following hypothesis: $H_0 : \mu = 0\, rad$ and $H_1 : \mu \neq 0\, rad$. The testing will evaluate whether the average head movement (μ) is more than $0\, rad$. A standard deviation is set at 0.1. The standard deviation is based on initial posturing variation of individuals during natural position. A significant head movement will indicate that during the facial expression of disgust, the neck movement is distinct and should be account as part of facial expression. A level of significant of $\alpha = 0.05$ is being used during the analysis to reject or accept H_1. To further emphasize the need for neck quantification, a graphical representation of head movement is rendered. The first representation illustrates the movement of head that incorporate the neck section as being proposed by this paper while the second representation uses fiducial points tracked using only points from the face area as proposed in work of Pantic and Pantras [22].

6   Results

Based on the experiment of recording neck movement, the resulting graph will be as the following Fig. 2. Looking at Fig. 2a, apparently the neck does not sit still during the manifestation of disgust. In analyzing the movement using two-tailed test, the *z-score* yield a value of 3.4 ($p - value = 0.0006$), this means that H_0 can be rejected and the average head movement definitely is not 0 *rad*.

Further inspection reveals that many samples recorded a trough like pattern during the early phase or towards the end of the movement. This is consistent with the expression of disgust where a person will pull back his head and swaying it back to its original condition. Figure 2b shows a backward movement of some of the samples for clearer illustrations. While the trough varies from one person to another, the trough existence is consistent in almost all of the samples taken. Assuming that the effect of roll, pitch and yaw is minimal, displaying disgust results in the head to be slightly pulled back causing the neck to be tightened as in Fig. 3.

The graphical representation of this reaction can be made possible by usage of DPHT where neck displacement information is available. Figure 4 shows the different situation of the head movement.

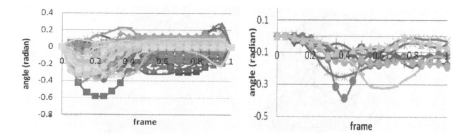

Fig. 2 (**a**) Neck movement for S1 to S80. (**b**) Neck movement for some of the samples

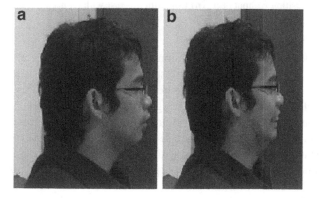

Fig. 3 (**a**) Initial head position. (**b**) Face depicting disgust

a   b   c

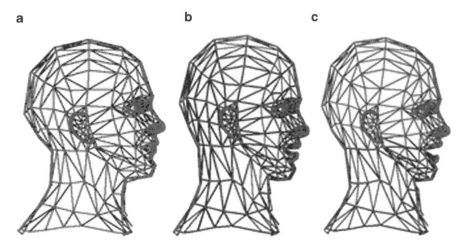

Fig. 4 (**a**) Initial position. (**b**) Depicting disgust – feature points do not include the neck area.
(**c**) Depicting disgust (DPHT) – feature points includes the neck area

a   b

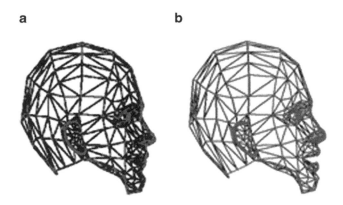

Fig. 5 (**a**) Initial position without neck. (**b**) Depicting disgust without neck

Figure 4a depicts the initial condition of the head prior to any expression.
Figure 4b displays the expression manifestation using only fiducial points as proposed by Pantic and Pantras in [22]. Figure 4c on the other hand illustrates the disgust using DPHT system. Comparing Fig. 4c and b, it is apparent, without the
neck movement information, the expression is less expressive. This is due to the
movement mainly comes from the face section only. Figure 4c carries a more in-
tense manifestation with the neck leaning to the back.

Figure 5a and b show the neutral and expression state without the usage of neck.
As oppose to what can be illustrated in Fig. 4c, the depiction of expression in Fig. 5b
lacks manifestation of intensity and above all ignores the role being played by the
neck during expressing an expression such as disgust. This example is yet another

reason for neck to be treated as an integral element in facial expression study as it carries some weight in showing the intensity of expression.

7   Conclusions

Based from the experiment results and visualization, it is apparent that the head movement inclusive of the neck contributes greatly towards understanding facial expression especially during expression such as disgust. Statistical analysis shows that neck exhibits significant movements during disgust which can possibly indicate the intensity of the expression. Therefore any face expression analysis will be more complete if head movement analysis similar to the proposed method is included. Most importantly, this information enables more realistic computer graphics rendering of facial expression movements. Usage of two pivot points in DPHT specifically at the base of the neck and ear hole will make the temporal modeling to be smoother because apart from pitch, yaw and roll, additional movement of the head such as pullback and push forward that are associated with the neck movement will also be defined. More importantly, the description of the /additional movement will make the head movement definition to be more complete and accordance to the head AU as proposed by Paul Ekman in his FACS [8].

References

1. Ansari, A., & Abdel-Mottaleb, M. (2005). Automatic facial feature extraction and 3D face modeling using two orthogonal views with application to 3D face recognition. *Pattern Recognition, 38*(2005), 2549–2563.
2. Arya, A., & DiPaola, S. (2007). Face modeling and animation language for MPEG-4 XMT framework. *IEEE Transactions on Multimedia, 9*(6), 1137–1146.
3. Bartlett, M.A., Hager, J.C., Ekman P., & Sejnowski, T. (Mar 1999). Measuring facial expressions by computer image analysis. *International Journal of Psychophysiology, 36*(2), 253–263.
4. Cohn, J.F., Schmidt, K., Gross, R., & Ekman, P. (2002). Individual differences in facial expression: Stability over time, relation to self-reported emotion and ability to inform person identification. In *IEEE International Conference on Multimodal Interfaces (ICMI 2002)*, pp. 491–496. PA: Pittsburgh.
5. Devernay, F., & Faugeras, O. (2001). Straight lines have to be straight. *Machine vision and application* (pp. 14–24). Springer.
6. Dornaika, F., & Ahlberg, J. (2004). Face and facial feature tracking using deformable models. *International Journal of Image and Graphics (IJIG), 4*(3), 499–532.
7. Efford, N. (2000). *Digital image processing – a practical introduction using Java*. Essex, England: Addison Wesley.
8. Ekman, P., & Friesen, W.V. (1978). *Facial action coding system – a technique for the measurement of facial movement*. Palo Alto California: Consulting Psychologist Press.
9. Ekman, P., & Rosenberg, E. (1997). *What face reveals – basic and applied studies of spontaneous expression using the facial action coding system (FACS)*. New York: Oxford University Press.
10. Ekman, P., Friesen, W.V., Joseph, C. (2002). *Facial action coding system – the manual on CD ROM*. Utah, USA: Research Nexus Division of Network Information Research Corporation.

11. Essa, I.A., & Pentland, A.P. (1995). Facial expression recognition using a dynamic model and motion energy. In *Proceedings of the Fifth International Conference on Computer Vision*, pp. 360–367. USA.
12. Fua, P., Plankers, R., & Thalmann, D. (1999). From synthesis to analysis: Fitting human animation models to image data. In *Proceedings of the Computer Graphics International*, pp. 4–11.
13. Hsu, S.H., Kao, C.K., & Wu, M. (2009). Design facial appearance for roles in video games. *Expert System with Applications, 36*(3), 4929–4934, Pergamon Press.
14. Jones, A., & Bonney, S. (2000). *3D Studio Max 3 – Professional Animation*. New Riders, Indiana, USA.
15. Kapoor, A., & Picard, R.W. (2001). A real-time head nod and shake detector. In *Proceedings of the 2001 workshop on perceptive user interfaces*, pp. 1–5. Orlando, Florida, USA.
16. Lee, S., & Terzopoulos, D. (2006). Heads up! Biomechanical modeling and neuromuscular control of the neck. *ACM SIGGRAPH 2006 papers* (pp. 1188–1198). ACM.
17. Lim, I.S., & Thalmann, D. (2002). Construction of animation models out of captured data. In *Proceedings of the IEEE International Conference on Multimedia and Expo*, pp.829–832.
18. Mao, C., Qin, S.F., & Wright, D. (2006). Sketching-out virtual humans: From 2D storyboarding to immediate 3D character animation. In *Proceedings of the International Conference on Advances in Computer Entertainment Technology*, pp. 61.
19. Marieb, E. (2006). *Essentials of human anatomy and physiology* (pp. 150–151). San Fransisco, USA: Pearson.
20. Osipa, J. (2007). *Stop staring – face modeling and animation done right*. Indiana, USA: Sybex, Wiley.
21. Pandzic, I.S., & Forenheimer R. (Ed.) (2002). *MPEG-4 facial animation – the standard, implementation and application*. England: Wiley.
22. Pantic, M., & Patras, I. (2004). Temporal modeling of facial actions from face profile image sequences. In *IEEE International Conference on Multimedia and Expo*, pp. 49–52.
23. Pantic, M., & Patras, I. (2005). Detecting facial actions and their temporal segments in nearly frontal-view face image sequences. In *IEEE International Conference on Systems, Man and Cyberneticsi* pp. 3358–3363.
24. Park, I.K., Zhang, H., & Vezhevets, V. (2005). Image-based 3D face modeling system. *EURASIP Journal on Applied Signal Processing, 13*, 2072–2090.
25. Parkinson, B. (1995). *Ideas and realities of emotion*. London, U.K.: Routledge.
26. Ralf, P., Fua, P., Apuzzo, N.D. (1999). Automated body modeling for video sequences. In *Proceedings of the IEEE International Workshop on Modelling People*, pp. 45–52.
27. Raouzaiou, A., Tsapatsoulis, N., Karpouzis K., & Kollias, S. (2002). Parameterized facial expression synthesis based on MPEG-4 EURASIP. *Journal on Applied Signal Processing, 10*, 1021–1038.
28. Ratner, P. (2003). 3-D *Human modeling and animation*. New Jersey, USA: Wiley.
29. Reisenzein, R., Bördgen, S., Holtbernd, T., & Matz, D. (2006). Evidence for strong dissociation between emotion and facial displays: The case of surprise. *Journal of Personality and Social Psychology, 91*(2), 295–315, American Psychological Association, United States.
30. Roberts, S. (2004). *Character animation – 2D skills for better 3D*. Massachusetts, USA: Focal.
31. Sato, H., Ohya, J., & Terashima, N. (2004). Realistic 3D facial animation using parameter-based deformation and texture mapping. In *Proceedings of the 6th IEEE International Conference on Automatic Face and Gesture Recognition*, pp. 735–742.
32. Seyedarabi, H., Aghagolzadeh, A., & Khanmohammadi, S. (2004). Facial expressions animation and lip tracking using facial characteristic points and deformable model. *International Journal of Information Technology, 1*(4), 165–168.
33. Tian, Y., Kanade, T., & Cohn, J.F. (2001). Recognizing AU fro facial expression analysis. *IEEE Transactions on Pattern Analysis and Machine Intelligence, 23*(2), 97–115.
34. Utsumi, A., Kawato, S., & Abe, S. (2005). Attention monitoring based on temporal-signal behavior structures. In *Proceedings of the IEEE International Workshop on Human-Computer Interaction*, pp. 100–109.

35. Valstar, M., & Pantic, M. (2006). Fully automatic facial action unit detection and temporal analysis. In *Proceedings of IEEE Conference on Computer Vision and Pattern Recognition Workshop*, pp. 149–156.
36. Yip, B., & Jin, J.S. (2004). Viewpoint determination and pose determination of human head in video conferencing based on head movement. In *Proceedings of the 10th International Multi-Media Modeling Conference*, pp. 130–135. Brisbane, Australia.
37. Yusoff, F., Rahmat, R., Sulaiman, M., Shaharom, M., & Majid, H. (2009). 3D based head movement tracking for incorporation in facial expression system. *International Journal of Computer Science and Network Security, 9*(2), 417–424.

Chapter 11
Enhanced Audio-Visual Recognition System over Internet Protocol

Yee Wan Wong, Kah Phooi Seng, and Li-Minn Ang

Abstract In this chapter, we extend the research work in Wong [1] and enhance the audio-visual recognition system over internet protocol. The multiband feature fusion method is presented to solve the illumination problem. Then a radial basis function neural network with a new orthogonal least square algorithm is proposed to improve the generalization of the radial basis function neural network with conventional orthogonal least square algorithm. Result shows that the proposed neural network achieves higher recognition accuracy with lesser number of neurons as compared to the neural network with conventional orthogonal least square algorithm. With this neural network, the recognition accuracy of the audio-visual recognition system is improved as compared to the audio-visual recognition system in Wong [1]. Then the audio-visual recognition system over internet protocol is developed where the multiband feature fusion and the proposed neural network are implemented.

Keywords Audio-visual recognition system · Neural network · Illumination variation · Internet protocol · Multiband feature fusion

1 Introduction

Audio-Visual (AV) recognition system is an automatic system that recognizes a person's identity using audio and facial data. Figure 1 depicts a typical AV recognition system. The feature extraction module is important to extract the facial and audio features from the raw data. These features are fed into the classifiers for person identification. The AV fusion combines the outputs from the classifiers and produces the final output. The recognition performance of the AV system is influenced by the robustness of the system to facial illumination variation and the classification accuracy of the classifier.

Y.W. Wong (✉), K.P. Seng, and L.-M. Ang
The University of Nottingham Malaysia Campus, Jalan Broga, 43500 Semenyih, Selangor, Malaysia
e-mail: yeewan.wong@nottingham.edu.my; jasmine.seng@nottingham.edu.my; kenneth.ang@nottingham.edu.my

S.-I. Ao et al. (eds.), *Intelligent Automation and Computer Engineering*,
Lecture Notes in Electrical Engineering 52, DOI 10.1007/978-90-481-3517-2_11,
© Springer Science+Business Media B.V. 2010

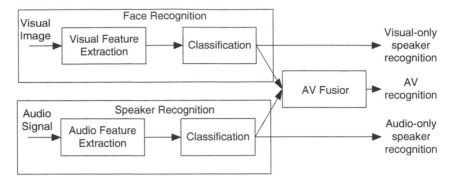

Fig. 1 A typical audio-visual recognition system

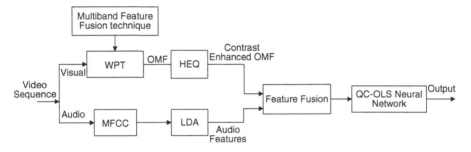

Fig. 2 The proposed audio-visual recognition system

In Wong [1], we presented the AV recognition system which is insusceptible to facial illumination variation. In this chapter, we extend this work by proposing a new neural network to the AV recognition system. Figure 2 shows the block diagram of the AV recognition system. The face image is decomposed by the wavelet packet transform (WPT) [2] and the multiband feature fusion method in Wong [3] extracts the facial features named optimal multiband feature (OMF). The contrast of extracted facial feature is enhanced by the histogram equalizer (HEQ). Mel-frequency cepstrum coefficient (MFCC) [4] is used to extract the audio features and the linear discriminant analysis (LDA) [13] is applied as a feature selection technique. Intra-modal feature fusion [5] combines both features. The radial basis function (RBF) neural network with the new proposed algorithm, quadratic-constant orthogonal least square (QC-OLS) is proposed to perform the classification. The AV recognition system is then implemented over internet protocol (IP) to enable long distance access.

This chapter is organized as follows. Section 2 discusses the wavelet packet transform and the multiband feature fusion method. Section 3 present the detail of the radial basis function (RBF) neural network with quadratic-constant orthogonal least square. Section 4 is specific to the important components for the implementation of the developed AV recognition system over IP. In Section 5, experimental results for the proposed neural network are presented. The compression and packet loss effects of sending the audio and video data over IP on recognition performance are investigated in this section. Section 6, the conclusion, includes some important remarks of this chapter.

2 Background

The recognition performance of the AV recognition system is degraded by face illumination variation. Research efforts [1, 3, 8, 9] were proposed to solve the illumination problem. The multiband feature fusion method in Wong [1, 3] is one of the methods to solve the illumination problem by extracting the illumination invariant facial features from the image. In this method, the WPT [2] is used to decompose image into various frequency subbands.

2.1 Discrete Wavelet Transform

To understand about WPT [2], the discrete wavelet transform (DWT) needs to be discussed first. The DWT is a process to compute the set of coefficients. For a two-dimensional image, three two-dimensional wavelets ψ and a two-dimensional scaling function φ are used. With the image $f(x, y)$, for the first stage decomposition, we have $c_{-1,kx,ky}^{LL}$ is the first subband of approximation coefficients at a coarse resolution. At each stage of transform, the image is decomposed into four quarter-size images (subband), each of the subband is formed by down sampling operation in x and y by a factor of 2. The other three subbands $d_{-1,kx,ky}^{LH}$, $d_{-1,kx,ky}^{HL}$, and $d_{-1,kx,ky}^{HH}$ are the detail coefficient at coarse resolution of 2^{-1}. The two-dimensional DWT can be viewed as a one-dimensional DWT along x and y directions. LL represents the low-frequency component in both x and y directions. LH for example, represents the low-frequency component in x and high-frequency component in y direction. In DWT, the approximation subband LL can be further decomposed to lower resolution approximation subband and detail subbands.

Wavelet packet [2] allows finer and adjustable resolution of frequencies at high frequencies. It gives a richer structure that allows adaptation to particular signals. The decomposition of WPT includes not only low frequency subband but also high frequency subbands. In this chapter, the Haar filter bank is employed for the decomposition.

2.2 Multiband Feature Fusion Method

The main goal of our research work on the multiband feature fusion method in Wong [1,3] is to select frequency subbands that carry illumination invariant features.

In order to extract the illumination invariant features from multiscale space, WPT decomposes the image into more compact frequency subbands. Since the selected subbands are expected to be invariant to illumination, we require these subbands to satisfy the following conditions:

Condition 1. The similarity between classes should be at minimum.

The similarity matrix $\rho(i, j)$ with the data size $N \times N$ where N refers to number of classes in the database. It records the similarity between image i and image j. For a good representation, $\rho(i, j)$ should be closed to one if $i = j$ and $\rho(i, j)$ should be close to zero if $i \neq j$. The Average Unmatched Similarity Value (AUMSV) [12] is defined as below,

$$AUMSV = \frac{1}{(N^2 - N)} \sum_{i=1}^{N} \sum_{j=1}^{N} \rho(i, j) \tag{1}$$

to give a single numerical value to the similarity performance of the subband. It ranges from 0 to 1, which means the higher the discriminatory power, the smaller the AUMSV value.

Condition 2. The ratio for the between-class distance and within-class distance should be maximized.

The ratio for the between-class distance and within-class distance gives an idea of the scatter of the subband representation, which means the higher the ratio, the better the classification performance. The two frequency subbands that achieve the lowest *AUMSV* and highest *BWR* will be selected and they will be concatenated to form the optimal multiband feature set (OMF). After the OMF is found the histogram equalizer (HEQ) is applied on to the subband in OMF to enhance the contrast of the detail images.

The recognition performance of the multiband feature fusion method is compared with the recognition performance of PCA [14], PCA w/o 3^{15} and Independent Component Analysis (ICA) [15] in the Wong [1]. The result obtained shows that the OMF + HEQ outperformed other algorithms tested in the Extended Yale B [10], AR [16] and CUAVE [17] databases.

3 Radial Basis Function Neural Network

The radial basis function (RBF) neural network [19] that has a Gaussian function as the function in the hidden neurons solves the classification problems using much lesser number of hidden neurons as compared to Multilayer perceptron (MLP) [18]. The performance of the RBF neural network is critically influenced by the centers. To choose a better set of centers, S. Chen et al. [21] employed the orthogonal least square (OLS) method to select the RBF centers from all the data points. Realizing the importance of network generalization, S. Chen et al. [22] combines the quadratic weight decay function with the OLS algorithm to derive a regularized OLS (ROLS) for RBF network to avoid the model to be overfitting to the noise. Unlike the ROLS algorithm which only provides weight decay stabilization effect only, we introduce a quadratic-constant OLS (QC-OLS) algorithm which provides weight decay and pruning effect to improve the generalization ability of the conventional ROLS algorithm. The RBF neural networks with the OLS, ROLS and the proposed QC-OLS algorithms will be discussed in the next sections.

3.1 RBF Neural Network with Orthogonal Least Square

In order to understand how OLS [21] works, it is important to relate RBF network and the linear regression model

$$d(t) = \sum_{i=1}^{M} p_i(t)\theta_i + \varepsilon(t) \tag{2}$$

where $d(t)$ is the desired output, the θ_i are the parameters which refers to weight in neural network, and the $p_i(t)$ are known as the regressors which are some fixed functions of $x(t)$. The error signal $\varepsilon(t)$ is assumed to be uncorrelated with the regressors $p_i(t)$. A fixed center c_i in a nonlinear function $\phi(.)$ in RBF neural network corresponds to the regressors $p_i(t)$ in the linear regression model. The RBF networks with n_H hidden neurons can be defined as

$$f_r(x) = \sum_{i=1}^{n_H} \theta_i \phi\left(\|x(t) - c_i\|\right) \tag{3}$$

where $x = [x_1 \ldots x_m]^T$ is the input vector with m inputs, θ_i are the weights, $c_i = [c_{1,i} \ldots c_{m,i}]^T$ are the RBF centers, $\|.\|$ is the Euclidean norm and $\phi(.)$ is the nonlinear function in the hidden neurons. If the training set has samples and we use every training samples $x(t)$ as center, for $1 \leq i \leq N$. The desired output can be expressed as

$$d(t) = \sum_{i=1}^{N} \theta_i \phi_i(t) + e(t), \quad 1 \leq t \leq N \tag{4}$$

where $\phi_i = (\|x(t) - c_i\|)$ for simplification and $e(t)$ is the error between the desired and the actual network output (4) can be expressed in term of matrix form as

$$d = \Phi\Theta + e \tag{5}$$

In order to show how an individual regressor contributes to the output energy, the OLS method transforms the set of Φ_i into a set of orthogonal basis vectors. Let an orthogonal decomposition of the regression matrix Φ be $\Phi = WA$ where A is an $N \times M$ triangular matrix with 1's on the diagonal and 0's below the diagonal and

W is an $N \times M$ matrix with orthogonal columns^{wi}. The space spanned by the set of the orthogonal basis vectors^{wi} is the same space spanned by the set of and (5) can be rewritten as

$$d = Wg + e \tag{6}$$

The orthogonal weight vector $g = [g_1 \ldots g_N]^T$ and the original weight vector Θ satisfy the triangular system

$$A\Theta = g \tag{7}$$

The modified Gram–Schmidt (MSG) [21] procedure is used to compute the orthogonalization. To derive the ROLS [22], it is important to consider the zero-order regularized error criterion

$$e^T e + \lambda g^T g \tag{8}$$

where $\lambda \geq 0$ is the regularization parameter. The term $\lambda g^T g$ (quadratic weight decay function) penalizes large g_i. The g with regularization has changed to

$$g = w^T d \big/ \left(w^T w + \lambda \right) \tag{9}$$

The error reduction ratio due to w_i is

$$[rerr]_i = \left(w_i^T w_i + \lambda \right) g_i^2 \big/ d^T d \tag{10}$$

This ratio seeks a subset of significant regressors in a forward-regression manner. The selection is terminated at the M_s th stage when

$$1 - \sum_{k=1}^{M_s} [err]_s < \varepsilon \tag{11}$$

is satisfied, where $0 < \varepsilon < 1$ is a chosen tolerance or terminated at a selected number of neurons.

3.2 RBF Neural Network with Quadratic-Constant Orthogonal Least Square

In Section 3.1, we have seen how the ROLS control the model complexity by involving a penalty term (weight decay function) to the error function to improve the generalization of the OLS. Specifically, the ROLS can be combined with other weight decay functions to construct compact network [20]. Hence, we proposed to implement a new weight decay function named quadratic-constant (QC) function to the adaptive OLS neural network. The QC weight decay function is given by

$$E_W = \sum_i \left(w_i^2 + abs\,(w_i) \right) \tag{12}$$

The difference between QC decay function with the quadratic weight decay function is that QC has much narrower weight distribution than the former. Hence, the generalization ability of the network trained with this decay function can be improved. We will look at the Bayesian's theory on this function. The prior distribution for the weight reflects any prior knowledge about the form of network we expect to find. In general, this distribution can be written as an exponential of the form [10]

$$p(w) = \frac{1}{Z_w(\alpha)} \exp(-\alpha E_w) \tag{13}$$

where $Z_w(\alpha)$ is a normalization factor given by

$$Z_w(\alpha) = \int \exp(-\alpha E_w) \, dw \tag{14}$$

To encourage smooth network mapping, the network weights is preferably small. The prior distribution of the QC and quadratic weight decay function are in (15) and (16) respectively.

$$p_1(w) = \frac{1}{Z_w(\alpha)} \exp\left(-\alpha \left\| w^2 + abs(w) \right\|\right) \tag{15}$$

$$p_2(w) = \frac{1}{Z_w(\alpha)} \exp\left(-\frac{\alpha}{2} \left\| w^2 \right\|\right) \tag{16}$$

Thus, when $\|w\|$ is large, E_w is large, and $p(w)$ is small, and so this choice of prior distribution says that we expect the weight values to be small rather than big. If we compare the equation from (15) and (16), we can see that QC weight decay function achieves the lowest of all.

Besides, it is important to compare the effect of the weight decay functions to the minimum condition of the overall cost function. The training cost function is given by

$$C = E_D + \lambda E_w \tag{17}$$

where $\lambda \geq 0$ is the regularization parameter, and E_D is the training data error function. At the minimum of C, the quadratic weight decay function is

$$\left| \frac{\partial E_D}{\partial w_i} \right| = \frac{\alpha}{\beta} |w_i| \tag{18}$$

The sensitivity of the data misfit to a given weight is proportional to the amplitude of the weight and is unequal for different weights. Therefore, it can be seen that if the weight is zero, then $\partial E_D / \partial w_j = 0$. This is same as the unregularised network where quadratic weight decay contributes nothing towards network pruning in strict case [11]. Now if we compare (18) to the proposed QC weight decay function, as we shall see, that

$$\left|\frac{\partial E_D}{\partial w_j}\right| = \frac{\alpha}{\beta}|w_j| + \frac{\alpha}{\beta} \quad if \ |w_j| > 0 \tag{19}$$

$$\left|\frac{\partial E_D}{\partial w_j}\right| < \frac{\alpha}{\beta} \quad if \ |w_j| = 0 \tag{20}$$

(19) refers to the condition that same with the quadratic weight decay when the weight is non-zero. However, when the weight is equal to zero as refer to (20), weights that fail to achieve the sensitivity of α/β can be pruned [11]. This shows that the proposed weight decay function combined both weight decay and pruning effects. By implementing QC weight decay function to the OLS neural network, the regularized error criterion of the adaptive OLS neural network is

$$e^T e + \lambda g^T g + \lambda g \tag{21}$$

After some derivation, the g is as below

$$g = w^T d - \lambda /(w^T w + \lambda) \tag{22}$$

The regularization error criterion (21) can be expressed as

$$e^T e + \lambda g^T g + \lambda g = d^T d - \sum_{i=1}^{N} \left(w_i^T w_i + \lambda \right) g_i^2 - \sum_{i=1}^{N} g_i \lambda \tag{23}$$

Normalizing (23) by $d^T d$ yields the error reduction ratio due to w_i is

$$[rerr]_i = \left(\left(w_i^T w_i + \lambda \right) g_i^2 + g_i \lambda \right) / d^T d \tag{24}$$

The RBF neural network with the proposed QC-OLS algorithm is then implemented in the AV recognition system to classify the combination of audio and facial features.

4   AV Recognition System over IP

Figure 3 depicts architecture of video and audio streaming over network for the enhanced AV recognition system. At the client side, the raw video and audio signals will be first compressed by H.263 and G.723 encoder respectively. The bit-stream will be packetized and sent over the internet by RTP. Packets may be dropped or experience delay inside the network depending on the network congestion. For packets that are delivered to the server successfully, they are passed through the transport protocols and being depacketized to bit-streams before being decoded at the video and audio decoder. At the server side, the received packets

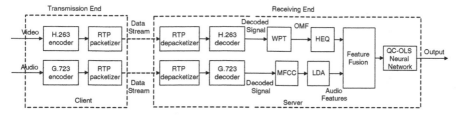

Fig. 3 Block diagram of the developed AV recognition over IP

will be depacketized and passed to the audio and video decoder. The decoded image frames will be decomposed by the WPT and the OMF is extracted. HEQ is then applied onto the OMF to enhance the image contrast. The audio signal is the MFCC for audio feature extraction and LDA for feature selection. Both the audio and facial features will be fused by the intra-modal feature fusion method [5] and the QC-OLS neural network performs classification.

5 Experiment and Result

There are two experiments in this section. The first experiment depicts the recognition performance of the QC-OLS neural network in audio-visual recognition systems. The second experiment shows the recognition performance of the developed AV recognition system over IP.

5.1 AV Recognition System with RBF Neural Network

This experiment evaluates the classification performance of the proposed RBF network with QC-OLS algorithm in face recognition system, audio recognition system and audio-visual recognition system. The OMF + HEQ facial features and MFCC and LDA audio features are fed to the QC-OLS neural network for classification. The recognition performance of the proposed neural network is compared with the recognition performance of the ROLS and MLP neural networks. The regularization parameter is set as 0.01 for QC-OLS and 0.1 for ROLS. The widths of the neurons for both the networks are set as 30. The CUAVE AV database is used in this experiment.

Three training sample images are used for training and three testing images are used to show the generalization performance of the proposed network in face recognition system. Figure 4 shows the recognition error rate of the QC-OLS neural network, ROLS and MLP neural networks against the number of neurons. The proposed network achieves the lowest recognition error rate at 13% with 60 neurons.

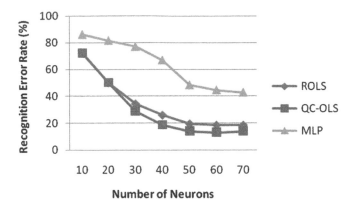

Fig. 4 Recognition error rate against number of neurons for ROLS, QC-OLS and MLP neural network in face recognition system

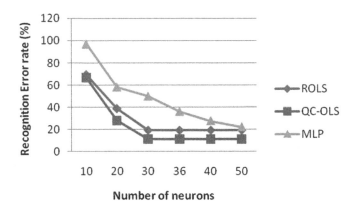

Fig. 5 Recognition error rate against number of neurons for ROLS, QC-OLS and MLP neural network in audio recognition system

The result illustrates that QC-OLS outperforms the ROLS and MLP in term of recognition accuracy in face recognition system.

The classification performance of QC-OLS network in audio recognition system is shown in Fig. 5. 10 dB signal to noise ratio is used in this experiment. Three training samples and one testing samples are used for the network training and testing purposes. Result shows that the QC-OLS achieves lower recognition error rate than ROLS and MLP across all the number of neurons.

The recognition performance of the QC-OLS network in AV recognition system is shown in Fig. 6. Three training images are used as the training samples and one image is used as the testing sample. The QC-OLS achieves lower recognition error rate as compared to ROLS and MLP neural networks against the 20–50 neurons. The QC-OLS achieves the lowest recognition error rate at 2.8% with 50 neurons. The results show that the proposed QC-OLS achieves higher recognition accuracy

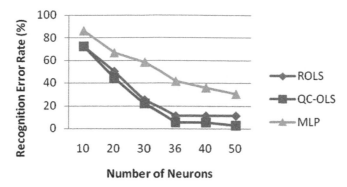

Fig. 6 Recognition error rate against number of neurons for ROLS, QC-OLS and MLP neural network in audio-visual recognition system

with lesser number of neurons as compared to ROLS and MLP in AV recognition system.

5.2 AV Recognition System over IP

The multiband feature fusion and QC-OLS neural network are implemented in the AV recognition system over IP as depicted in Fig. 3. The JM studio [6] is used to transmit and receive the data. In this experiment, we evaluate the recognition performance of AV recognition system under varying bandwidth. Bandwidth Limiter Enterprise [7] is employed to control the link capacities. The link capacity at the client side is fixed at 10 Mbits/s and the link capacity at the server sides varies from 160,400 kbits/s, and 8 Mbits/s. G.723 with 16 kbits/s is selected as the codec in this experiment. Three training samples per subject are used for the QC-OLS neural network training. The width of the neuron is 30 and the number of neuron is 50.

Table 1 shows the recognition performance of the system under low link capacity (160 kbits/s). Due to the limited link capacity, the packet loss ratio for CIF is the highest of 65.4% and causes the highest recognition error rate of 41.7%. When QCIF and SQCIF are at the same limited link capacity, the error rate is 11.1% and 5.6% respectively. Table 2 shows that only CIF encounters packet loss at 400 kbits/s link capacity. Smaller video formats QCIF and SQCIF achieve a constant error rate at 2.8% and 5.6% across link capacity of 400 kbits/s to 8 Mbits/s as shown in Tables 2 and 3. At larger link capacity 8 Mbits/s, there is no packet loss for CIF and it achieves error rate of 2.8%. The results also show that the recognition performance of QCIF is the most promising one. This is because the recognition performance of QCIF is less affected by the link capacity variations. If we compare the recognition performance of the enhanced AV recognition system over IP and the AV recognition system over IP in Wong [1], we can conclude that the enhanced system outperformed the system proposed in Wong [1] in term of recognition accuracy.

Table 1 Packet loss ratio and recognition error rate of the AV recognition system over 160 kbits/s link capacity

Video dimension	Audio bit-rate	Packet loss ratio (%)	Error rate (%)
Standalone (720 × 480)	Standalone (16 kbit/s)	0	2.8
Over IP CIF (352 × 288)	Over IP G.723 (16 kbit/s)	65.4	41.7
Over IP QCIF (176 × 144)	Over IP G.723 (16 kbit/s)	11.8	11.1
Over IP SQCIF (128 × 96)	Over IP G.723 (16 kbit/s)	0	5.6

Table 2 Packet loss ratio and recognition error rate of the AV recognition system over 400 kbits/s link capacity

Video dimension	Audio bit-rate	Packet loss ratio (%)	Error rate (%)
Standalone (720 × 480)	Standalone (16 kbit/s)	0	2.8
Over IP CIF (352 × 288)	Over IP G.723 (16 kbit/s)	22.6	25.0
Over IP QCIF (176 × 144)	Over IP G.723 (16 kbit/s)	0	2.8
Over IP SQCIF (128 × 96)	Over IP G.723 (16 kbit/s)	0	5.6

Table 3 Packet loss ratio and recognition error rate of the AV recognition system over 8 Mbits/s link capacity

Video dimension	Audio bit-rate	Packet loss ratio (%)	Error rate (%)
Standalone (720 × 480)	Standalone (16 kbit/s)	0	2.8
Over IP CIF (352 × 288)	Over IP G.723 (16 kbit/s)	0	2.8
Over IP QCIF (176 × 144)	Over IP G.723 (16 kbit/s)	0	2.8
Over IP SQCIF (128 × 96)	Over IP G.723 (16 kbit/s)	0	5.6

6 Conclusion

In this chapter, the audio-visual recognition system based on the multiband feature fusion technique and RBF neural network with QC-OLS algorithm over internet protocol is presented. Recognition performance of the multiband feature fusion technique achieves high recognition accuracy in face recognition system. The QC-OLS neural network is proposed to further increase the recognition accuracy of the system. This neural network achieves higher recognition accuracy than ROLS and MLP. The multiband feature fusion and QC-ROLS neural network are implemented in the audio-visual recognition system over internet protocol. The QCIF is shown to be most suitable for AV recognition system over IP because it achieves promising recognition performance in both limited and unlimited link capacities. The enhanced AV recognition system outperforms the AV recognition system in Wong [1] in term of recognition accuracy.

References

1. Wong, Y.W., Seng, K.P., & Ang, L.-M. (2009). Audio-visual recognition system insusceptible to illumination variation over Internet Protocol. *IAENG Internation Journal of Computer Science, 36*(2), IJCS_36_2_08.
2. Primer, A. (1998). *Introduction to wavelet and wavelet transform*. Prentice Hall.
3. Wong, Y.W., Seng, K.P., & Ang, L.-M. (Mar 2009). Audio-visual authentication system over the Internet Protocol. *Proceedings of the International MultiConference of Engineers and Computer Scientists 2009* (IMECS 2009). The 2009 IAENG *International Conference on Imaging Engineering* (ICIE'09), I, 938–943.
4. Campbell, J.P., Jr. (Sept 1997). Speaker recognition: A tutorial. *Proceedings of IEEE, 85*(9), 1437–1462.
5. Wong, Y.W., Seng, K.P., Ang, L.-M., Khor, W.Y., & Liau, H.F. (Dec 2007). Audio-visual recognition system with intra-modal fusion. *2007 International Conference on Computational Intelligence and Security*, pp. 609–613.
6. JM studio: java.sun.com/javase/technologies/desktop/media/jmf/2.1.1/samples/samplecode.html.
7. Bandwidth controller enterprise: wareseeker.com/Network-Tools/bandwidth-controller-enterprise-1.18.zip/280994.
8. Belhumeur P., & Kriegman, D. (1998). What is the set of images of an object under all possible lighting conditions. *International Journal of Computer Vision, 28*, 245–260.
9. Lee, K.-C., Ho, J., & Kriegman, D.J. (May 2005). Acquiring linear subspaces for face recognition under variable lighting illuminations. *IEEE Transactions on Pattern Analysis and Machine Intelligence, 27*(5), 684–698.
10. Christopher M. Bishop (2003). *Neural networks for pattern recognition*. Oxford University Press, pp. 116–163.
11. Williams P.M. (1995). Bayesian regularisation and pruning using a Laplace prior. *Neural Computation, 7*(1), 117–143.
12. Feng, G.C., Yuen, P.C., & Dai, D.Q. (Apr 2000). Human face recognition using PCA on wavelet subband. *Journal of Electronic Imaging, 9*(2), 226–233.
13. Swets D.L., & Weng, J. (Aug 1996). Using discriminant eigenfeatures for image retrieval. *IEEE Transactions on Pattern Analysis and Machine Intelligence, 18*(8), 831–836.
14. Turk, M., & Pentland, A. (1991). Eigenfaces for recognition. *Journal of Cognitive Neuroscience 3*(1), 71–86.
15. Bartlett, M.S., Movellan, J.R., & Sejnowski, T.J. (Nov 2002). Face recognition by independent component analysis. *IEEE Transactions on Neural Networks, 13*, 1450–1464.
16. Martinez A.M., & Benavente, R. (1998). The AR face database. CVC Tech. Report #24.
17. Patterson, E.K., Gurbuz, S., Tufekci, Z., & Gowdy, J.N. (2002). CUAVE: A new audio-visual database for multimodal human-computer interface. In *Proceedings ICASSP*, pp. 2017–2020.
18. Jain, L.C., Halici, U., Hayashi, I., Lee, S.B., & Tsutsui, S. (1999). *Intelligent biometric techniques in fingerprint and face recognition*. Boca Raton, FL: CRC Press,.
19. Er, M.J., Wu, S., Lu, J., & Toh, H.L. (May 2002). Face recognition with radial basis function (RBF) neural networks. *IEEE Transactions on Neural Networks, 13*(3), 697–710.
20. Wu, L., & Moody, J. (1996). A smoothing regularizer for feedforward and recurrent neural networks. *Neural Computers, 8*, 461–489.
21. Chen, S., Cowan, C.C.N., & Grant, P.M. (1991). Orthogonal least squares for radial basis function networks. *IEEE Transactions on Neural Network, 2*(2), 302–309.
22. Chen, S., Chng, E.S., & Alkadhimi, K. (1996). Regularized orthogonal least squares algorithm for constructing radial basis function networks. *Internetional Journal of Control, 64*(5), 829–837.

Chapter 12
Lossless Color Image Compression Using Tuned Degree-K Zerotree Wavelet Coding

Li Wern Chew, Li-Minn Ang, and Kah Phooi Seng

Abstract This chapter presents a lossless color image compression scheme using the degree-k zerotree coding technique. From studies carried out on the degree-k zerotree coding, it has been found that at lower bit-rates, a higher degree zerotree coding gives a better coding performance whereas at higher bit-rates, coding with a lower degree zerotree is more efficient. Hence, the degree of zerotree tested is tuned in each encoding pass in the proposed Tuned Degree-K Zerotree Wavelet (TD-KZW) coding to obtain an optimal compression performance. Since the TDKZW coder uses the set-partitioning approach similar to the Set-Partitioning in Hierarchical Trees (SPIHT) coder, it allows embedded coding and also enables progressive transmission to take place. In addition, a new spatial orientation tree (SOT) structure for color coding is also proposed here for low memory implementation of the TD-KZW coder (LM-TDKZW). Simulation results on standard test images show that the proposed TDKZW coder gives a better lossless color image compression performance than the SPIHT coder. The results also show that the proposed LM-TDKZW coder not only requires as little as 6.25% of the memory needed by the SPIHT coder, it is able to achieve an almost equivalent lossless compression performance as the SPIHT coder.

Keywords Color image compression · Degree-k zerotree coding · Lossless compression · Spatial orientation tree structure · Wavelet-based image coding

1 Introduction

Wavelet-based image compression which applies the entropy coding in the discrete wavelet transform (DWT) domain began to gain attention in the field of image coding since the 1990s. It has been shown to outperform the conventional image

L.W. Chew (✉), L.-M. Ang, and K.P. Seng
Department of Electrical and Electronic Engineering, The University of Nottingham,
43500 Selangor, Malaysia
e-mail: eyx6clw@nottingham.edu.my; kezklma@nottingham.edu.my;
kezkps@nottingham.edu.my

S.-I. Ao et al. (eds.), *Intelligent Automation and Computer Engineering*,
Lecture Notes in Electrical Engineering 52, DOI 10.1007/978-90-481-3517-2_12,
© Springer Science+Business Media B.V. 2010

compression standard, the Joint Photographic Experts Group (JPEG) [1] which uses the discrete cosine transform (DCT) as its base. Since DCT causes blocking artifacts which reduces the quality of the reconstructed image, the wavelet-based coding is generally preferred over JPEG due to its ability to overcome blocking artifacts and to provide embedded coding [2, 3].

Wavelet-based image coding can be categorized into zerotree coding and subband coding. Zerotree coding techniques include the Embedded Zerotree Wavelet (EZW) [2] and the Set-Partitioning in Hierarchical Trees (SPIHT) [3] algorithms whereas the Embedded Block Coding with Optimized Truncation (EBCOT) [4] which is incorporated into the JPEG2000 [5] is categorized under subband coding technique.

Although JPEG2000 provides a higher compression efficiency as compared to EZW and SPIHT, its multi-layered coding procedures are very complex and computationally intensive. Also, the need for multiple coding tables for arithmetic coding requires extra memory allocation which makes the hardware implementation of the coder more complex and expensive [6, 7].

The EZW and SPIHT coders apply the concept of bit-plane coding. Their embedded property of being able to generate bit-streams according to the order of importance allows variable bit-rate coding and decoding. Thus, an exact targeted bit-rate or tolerated distortion degradation can be achieved since the encoder or decoder can stop at any point in the bit stream. In spite of this, a full high quality image can still be obtained. Both EZW and SPIHT coders also allow for progressive transmission [2, 3].

Comparing EZW and SPIHT, the latter not only has all the advantages that EZW possesses, it has also been shown that SPIHT provides a better compression performance than EZW coding. In addition, SPIHT algorithm has straight forward coding procedures and does not require any coding table since arithmetic coding is not essential in SPIHT coding [3].

In this chapter, the theory and coding methodology of degree-k zerotree coding is first presented. In general, a degree-k zerotree coder codes the significance of a spatial orientation tree (SOT) up to k levels and an SOT is referred to as a degree-k zerotree if all the nodes are insignificant with respect to a threshold except for the nodes in the top k levels [8, 9]. From studies carried out, it has been found that for bit-plane coding, a higher degree zerotree coding performs better at lower bit-rates whereas a lower degree zerotree coding provides a better coding performance at higher bit-rates. Based on this finding, a Tuned Degree-K Zerotree Wavelet (TD-KZW) [10] coding is proposed.

As its name implies, the degree of zerotree tested is tuned in each encoding pass to obtain an optimal compression performance in the proposed TDKZW coding. In addition, a new SOT structure for color image compression is also proposed for the low memory implementation of the TDKZW coder. For the sake of brevity, the proposed low memory work is denoted by LM-TDKZW. The new SOT takes the 1×4 nodes as its four offsprings in certain wavelet subbands and uses a partially-linked structure to connect the no-descendant roots in the low-pass subband in the luminance plane to the wavelet coefficients in the chrominance plane. This is carried out to ensure that fewer zerotrees are obtained and the coding efficiency of the LM-TDKZW coder at lower bit-rates is increased, especially when the dimension of the low-pass subband is large [11]. Lastly, the lossless performance of both the proposed TDKZW and LM-TDKZW coders are evaluated on color images and are compared to the traditional SPIHT coder.

The remaining sections of this chapter are organized as follows: Section 2 presents the degree-k zerotree coding. This includes the EZW and SPIHT coding techniques. In addition, the strengths and weaknesses of both the higher and lower degree zerotree coding are illustrated using examples. Section 3 describes the proposed TDKZW coding scheme. Here, an analysis is first carried out on a degree-1 to a degree-7 zerotree coding on six standard color test images to obtain a suitable tuning parameter for our proposed TDKZW coder. A new SOT structure for color image compression is then proposed for our LM-TDKZW coder followed by a discussion on the memory requirements for the proposed works as compared to the SPIHT coder. In Section 4, the lossless compression performance of our proposed TDKZW and LM-TDKZW coders on color images are evaluated using software simulations and are compared to the performance of the SPIHT coder. Finally, Section 5 concludes this chapter.

2 Degree-K Zerotree Coding

Zerotree coding schemes operate by exploiting the relationships among the wavelet coefficients across the different scales at the same spatial location in the wavelet subband. The coding algorithms are developed based on the hypothesis that "In an SOT, if the parent node is found to be insignificant with respect to a threshold T, then all of its descendants are likely to be insignificant with respect to T" [2]. To achieve compression, this SOT is encoded using a single symbol which indicates that all the nodes in the SOT are insignificant and there is no need to code these insignificant nodes.

A degree-k zerotree coder [8, 9] was recently proposed to quantify the coding performance of zerotrees for wavelet-based image compression algorithms such as the EZW and SPIHT. In general, a degree-k zerotree is an SOT when all its nodes are insignificant with respect to a threshold value except for the nodes in the top k levels. Figure 1 shows an example of a degree-1 to a degree-3 zerotree structure. A degree-k zerotree coder performs significance tests on a degree-1 zerotree up to a degree-k zerotree for each wavelet coefficient [8, 9].

In this section, the coding methodologies of EZW and SPIHT coders are first presented. Next, the coding performance of a degree-1 to a degree-3 zerotree coder is investigated through given examples followed by a discussion on the strengths and weaknesses of both higher and lower degree zerotree coding.

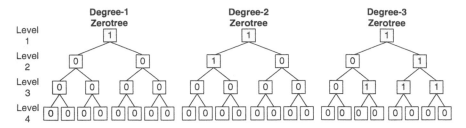

Fig. 1 Spatial orientation tree structure for degree-1, degree-2 and degree-3 zerotree

2.1 Embedded Zerotree Wavelet Coding

Due to the nature of wavelet decomposition, higher scale subbands will contain more energy than the lower scale subbands. Thus, the EZW [2] encoding starts from a higher to a lower scale subband with an initial threshold, $T_0 > |X_j|/2$ where X_j is the largest coefficient found in the wavelet-transformed image. The encoding is then continued under a sequence of thresholds $T_0, T_1, T_2, T_3, \ldots T_{N-1}$ where $T_i = (T_{i-1}/2)$. This coding methodology is referred to as bit-plane encoding.

The EZW coding scheme encodes the wavelet coefficients starting with a degree-0 zerotree, which is an SOT where its root node and all its descendants are insignificant with respect to a threshold T. This degree-0 zerotree is then coded as zerotree root (ZTR) if it is not the descendant of a previously found ZTR for that threshold T. However, if a coefficient is insignificant with respect to T but has some significant descendants, it is then coded as isolated zero (IZ). For a coefficient that is found to be significant with respect to T, it is coded as positive significant (POS) or negative significant (NEG) depending on the sign of the coefficient.

Two lists need to be maintained during the EZW coding: a dominant list which contains the coordinates of the coefficients that have not yet been found to be significant and a subordinate list which contains the magnitudes of those coefficients that have already been found to be significant. For each threshold, all coefficients in the dominant list are scanned and coded. Those coefficients that are coded as significant are then transferred to the subordinate list. In the subordinate pass, each of the coefficients in the list is refined to an additional bit of precision. This encoding process is repeated until all the wavelet coefficients are coded or the target rate has been met. The encoded symbols stream that contains a mixture of four symbols (POS, NEG, ZTR and IZ) and the refinement bit '1' or '0' is then arithmetically coded for transmission.

The inherent ability of embedded coding to generate the encoded bit stream in the order of importance allows the EZW encoder to stop encoding at any point when a target rate or target distortion metric is met. Similarly, at any given bit stream, the decoder can cease decoding at any point and yet provides a full image reconstruction. This allows signal-to-noise ratio (SNR) scalability as the reconstructed image quality is dependent on the number of bits that have been decoded. It also allows progressive transmission as the image is decompressed with increasing accuracy.

2.2 Set-Partitioning in Hierarchical Trees Coding

The SPIHT [3] coding scheme, which is the improved version of EZW, not only
has all the advantages that EZW possesses, it has also been shown to provide a
better compression efficiency than EZW even without the arithmetic coding. This is
because compared to the EZW which is a degree-0 zerotree coder, SPIHT which is
a degree-2 zerotree coder provides more levels of descendant information for each
coefficient tested. This hypothesis has been justified in works proposed by Cho [8]
and Cicala [9].

SPIHT coding adopts a set-partitioning approach where the significance test result is binary. A magnitude test is carried out on every wavelet coefficient c, at
coordinate (i, j) to indicate the significance of the coefficient. A coefficient is encoded as significant i.e. $Sn(i, j)$ = '1' if its value is larger than or equal to the
threshold T, or as insignificant i.e. $Sn(i, j)$ = '0' if its value is smaller than T.

The order of subsets which are tested for significance is stored in three ordered
lists: (i) List of significant pixels (LSP), (ii) List of insignificant pixels (LIP), and
(iii) List of insignificant sets (LIS). LSP and LIP contain the coordinates of individual pixels whereas LIS contains either the set $D(i, j)$ or $L(i, j)$ which is referred
to as type A and type B respectively. Similar to EZW, SPIHT coding first defines
a starting threshold $T_0 > |X_j|/2$ where X_j is the maximum value of the wavelet
coefficients. The LSP is set as an empty list and all the nodes in the highest subband
are put into the LIP. The root nodes with descendants are put into the LIS.

Two encoding passes which are the sorting pass and the refinement pass are then
performed in the SPIHT coder. During the sorting pass, a significance test is performed on the coefficients according to the order in which they are stored in LIP.
Elements in LIP that are found to be significant with respect to the threshold are
moved to the LSP. A significance test is then performed on the sets in the LIS. Here,
if a set in LIS is found to be significant, the set is removed from the list and is partitioned into four single elements and a new subset. This new subset is added back to
LIS and the four elements are then tested and moved to LSP or LIP, depending on
whether they are significant or insignificant with respect to the threshold.

Refinement is then carried out on every coefficient that is added to the LSP except
for those that are just added during the sorting pass. Each of the coefficients in the
list is refined to an additional bit of precision. Finally, the threshold is halved and
SPIHT coding is repeated until all the wavelet coefficients are coded for lossless
compression coding or until the target rate is met for lossy compression.

2.3 Set-Partitioning Coding with Degree-K Zerotree

In this sub-section, a new family of set-partitioning coding schemes with a degree-k
zerotree is presented. The proposed coding schemes perform significance tests on
a degree-1 zerotree up to a degree-k zerotree for a wavelet coefficient (i, j) using
the set-partitioning approach similar to SPIHT. The notations used in the proposed

coding scheme for node (i, j) at level L are summarized as follows: SIG(i, j) is the significance of node (i, j); DESC(i, j) is the significance of descendants of node (i, j); GDESC(i, j) is the significance of grand descendants of node (i, j); and G_NDESC(i, j) is the significance of descendant nodes at level (L + N) onwards for node (i, j).

For a degree-1 zerotree coding, significance tests are performed on the individual coefficient (i, j) and on the descendants of the coefficient, i.e. SIG(i, j) and DESC(i, j) are determined in this coding. For a degree-2 zerotree coding, significance tests are performed to determine the SIG(i, j), DESC(i, j) and GDESC(i, j), which is equivalent to the SPIHT coding. In degree-3 zerotree coding, a significance test is also carried out on the G_2DESC(i, j) in addition to the significance tests on the SIG(i, j), DESC(i, j) and GDESC(i, j). An illustration of a degree-1, degree-2 and degree-3 zerotree coding with two different thresholds is shown in Fig. 2. This example shows the number of output bits needed to encode all the nodes in the SOT at different threshold values, T.

As shown in Fig. 2a, at T = 16; SPIHT coding performs better than the degree-1 zerotree coding since it generates fewer bits for the encoding of the SOT whereas the degree-3 zerotree coding requires even fewer encoding bits as compared to the SPIHT coding. In degree-3 zerotree coding, since the G_2DESC(i, j) is found to be insignificant, significance tests on GDESC(k, l) where (k, l)s are the offsprings of O(i, j) can be omitted. Therefore, when compared to SPIHT coding, with just one extra G_2DESC(i, j) bit transmitted in a degree-3 zerotree coding, a few GDESC(k, l) bits can be saved if G_2DESC(i, j) is shown to be insignificant. On the other hand, as shown in Fig. 2b, when T = 4; a degree-1 zerotree coding is found to be most efficient in encoding the SOT. Here, it is anticipated that the SOT is likely to have significant nodes and is subsequently partitioned into sub trees after performing the significant test on DESC(i, j).

From our studies carried out on the degree-k zerotree coding, it has been found that at lower bit-rates i.e. at a higher threshold, where most of the coefficients are insignificant, a higher degree zerotree coding gives a better coding performance. In contrast, at higher bit-rates i.e. at a lower threshold, a higher degree zerotree coding is less efficient since the wavelet coefficients are more likely to be significant as the number of planes encoded is increased [10]. Hence, to obtain an optimal compression performance, the degree of zerotree tested is tuned in each encoding pass in the proposed TDKZW coding. Section 3 describes the coding methodology of the proposed work.

3   Tuned Degree-K Zerotree Wavelet Coder

Similar to other wavelet-based color image coders, the proposed TDKZW coding scheme is divided into four main stages as shown in Fig. 3. A color image in Red-Green-Blue (RGB) format is first transformed into the luminance and chrominance components using the reversible luminance-offset orange-offset green (YCoCg)

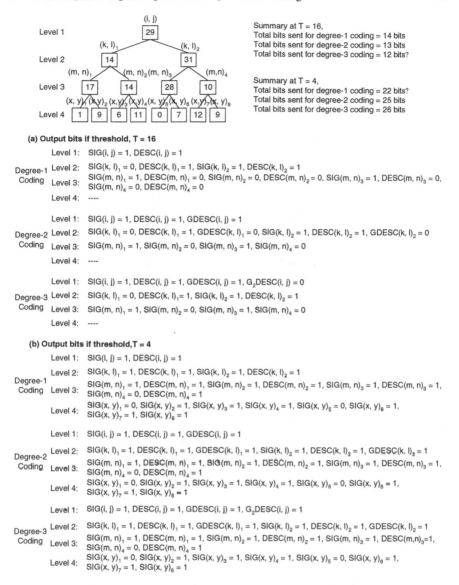

Summary at T = 16,
Total bits sent for degree-1 coding = 14 bits
Total bits sent for degree-2 coding = 13 bits
Total bits sent for degree-3 coding = 12 bits?

Summary at T = 4,
Total bits sent for degree-1 coding = 22 bits?
Total bits sent for degree-2 coding = 25 bits
Total bits sent for degree-3 coding = 26 bits

(a) Output bits if threshold, T = 16

Level 1: SIG(i, j) = 1, DESC(i, j) = 1

Degree-1 Level 2: SIG(k, l)$_1$ = 0, DESC(k, l)$_1$ = 1, SIG(k, l)$_2$ = 1, DESC(k, l)$_2$ = 1
Coding Level 3: SIG(m, n)$_1$ = 1, DESC(m, n)$_1$ = 0, SIG(m, n)$_2$ = 0, DESC(m, n)$_2$ = 0, SIG(m, n)$_3$ = 1, DESC(m, n)$_3$ = 0,
            SIG(m, n)$_4$ = 0, DESC(m, n)$_4$ = 0
      Level 4: ----

Level 1: SIG(i, j) = 1, DESC(i, j) = 1, GDESC(i, j) = 1

Degree-2 Level 2: SIG(k, l)$_1$ = 0, DESC(k, l)$_1$ = 1, GDESC(k, l)$_1$ = 0, SIG(k, l)$_2$ = 1, DESC(k, l)$_2$ = 1, GDESC(k, l)$_2$ = 0
Coding Level 3: SIG(m, n)$_1$ = 1, SIG(m, n)$_2$ = 0, SIG(m, n)$_3$ = 1, SIG(m, n)$_4$ = 0
      Level 4: ----

Level 1: SIG(i, j) = 1, DESC(i, j) = 1, GDESC(i, j) = 1, G$_2$DESC(i, j) = 0

Degree-3 Level 2: SIG(k, l)$_1$ = 0, DESC(k, l)$_1$ = 1, SIG(k, l)$_2$ = 1, DESC(k, l)$_2$ = 1
Coding Level 3: SIG(m, n)$_1$ = 1, SIG(m, n)$_2$ = 0, SIG(m, n)$_3$ = 1, SIG(m, n)$_4$ = 0
      Level 4: ----

(b) Output bits if threshold, T = 4

Level 1: SIG(i, j) = 1, DESC(i, j) = 1

         Level 2: SIG(k, l)$_1$ = 1, DESC(k, l)$_1$ = 1, SIG(k, l)$_2$ = 1, DESC(k, l)$_2$ = 1
Degree-1 Level 3: SIG(m, n)$_1$ = 1, DESC(m, n)$_1$ = 1, SIG(m, n)$_2$ = 1, DESC(m, n)$_2$ = 1, SIG(m, n)$_3$ = 1, DESC(m, n)$_3$ = 1,
Coding   SIG(m, n)$_4$ = 0, DESC(m, n)$_4$ = 1
         Level 4: SIG(x, y)$_1$ = 0, SIG(x, y)$_2$ = 1, SIG(x, y)$_3$ = 1, SIG(x, y)$_4$ = 1, SIG(x, y)$_5$ = 0, SIG(x, y)$_6$ = 1,
                   SIG(x, y)$_7$ = 1, SIG(x, y)$_8$ = 1

Level 1: SIG(i, j) = 1, DESC(i, j) = 1, GDESC(i, j) = 1

         Level 2: SIG(k, l)$_1$ = 1, DESC(k, l)$_1$ = 1, GDESC(k, l)$_1$ = 1, SIG(k, l)$_2$ = 1, DESC(k, l)$_2$ = 1, GDESC(k, l)$_2$ = 1
Degree-2 Level 3: SIG(m, n)$_1$ = 1, DESC(m, n)$_1$ = 1, SIG(m, n)$_2$ = 1, DESC(m, n)$_2$ = 1, SIG(m, n)$_3$ = 1, DESC(m, n)$_3$ = 1,
Coding   SIG(m, n)$_4$ = 0, DESC(m, n)$_4$ = 1
         Level 4: SIG(x, y)$_1$ = 0, SIG(x, y)$_2$ = 1, SIG(x, y)$_3$ = 1, SIG(x, y)$_4$ = 1, SIG(x, y)$_5$ = 0, SIG(x, y)$_6$ = 1,
                   SIG(x, y)$_7$ = 1, SIG(x, y)$_8$ = 1

Level 1: SIG(i, j) = 1, DESC(i, j) = 1, GDESC(i, j) = 1, G$_2$DESC(i, j) = 1

         Level 2: SIG(k, l)$_1$ = 1, DESC(k, l)$_1$ = 1, GDESC(k, l)$_1$ = 1, SIG(k, l)$_2$ = 1, DESC(k, l)$_2$ = 1, GDESC(k, l)$_2$ = 1
Degree-3 Level 3: SIG(m, n)$_1$ = 1, DESC(m, n)$_1$ = 1, SIG(m, n)$_2$ = 1, DESC(m, n)$_2$ = 1, SIG(m, n)$_3$ = 1, DESC(m,n)$_3$=1,
Coding   SIG(m, n)$_4$ = 0, DESC(m, n)$_4$ = 1
         Level 4: SIG(x, y)$_1$ = 0, SIG(x, y)$_2$ = 1, SIG(x, y)$_3$ = 1, SIG(x, y)$_4$ = 1, SIG(x, y)$_5$ = 0, SIG(x, y)$_6$ = 1,
                   SIG(x, y)$_7$ = 1, SIG(x, y)$_8$ = 1

Fig. 2 An illustration of degree-1, degree-2 and degree-3 zerotree coding with two different thresholds, T = 16 and T = 4

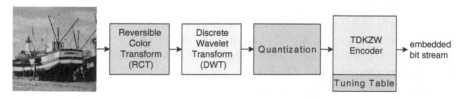

Fig. 3 The proposed Tuned Degree-K Zerotree Wavelet (TDKZW) coding scheme

color space transformation [12]. The transformed image data is then fed into the DWT module for wavelet decomposition. Next, quantization is then carried out followed by the TDKZW coding. It should be noted that if lossless compression is expected, the quantization step is set to one. At the end of encoding, an embedded bit-stream is obtained and is either transmitted or stored for future use. The decoding process is simply the reverse of the encoding process.

It is also shown in Fig. 3 that a tuning table is incorporated as part of the TDKZW coding. This tuning table holds the tuning parameters which indicate which degree-k zerotree coding provides the best compression result at each encoding pass. Based on these tuning parameters, the degree of zerotree tested is tuned in each encoding pass during the TDKZW coding to achieve optimal compression efficiency.

The proposed TDKZW coding uses the set-partitioning approach similar to the SPIHT coding where all the entries added to the end of the LIS are evaluated in the same sorting pass. Since they are partitioned, moved and added to the LIS based on the set-partitioning decision and are evaluated according to their order of importance, the bit streams generated have properties that are similar to that of embedded coding. Besides this, it also makes progressive transmission possible.

In terms of complexity, the proposed algorithm does not significantly increase the coder complexity as the fast technique used to identify zerotrees [13] can be incorporated into the proposed TDKZW coding scheme. Here, zerotrees for all sorting passes are identified prior to encoding. This increases the processing speed of zerotree coding by eliminating the need for recursively checking the zerotrees in the sorting pass.

Due to the nature in which the order of wavelet decomposition is processed, coefficients in a lower level of an SOT are available before those coefficients in the higher levels. Thus, the zerotree information at the N level can be determined by performing a bitwise-OR operation on the zerotree information at the $(N + 1)$ level with the parent node at N level [13]. Since only bitwise-ORing operation is needed, the increase in complexity of the coder due to the significance tests on higher degree zerotrees becomes insignificant.

3.1 Tuning Table

In SPIHT implementation [14], a six-scale wavelet decomposition and a further one scale virtual decomposition i.e. DWT-6, SOT-7, are carried out on an image of size 512×512 pixels prior to the entropy coding. A similar decomposition level is used in our proposed TDKZW coder. Thus, the maximum level of degree-k zerotree coding that can be achieved is seven. In other words, the proposed TDKZW coder is tuned from a maximum degree of $K = 7$ to a minimum degree of $K = 1$. Analyses were performed on six standard color test images to determine the best tuning table and the results are presented in Table 1. Each tuning parameter obtained indicates the degree-k zerotree coding scheme that encodes the image with the fewest number of bits i.e. the degree-k zerotree coding that gives the best compression performance.

Table 1 Predefined tuning table for TDKZW coding using bit-plane coding methodology

Degree-k zerotree coding scheme that gives the fewest number of encoding bits at each bit-plane. Test images used are in 24-bits RGB format with size 512×512 pixels

Bit-plane	Airplane	Baboon	Lenna	Peppers	Sailboat	Tiffany
1	7	2	7	7	7	2
2	7	2	7	7	7	2
3	7	2	7	7	7	2
4	7	2	7	6	7	2
5	7	2	7	6	7	2
6	7	2	7	5	7	2
7	7	2	7	2	6	2
8	6	2	4	2	2	2
9	5	2	2	2	2	2
10	2	1	2	1	1	1
11	2	1	1	1	1	1
12	1	1	1	1	1	1
13	1	1	1	1	1	1

As shown in Table 1, for Airplane, Lenna, Peppers and Sailboat, a higher degree-k zerotree coding performs better at lower bit-rates than SPIHT coding which is a degree-2 zerotree coding. However, for Baboon and Tiffany, SPIHT coding is more efficient at lower bit-rates as compared to the higher degree-k zerotree coding. It can also be seen from Table 1 that a degree-1 zerotree coding gives a better compression performance than both SPIHT and higher degree-k zerotree coding at higher bit-planes for all the six test images.

3.2   Spatial Orientation Tree Structure

An SOT structure with partially-linked nodes has been proposed for color image coding by Kassim [11]. A similar SOT structure is applied in our proposed work. Figure 4a shows the SOT structure used in the TDKZW coder. As can be seen, the no-descendant root in the low-low (LL) subband in luminance plane is used to link the coefficients in the chrominance plane to form a longer zerotree. During the initialization stage of TDKZW coding, only the luminance nodes are added to the LIP and LIS lists. The chrominance nodes will only be added to the lists when they are found to be significant through their luminance parent nodes in LIS [11]. This SOT structure with partially-linked nodes is applied to increase the coding performance of the color compression especially at lower bit-rates and when the dimension of the LL subband is large.

Recently, a strip-based coding [15] for low memory implementation of the wavelet-based image coder has been proposed. In strip-based coding, an image is

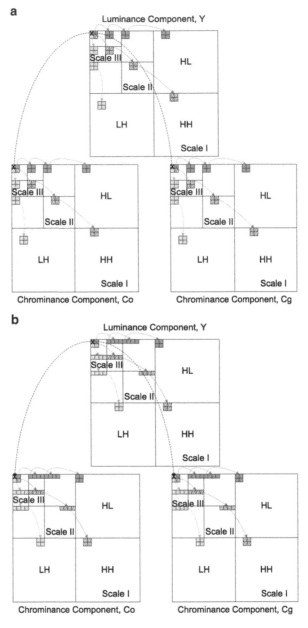

Fig. 4 Spatial orientation tree structure for color image compression. (**a**) The traditional SOT structure for TDKZW coder. (**b**) The modified SOT-C structure for LM-TDKZW

acquired in raster scan format from an image sensor and the computed wavelet coefficients are buffered in a strip-buffer. Entropy coding is then carried out on the strip-buffer which holds a few lines of wavelet coefficients from all subbands that belong to the same spatial location. Once the coding is done for a strip-buffer, it is released and is ready for the next set of data lines. Since only a portion of the full wavelet decomposition subband is encoded at a time, it is not necessary to wait for the full transformation of the image and coding can be performed once a strip is fully buffered. This enables the image coding to be carried out rapidly and also greatly reduces the memory requirement for the coder implementation.

Here, a low memory implementation of the TDKZW coder i.e. LM-TDKZW, using the strip-based approach is introduced. New tree structures proposed by Chew [16] have been taken into consideration. The proposed LM-TDKZW coder uses a lower scale of wavelet decomposition with the new tree structure SOT-C. As shown in Fig. 4b, in contrast to the traditional SPIHT SOT where each node has 2×2 adjacent pixels of the same spatial orientation as its descendants, the proposed SOT-C take the next four pixels of the same row as its children for nodes in the two highest wavelet subbands.

3.3   Memory Requirements

Table 2 shows the memory requirement for the proposed TDKZW and LM-TDKZW coders using the strip-based approach as compared to the traditional strip-based SPIHT coder.

In SPIHT and TDKZW coding, a six-scale wavelet decomposition and a seven-scale SOT decomposition are performed on test image of size 512×512 pixels. Here, the traditional 2×2 SOT structure with partially-linked nodes is used. Since the proposed TDKZW coder and SPIHT coder applies the same amount of wavelet decomposition and uses a similar SOT structure, they require the same amount of memory lines for its strip-based implementation.

On the other hand, the LM-TDKZW coder applies a four-scale wavelet decomposition and a five-scale SOT decomposition on test image of size 512×512 pixels. As shown in Table 2, the LM-TDKZW only requires 1/16 of the memory needed as compared to both the SPIHT and TDKZW coders. This is achieved by using a modified SOT-C structure in conjunction with a lower scale of DWT decomposition.

Table 2   Memory requirements for the strip-based implementation of SPIHT coding and our proposed TDKZW and LM-TDKZW coders

Coding scheme	DWT scale	SOT scale	Memory lines needed (DWT/SOT)	Type of spatial orientation tree (SOT) structure
SPIHT	6	7	32/128	Original 2×2 structure
TDKZW	6	7	32/128	Original 2×2 structure
LM-TDKZW	4	5	8/8	SOT-C

Table 3 Lossless color
image compression efficiency
represented by bit-per-pixel
(Bpp) of the proposed
TDKZW and LM-TDKZW
coders as compared to the
SPIHT coder

Test image	TDKZW	SPIHT	LM-TDKZW
Airplane	10.91	10.95	10.96
Baboon	18.10	18.12	18.14
Lenna	13.35	13.40	13.46
Peppers	14.89	14.93	14.96
Sailboat	15.88	15.90	15.94
Tiffany	13.23	13.27	13.28
Average	14.39	14.43	14.46

4 Performance Evaluation

The lossless compression performance of the proposed TDKZW and LM-TDKZW
image coding schemes were evaluated by software simulations using MATLAB.
These simulations were carried out using the reversible Le Gall 5/3 wavelet filter
and without arithmetic coding. Six standard color test images used are Airplane,
Baboon, Lenna, Peppers, Sailboat and Tiffany. Each of them is represented by a
24-bits RGB model. Tuning parameters given in Table 1 are used in our proposed
works.

Table 3 presents the lossless color image compression efficiency obtained from
the simulations carried out in terms of bit-per-pixel (bpp). From the simulation results, it can be seen that the proposed TDKZW coder performs better than the SPIHT
coder. This is because in our proposed work, the degree of zerotree tested is tuned
in each encoding pass to achieve optimal compression performance. It can also be
seen that the proposed LM-TDKZW coder which uses the modified SOT-C structure gives an almost equivalent lossless compression performance as the SPIHT
coder despite needing only 6.25% of its memory requirement. In addition, with a
lower scale of wavelet decomposition, the complexity in the implementation of the
LM-TDKZW coder is significantly reduced.

5 Conclusion

Based on the analysis carried out on the degree-k zerotree coding, it has been found
that at low bit-rates, a higher degree-k zerotree coding gives a better compression
performance whereas at higher bit-rates, a lower degree-k zerotree coding is more
efficient. In view of this, the degree of zerotree tested is tuned in each encoding
pass in our proposed TDKZW coder to obtain optimal compression performance.
A low memory implementation of the TDKZW coder known as the LM-TDKZW
coder which uses a modified SOT-C structure with partially-linked nodes in conjunction with a lower scale of wavelet decomposition is also proposed here. Simulation
results on color test images show that our proposed TDKZW coder gives a better
lossless compression efficiency than the SPIHT coder. The results also show that

the proposed LM-TDKZW coder is able to achieve a 93.75% saving in memory requirement as compared to the SPIHT coder without having any appreciable negative impact on its compression performance. Besides, with a lower scale of DWT composition, this not only reduces the complexity and memory needed, the cost for the hardware implementation of our proposed LM-TDKZW coder is also significantly reduced.

References

1. Christopoulos, C., Skodras, A., & Ebrahimi, T. (2000). The JPEG still image coding system: an overview. *IEEE Transactions on Consumer Electronics, 46*, 1103–1127.
2. Shapiro, J.M. (1993). Embedded image coding using zerotrees of wavelet coefficients. *IEEE Transactions on Signal Processing, 41*(12), 3445–3462.
3. Said, A., & Pearlman, W.A. (1996). A new fast/efficient image codec based on set partitioning in hierarchical trees. *IEEE Transactions on Circuits and System Video Technology, 6*(12), 243–250.
4. Taubman, D. (2000). High performance scalable image compression with EBCOT. *IEEE Transactions on Image Processing, 9*(7), 1158–1170.
5. Taubman, D., & Marcellin, M.W. (2002). *JPEG 2000 image compression: fundamentals, standards and practice*. Norwell, MA, USA: Kluwer Academic Publishers.
6. Huang, W.B., Su, W.Y., & Kuo, Y.H. (2006). VLSI implementation of a modified efficient SPIHT encoder. *IEICE Transactions on Fundamentals of Electronics, Communication and Computer Science, E89-A*(12), 3613–3622.
7. Jyotheswar, J., & Mahapatra, S. (2007). Efficient FPGA implementation of DWT and modified SPIHT for lossless image compression, *Elsevier – Journal of Systems Architecture, 53*(7), 369–378.
8. Cho, Y., & Pearlman, W.A. (2007). Quantifying the coding performance of zerotrees of wavelet coefficients: Degree-k zerotree. *IEEE Transactions on Signal Processing, 55*(6), 2425–2431.
9. Cicala L., & Poggi, G. (2007). A generalization of zerotree coding algorithms, *Picture Coding Symposium 2007*, Lisboa, Portugal.
10. Chew, L.W., Ang, L-M., & Seng, K.P. (2009). Lossless image compression using tuned Degree-K zerotree wavelet coding. *Proceedings of the International MultiConference of Engineers and Computer Scientists (IMECS 2009), 1*, 779–782.
11. Kassim, A.A., & Lee, W.S. (2003). Embedded color image coding using SPIHT with partially linked spatial orientation trees. *IEEE Transactions on Circuits and Systems for Video Technology, 13*(2), 203–206.
12. Malvar, H., & Sullivan, G. (2003). YCoCg-R: A color space with RGB reversibility and low dynamic range, ISO/IEC JTC1/SC29/WG11 and ITU-T SG16 Q.6.
13. Shapiro, J.M. (1996). A fast technique for identifying zerotrees in the EZW algorithm, *ICASSP-96, 3*, 1455–1458.
14. Said, A., & Pearlman, W.A. (January 10, 2009) SPIHT image compression; http://www.cipr.rpi.edu/research/SPIHT/spiht3.html.
15. Bhattar, R.K., Ramakrishnan, K.R., & Dasgupta, K.S. (2002). Strip based coding for large images using wavelets. *Elsevier – Signal Processing: Image Communication, 17*, 441–456.
16. Chew, L.W., Ang, L-M., & Seng, K.P. (2008). New virtual SPIHT tree structures for very low memory strip-based image compression. *IEEE Signal Processing Letters, 15*, 389–392.

Chapter 13
Motion Estimation Algorithm Using One-Bit-Transform with Smoothing and Preprocessing Technique

Wai Chong Chia, Li Wern Chew, Li-Minn Ang, and Kah Phooi Seng

Abstract A high performance 2D one-bit-transform (1BT) motion estimation algorithm with smoothing and preprocessing (S + P) is introduced in this paper. The 1BT technique is used to transform an 8-bit image into a 1-bit representation image (1BT image). In the 1BT motion estimation algorithm, the 8-bit current frame (c frame) and reference frame (p frame) are first transformed into their 1BT image respectively, before calculating the Sum of Absolute Difference (SAD) and performing the search operations using the Full Search Block Matching Algorithm (FSBMA). In our proposed algorithm, a smoothing threshold (Threshold$_S$) is incorporated into the filtering kernel, which is used to perform the transformation from 8-bit image into the 1BT image. The smoothing technique can greatly reduce the scattering noise created in the 1BT image. This will help to improve the accuracy when performing the search operations. After the transformation, the 1BT image for the c frame and p frame is divided into number of macroblocks. The macroblock in the c frame will be first compared to the macroblock at the same position in the p frame. If the SAD is below the preprocessing threshold (Threshold$_P$), the macroblock is considered to have negligible movement and search operation is not required. This preprocessing technique can greatly reduce the total number of search operations. Simulation results show that an improvement up to 0.65 dB, with reduction in search operation up to 95.07% is achieved. Overall, the proposed S + P technique is very suitable to be used in applications such as video conferencing and monitoring.

Keywords Full search block matching algorithm (FSBMA) · Motion estimation · One-bit-transform

W.C. Chia (✉), L.W. Chew, L.-M. Ang, and K.P. Seng
School of Electrical and Electronic Engineering, The University of Nottingham, Malaysia
e-mail: keyx7cwc@nottingham.edu.my; eyx6clw@nottingham.edu.my; kezklma@nottingham.edu.my; kezkps@nottingham.edu.my

S.-I. Ao et al. (eds.), *Intelligent Automation and Computer Engineering*,
Lecture Notes in Electrical Engineering 52, DOI 10.1007/978-90-481-3517-2_13,
© Springer Science+Business Media B.V. 2010

1 Introduction

Motion estimation is a common video compression technique used to exploit the correlated information between the current frame and the reference frame. The basic idea is to first divide the current frame into number of macroblocks as shown in Fig. 1, and search for the similar macroblock in the reference frame. In this case, it is only required to transmit the motion vector that shows the new location of the macroblock. Hence, the amount of information to be transmitted is very much lower than transmitting the entire new frame. Among the various types of motion estimation algorithm, the most popular algorithm that provides optimal performance is the Full Search Block Matching Algorithm (FSBMA).

In FSBMA, search operation is performed on every macroblock in the current frame. For each macroblock, searching is conducted within the search window in the reference frame to determine the best matching macroblock. The degree of matching is commonly evaluated by using the Sum of Absolute Difference (SAD), due to its simplicity in implementation. For a block with size of N × N pixels, the SAD can be calculated by Eq. 1, whereby c represents the current frame, p represents the reference frame, and i and j represents the coordinate of the image.

$$SAD(x, y) = \sum_{i=1}^{N-1} \sum_{j=1}^{N-1} |c(i, j) - p(i + x, j + y)| \tag{1}$$

Although FSBMA provides optimal performance, it is computationally intensive due to the high number of search operations required [1]. According to [5, 8], 50% to 80% of the video processing time is consumed by the FSBMA. This creates a

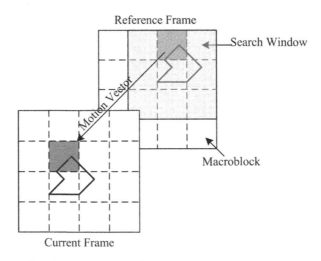

Fig. 1 The basic of motion estimation for video compression

problem in the era where power consumption has becoming an important design constraint, especially for embedded systems [2, 3].

In order to overcome this problem, many fast motion estimation algorithms such as the New Three Step Search [4], Four Step Search [5], 2D-Logarithmic Search [6] and Conjugate Direction Search [7] have been proposed. These algorithms reduce the number of search candidates and hence reduce the number of search operations required. However, the performance of these algorithms is only sub-optimal, because there is always a possibility for these algorithms to miss the best matching candidate.

Other than reducing the number of search operation, simplifying the computation of degree of matching is another way to resolve the problem. The algorithm reported in [8, 9] uses a technique called one-bit-transform (1BT) to transform an 8-bit image into its 1-bit representation image. Then, the FSBMA and SAD calculations are carried out on the 1BT image. Since the 1BT image contains only either logic '0' or '1', this help to simplify the calculation process of SAD down to simple exclusive-OR (XOR) operation as shown in Eq. 2, and hence reduces the overall complexity. Other related algorithms are also reported in [10–12].

$$SAD(x, y) = \sum_{i=1}^{N-1} \sum_{j=1}^{N-1} |c(i, j) \oplus p(i + x, j + y)| \qquad (2)$$

In this paper, a smoothing and preprocessing (S + P) technique is added to the Multiplication-Free 1BT (MF-1BT) motion estimation algorithm that was first presented in [8]. Although another algorithm based on 1BT is also introduced in [9], it involves floating-point multiplication which is usually slow in hardware and software implementation [9]. Hence, the MF-1BT algorithm is chosen to be the basis of our proposed algorithm. The proposed S + P technique not only maintains the simplicity in the matching error calculations, but also reduces the number of search operations and provides better performance.

Firstly, scattering noise which is a common problem in the 1BT image is reduced by using the proposed smoothing technique. The scattering noise is one of the reasons that lead to performance degradation, since it affects the accuracy in finding the best matching block. Secondly, the large number of search operations in the FSBMA is reduced by using the proposed preprocessing technique. Reducing the number of search operations will shorten the video processing time. This can either help to improve the frame rate or reduce the overall power consumption.

The paper is organized in the following manner. First, a brief overview on the MF-1BT motion estimation algorithm is given in Section 2. Then, the proposed S + P technique will be separately explained in Sections 3 and 4 respectively. This is followed by the simulation results that were presented and discussed in Section 5. Finally, the chapter is concluded in Section 6.

2 Multiplication-Free 1BT (MF-1BT)

The overall system block diagram of the MF-1BT motion estimation algorithm is shown in Fig. 2. First of all, the current (I_X) and the reference (I_Y) 8-bit video frames are filtered by using the convolutional kernel (K) proposed in [8]. Then, the current (I_X) and reference (I_Y) 8-bit video frames are compared with their filtered version ($I_{F.X}$ and $I_{F.Y}$) respectively. The 1BT image is generated based on the thresholding decision shown in Eq. 3, whereby B represents the output 1BT image, I represents the input 8-bit image, I_F represents the filtered version of the input image, and i and j represent the coordinate of the image.

$$B(i, j) = \begin{cases} 1, \text{If } |I(i, j)| \geq |I_F(i, j)| \\ 0, \text{Otherwise} \end{cases} \tag{3}$$

It can be seen that the 1BT image is purely a binary image which contains only the logic value of '0' and '1'. Next, the 1BT image of the current (B_X) and the reference (B_Y) frame is divided into number of macroblocks as shown in Fig. 1 respectively, and FSBMA is performed on the 1BT image to determine the motion vector for each of the macroblocks. Finally, the motion vector for each of the macroblocks will be transmitted as output.

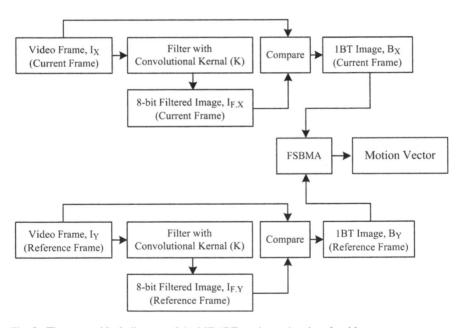

Fig. 2 The system block diagram of the MF-1BT motion estimation algorithm

3   The Smoothing Technique

The proposed smoothing technique is incorporated into the convolutional kernel (K) module that is shown in Fig. 2. The purpose of smoothing is to remove the scattering noise that is created in the 1BT image. Figure 3 shows some sample frames from the

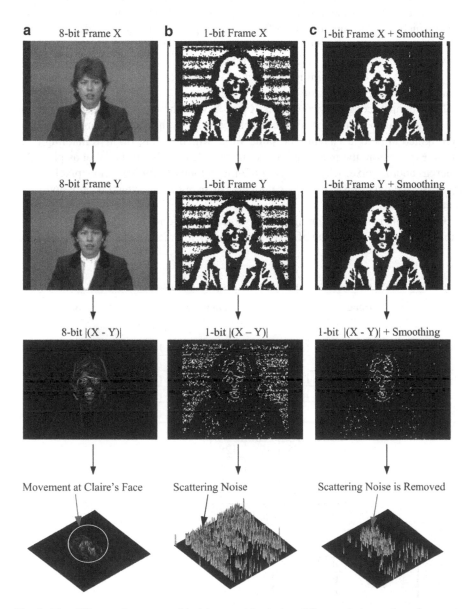

Fig. 3 The difference between residual images (absolution difference between the subsequent two frames) obtained from (**a**) original 8-bit frames, (**b**) 1BT frames, and (**c**) 1BT frames with smoothing

video sequence *Claire* that explains the effect of smoothing. From the 8-bit residual image which represents the absolute difference between the subsequent two frames as shown in the third and fourth row of Fig. 3a, notice that the background is very smooth, and the moving object (*Claire*'s face) that is concentrated at the middle of the frame can be seen clearly.

By comparison, the moving object is difficult to be seen in the 1BT residual image shown in the third and fourth row of Fig. 3b, due to the scattering noise (white pixel which indicates an absolute difference of logic '1') that surrounded the moving object. The scattering noise is created by the small difference in pixel values between the original video frame and its filtered version. Although the difference is very small, it can have a significant effect on the SAD calculation in the 1BT image which will be explained as followed.

For example, consider two images X and Y with the configuration as shown in Table 1. It can be seen that all the pixels other than the two with its magnitude highlighted in bold are having the same value which is 128. In the remaining part of the explanation, the pixel having the magnitude of 127 will be label as pixel A, whereas another pixel which carrying the magnitude of 129 is label as pixel A'. In this case, the overall SAD can be calculated by using Eq. 1, which is only approximately 0.049%.

On the other hand, assuming that the filtered version of image X and Y are as shown in Table 1. The value of pixel A is smaller than its filtered version, whereas

Table 1 The effect of small variation in pixel value on the calculation of SAD

	Image X				Image Y				SAD
8-bit Image	128	128	128	128	128	128	128	128	$= \dfrac{129-127}{4\times4\times255}$
	128	128	128	128	128	128	128	128	
	128	128	**127**	128	128	128	**129**	128	
	128	128	128	128	128	128	128	128	$= 0.049\ \%$
Filtered Version				128	128	128	128		
				128	128	128	128		
				128	128	128	128		
				128	128	128	128		
1-bit Image	0	0	0	0	0	0	0	0	$= \dfrac{1-0}{4\times4\times1}$
	0	0	0	0	0	0	0	0	
	0	0	**0**	0	0	0	1	0	
	0	0	0	0	0	0	0	0	$= 6.25\ \%$

the value of pixel A' is larger than its filtered version. According to the thresholding decision shown in Eq. 3, pixel A and pixel A' will be assigned with logic value '0' and '1' in the 1BT image of X and Y respectively. Under the same condition applied to the 8-bit image, the overall SAD which can be calculated using Eq. 2, is increased to 6.25%. It should be noted that the maximum difference of pixel value in 1BT image is 1, instead of 255 in 8-bit image. This shows that a small difference in the pixel value can affect the SAD significantly.

In order to overcome this problem, it is necessary to prevent the small changes in pixel values from affecting the decision output of the 1BT image. This is the main reason that leads to the introduction of the smoothing threshold (Threshold$_S$) into the convolutional kernel (K), which is used to transform an 8-bit image into its 1BT image. By adding the smoothing threshold into (3), the thresholding decision is altered and shown in (4).

$$B(i, j) = \begin{cases} 1, \text{If } |I(i, j)| \geq |I_F(i, j)| + \text{Threshold}_S \\ 0, \text{Otherwise} \end{cases} \tag{4}$$

The effect of applying the smoothing technique can be seen clearly from the 1BT residual image shown in Fig. 3c. Notice that the scattering noise in the background is greatly reduced. Furthermore, it is now easier to locate the moving object within the video frame. By reducing the scattering noise, the accuracy in searching for the best matching macroblock can be improved.

4 The Preprocessing Technique

The preprocessing technique is added to the FSBMA module that is shown in Fig. 2 to reduce the number of search operations. In many situations, the background or certain object in a video sequence is usually static or has negligible movement. For example, it can be seen that the movement in video sequence *Claire* is mostly concentrated on the face and shoulder only. Therefore, it is a waste to perform the FSBMA for the entire video frame, due to the high probability that the best matching block is located at the same position in the reference frame.

The basic idea of preprocessing is to prevent FSBMA being performed on macroblocks which have negligible movement. This will helps to shorten the video processing time. In addition to this, the effect of scattering noise in affecting the performance can also be reduced. For example, consider an example shown in Fig. 4 which assumes that the macroblock (shaded in dark gray) only covers a portion of a static background in the original 8-bit frame. If FSBMA is carried out on this macroblock, it can cause a false matching due to the similar artifact pattern created by the scattering noise that located in other position. The false matching can lead to degradation in performance, and it can be avoided by preventing the FSBMA be-

Fig. 4 An example of the
effect of scattering noise in
affecting the performance

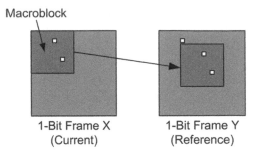

ing performed on the macroblock using the preprocessing technique. Algorithm I describes the preprocessing steps that will be carried out before FSBMA is performed on a macroblock.

Algorithm I

- For each of the macroblock in the reference frame:

    - Compute SAD (0, 0) using Eq. 2
    - If SAD (0, 0) $\leq$ Threshold$_P$
        - Halt
    - Else
        - Perform FSBMA

For each of the macroblocks in the current frame, it will be first compared with the macroblock at the same position in the reference frame. Then, the SAD is computed using (2) and compared to the preprocessing threshold (Threshold$_P$). If the SAD is smaller than or equal to Threshold$_P$, it is presumed that the macroblock has negligible movement and no search operation is required. Otherwise, FSBMA is performed to find the best matching block within the predefined search window in the reference frame. This process will be repeated for all macroblocks.

5 Simulation Results and Discussions

Simulation of the proposed S + P technique is performed on ten video sequences with different characteristics, which include *Claire*, *Miss America*, *Akiyo*, *Carphone*, *Container*, *Foreman*, *Football*, *Salesman*, *Tennis* and *Mobile*. Generally, the video sequences can be categorized into four categories as shown in Table 2.

It should be noted that only the luminance (Y) component is considered in the simulation. The block size of 16 × 16 pixels with a search window of 16 pixels and step size of 1 pixel is adopted. Finally, the motion vectors generated are used to reconstruct the video frame, and its quality is evaluated by calculating the peak signal-to-noise-ratio (PSNR) which measured in decibel (dB).

Table 2 The characteristics of the ten video sequences adopted for the simulation

Category	Characteristics	Video sequence
Video conferencing	Imitate a video conferencing environment by having the object sitting in a fix position with limited motion and static background	Claire Miss America Akiyo Salesman
Monitoring	Imitate an object monitoring environment with static background, and object moving from one position to another	Container
Cell phone video	Imitate the environment whereby the video is captured by using a cell phone	Carphone Foreman
High motion	Imitate the environment which contains object with high motion or rapid changes in the background	Football Tennis Mobile

5.1 Determination of Threshold$_S$ and Threshold$_P$

Before the simulation is performed, it is necessary to first determine the optimum value of Threshold$_S$ and Threshold$_P$. Hence, an analysis is carried out to determine the suitable value of Threshold$_S$ and Threshold$_P$. In this analysis, the proposed S + P algorithm is applied to the first 20 frames of all the ten video sequences. Then, the values of Threshold$_S$ and Threshold$_P$ are varied from 0 to 10 and 2 to 16 respectively. The PSNR achieved by different video sequences at different value of Threshold$_S$ and Threshold$_P$ are recorded. Figure 5 shows the plot of the average PSNR achieved by the ten video sequences versus Threshold$_S$ for different value of Threshold$_P$.

From the plot, it was found that the best choice for Threshold$_S$ and Threshold$_P$ are 3 and 12 respectively. The performance starts to degrade when further increasing the value. In fact, the optimum threshold for each video sequence is about ± 1 to ± 2 from the value we have chosen. When the value of Threshold$_P$ is range from 10 to 16 while the value of Threshold$_S$ is 3, the difference in the average PSNR is not very significant. The main criteria is to select a value of Threshold$_S$ and Threshold$_P$ that can work for most of the situation.

However, it is important to note that there is a possibility that a better performance can be achieved for different video sequences with other threshold values. By assuming that the remaining frames are following the same trend as the first 20 frames, these two threshold values are applied to the entire sequence.

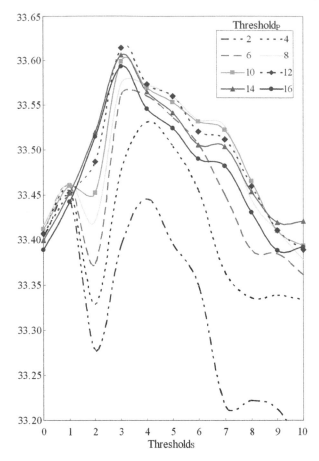

Fig. 5 Average PSNR (db) for the first 20 frames of several video sequences by varying the value of Threshold$_S$ and Threshold$_P$

5.2 Simulation Results

The average PSNR for each video sequence is recorded in Table 3. The proposed algorithm performs very well in lower motion video such as *Claire, Miss America, Akiyo, Salesman,* and *Container.* Most significantly, the average PSNR achieved in these five video sequences is very close to the optimal performance of 8-bit FS-BMA. The average PSNR is just 0.1 to 0.4 dB lower than 8-bit FSBMA. But for the MF-1BT, the average PSNR is 0.08 to 1.05 dB lower than 8-bit FSBMA. Overall, the proposed algorithm achieved an improvement up to 0.65 dB.

On the other hand, the improvement for moderate motion video such as *Foreman,* and high motion video such as *Football, Tennis* and *Mobile* is not as significant as lower motion video. For *Foreman,* an average gain of 0.22 dB is achieved. A sample

Table 3 Average PSNR (dB) of several video sequences reconstructed using different motion estimation algorithm. The block size of 16 × 16 pixels with search window of 16 pixels and step size of 1 pixel is adopted. Only the luminance (Y) component is considered

Video sequence	Frame size	Sequence length	Methods			
			8-bit FSBMA	1BT [9]	MF-1BT [8]	Proposed S + P
Claire	176 × 144	493	42.90	42.08	42.09	**42.71**
Miss America	176 × 144	149	41.34	40.44	40.29	**40.94**
Akiyo	176 × 144	299	44.30	44.20	**44.22**	44.21
Carphone	176 × 144	299	32.10	30.60	30.61	**30.70**
Container	176 × 144	299	43.09	42.77	42.76	**42.89**
Tennis	352 × 240	149	29.92	28.81	**28.81**	28.78
Football	352 × 240	125	22.89	**21.80**	21.79	21.70
Foreman	352 × 288	399	31.55	29.85	29.92	**30.11**
Mobile	352 × 288	299	24.59	24.20	**24.24**	24.19
Salesman	352 × 288	448	27.12	26.60	26.58	**26.96**

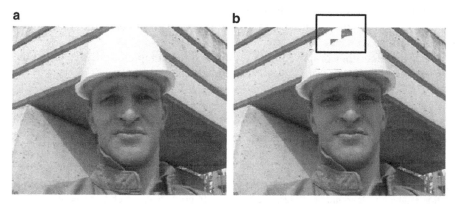

Fig. 6 The reconstruction of a sample frame from video sequence Foreman (**a**) with and (**b**) without using the proposed S + P technique

reconstructed frame shown in Fig. 6 shows that the visual quality is improved by the proposed S + P technique. As been highlighted by the black box in Fig. 6, it can be seen that the "helmet" is better reconstructed.

But for *Football, Tennis* and *Mobile*, the proposed algorithm is 0.09, 0.03 and 0.05 dB lower than the MF-1BT algorithm. However, the degradation is not significant when compared to the gain achieved in other video sequences. Other than performance, the reduction in number of search operations achieved by the proposed algorithm should also be taken into account.

5.3 Reduction in Number of Search Operations

The reduction in the number of search operations for each video frame is calculated by using Eq. 5, whereby N_{FSBMA} represents the total number of search operations required in FSBMA, and N_{S+P} represents the number of search operations carried out when the proposed S + P technique is used. First, the reduction in the number of search operations for all the frames is computed. Then, it is averaged up to obtain the final search reduction for the entire video sequence that is recorded in Table 4.

$$\text{Reduction}(\%) = \frac{N_{FSBMA} - N_{S+P}}{N_{FSBMA}} \times 100\% \tag{5}$$

On the other hand, the reduction in number of search operations for fast motion video such as *Football, Tennis* and *Mobile* are 2.03%, 11.08% and 5.11% respectively. Although the proposed S + P technique causes a minor degradation in performance for *Football, Tennis* and *Mobile*, but in return the number of search operations is reduced for these three video sequences. By reducing the number of search operations required, the video overall processing time and power consumption can also be reduced.

Since the algorithm proposed in [8, 9] applied the complete FSBMA on all the macroblocks in a video frame, the search reduction is zero when compared to the 8-bit FSBMA. For lower motion video sequences such as *Claire, Miss America, Akiyo*, and *Container*, the total number of search operations being performed are significantly reduced by 93.91%, 63.59%, 95.07% and 77.77% respectively, when the proposed S + P technique is incorporated.

Table 4 Average search reduction (%) of several video sequences reconstructed using different motion estimation algorithm. The block size of 16 × 16 pixels with search window of 16 pixels and step size of 1 pixel is adopted. Only the luminance (Y) component is considered

Video sequence	Frame size	Sequence length	Methods 8-bit FSBMA	1BT [9]	MF-1BT [8]	Proposed S + P
Claire	176 × 144	493	–	0	0	93.91
Miss America	176 × 144	149	–	0	0	63.59
Akiyo	176 × 144	299	–	0	0	95.07
Carphone	176 × 144	299	–	0	0	44.01
Container	176 × 144	299	–	0	0	77.77
Tennis	352 × 240	149	–	0	0	13.65
Football	352 × 240	125	–	0	0	3.37
Foreman	352 × 288	399	–	0	0	14.48
Mobile	352 × 288	299	–	0	0	7.36
Salesman	352 × 288	448	–	0	0	29.16

6 Conclusion

Implementation of the S + P technique into the MF-1BT motion estimation algorithm produces good results for lower motion video such as *Claire, Miss America, Salesman, Container* and *Akiyo*. The performance is approaching the optimal 8-bit FSBMA. Furthermore, the number of search operations is greatly reduced, and it also helps to reduce the video processing time and power consumption. Simulation results show that the proposed S + P method can improve the performance up to 0.65 dB, and reduce the number of search operations up to 95.07%. Overall, the proposed S + P method provides an efficient solution for the implementation of motion estimation algorithms under hardware constraint environments. This S + P technique is suitable for video conferencing and monitoring application which involves lower motion and static background.

References

1. Do, V.L., & Yun, K.Y. (Aug 1998). A low-power VLSI architecture for full-search block-matching motion estimation algorithm. *IEEE Transactions on Circuits and Systems for Video Technology, 8*(4), 393–398.
2. Piguest, C. (2005). *Low-power electronics design* (pp. 36–15). CRC.
3. Sayood, K. (2005). *Lossless compression handbook* (pp. 407). Academic.
4. Li, R., Zeng, B., & Liou, M.L. (Aug 1994). A new three-step search algorithm for block motion estimation. *IEEE Transactions on Circuit Systems and. Video Technology, 4*(4), 438–442.
5. Po, L.M., & Ma, W.C. (June 1996). A novel four-step search algorithm for fast block motion estimation. *IEEE Transactions on Circuit Systems and Video Technology, 6*(3), 313–317.
6. Jain, J.R., & Jain, A.K. (Dec 1981). Displacement measurement and its application in inter-frame image coding. *IEEE Transactions on Communication, COM-29*(12), 1799–1808.
7. Srinivasan, R., & Rao, K.R. (Aug 1985). Predictive coding based on efficient motion estimation. *IEEE Transactions on Communication, COM-33*(8), 888–896.
8. Erturk, S. (Feb 2007). Multiplication-free one-bit transform for low-complexity bock-based motion estimation. *IEEE Signal Processing Letter, 14*(2), 109–112.
9. Natarajan, B., Bhaskaran, V., & Konstantinides, K. (Aug 1997). Low-complexity block-based motion estimation via one-bit transforms. *IEEE Transactions on Circuit Systems and Video Technology, 7*(4), 702–706.
10. Erturk, A., & Erturk, S. (Jul 2005). Two-bit transform for binary block motion estimation. *IEEE Transactions on Circuits and Systems Video Technology, 15*(7), 938–946.
11. Feng, J., Lo, K.T., Mehrpour, H., Karbowiak, A.E. (1995). Adaptive block matching motion estimation algorithm using bit-plane matching. In *IEEE International Conference on Image Processing* (pp. 496–499). Washington, DC.
12. Mizukim, M.M., Desai, U.Y., Masaki, I., Chandrakasan, A. (1996). A binary block matching architecture with reduced power consumption and silicon area requirements. In *IEEE ICASSP-96* (pp. 3248–3251), vol. 6. Atlanta.

Chapter 14
Configuration of Adaptive Models in Arithmetic Coding for Video Compression with 3DSPIHT

Wai Chong Chia, Li-Minn Ang, and Kah Phooi Seng

Abstract The 3D Set Partitioning In Hierarchical Trees (SPIHT) for video compression is an extension of the SPIHT algorithm, which is initially introduced by A. Said and W. Pearlman for image compression. Previous works have shown that the performance of 3DSPIHT with Arithmetic Coding (AC) is comparable to H.263 and MPEG-2. Moreover, the output bit stream of 3DSPIHT is inherently embedded and scalable in rates. It is also relatively easy to make the bit stream become scalable in resolution with some minor changes. Although all these features are very attractive for certain applications that required progressive transmission or heterogeneous network, the configuration of AC can be tedious and remains as a challenging task. The changeable parameters in AC include the type (fixed or adaptive) of models, number of models, and maximum frequency to reset the models. This work presents a configuration of adaptive models in AC, which can help to improve the coding efficiency of AC for 3DSPIHT, and thus achieve better performance in terms of Peak Signal-to-Noise Ratio (PSNR). The adaptive models are used to store the probability distribution of all the symbols that appear in a system. In the proposed configuration, each type of output bits in 3DSPIHT is assigned with a separate set of adaptive models. This proposed configuration takes into account the different probability patterns which exist in each type of output bits. The maximum frequency used to reset the adaptive models is also investigated. It will not only affect the adaptation rate which directly relates to the coding efficiency of AC, but also the memory requirement. The simulation results show that the proposed configuration can improve the mean PSNR for various video test sequences in QCIF and SIF formats.

Keywords Adaptive models · Arithmetic coding · 3DSPIHT · SPIHT · Video compression

W.C. Chia (✉), L.-M. Ang, and K.P. Seng
School of Electrical and Electronic Engineering, The University of Nottingham Malaysia, Jalan Broga, Semenyih, 43500, Selangor Darul Ehsan, Malaysia
e-mail: keyx7cwc@nottingham.edu.my; kezklma@nottingham.edu.my; kezkps@nottingham.edu.my

S.-I. Ao et al. (eds.), *Intelligent Automation and Computer Engineering*,
Lecture Notes in Electrical Engineering 52, DOI 10.1007/978-90-481-3517-2_14,
© Springer Science+Business Media B.V. 2010

1 Introduction

Due to its superior performance in image compression, many research works have been carried out to extend and improve the Set Partitioning In Hierarchical Trees (SPIHT) algorithm [1]. The 3DSPIHT algorithm is one of the extensions that attain remarkable performance in video compression. When the Arithmetic Coding is incorporated, the performance of 3DSPIHT is comparable to H.263 and MPEG-2 without using any of the motion compensation technique. In fact, it is also one of the advantages of 3DSPIHT, which reduces the computational complexity by replacing the motion compensation technique with wavelet transform. Moreover, the output bit stream of 3DSPIHT is inherently embedded and scalable in rates. It is also relatively easy to make the bit stream become scalable in resolution with some minor changes.

Although the features stated are attractive for applications that require progressive transmission or heterogeneous network, the configuration of AC can be tedious and remains as a challenging task. The changeable parameters in AC include the type (fixed or adaptive) of models, number of models, and maximum frequency to reset the models. Several 3DSPIHT algorithms for video compression have been proposed in [2–4]. Even though the AC is adopted in all these works to further enhance the performance, the configuration of AC is not well reported. This creates some difficulties in replicating the original algorithm. From our point of view, the configuration of AC can be further optimized to enhance the performance.

In this work, a configuration of the adaptive models in AC for 3DSPIHT is presented. The adaptive models are used to store the probability of occurrence for all the symbols that can appear in the system, which strongly affects how efficient that the symbol can be coded using AC. This proposed configuration will takes into account the different probability patterns which exist in different type of output bits in 3DSPIHT. In addition, the maximum frequency that is used to reset the adaptive models is also investigated. The maximum frequency will not only affect the adaptation rate which directly relates to the coding efficiency of AC [5], but also the amount of memory required to setup the adaptive models. The simulation results show that the proposed configuration can improve the mean PSNR for various video test sequences in the QCIF and SIF formats.

The chapter is organized as follows. A brief overview on the ordinary SPIHT algorithm and some properties of 3DSPIHT will be first described in Sections 2 and 3 respectively. The proposed configuration of the adaptive models is explained in detail in Section 4. This is followed by the simulation results and discussion which presented in Section 5, and conclusion of the chapter in Section 6.

2 The SPIHT Algorithm

Since the SPIHT algorithm is the basis of the 3DSPIHT, a brief overview is presented in this section. The type of output bits in the coding process is clarified during

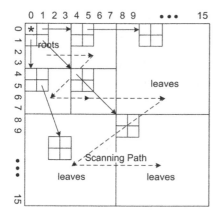

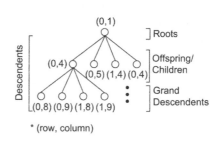

Fig. 1 The 2D SOT structure adopted by SPIHT

the explanation of the algorithm. All these clarification are very important and will be used to explain the proposed configuration of the adaptive models.

Before using the SPIHT algorithm to encode an image, the 2D Discrete Wavelet Transform (DWT) is performed to decompose the image into multiple number of subbands. With the coefficients in the lowest frequency subband serving as the root, the SPIHT algorithm uses the Spatial Orientation Tree (SOT) structure to encode the wavelet coefficients. In this case, a node in the tree has either four offspring or no offspring. The offspring is formed by grouping the wavelet coefficients in 2×2 adjacent pixels. Figure 1 illustrates the parent-offspring relationship of the SOT structure.

The SPIHT algorithm uses three lists which are called the List of Insignificant Pixels (LIP), List of Insignificant Sets (LIS), and List of Significant Pixels (LSP) to identify the status and control the coding process of wavelet coefficients. Initially, all the coordinates of coefficients located in the lowest frequency subband are added to the LIP. The entries in the LIS are similar to the entries in the LIP, except that the coordinates of coefficients denoted with "*" in Fig. 1 are excluded. In addition to this, all the entries in the LIP are marked as Type A entries. On the other hand, the LSP is initially left empty as it is used to store the coefficients which are tested to be significant.

The encoding process of SPIHT algorithm is divided into two phases which are called the sorting phase and refinement phase. In the sorting phase, each entry in the LIP is tested against a predefined threshold 2^n, where n is the level of significance. If the value of the entry is larger than the threshold 2^n, the entry is considered as significant at this level. In this case, a '1' bit and its sign bit are sent. Otherwise, it is considered as insignificant and only a '0' bit is sent. The entry which is tested to be significant is moved to the LSP before testing another entry. Throughout the chapter, the bit which defines the significance of the entry in LIP is labeled as *LIP SIG* bit, and the sign bit is labeled as *LIP SIGN* bit.

On the other hand, the testing process of the entry in the LIS is slightly different from that in the LIP. In the case for the entry in the LIS, all the descendents of the entry are tested against the threshold 2^n. If any one of the descendents is larger than the threshold, the entry will be marked as significant. A '1' bit is sent for the case of significant, and a '0' bit is sent for the case of insignificant. This bit which defines the significance of the descendents will be labeled as *DESC* bit.

When the entry in the LIS is marked as significant, the four offspring of the entry will be tested in the same way as the entry in LIP. This means that a '1' bit and a sign bit are sent for offspring which is tested to be significant, and a '0' bit is sent for the opposite case. The coordinate of the significant offspring will be added to the end of the LSP whereas the insignificant offspring will be added to the end of the LIP. In this case, the bit which defines the significance of an offspring is labeled as *LIS SIG* bit, and the sign bit is label as *LIS SIGN* bit.

After the testing process of the four offspring, the entry of the LIS which is initially marked as Type A will be shifted to the back of the LIS and marked as Type B. When coding a Type B entry in the LIS, all the descendents excluding the 4 direct offspring will be tested against the threshold 2^n. If any one of the remaining descendents is larger than the threshold, the entry will be marked as significant. A '1' bit is sent for the case of significant and a '0' bit is sent for the case of insignificant. In addition to this, the coordinates of the four direct offspring of the Type B entry are added to the end of the LIS as new Type A entries. This bit which defines the significance of the Type B entry is labeled as *GDESC* bit throughout the chapter. The definition of all the type of output bits is illustrated in Fig. 2.

In the refinement phase, the n bit of each entry in the LSP which appears in the previous pass is sent. This bit will be labeled as *REF* bit in the rest of the chapter. In this case, no bit is sent during the first pass of the encoding process. After the refinement phase is completed, the n value is decremented for the next pass where the entire process described above is performed again. The decoding process of SPIHT algorithm is just the reverse of the encoding process.

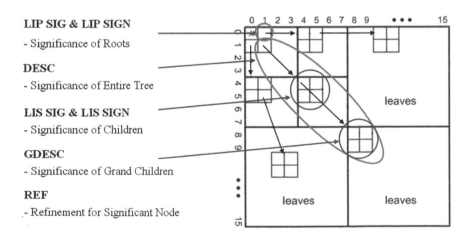

Fig. 2 Definition and notation of different type of output bits in SPIHT

3   3DSPIHT for Video Compression

Generally, the encoding and decoding processes of the 3DSPIHT for video
compression are totally same as the SPIHT for image compression. The main
difference is the DWT and the tree structure that have been changed from 2D to 3D.
It is important to note that these do not affect the clarification of the type of output
bits which is stated previously.

Before the 3DSPIHT is adopted, it is necessary to first form a 3D data structure
from Group of Frames (GOF) as shown in Fig. 3. Then, the 3D DWT which consists
of performing a 2D DWT in the spatial domain of each frame in the GOF and
followed by a 1D DWT across the temporal domain of all the frames in the GOF
can be carried out.

The number of subband created by the 3D DWT can be very different, depending
on the configuration of its 2D DWT and 1D DWT that are applied to the spatial
and temporal domain respectively. Generally, there are two types of 3D DWT. One
is called the symmetry 3D DWT as shown in Fig. 4a, while another is called the
decoupled or wavelet packet 3D DWT as shown in Fig. 4b. The structure shown in
Fig. 3 uses 2 decomposition levels for both the spatial and temporal domains. In this
case, the numbers of subbands in the symmetry 3D DWT and the decoupled 3D
DWT are 15 and 21 respectively.

Since the data structure has changed from 2D to 3D, it is necessary to redefine
the parent-offspring relationship of the tree structure. The 3DSPIHT algorithm is
not affected by the dimension, as long as the parent-offspring relationship of the
tree structure is clearly defined.

Initially, the 3D SOT adopted in [6] is a direct extension of the 2D SOT. This
3D SOT is considered as a type of symmetry tree structure, since all the trees are
symmetrically extended in both spatial and temporal domains. In the symmetry 3D
SOT, a node is has either 8 offspring which are formed by a group of $2 \times 2 \times 2$
adjacent pixels or no offspring (leaves). Similar to the 2D SOT, the nodes denoted
with "*" have no offspring. The parent-offspring relationship of the symmetry 3D
SOT is shown in Fig. 5.

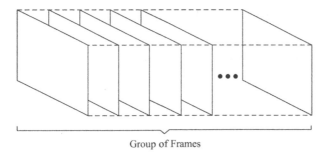

Group of Frames

Fig. 3 Forming a 3D data structure from a series of n frames

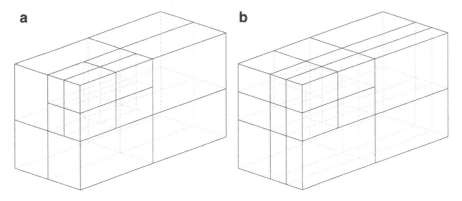

Fig. 4 The structure of (**a**) symmetry 3D DWT and (**b**) decoupled 3D DWT

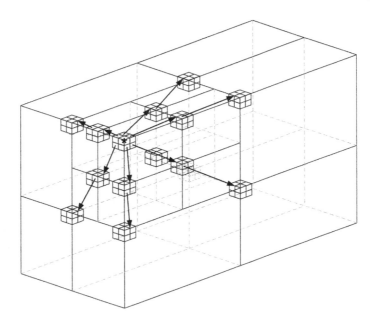

Fig. 5 The 3D SOT structure adopted by 3DSPIHT

4 Configuration of Adaptive Models

The AC introduced in [5] is adopted for our proposed configuration. This AC is also adopted by the previous works presented in [2–4]. The main advantage of this AC is the isolation of adaptive model from the coding algorithm. Hence, the configuration of the adaptive models can be modified easily.

The adaptive model is used to store the probability of occurrence for all the symbols that can appear in the system. It is updated after a symbol is encoded. The coding efficiency of AC is very much affected by the probability distribution in the adaptive model. When the frequency is skewed to certain symbols, the compression rate of AC will increase.

In 3DSPIHT, the information of the eight offspring is usually encoded with AC as a single symbol to exploit the local correlation. The amount of information to be encoded depends on the number of insignificant pixels m in the group. In the previous works, several different adaptive models each with 2^m symbols, where $m \in \{1, 2, 3, 4, 5, 6, 7, 8\}$, are used to encode the information of the 8 offspring [2,3].

From here, several questions may arise. Firstly, how many adaptive models were used and how they were configured. Secondly, how each type of output bits is assigned with a separate set of adaptive models or the adaptive models were shared by few types of output bits. Thirdly, how were the sign bits encoded. Fourthly, what is the maximum frequency used to reset the adaptive models.

According to [1], the gain in coding the sign bits with AC is very little. It is true when the sign bits are viewed in terms of bit by bit. In this case, the numbers of positive and negative coefficients are close to each other. But from our observation, when the sign bits of the 8 offspring are coded as one symbol, certain patterns can occur more frequently than others. Due to this reason, it is possible to achieve some gain in coding the sign bits as a single symbol with AC in 3DSPIHT.

Other than the probability distribution, the maximum frequency also has significant effect on the coding efficiency of AC [5]. This maximum frequency will determine the adaptation rate of the AC. Furthermore, it also affects the amount of memory that is required to store the frequency for a symbol in the adaptive model. In our new configuration, the maximum frequency is set to 64, which is different from the value reported in [6]. This value is obtained from an analysis that will be explained in Section 4.

The proposed configuration of the adaptive models is summarized in Table 1. Even though the configuration of adaptive models for the *LIP SIG*, *LIP SIGN*, *DESC*, and *LIS SIGN* bits are the same, they are not sharing the same set of adaptive models. Each of them is assigned with a separate set of adaptive models with

Table 1   The proposed configuration of adaptive models for different type of output bits in 3DSPIHT

Output bits	Number of adaptive models with 2^m symbols							
	$m=1$	$m=2$	$m=3$	$m=4$	$m=5$	$m=6$	$m=7$	$m=8$
LIS SIG	1	1	1	1	1	1	1	1
LIS SIGN	1	1	1	1	1	1	1	1
DESC	1	1	1	1	1	1	1	1
LIS SIG	0	0	0	0	0	0	0	1
LIS SIGN	1	1	1	1	1	1	1	1
GDESC	1	0	0	0	0	0	0	0
REF	1	0	0	0	0	0	0	0
Total (T_{AM^m})	6	4	4	4	4	4	4	5

2^m symbols, where $m \in \{1, 2, 3, 4, 5, 6, 7, 8\}$. The adaptive model with different number of symbols is required to adapt for the case, where only the number of insignificant pixels m in the group are encoded. For example, consider the case where we are required to encode a cube with 8 coefficients. In this case, a adaptive model with 256 symbols is needed. If two of the coefficients are significant in the current bit plane, it is not necessary to encode them in the next bit plane. Hence, only 6 coefficients are left and we can use an adaptive model with 64 symbols.

This is different for the *LIS SIG* bit because it always comes with 8 offspring at a time. Hence, only one adaptive model with 256 symbols is needed. On the other hand, the *GDESC* and *REF* bits are coded bit by bit separately with adaptive model of 2 symbols. Degradation in performance is observed during an attempt to code these bits in the same way as others. Algorithm I shows the detail coding sequence using the proposed configuration.

Algorithm I

1. Initialization phase

- Add all the roots to the LIP in a group of $2 \times 2 \times 2$ adjacent pixels.
- Add all the roots with descendents to the LIS in a group of $2 \times 2 \times 2$ adjacent pixels.
- Initialize the threshold value.

2. Sorting phase

- For each set (p, q) in the LIP that contains a group of $2 \times 2 \times 2$ adjacent pixels:
  - Determine number of entry (i, j) $\in$ set (p, q), W, and number of sign bits, V.
  - For each of the entry (i, j) $\in$ set (p, q):
    - Perform magnitude test on entry (i, j) to determine the value of LIP SIG bit and LIP SIGN bit (if any).
    - If the entry (i, j) is significant, shift it to the LSP.
  - Use the 2^W symbols adaptive model to encode the group of LIP SIG bits and 2^V symbols adaptive model for the LIP SIGN bits (if any).
- For each Type A set (p, q) in the LIS:
  - Determine number of entry (i, j) $\in$ Type A set (p, q), S.
  - For each of the entry (i, j) $\in$ Type A set (p, q):
    - Check all descendents of entry (i, j) to determine the value of DESC bit.
  - Use the 2^S symbols adaptive model to encode the group of DESC bits.
  - For each of the entry (i, j) $\in$ Type A set (p, q):
    - If DESC bit for entry (i, j) = 1:
    - For each offspring (k, l) $\in$ entry (i, j):
      - Perform magnitude test on offspring (k, l) to determine the value of LIS SIG bit and LIS SIGN bit (if any).
      - If the offspring (k, l) is significant, shift it to the LSP. Otherwise, shift it to LIP.
    - Determine the number of sign bits, R.

- Use a 2^8 symbols adaptive model to encode the group of LIS SIG bits and 2^R symbols adaptive model to encode the LIS SIGN bits (if any).
- If any of the offspring (k, l) is not leaves, add entry (i, j) to the back of LIS and mark as new Type B entry.
- Removed entry (i, j) from Type A set (p, q).
- For each Type B entry (i, j) in the LIS:
  - Check all grand descendents of entry (i, j) and encode the GDESC bit using a 2 symbols adaptive model.
  - If GSEDC = 1:
  - Add the offspring (k, l) of entry (i, j) to the end of LIS in set (p, q) of $2 \times 2 \times 2$ adjacent pixels.
  - Remove the Type B entry (i, j) from LIS.

3. Refinement phase

- Except for new entries in the LSP, encode the REF bits for each entry (i, j) in the LSP using a 2 symbols adaptive model.

4. Update

- Divide the threshold value by 2 and repeat the above steps until the desired bit rate is achieved.

5   Simulation Results and Discussions

The proposed configuration is applied to the 3DSPIHT with AC and tested on various video sequences in QCIF and SIF formats. The decoupled 3D DWT and 3D SOT shown in Figs. 4b and 5 respectively are adopted in the simulation. For the 3D DWT, 3 level of decomposition is adopted in both spatial and temporal domain for the QCIF and SIF video sequences. The resolution of all the QCIF video sequences is 176×144, whereas the resolution of the SIF video sequences can be divided to 352×240 and 352×288. The 9/7 tap [7] and Haar wavelet filter is adopted for the spatial domain and temporal domain respectively. It has been mentioned in [4] that the use of Haar wavelet filter in the temporal domain can slightly improved the performance and reduce the computational complexity when compared to the 9/7 tap wavelet filter. It should be noted that the DWT is first carried out in the spatial domain and followed by the temporal domain.

5.1   QCIF Video Sequences

The simulation is performed with 10 fps by coding every third frame. Only the luminance component (Y) of frame 0 to 285 is considered in the simulation, and the

GOF size of 16 frames is chosen. Since the simulation is performed by coding every third frame, the actual number of frames encoded is 96. The result of the 3DSPIHT with AC that uses the proposed configuration (Proposed) is compared to the result generated from [8]. It should be noted that only the result for encoded frames are shown in Fig. 6.

5.2 SIF Video Sequences

Unlike the setting applied to QCIF video sequences, the simulation for SIF video sequences is performed with 30 fps by coding every frame. Only the luminance component (Y) of frame 0 to 47 is considered. The GOF size remains unchanged as 16 frames. In this case, the total number of frames to be coded is 48. All the simulation results are shown in Fig. 7.

5.3 Adaptation of the Adaptive Models

Generally, the mean PSNR achieved by the proposed configuration is 0.2 to 1.66 dB higher than the original configuration adopted by the 3DSPIHT. By using different sets of adaptive models for different types of output bits, better adaptation can be achieved as different types of output bits have different probability patterns. This also prevents the probability pattern of one type of output bits from affecting the others. For example, assuming that there are two models with one of the probability patterns skewed to symbol a while another skewed to symbol b. If the two models are combined, the combined model may have two equal probable symbols when the frequency of symbol a and symbol b is very close to each other.

On the other hand, the maximum frequency can also significantly affects the adaptation rate of the adaptive models [5]. In order to determine the optimum value of maximum frequency, an analysis is performed to observe the number of output bits for all the video sequences when different value of maximum frequency is adopted. The average number of output bits for QCIF and SIF video sequences is computed and summarized in Table 2.

It can be seen that by using the maximum frequency of 64, few hundred bits can be saved in the later bit plane for both QCIF and SIF video sequences. Furthermore, it also helps to reduce the memory required to store the frequency of each symbol in the adaptive model.

5.4 Memory Requirement

The main drawback of the proposed configuration is the amount of memory that is required by the adaptive model, since different types of output bits are now assigned

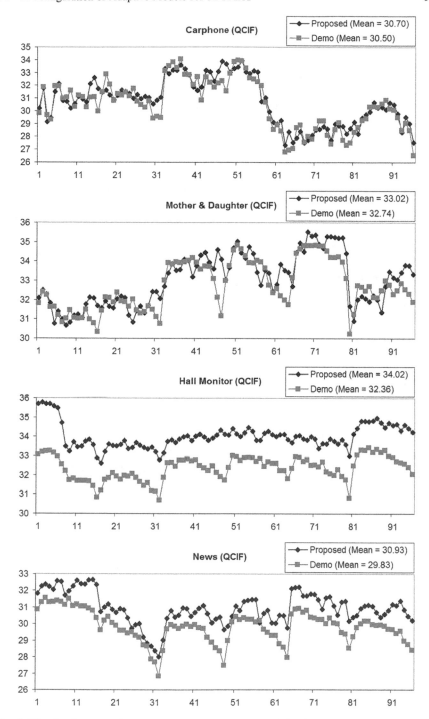

Fig. 6 The simulation results for various video sequences in QCIF format

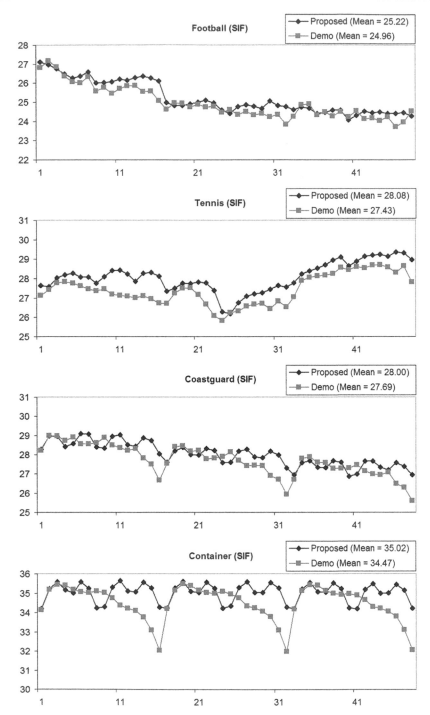

Fig. 7 The simulation results for various video sequences in SIF format

Table 2 The average number of output bits for all the QCIF and SIF video sequences at different bit plane under different value of maximum frequency

Format	QCIF video sequence				SIF video sequence			
Maximum frequency	32	64	128	256	32	64	128	256
Bit plane 1	584	584	584	584	1,318	1,266	1,249	1,387
Bit plane 2	1,272	1,272	1,272	1,300	2,977	2,758	1,680	3,132
Bit plane 3	2,670	2,686	2,700	2,776	5,610	5,294	5,244	6,117
Bit plane 4	6,229	6,251	6,282	6,384	11,128	10,837	10,891	12,124
Bit plane 5	13,977	13,982	14,048	14,151	24,878	24,610	24,779	26,107
Bit plane 6	30,624	30,555	30,700	30,789	68,745	68,465	68,784	70,062
Bit plane 7	65,595	65,377	65,641	65,753	180,612	180,198	180,693	181,849

with a separate set of model. It is more than sufficient to use one byte of memory to store the frequency for one symbol, because the maximum frequency can only go up to 64. If only the memory required to store the frequency of all the symbols is taken into consideration, the memory requirement can be calculated by using (1), whereby $T_{AM}{}^m$ represents the total number of adaptive models with 2^m symbols.

$$\text{Memory(Symbol)} = \left\lceil \sum_{m=1}^{m=8} T_{AM}{}^m \times 2^m \times \log_2(64) \right\rceil \text{Bit} \tag{1}$$

By referring to the proposed configuration shown in Table 1, the memory requirement is approximately 2.3 kilo Byte (kB). For the case where different types of output bits are sharing the same set of adaptive model with m symbols, the memory requirement is only approximately 0.5 kB.

For practical implementation, a cumulative version [5] of the frequency for each symbol in the models is usually computed to increase the processing speed of AC. Hence, the number of memory blocks need to be allocated is also proportional to the number of symbols in the models. For the worst scenario, the number of bits required for each memory block is equivalent to the number of bits required to represent the maximum cumulative frequency.

$$\text{Memory(Cumulative)} = \left\lceil \sum_{m=1}^{m=8} T_{AM}{}^m \times 2^m \times \log_2(2^m \times 64) \right\rceil \text{Bit} \tag{2}$$

In this case, the amount of memory required to store the cumulative version of the frequency for each symbol is approximately 3.78 kB. In total, about 6 kB is needed. But as compared to the amount of memory used to store the video frame or implement the 3DSPIHT algorithm which measure from few hundred to about a thousand kB [8], 6 kB of memory is almost negligible. Hence, the increment in memory requirement is not excessive. However, it should be noted that the integration of AC can significantly increase the computational complexity of the overall system. Therefore, it might not be suitable for implementation under a hardware constraint environment.

6 Conclusion

The simulation results show that applying the proposed configuration of the adaptive models into 3DSPIHT with AC can improve the performance in both QCIF and SIF video sequences by approximately 0.2 to 1.6 dB. The main drawback of the proposed configuration is the increase in memory requirement introduced by the adaptive models. However, the increment is almost negligible as compared to the amount of memory required to store the video frame and implement the 3DSPIHT algorithm, which can be ranging from hundred to thousand kB. By separating the adaptive models for each type of output bits, better adaptation can be achieved. This is due to the different probability patterns that exist in different type of output bits.

References

1. Said, A., & Pearlman, W.A. (June 1996). A new, fast, and efficient image codec based on set partitioning in hierarchical trees. *IEEE Transactions on Circuits and Systems for Video Technology, 6*(3), 243–250.
2. Kim, J., Xiong, Z., & Pearlman, W.A. (Dec 2000). Low bit-rate scalable video coding with 3-D set partitioning in hierarchical trees (3-D SPIHT). *IEEE Transactions on Circuits and Systems for Video Technology, 10*(8), 1374–1387.
3. Kim, J., & Pearlman, W.A. (Mar 1997). An embedded wavelet video coder using three-dimensional set partitioning in hierarchical trees. In *Proceedings of the DCC*, pp. 251–260.
4. He, J., Dong, Y.F., Zheng, & Gao, Z. (Oct 2003). Optimal 3-D coefficient tree structure for 3-D wavelet video coding. *IEEE Transactions on Circuits and Systems for Video Technology. 13*(10), 961–972.
5. Witten, H., Neal, R., & Cleary, J.G. (Jun 1987). Arithmetic coding for data compression. *Communications ACM, 30*, 520–540.
6. Shapiro, J.M. (Dec 1993). Embedded image coding using zerotrees of wavelet coefficients. *IEEE Transactions on Signal Processing, 41*(12), 3445–3462.
7. Chow, J., & Lee, C. (2002). Memory-efficient implementation of 3-dimensional zerotree video coding. In *Proceedings of the International Conference on Consumer Electronics*, pp. 350–351.
8. Kim, B.J., Pearlman, W.A., & Said, A. Three dimensional set partitioning in hierarchical trees compression, [Online] Available: http://www.cipr.rpi.edu/research/S.

Chapter 15
Ad Hoc In-Car Multimedia Framework

Hemant Sharma, Kamal Sharma, and A.K. Ramani

Abstract The chapter provides the description of framework architecture for ad hoc pervasive multimedia services that addresses the needs for automotive applications. It includes mechanisms to access multimedia content as part of contextual information. It is composed of core service components and augmented by several enhanced components that comprise a distributed service enabler space.

Keywords In-car multimedia · Ad hoc network · Bluetooth · Pervasive computing

1 In-Vehicle Systems and Bluetooth

The proliferation of multimedia capable mobile devices such as novel multimedia enabled cellular phones has encouraged users inside a vehicle to consume multimedia content and services while on move [1–7]. However, selecting and enabling the presentation of the most appropriate content for the given device is rather complicated due to the vast amount of multimedia data available and the heterogeneity of systems available. Appropriate software infrastructure is required and being developed to enable automatic discovery and efficient presentation of multimedia content. Context information can be used to guide the applications and systems in selecting the proper content and format for a given user (passenger or driver) at a certain time, place and under a specific context.

Recent advances in communication and multimedia technology in the past decade have greatly shaped the evolution of in-vehicle ubiquitous multimedia environment. On-board infotainment system is able to access multimedia content from a CD changer, iPod or Bluetooth enabled Cellular Phone. However, there are still many challenges in this type of environment. One basic problem of this

H. Sharma (✉)
Delphi Delco Electronics Europe GmbH, Germany
e-mail: hemant.sharma@delphi.com

K. Sharma and A.K. Ramani
ICSIT, Devi Ahilya University, Indore, India

S.-I. Ao et al. (eds.), *Intelligent Automation and Computer Engineering*,
Lecture Notes in Electrical Engineering 52, DOI 10.1007/978-90-481-3517-2_15,
© Springer Science+Business Media B.V. 2010

multimedia service model is the dynamic heterogeneity caused by the physical attributes of diversified computing devices and vehicle networks. Device and user mobility, another important problem of in-vehicle ubiquitous computing, posses a great challenge to the management of multimedia services.

The evolutionary development of in-car electronic systems has lead to a significant increase of the number of connecting cables within a car. To reduce the amount of cabling and to simplify the interworking of dedicated devices, currently appropriate wired bus systems are being considered. These systems are related with high costs and effort regarding the installation of cables and accessory components. Thus, wireless systems are a flexible and very advanced alterative to wired connections. However, the realization of a fully wireless car bus system is still far away. Fiber optic connections for multimedia bus systems offer advantages regarding costs and bandwidth, and the demanded reliability of mission-critical networks still require wired connections to ensure the safe operation of the car. Though, cost-effective wireless subsystems which could extend or partly replace wired bus systems are already nowadays conceivable. A very promising technology in this context is specified by the recent Bluetooth 3.0 standard.

2 Pervasive In-Vehicle Multimedia

In-vehicle entertainment systems offer a new generation of ultra-compact embedded computing and communication platforms, providing occupants of a vehicle with the same degree of connectivity and the same access to digital media and data that they currently enjoy in their home or office. These systems are also designed with an all-important difference in mind: they are designed to provide an integrated, upgradeable control point that serves not only the driver, but the individual needs of all vehicle occupants.

Innovative key factor for improving multimedia applications and services, comfort and safety, and vehicle management are represented by pervasive networked technologies in the automotive sector. This networking enables:

- Innovation of in-vehicle multimedia system and software architectures
- Enhancement of the application and services offered to end users

Figure 1 below presents typical in-vehicle pervasive network along with the physical components and devices that could part of the network.

The network contains a pure multimedia part, represented by Multimedia Bus, and a control and external communication part, represented by Networking for Automotive. Multimedia Bus represents the wired multimedia resources. External multimedia resources can form an ad hoc network via the communication interfaces of the vehicle network.

One commonly available technology that can support ad hoc entertainment services inside a car is Bluetooth, a low power, short range radio technology mainly used to support wireless devices. Bluetooth radios on typical mobile devices can

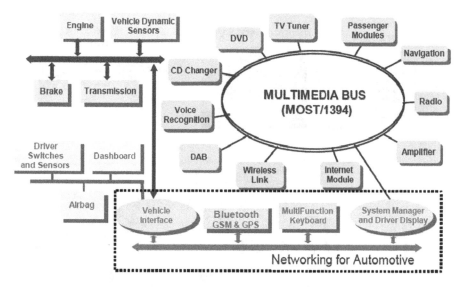

Fig. 1 Typical pervasive vehicle network with multimedia devices

support a data rate of up to 24 Mbps over a range of ten meters. Bluetooth supports a wide range of use cases, including file transfer and personal area networking. The Bluetooth standard defines a number of profiles, standard interfaces for particular broad categories of use cases. While the specific Bluetooth capabilities of a mobile phone vary based on the manufacturer, the model, and the needs and desires of the mobile operator, many do support Bluetooth profiles beyond the simple pairing of phone and headset. In addition, a number of mobile phones expose their Bluetooth interfaces to third party application developers via standard application programming interfaces.

Widespread automotive availability, strong consumer demand and regulatory push have also provided a strong platform for rapid growth in automotive Bluetooth option take rates, but there are significant underdeveloped automotive Bluetooth opportunities. In order to exploit the services, provided by available multimedia resources in a vehicle and over wireless network, a software framework is necessary that can seamlessly provide ad hoc network resources to entertainment applications.

3   Ad Hoc Framework Architecture

3.1   System Overview

An ad hoc pervasive network in side a car and participating devices are shown in Fig. 2. The diagram shows a smart phone, infotainment system, and rear seat

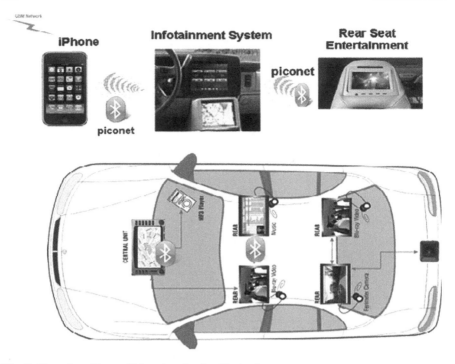

Fig. 2 Pervasive ad hoc vehicle network using bluetooth

entertainment system. All the three devices are capable of establishing a Bluetooth connection. A piconet can be established between Infotainment system and iPhone when the driver or any other occupant of the vehicle pairs the phone. Similarly a piconet could be established between Infotainment system and rear seat entertainment system. The piconet shall enable sharing of iPhone contents or access to contents from internet, if appropriate application framework is available at infotainment system.

3.2 Framework Design Concepts

In order to address various design challenges, the software architecture presented here incorporates the reflective techniques into its design, in a sense that the system can reason about and modify itself on a meta-object level. The architecture considers the following design concepts:

- *Open*: the term has two-fold meaning in the context of this system. First of all, the internal semantics of the system objects are exposed to the applications on top. Secondly, the system is open to the modification or replacement of its constructing components.

- *Component-based*: the system is decoupled into different sets of components, each having individual functionalities. As the building blocks of the whole architecture, they can be loaded or removed dynamically.
- *Active*: the system environment updates and application behavior changes, and has certain self administrative power to accomplish the system functionality.
- User-centric: we believe that user, not the application, will be the ultimate client of pervasive multimedia service delivery.

In the next section, we present the architecture of the multimedia framework based on these design concepts. We name the framework as mCAR.

4 *m*CAR Architecture

This section presents a high level overview of mCAR, the context aware multimedia framework architecture, describing briefly some of the design decisions taken for each part. The framework is divided in three main modules; the mCAR Kernel, the Communication Subsystem (CS) and the Content Module (CM). The user of mCAR provides a set of decision modules and multimedia protocol software to resolve the communication with the vehicle infrastructure.

The Kernel defines a set of software interfaces that are used by the decision modules for accessing both the CS and the CM. This module also provides a set of mechanisms for handling concurrent access to the data. The Kernel initializes the server and the decision modules. To maintain low coupling between the Framework and the decision modules, an extensible event system has been developed. This low coupling approach allows dynamically adding and removing listeners without changing the predefined internal behavior of the Framework.

The decision modules and their dependencies are specified in a configuration file that contains the identifier, the class that implements the decision module and their execution dependencies. This approach allows a declarative way for specifying the logic components used by the system. The execution dependencies are embedded as a list of decision modules that run before the current one is executed. This defines a direct acyclic graph of executing dependencies between all the decision modules. A topologic sort over this graph provides the execution order for each module. As previously stated, the modules are executed when an event occurs, but it is also possible that the decision module runs in parallel to the framework. The decision modules are loaded in runtime, allowing the system to replace or modify certain logic without stopping the execution of the multimedia application. The architecture and components that integrate a multimedia service implemented in the framework are presented in Fig. 3.

The Communication System implements asynchronous communication for the framework. It abstracts the network protocol and the network link (wired or wireless). Thus, the user of the framework defines the communication protocol used at the application level. Applications open a data connection with the framework and

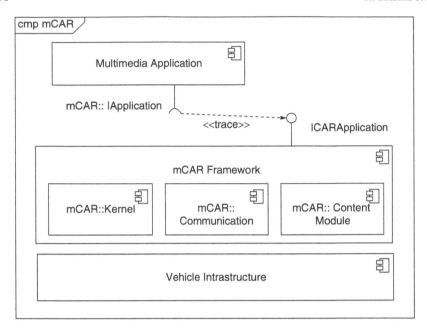

Fig. 3 *mCAR* framework components

for each new connection established the CS creates a content worker thread for handling the communication with the specific multimedia resource. For sending data to a resource, each worker contains a local queue and the worker sends data messages stored in this queue to the specific resource. The worker stores data received from the resource in a general queue that is part of the CS.

The information of this queue is processed by certain number of dispatcher threads, each dispatcher pass the data message to the Multimedia Protocol component. The Multimedia Protocol component is loaded at the startup of Framework. This component processes the incoming messages of the resources and may propagate events to the decision modules (or other listeners) each time a new message arrives. Different types of events can be defined and propagated by the user through this component. The user application can provide different implementations of the *Multimedia Protocol* (only one at a time is used). The latter allows different application-level protocols to operate with the Framework. The internal composition of Communication System is presented in Fig. 4.

The proposed queue system, which is based on a producer-consumer model, makes possible to achieve certain level of asynchronous application – resource communication. It is also possible to distribute the work of the dispatchers and the workers under several resources, improving the scalability of the system.

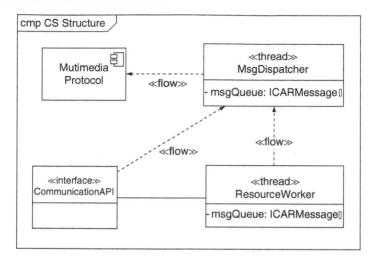

Fig. 4 *mCAR* communication system

5   Framework Interfaces

This section introduces the class and interface designs of the Event System, Decision Modules and the Multimedia Protocol. The interfaces are described using UML. The user of the framework provides some of these interfaces to implement different system behaviors. The framework is currently implemented in C++.

5.1   Event System

The event system used in the framework is defined using three interfaces, the ICAREventProvider, ICAREventListener and ICAREvent, the UML class diagram for these interfaces is presented in Fig. 5.

ICAREventListener instances are registered to listen events from a specific event source (*ICAREventProvider* instances). Each *ICAREventListener* instance provides the list of dependencies that should be executed before itself, defining a partial execution order for all the listeners. *ICAREventProvider* instances are active components that propagate certain mCAR or user defined events (*ICAREvent* instances). Event source instances can be linked in a chain of event propagation, using the appropriate method of *ICAREventProvider*. When an event is propagated, the following protocol steps should be respected by each event source:

- Executes default action event for the current event source.
- Executes the *listenEvent* method for all registered listeners of the current event source, respecting the predefined order.
- Propagates the event to the parent of the current event source (repeat the process from step 1).

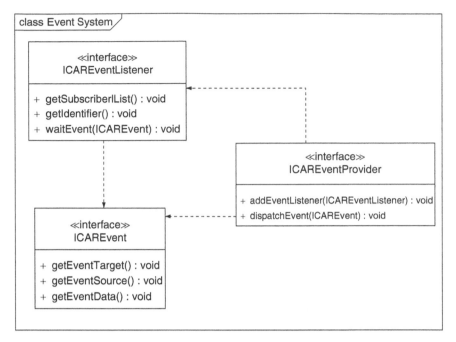

Fig. 5 Event system interfaces

ICAREvent instances are the event objects that contain the event source target and some data related to the event. The event source target is the logic element of the framework that generates the event, for example, if a vehicle logged in event is propagated then the event source target is the resource identifier.

5.2 Decision Module

The framework instantiates decision modules using a factory configurable by an XML configuration file. A default implementation for the factory is provided by the framework; this factory loads the XML information, resolves the dependencies and allows the runtime instantiation of each decision module. The UML class diagram for the decision modules is presented in Fig. 6.

Each declared decision module in the XML is an instance of ICARDecision-Module interface. All the decision modules are initialized with the Kernel interface (API to the framework), including the decision module name and dependencies list (information declared in the XML). The ICARDecisionModule instance may run in parallel of the Framework (for example as a proxy to other legacy system), or be activated and executed only under certain events propagated.

Fig. 6 Decision module
interfaces

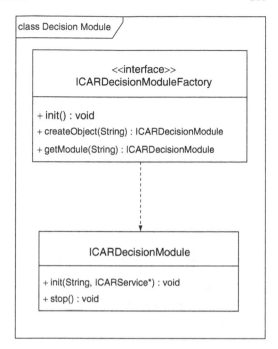

5.3 Multimedia Protocol

The Multimedia Protocol component processes the communication messages received from the vehicles, propagating the corresponding events in each case. This component is divided in three interfaces: *ICARMessage*, *ICARMessageFactory* and *ICARMultimediaProtocol*.

The *ICARMessage* interface represents a message to exchange with the resources. The implementation of this interface provides the methods marshal/unmarshal which receive Input and Output Streams respectively. These methods allows the message to be serialized and deserialized using different formats provided by the application, making transparent to workers and dispatchers, the real stream format exchanged over the network protocol.

The *ICARMessageFactory* instance, provided as part of the Multimedia Protocol component by the application, will be the creator of the *ICARMessage* instances. The *ICARMultimediaProtocol* instance is responsible for processing the messages received from the resources, before different events may be propagated. A UML class diagram of the Multimedia Protocol interfaces is presented in Fig. 7.

5.4 Application Model

The Application Model of the mCAR framework (Fig. 8) provides support for building and running pervasive multimedia applications on top of the framework kernel.

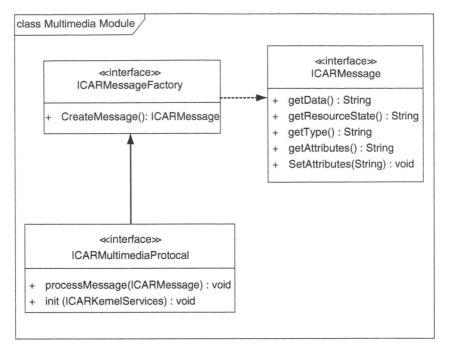

Fig. 7 Multimedia protocol interfaces

The application model shall guide modeling of multimedia application components with the use of interfaces of component of the framework in an efficient manner.

The applications access the framework functionality through an *IApplication* interface. Each time an application is started, an *ApplicationSkeleton* is created to allow interactions between the framework and the application itself. In particular, application interfaces allow applications to request services from the underlying framework, and to access their own application configuration profile through a well-defined reflective meta-interface. The multimedia application is realized as composition of components based on the component model of the framework.

6 Framework QoS Challenges

The provisioning of multimedia streaming applications in the ad hoc networks requires managing differentiated Quality of Service (QoS) levels depending on service/user/device requirements in order to properly allocate network bandwidth, especially the limited one available in the wireless last-meter. In particular, the Bluetooth specification offers limited support to QoS differentiation, by allowing to choose which of the three kind of logical transports to exploit and to statically configure QoS requirements for ACL ones. In addition, current implementations

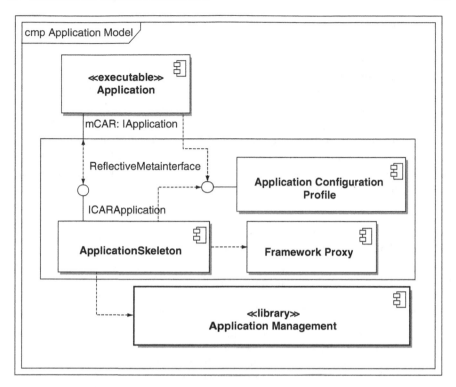

Fig. 8 Application model for the framework

of the Bluetooth software stack do not allow applications to exploit the limited QoS functions included in the specification in a portable way. The result is that the development of QoS-related Bluetooth operations in wireless Internet applications currently depends on specific implementation details of the target Bluetooth hardware/software stacks. This relevantly complicates service design and implementation, limits the portability of developed applications, and calls for the introduction of novel framework supports for QoS management.

The provisioning of multimedia streaming applications in the wireless in-car entertainment environment requires addressing novel and peculiar characteristics of the scenario, e.g., intermittent connectivity, resource-limited terminals, and proper allocation of the limited bandwidth available especially in the wireless last-meter. The above solutions have achieved significant results in the wired infotainment scenario, but do not fit well with the ad hoc automotive environment. On the one hand, network-layer solutions has been designed to perform traffic shaping and prioritization on wired networks and usually require updating/replacing the installed best effort vehicle infrastructure in order to support the proposed low-level protocols. On the other hand, middleware solutions for the wired network do not face the dynamicity issues of mobile provisioning environment and do not provide specific countermeasures for the relevant resource discontinuities inside the vehicle.

7 Summary

This chapter presented the software architecture of mCAR. The architecture is based on the design concepts outlined in the chapter to meet the multimedia service deployment and management challenges. The application model for development of multimedia applications on the top of mCAR shall enable rapid design of multimedia applications. The applications shall use the provided interfaces of mCAR to use the framework services. The framework behavior is configurable via application specific configuration file. Specification of application specific decision module makes the framework open and scalable to accommodate support for wide range application and multimedia resources.

References

1. Bluetooth 3.0 Specifications, www.bluetooth.com
2. Henricksen, K., Indulska, J., & Rakotonirainy, A. (Aug 26–28, 2002). *Modeling context information in pervasive computing systems, 1st International Conference on Pervasive Computing*, pp. 167–180. Zurich, Switzerland: Springer.
3. Zhang, Q., Zhu, W., & Zhang, Y-Q. (2005). End-to-end QoS for video delivery over wireless internet. *Proceedings of the IEEE 03 1*, 123–134.
4. Sharma, K.K., Sharma, H., & Dr. Ramani, A.K. (2009). mCAR: software framework architecture for in-vehicle pervasive multimedia services. *IMECS 2009, I*, 1100–1105.
5. Misic, V.B., Ko, E.W.S., & Misic, J. (2004). Load and QoS-adaptive scheduling in bluetooth piconets. *Proceedings of the 37th Hawaii International Conference on System Sciences*.
6. Chen, G., & Kotz, D. (June 2002). Context aggregation and dissemination in ubiquitous computing systems. In *4th IEEE Workshop on Mobile Computing Systems and Applications*. Callicoon.
7. Judd, G., & Steenkiste, P. (Mar 2003). Providing contextual information to pervasive computing applications. In *1st IEEE Conference on Pervasive Computing and Communications*, pp. 133–142. Fort Worth.

Chapter 16
Reliable Routing Protocol for Wireless Sensor Network

Mohammad S.I. Alfares, Zhili Sun, and Haitham Cruickshank

Abstract Wireless Sensor Network (WSN) is one of the major research areas in computer network field today. The function of WSN in this chapter is to provide sensing services in an un-attended harsh environment. Sensed data need to be delivered to the base station and to cope with the network unreliability problem. Few routing protocol takes into consideration of this problem. It is a great challenge of the hierarchical routing protocol to provide network survivability through redundancy features. In this chapter, a short literature review of the existing routing protocol is carried out. Then a novel hierarchical routing protocol, which addresses network survivability and redundancy issues, is introduced. Initial analysis shows promising results of the proposed protocol comparing with OEDSR and LEACH, which is a well known protocol as benchmark. Finally, conclusion was drawn based on the research and future direction for further research is identified.

Keywords Wireless sensor network · Hierarchical · Routing protocol · Reliability · Redundancy · Survivability

1 Introduction

Wireless Sensor Network (WSN) is one of the major research areas in computer network field today. It is considered as one of ten emerging technologies that will bring far-reaching impacts on the future of humanity lives [1]. Also, the importance is due to that numerous applications can benefit from the WSN, such as healthcare, environmental, forest fire, and military applications, etc. [2–5]. As a new challenging research field, WSN now is undergoing an intensive research to overcome its complexity and constraints [5]. Such constraints are:

- Power source
- Communication and bandwidth

M.S.I. Alfares (✉), Z. Sun, and H. Cruickshank
Centre for Communication Systems Research (CCSR), University of Surrey, Guildford, GU2 7XH, Surrey, UK
e-mail: m.fares@surrey.ac.uk; z.sun@surrey.ac.uk; H.Cruickshank@surrey.ac.uk

S.-I. Ao et al. (eds.), *Intelligent Automation and Computer Engineering*,
Lecture Notes in Electrical Engineering 52, DOI 10.1007/978-90-481-3517-2_16,
© Springer Science+Business Media B.V. 2010

- Mobility
- Processors and memory
- Network density and data aggregation

Because of these concerns, existing routing protocols cannot be deployed directly in WSN. Moreover, one of the difficult challenging features to be offered by the routing protocol in WSN is to provide reliable network connectivity in the presence of harsh environment (resist to link and sensor nodes failures). The proposed routing algorithm presented in this chapter is aimed to the forest fire monitoring and similar applications. The major specifications of these kinds of applications are the large area of deployment (about 20×20 km), heterogeneous in sensed data and in Sensor Nodes (SN) types, and harsh environment (goes behind the high probability of node failure). Previous studies [6–8] have shown that link connectivity greatly affects the performance of routing protocols in WSN. However, the impact of unreliable link/node on the connectivity of WSN is not fully tackled in preceding research.

This chapter presents a Self Organizing Network Survivability routing protocol (SONS). It is designed to cope with the large area of deployment, link or node failures and heterogeneous network in forest fire monitoring and similar applications.

2 Application Descriptions

This section gives a brief description of the forest fire and similar scenarios (for further information refer to [9, 10]). Forest fire size is classified to small, medium, and large. The large size could reach more than 1 million acres [11]. The speed of fire front line is about 3–8% of the wind speed (depends on the density and type of vegetation, and slope).

Most of the large forest fires are contained in several weeks. The area temperature is high (reach more than 92°C). Forest fire needs massive resources to fight it. It needs many fire crews, many helicopters, water tanks and pumps, trucks, civilians support and many more. All these resources can be organized and managed by the commander centre with the support of sensing data (see Fig. 1).

The huge damage caused by forest fire (reach multi-billion US$), makes developing a new techniques or equipments a vital task. Based on the above, the WSN solution should work in a harsh environment with a heterogeneous data and devices.

3 Related Work

Many routing protocols in WSN have been proposed to take into account of the inherent features of WSNs, along with the application and architecture requirements. These routing protocols can be classified according to the network structure, protocol operation, resource utilization, or routing protocols [2, 5]. The proposed routing

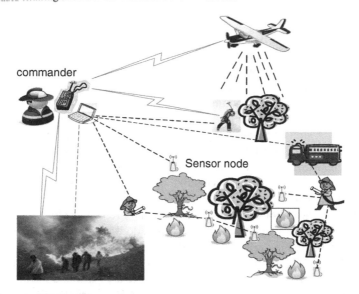

commander

Sensor node

Fig. 1 Scenario of WSN in forest fire application

protocol is classified under hierarchical network structure, and hybrid protocol operation that provides network connectivity and survivability features.

Low Energy Adaptive Clustering Hierarchy (LEACH) Protocol [12] and LEACH-Centralize (LEACH-C) [13] are the well known routing protocols in WSN based on hierarchical structure. In general LEACH and LEACH-C divides the network into clusters (sections), and in each cluster there is an elected sensor node to act as head of the cluster identified as Cluster Head (CH). The cluster head task is to manage communication among member nodes of the cluster, data processing, and relay processed sensed data to the Base Station (BS) directly. LEACH and LEACH-C outperform flat network protocols in terms of network life time. However, LEACH and LEACH-C are not suitable to be deployed in a large area because of direct communication between CH and BS. Moreover, it does not take into account how the network resists the link and SN failure [5].

Optimized Energy-Delay Sub-network Routing (OEDSR) protocol [14] is a hierarchical based structure protocol. It is an extension to Optimized Energy-Delay Routing (OEDR) protocol [15]. Where, only sub-networks are formed around an event/fault and elsewhere in the network nodes are left in sleep mode. OEDSR borrows the concept of relay-nodes (next hop node) selection from OEDR. In OEDR, relay nodes selection is based on maximizing the number of two hop neighbours. Whereas, the selection in OEDSR is based on maximizing the link cost factor.

OEDSR assumes that the BS has a sufficient power supply, thus a high power beacon from the BS is sent to all SNs on the network. This assumption makes all SNs know their distance to the BS, which the link cost factor formula rely on this assumption.

Another hierarchical routing protocol is Self-Organizing Protocol (SOP) [16]. SOP routing protocol can be used with stationary or mobile sensor nodes. The protocol works simply by selecting a sensor nodes to act as router (relay messages), and these routers are to be stationary which form the backbone of the network to carry data to the BS. Every sensor node should be able to reach a router in order to be part of the network. A routing architecture that requires addressing of each sensor node is achieved, by identifying the address of the router node to which they are connected. A similar enhanced idea is the Proactive Routing with Coordination (PROC) [17], in which both protocols are used in a continuous dissemination (not on-demand) network type. PROC detect faulty node or link when a certain number of consecutives data is not acknowledged, then the sender node removes the failed node from its neighbor table and resend the data to the best selected parent node from the updated neighbor table. On the other hand, PROC have long recovery time (the time it takes to recover the failed node), and does not have control scheme for the traffic flow pattern in the entire network (local solution rather than global).

A multi-hop hybrid routing protocol based on LEACH [18] is proposed by combining the clustering and multi-hop techniques. It uses a hierarchical multi-hop routing method to forward sensed data from CH to CH toward the BS. The protocol does not provide redundancy and recovery techniques in case of link or node failure, but for the next cycle of electing a new CH the fault will be resolved automatically (takes long time) for in-cluster communication. For intra-cluster communication if the intermediate CH or communication link failed then the CH will search for alternative path. In case of no alternative path exist, then the CH will send the data directly to the BS which is not feasible solution in large area of deployment.

4   Self-organizing Network Survivability Routing Protocol

In this section, we present a new routing protocol. One of the key points in the proposed routing protocol is to solve the link and node reliability problem. Most protocols described in Section 3 especially hierarchical do not address link/node reliability problem. It is a challenge for the hierarchical routing protocol to provide network survivability and redundancy features. Therefore, SONS routing protocol designed to cope with these features.

In general, SONS uses multi-hop hierarchy (to cover large area) and spanning tree (for fast routing and less overhead) as basic ideas to deal with large area of deployment issue. SONS is fully distributed that every CH choose the nearest parent-CH to the BS to forward data to in the normal situation. If the network is congested or the parent-CH is dead, then the CH chooses the next best parent with zero communication (because of the wireless broadcast communication nature). Furthermore, SONS takes high density SN deployment advantage, to provide network fault tolerance feature through introducing redundant CH. Mainly the operation of SONS can be divided into three parts. The start-up phase, the synchronization phase and the message exchange phase. Figure 2 illustrates the proposed algorithm.

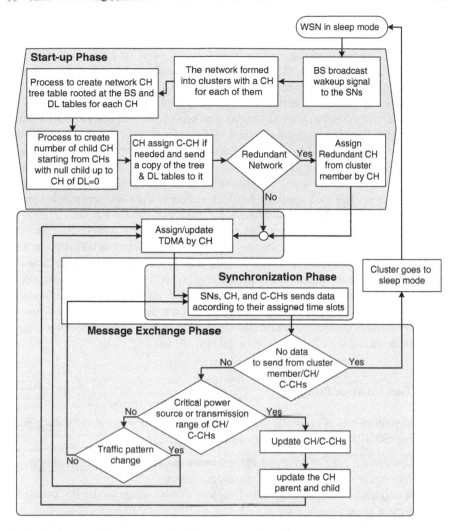

Fig. 2 Flow chart of the proposed algorithm

4.1 Start-Up Phase

This phase is responsible to build up the WSN in the network layer. When the SNs deployed in the area of interest, it will be initially in a sleep mode state. Then the start-up phase starts by firstly BS send a wakeup signal to the network. Then the network will be formed into clusters each with assigned elected CH (initially the same as in LEACH [12], and then afterwards remaining power source is the added factor). Subsequently, a tree is created from the BS as the root up to the leaves through only the CHs.

Based on the tree construction, there will be parents and children for each CH except for the leaf CH where it has no child. For every turn of electing CH, the new CH will send a control message to its parent and child(s) (if exist) to update the changes of the tree members. After that, the total number of children of each CH will be specified starting from leaf CH to its parents CH up to the root (BS). Subsequently, we can benefit of the network high density deployment and the in-expensive cost of the SNs, by helping the CH to relay and process data of other children clusters toward the BS. This procedure is via SN of cluster to be as Co-operative cluster head (C-CH), and assigned by the CH according to the next rank on the CH election. This procedure is applied to CH with a total number of children exceeds n (from simulation, optimal n will be chosen). This will help the network to avoid power drain from the CH. Especially for the CH near to the BS (energy holes problem [19]). In addition, the procedure provides more paths to the BS which helps in solving the data converge-cast problem. Converge-cast is a communication pattern, where the data flow is from a set of nodes to one node (many-to-one). This point will make the procedure simple (not complicated), that the code is small which helps in saving memory space and needs less processing power. Finally, redundancy option for the CH is specified by the BS to be either fully-redundant, or semi-redundant, or no-redundant. Fully-redundant is simply assigns a redundant CH node by means of the CH for each cluster. In semi-redundant, the CH near to the BS will assign a redundant node and the CH distant from the BS will not.

4.1.1 Tree Creation Process

The tree creation process initially starts from the BS. The process uses three-way messaging technique. The tree will be created as follow:

(i) BS (afterward parent CH) send a tree formation (TF) message to its neighbour CH(s) that are in the modified transmission range.
(ii) The neighbours CH(s) reply with tree join (TJ) message to the BS asking to add them to the tree.
(iii) A confirmation tree join grant (TJG) message is sent from the BS to the replied CH(s) associated with its distance level value. Then BS create a table of all answered back children CH(s). As well, the replied neighbour CH(s) will create a parent table that contain the parent and the BS with distance level of zero. The Distance Level (DL) is the number of hops from CH to reach the BS via intermediate parent CHs, and is equal to the $DL_{Parent} + 1$. Afterwards, CH may have many parents, and then forward data to the parent with minimum DL. This will help to transfer data to other route if one link is failed or congested, by maintaining the DLs in the parent table (increase DL of parent by 2 if it congested). Maintaining DLs of the parent table will achieve network reliability feature.

As soon as the CH join the tree, it will repeat the three ways messaging process (steps i to iii). If the CH(s) receives no reply then it will stop the operation and knows that it is the leaf CH of the tree.

4.1.2 Child Discovery Process

This process helps each CH to discover the expected load by find out number of clusters to be served (i.e. number of child CH(s) down to the leaf tree).

The leaf CH will start the process by sending a number of child (NC) message to its parent. Where the NC field in the message is set to zero (has no child). The parent will then update its child table. After that, the parent CH checks if all child reply with an NC message. If so, then it sends an NC message to its parent with the total number of grandchild CH. The process is continuing bottom to top of the tree until reaching the BS.

4.2 Synchronization Phase

In this phase, every CH will manage communication time schedule of its cluster SNs member. In every cycle, CH will assign communication time slot to every cluster members according to their needs. This allows SNs to identify when to wake up to send its sensed data to the CH to save energy (extend network life time).

4.3 Message Exchange Phase

This phase has the longest time interval compared to the start-up and synchronization phases. It is the event where the sensor nodes send their requested sensed data to the BS via CH/C-CH according to their time slots. As well, CH and C-CH relay processed data (remove duplicate data, enhance data accuracy, and compress data of similar priority) to the BS.

5 Results

Due to page limitation constraint, the analysis has been omitted (for further information refers to [20]). To study the proposed protocol in network connectivity and survivability features, we need to answer the question. How long does the network take to recover from faulty link or node? And how much power it consumes?

To be able to answer this question we use OMNeT++3.2 simulation tools with INET framework. We consider the same power model used in [12]. With the same parameters of $E_{elec} = 50\,nJ/bit$ and $e_{amp} = e_{fs} = e_{mp} = 100\,pJ/bit/m^2$. We assume that CH will send 400 bits/packet. The simulated area is $1,500 \times 1,000\,m$, with a random number of SNs deployed randomly. Also, the scenario is created using C++ to specify random failure by using Poisson's reliability SN distribution [21]. The assumptions for the network in our study are:

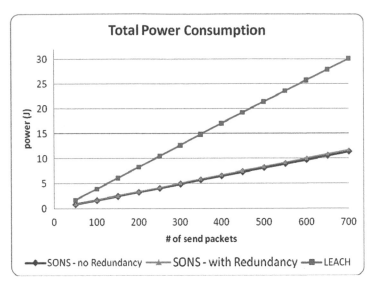

Fig. 3 Total power consumption

- Fixed location of BS, CHs, and redundant CH.
- The SNs are equipped with the power control facilities to vary their transmitter power.
- The clusters assumed to be circular and then the transmissions take a spherical shape.
- Assume worst case failure (node failure).

From the simulation, Fig. 3 shows a comparison in terms of total power consumption. It shows that both SONS types consume less energy than LEACH. Moreover, the difference in power consumption is increased as the number of sent packets increased. In Forest fire and similar applications, LEACH could not far communicate directly with the BS, but in simulation we assume it is possible (if new wireless technology found). Also, we use the short distance of P_{tx} as well, otherwise the power consumption in LEACH will be higher.

Additionally, the simulation measures the SONS performance in terms of average EED. Figure. 4 illustrates a comparison of the average EED against network size. It clearly shows, that SONS outperform OEDSR. Moreover, as network size increases the average EED of SONS is barely changed as in OEDSR and DSR.

Further studies on SONS response time in case of CH failure. The response time for the SONS in case of failure can be clearly revealed in Fig. 5. It shows end-to-end delay per packet with eight hops to reach the BS. Where, the CH failure occurs during sending packet 21. SONS with redundancy respond better and faster by about 1/3 less time than SONS with no redundancy.

One of the forest fire application requirements is to route heterogeneous sensed data (quality of service (QoS)). The data has been categorized using a priority field

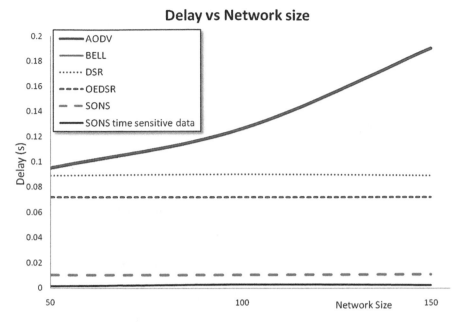

Fig. 4 Average end-to-end delay versus network size

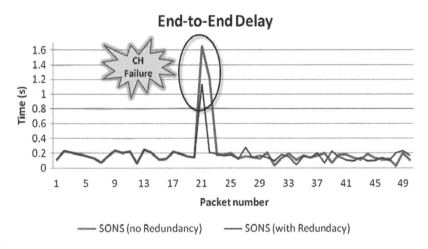

Fig. 5 SONS delivery time comparison

on the packet header. In which enable SONS to serve the packets with higher priority before the lower one. The classification comes from the application nature. This feature makes SONS to operate on heterogeneous WSN network.

By using simulation we investigate the SONS QoS behavior with respect to the end to end time delay. From the simulation, Fig. 6 shows the end-to-end time delay of heterogeneous data with a single path to the BS. The figure illustrates that the time sensitive data gets guaranteed service response even if the network is busy with

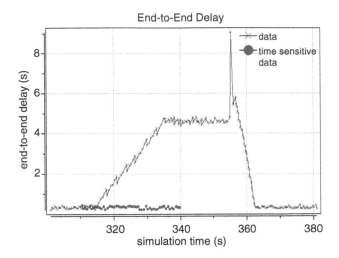

Fig. 6 End-to-end delay of heterogeneous data with a single path to the BS

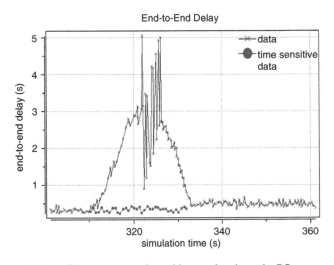

Fig. 7 End-to-end delay of heterogeneous data with several paths to the BS

normal data. Also, the figure is build up based on no alternative path to the BS is
available. Regardless to the rise of time delay for the normal data, SONS guarantee
data delivery to the BS. This helps in extending network life time by avoiding resend
the same data.

In addition, if one path is used from source to BS in multi-hop technique, it will
leads to extensive energy consumption to this path. With the time this results to an
un-even network energy distribution.

SONS routing protocol take care of this issue by distribute the traffic on the network if possible. This can be seen in Fig. 7, if the normal data congested with a time
sensitive data, then it try to search for alternative path. This will help in improving

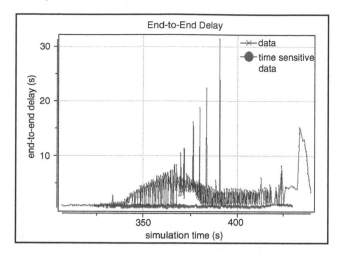

Fig. 8 End-to-end delay of heterogeneous data of the whole network

the end-to-end delay of the normal data and to distribute the traffic throughout the network.

Finally, Fig. 8 illustrates the SONS performance on the whole network with multiple senders. It can be seen that SONS guaranteed the priority service to the time sensitive data than the normal one. Additionally, it avoid congestion path using alternative path if available.

6   Conclusion and Future Directions

This chapter has shown the importance of network survivability and connectivity to overcome the link and node failure. The proposed protocol provides features of network survivability and redundancy. Initial analysis shows that SONS overcomes the weakness of the LEACH in recovery of link and node failures. In addition, simulation shows that SONS consumes less energy than LEACH. In comparison with OEDSR, DSR, and AODV in terms of EED, SONS has the superiority over these protocols. In addition, SONS scale well as network size is increase. And we demonstrate that SONS is capable to deal with link or node failure. Moreover, SONS has the feature to operate on heterogeneous WSN network.

Further investigation of the proposed protocol will be the topic of future work including simulation and application to real life case. Additionally, simulation tool will be used to measure the effect of **BS** mobility on the recovery time in harsh environment. Furthermore, network traffic pattern and shape will be investigated. Also, the proposed hierarchical routing protocol will be enhanced to be tested under mobility of both BS and SNs. Finally, latest QoS routing protocols in WSN will be coded (using VC++ on OMNeT++3.3) and compared against SONS.

References

1. MIT, (2004). 10 emerging technologies that will change your world. *Engineering Management Review, IEEE, 32,* 20–20.
2. Murthy, C.S.R., & Manoj, B. (2004). *Ad Hoc Wireless Networks: Architectures and Protocols.* Prentice Hall, PTR, USA.
3. Romer, K., & Mattern, F. (2004). The design space of wireless sensor networks. *Wireless Communications, IEEE, 11,* 54–61.
4. Akyildiz, I., Su, W., Sankarasubramaniam, Y., & Cayirci, E. (2002). A survey on sensor networks. *Communications Magazine, IEEE, 40,* 102–114.
5. Al-Karaki, J., & Kamal, A. (2004). Routing techniques in wireless sensor networks: A survey. *Wireless Communications, IEEE, 11,* 6–28.
6. LaI, D., Manjeshwar, A., Herrmann, F., Uysal-Biyikoglu, E., & Keshavarzian, A. (2003). Measurement and characterization of link quality metrics in energy constrained wireless sensor networks. *IEEE Global Telecommunications Conference, 2003. GLOBECOM'03.*
7. Gorce, J.M., Zhang, R., & Parvery, H. (2007). Impact of radio link unreliability on the connectivity of wireless sensor networks. *EURASIP Journal on Wireless Communications and Networking, 2007,* 1–16, doi:10.1155/2007/19196.
8. Zhou, G., He, T., Krishnamurthy, S., & Stankovic, J.A. (2004). Impact of radio irregularity on wireless sensor networks. *Proceedings of the 2nd International Conference on Mobile Systems, Applications, and Services,* pp. 125–138. NY, USA: ACM New York.
9. Hartung, C., Han, R., Seielstad, C., & Holbrook, S. (2006). FireWxNet: A multi-tiered portable wireless system for monitoring weather conditions in wildland fire environments. *Proceedings of the 4th International Conference on Mobile Systems, Applications and Services,* pp. 28–41. Uppsala, Sweden: ACM.
10. Doolin, D.M., & Sitar, N. (2005). Wireless sensors for wildfire monitoring. *Proceedings of SPIE,* pp. 477.
11. http://www.nifc.gov/fire_info/lg_fires.htm.
12. Heinzelman, W.R., Chandrakasan, A., Balakrishnan, H., & MIT, C. (2000). Energy-efficient communication protocol for wireless microsensor networks. *Proceedings of the 33rd Annual Hawaii International Conference on System Sciences,* pp. 10.
13. Heinzelman, W.B., Chandrakasan, A.P., Balakrishnan, H., & MIT, C. (2002). An application-specific protocol architecture for wireless microsensor networks. *IEEE Transactions on Wireless Communications, 1,* 660–670.
14. Ratnaraj, S., Jagannathan, S., & Rao, V. (2006). OEDSR: Optimized energy-delay sub-network routing in wireless sensor network. *Proceedings of the 2006 IEEE International Conference on Networking, Sensing and Control, 2006. ICNSC '06,* pp. 330–335.
15. Regatte, N., & Sarangapani, J. (2005). Optimized energy-delay routing in ad hoc wireless networks. *Proceedings of World Wireless Conference.* San Francisco, CA.
16. Subramanian, L., & Katz, R. (2000). An architecture for building self-configurable systems. *First Annual Workshop on Mobile and Ad Hoc Networking and Computing, 2000. MobiHOC. 2000,* pp. 63–73.
17. Macedo, D.F., Correia, L.H.A., dos Santos, A.L., Loureiro, A.A.F., & Nogueira, J.M.S. (2006). A rule-based adaptive routing protocol for continuous data dissemination in WSNs. *Journal of Parallel and Distributed Computing, 66,* 542–555.
18. Zhao, J., Erdogan, A., & Arslan, T. (2005). A novel application specific network protocol for wireless sensor networks. *IEEE International Symposium on Circuits and Systems, 2005. ISCAS 2005* (vol. 6), pp. 5894–5897.
19. Li, J., & Mohapatra, P. (2005). An analytical model for the energy hole problem in many-to-one sensor networks. *2005 IEEE 62nd Vehicular Technology Conference, 2005. VTC-2005-Fall,* pp. 2721–2725.
20. Alfares, M., Sun, Z., & Cruickshank, H. (2009). *A hierarchical routing protocol for survivability in wireless sensor network* (WSN) (pp. 262–268). *International MultiConference of Engineers and Computer Scientists,* 18–20 March, Hong Kong.
21. Garg, V.K. (2007). *Wireless communications and networking.* US: Elsevier Morgan Kaufmann.

Chapter 17
802.11 WLAN OWPT Measurement Algorithms and Simulations for Indoor Localization

Xinrui Wang and Tien-Fu Lu

Abstract This paper discussed novel algorithms for synchronizations and time resolution improvements of One Way Propagation Time (OWPT) measurements in the 802.11 Wireless Local Area Network (WLAN). In OWPT measurements, the Mobile Station (MS) records each 802.11 Beacon frame's arrival time. The Beacon frame's arrival time minus its Timestamp, which is recorded when the Beacon frame is transmitted from an Access Point (AP), is the Beacon frame's Propagation Time. The Propagation Time represents the distance between the MS and the AP. Rather than microseconds (μs) time resolution Timestamp, the MS could use its high precision clock to record the Beacon frame's arrival time in nanoseconds (ns). The first part of this paper proposes algorithms which can utilize the ns resolution arrival Time to improve the OWPT measurements time resolution from μs to ns and to highly synchronize the MS with all APs. These algorithms provide an opportunity to apply OWPT in 802.11 WLAN for highly accurate indoor localization. The second part discusses the possibility to utilize existing software and hardware platform to realize the proposed algorithms. At the end of this paper, the shortages of existing MS timing ability were raised and several options are provided for future researches to improve MS timing ability for OWPT application.

Keywords 802.11 WLAN · Synchronization · TOA · One way propagation time · Indoor localization · Time resolution improvement

1 Introduction

Indoor localization technologies have attracted considerable research interest over the last 20 years due to a wide range of application areas. Indoor localization technologies are the foundation of domestic robots navigation and mapping. The location information can also be used for tracking people, goods and equipment in

X. Wang (✉) and T.-F. Lu
School of Mechanical Engineering, University of Adelaide, SA, Australia, 5005
e-mail: xinrui.wang@adelaide.edu.au; tien-fu.lu@adelaide.edu.au

S.-I. Ao et al. (eds.), *Intelligent Automation and Computer Engineering*,
Lecture Notes in Electrical Engineering 52, DOI 10.1007/978-90-481-3517-2_17,
© Springer Science+Business Media B.V. 2010

buildings to benefit home entertainment, stock management and improving office work efficiency etc. The widely applied outdoor localization technology GPS, however, is hard to apply in indoor environments and dense urban areas. As the satellites' signals are reflected and/or diffracted by building structures, the receivers can not receive clear and strong enough satellite signals for localization purposes in these areas.

Many indoor localization technologies have been under continuous development in the previous decades to replace GPS for indoor localization. Generally, these localization technologies can be categorised by the type of sensors being used, including target sensors, distance proximity sensors, visual cameras and wireless sensors. The WLAN indoor localization technologies attracted considerable research interests recently, because the WLAN are widely deployed and localization technologies based on WLAN cost relatively low. WLAN localization technologies can be commonly further divided into two sub-categories, according to the different wireless signal features measured for localization. They are Received-Signal-Strength (RSS) and Time-of-Arrival (TOA).

RSS builds a signal strength fingerprint map of the working area based on real-time measurements [1] or wireless signal propagation models [2]. The signal strength fingerprint map consists of many sample points' signal strength and their coordinates refer to the working area. These sample points are distributed through the whole working area. When the MS is locating, it compares the receiving signal strength with the signal strength fingerprint map. The position of the sample point which has the most similar signal strength is assumed to be the MS position.

TOA measures the time lapsed when the signal transfers from AP to MS or from MS to AP then back to MS. These two different measurement patches are called One Way Propagation Time (OWPT) and Round Trip Time (RTT) respectively. The distance between the AP and the MS is calculated from OWPT or RTT. With three or more APs, these APs' positions and the distances between these APs and the MS can be used to calculate the MS's position with Trilateration.

Comparing these two different wireless signal measurement localization technologies, TOA is chosen for further developing. RSS demands considerable labour and computing ability to build the signal strength fingerprint map. But at the same time, the existing signal strength fingerprint map changes follow the changes of the indoor environment. Most of the former RSS localization deviations were between 5 and 10 m. The localization deviations are too loose to apply for accurate indoor localization applications. TOA measures the signal transferring time between an AP and an MS. The keystone of the TOA measurements is the time measurement accuracy and time resolution. The localization accuracy improves as the time measurement accuracy is enhanced.

In the rest part of the paper, Section 2 compares two different TOA measurements, Round Trip Time (RTT) and One Way Propagation Time (OWPT). In Section 3, the novel OWPT measurement algorithms are presented and discussed. These algorithms are proposed to synchronize AP and MS in OWPT measurement and to improve the OWPT measurement time resolution to ns. Section 4 describes the simulation model. The simulation results are analysed to evaluate the performance of the proposed algorithms. In Section 5, an experiment platform was build to verify the proposed algorithm. At the end of the paper is the conclusion.

2   Related Work

TOA measurement requires to accurately measuring the time when a wireless frame transfers between an AP and an MS. However, the resolution of all standard 802.11 time measurements is limited by the hardware and protocols to µs. As the wireless signals transfer in light speed, 1 µs time resolution leads to 300 m localization resolution, which makes it pointless for indoor localization applications. Besides the low time resolution, the synchronization errors between APs and MSs, and the processing time delays inside APs and MSs also cause enormous time measurement errors. To improve the time resolution and eliminate the synchronization errors and processing time delays, two TOA measurement algorithms, RTT and OWPT, are introduced.

2.1   Round Trip Time (RTT)

To improve the time measurement resolution and avoid synchronization errors between an AP and an MS, most researchers chose RTT to measure the time when a frame is transferred between an AP and an MS. Figure 1 demonstrates the process of RTT measurement. At first, the MS sends a Probe Request frame to an AP and records the sending time T_1. After the AP receives the request frame, a Probe Response frame is sent back to the MS and the MS records the Probe Response frame arrival time T_2. Ideally, half of the time lapsed between the time T_1 and the time T_2 is the propagation time T when the signal travels between the MS and the AP. Because all the time information is recorded by the MS clock, there's no need for synchronization between the AP and the MS. A ns time resolution clock on the MS could record both the time T_1 and T_2. With these two advantages, RTT could measure the propagation time T in ns resolution and eliminate the synchronization errors.

The shortcoming of the RTT is that the time measurement includes the unknown time delay in AP. In a wireless network, the MS is not the only client which sends requests or data to an AP. As shown in Fig. 1, when the AP receives the probe request, other requests and data frames which are sent by other clients may have already arrived in the AP. The AP processes these tasks based on time sequence. The MS Probe Request has to wait in the AP stack until the AP finishes all the prior tasks. After the AP processes the Probe Request, a Probe Response is sent out. As existing 802.11 protocols do not require the AP to record a frame's arrival time, the Probe Response only contains the AP's transmitting time in µs time resolution. ΔT,

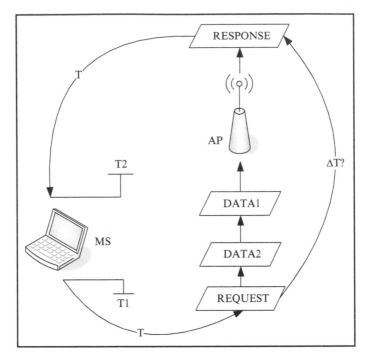

Fig. 1 RTT

which is the time delay in the AP, is unknown. ΔT would differ from time to time, depending on how busy the wireless network is. In a busy wireless environment, such as public places or offices, large volumes of data are continually being exchanged between the AP and other wireless devices, like laptops, computers, PDAs etc. In these situations, RTT's localization accuracy would be deteriorated by the enlarged ΔT.

Many former researchers noticed the time delays in AP causing RTT measurement errors. Several methods were employed to eliminate the time delays in AP. In [3], a calibration was implemented before RTT measurements. An MS and an AP were put together to set T equal to 0. The difference between T_2 and T_1 was the AP time delay ΔT. In calibration, a group of ΔT were collected. In one RTT measurement, a ΔT was randomly chosen from the calibration ΔT set. The chosen ΔT was extracted from the RTT measurement. Another method was described in the project of Gunther et al. [4] They utilized the WLAN feature that the AP and the MS send Acknowledgement frames (ACK) after they receive the Probe Requests and Probe Responses, respectively. ACK is a lower level frame which takes a much shorter time to process in an AP. They measured the time delay between the MS sent out Probe Request and received AP's ACK as the RTT. They also measured the time delay between the MS received the Probe Response till it sent out ACK. This is the processing delay in MS. The AP delay ΔT was replaced by the delay on MS and extracted from RTT to improve the RTT measurement accuracy.

However, neither of these two methods can record the AP delay, ΔT, directly. The localization deviations of the above two algorithms were unstable ranged from 2 m to more than 20 m. To more accurately measure the time delay ΔT, two Intel engineers, Mr. Golden and Mr. Bateman, customized an AP and an MS which can measure the time between when a frame arrived at an AP and when the AP sent the response. In this scenario, they measured the ΔT directly in every RTT measurement. In their experiment, highly accurate RTT measurement was achieved. Their results showed the RTT localization Root Mean Square Error (RMSE) to be between 1.1 and 5.5 m [5]. As mentioned before, however, existing 802.11 WLAN hardware and protocols do not support the RTT measurement proposed in [5]. The hardware and the protocols must to be modified to implement this highly accurate RTT measurement. To improve the TOA measurement accuracy in existing WLAN, OWPT measurement is explored in this paper.

2.2 One Way Propagation Time (OWPT)

Compared with RTT, OWPT measurement is quite simple. OWPT measures the time on both MS and AP sides. As shown in Fig. 2, when a frame transfers from an AP to an MS, the transmitting time T_1 and arrival time T_2 are recorded by AP and MS respectively. The signal transferring time T is calculated from the difference between T_1 and T_2. OWPT measurement does not contain the time delay in the AP or MS, giving OWPT a big advantage over the RTT measurement. However, in OWPT measurement, APs and MSs are required to be highly synchronized; otherwise the difference between T_1 and T_2 would be meaningless. To achieve sub-meter localization accuracy, the synchronization has to be on ns scale. Another challenge is that AP can only provide μs time resolution information. So, to reach acceptable indoor localization accuracy, the OWPT has to be specifically measured in ns resolution when the time information from AP is in μs resolution. Due to the strict synchronization and time resolution requirements, in the literature, so far no researcher has tried to apply the OWPT on WLAN. Only a ultrasonic localization system called Cricket System shares the similar OWPT mode to measure the transferring time of ultrasonic waves.

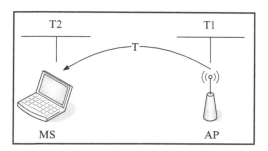

Fig. 2 OWPT

If OWPT is going to be applied on WLAN, efficient algorithms have to be developed to implement synchronization and OWPT measurement between APs and MSs in ns scale.

3 OWPT Synchronization and Time Measurement Resolution Improvement Algorithms

3.1 Introduction

Beacon frames are the most suitable frames for OWPT measurement in 802.11 WLAN. Every AP in WLAN periodically transmits Beacon frames with the broadcasting time, which is stored in a piece of the Beacon frame field and called Timestamp. Timestamp is generated by an AP clock and records the time when the first bit of the Beacon frame hits the Physical Layer for transmission. The time resolution of Timestamp is governed by 802.11 protocols as µs, unless customized APs, commercial APs can not provide higher than 1 µs time resolution Timestamp [6]. When an MS receives the Beacon frames, every Beacon frame's arrival time is recorded by the MS. Usually, a main part of an MS is a computer or a laptop. There are many high frequency microchips in an MS to process tasks, and manage the power etc. For example, a normal CPU working frequency is much higher than 1 GHz, corresponding to less than 1 ns time resolution. If one of the high frequency microchip is employed as the MS clock to record the Beacon frames arrival time, the Beacon frames arrival time should be recorded with less than 1 ns time resolution. When the OWPT is calculated from the Timestamp and arrival time, these two different time resolutions introduce the possibility of improving the OWPT time resolution to ns and synchronize the AP and the MS to ns scale.

3.2 OWPT Measurement Time Resolution Improvement

When the Beacon frames are received, the MS obtains the Beacon frames Timestamps in µs and arrival time in ns. Unfortunately, to improve the OWPT measurement time resolution to ns is not simply subtracting the Timestamp from the arrival time, even though the results are presented in ns time resolution. Because of loose time resolution, there is a time delay Δt between the Beacon frame Timestamp T and the Beacon frames broadcasting time t, as shown in Fig. 3. The Δt is inherited to the OWPT measurement.

Equation 1 shows the resolutions of different time elements which consist in the Beacon frames broadcasting time.

$$t \times 10^{-9} = T \times 10^{-6} + \Delta t \times 10^{-9} \tag{1}$$

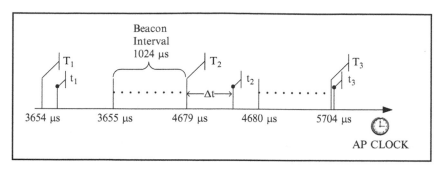

Fig. 3 Beacon frame broadcasting time

t: Beacon frame broadcasting time, ns

T: Beacon frame Timestamp, μs

Δt: Time delay between Timestamp and broadcasting time, ns

If $\Delta t \rightarrow 0$, Eq. 2 can be derived from Eq. 1:

$$t \times 10^{-9} = T \times 10^{-6} \qquad (2)$$

Equation 2 reveals that the μs time resolution Timestamp can replace the Beacon frame broadcasting time with ns time resolution if the time delay Δt is small enough.

According to 802.11 protocols, when a Beacon frame is transmitted, the time is copied from an AP clock into a Beacon frame Timestamp field. If a Beacon frame's Timestamp records the time T_1, the Beacon frame would be transmitted at any moment $T_1 + \Delta t_1$ within next 1 μs. The time delay Δt ranges from 0 to 999 ns. Randomly ranged Δt deteriorates the accuracy of OWPT measurement. To eliminate Δt, a Beacon frames selection algorithm is proposed to select a Beacon frame with the shortest Δt in a group of received Beacon frames. The selection algorithm utilizes the character of the randomly changed Δt, as illustrated in Fig. 4. In this section, the AP and the MS are assumed to be perfected synchronized to simplify the question and focus on how to select the Beacon frame.

Beacon frames are transmitted after every Beacon Interval. If the Beacon Interval is set to 1 time unit, an AP transmits a Beacon frame every 1,024 μs. An MS can receive 1,000 Beacon frames in approximately 1 s. If the MS is assumed to stay at the same location for about 1 s, the 1,000 received beacon frames would have the same propagation time ΔT. The MS, however, can only obtain the Beacon frames' Timestamp T_n and arrival time t_n. The pseudo propagation time $\Delta T'$ is calculated in Eq. 3:

$$\Delta T_n{'} = t_n - T_n = \Delta t_n + \Delta T, \; n \in [1,1000] \qquad (3)$$

Since the existence of random time delay Δt, the pseudo propagation time $\Delta T'$ is also randomly changed. As shown in Fig. 4, the propagation time $\Delta T_1{'}$ is shorter than the propagation time $\Delta T_2{'}$ but larger than the propagation time $\Delta T_3{'}$. If a

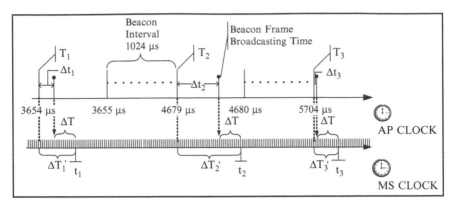

Fig. 4 Beacon frame transferring time measurement resolution improvement

group of 1,000 Beacon frames are collected, one Beacon frame with shortest Δt can be chosen to calculate the most accurate OWPT of this group.

Every captured Beacon frames were calculated its pseudo propagation time $\Delta T'$. From the second captured Beacon frame, its pseudo propagation time $\Delta T_n'$ is compared with last pseudo propagation time $\Delta T_{n-1}'$. Two results can be expected: $\Delta T_n' < \Delta T_{n-1}'$ and $\Delta T_n' >= \Delta T_{n-1}'$. If $\Delta T_n' < \Delta T_{n-1}'$, it means that the *n-th* Beacon frame contains smaller time delay Δt than *(n − 1)th* Beacon frame and so it's closer to the actually propagation time ΔT, vice versa. As a measurement group, the group's propagation time always chooses the smallest pseudo propagation time. After received 1,000 Beacon frames and the selection process finished, the measurement group's propagation time should be just couple ns different from ΔT, if it's not as the same as ΔT.

3.3 OWPT Synchronization Between AP and MS

A calibration is applied to synchronize APs and MSs before localization. The calibration process is developed from the OWPT Measurement Time Resolution Improvement algorithm which likes the other side of the same coin. The Timestamp of the Beacon frame comes from a 64-bit counter. It counts every μs after the AP turn-on. When the counter reaches its maximum value, it wraps around. As it is a 64-bit counter, it will take 580,000 years to reach its maximum value. In the MS, a same 64-bit counter is used to record Beacon frames arrival time. So once the OWPT measurement starts, there is a certain synchronization bias between AP and MS, as shown in Fig. 5.

In Fig. 5, the synchronization bias τ is considered in the OWPT measurement. The MS calculates the Beacon frames' pseudo transferring time in Eq. 4, and the MS is assumed to be static during calibration:

$$\Delta T_n' = t_n - T_n = (\Delta T + \tau) + \Delta t_n \tag{4}$$

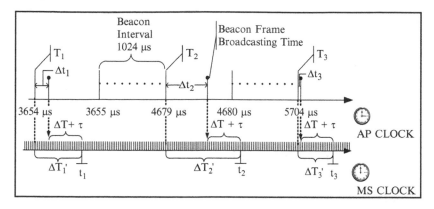

Fig. 5 Synchronization between APs and MSs

To calculate the synchronization bias τ, Eq. 4 is converted to Eq. 5:

$$\tau = t_n - T_n - \Delta T - \Delta t_n \qquad (5)$$

T_n and t_n are the Timestamp and the arrival time of the Beacon frame; Δt is the time delay between the Timestamp and broadcasting time of this Beacon frame; ΔT is the Beacon frame propagation time. In calibration, an AP and an MS are placed within a measured distance, which determines the ΔT in ns. The positions of the MS and the AP are chosen in a Line of Sight (LOS) situation to avoid any Multipath influences on the Beacon frames' propagation time measurement. The synchronization bias τ which is calculated directly from Eq. 5 contains a random error of Δt. This random error ranges from 0 to 999 ns. Hence, the Δt elimination algorithm which was described earlier, is applied in Eq. 5 to remove the Δt in the calibration. Unlike the OWPT Measurement Time Resolution Improvement scenario, the calibration has no strict processing time requirement. The calibration can take 1 min and it is still acceptable. For 1 min, n will be 60,000. That allows much more Beacon frames can be received in calibration. With a much bigger n, there is a greater possibility to reduce the Δt to 0. When $\Delta t \rightarrow 0$, Eq. 6 is derived from Eq. 5:

$$\tau = t_n - T_n - \Delta T \qquad (6)$$

After the calibration, τ can be accurately measured in ns time precision. Once the calibration is finished, the synchronization bias τ can be used for the following OWPT measurements.

Another benefit of the calibration synchronization is that an MS can synchronize with all the APs at the same time. When an MS is placed on the chosen calibration spot, the distances between the MS and all the APs are known. The calibration processes can be applied on every AP. APs broadcast Beacon frames in sequence in a Basic Service Set (BSS), so even though the Beacon Interval of every AP is $1,024\,\mu s$, the MS can receive two APs' Beacon frames adjacently in several μs.

After the 1 minute calibration, an MS can receive enough Beacon frames from all the APs in the working area to synchronize with each of them.

The calibration synchronization processes overcome one of the biggest challenges of OWPT measurement. With synchronized APs and MSs, as well ns time resolution measurement, OWPT can provide accurate Beacon frames propagation time measurement in WLAN.

4 OWPT Algorithms Simulation

4.1 Simulation Model

An AP-Channel-MS simulation Model was built to test the proposed OWPT Measurement Time Resolution Improvement algorithm and synchronization algorithm. The AP-Channel-MS simulation model was a simplified version of 802.11 g wireless network. The model consisted with three function blocks: AP, Channel, and MS, as shown in Fig. 6.

Instead of simulating all the WLAN frames and real network working situations, the simplified model only simulated the scenario that an AP broadcasted Beacon frames; an MS received Beacon frames after Beacon frames propagated through the Channel. The Beacon frame was also simplified. In the OWPT measurement, the most critical parameter was time. Therefore, in the model, the Beacon frame only contained the Timestamp and the Basic Service Set Identifier (BSSID). BSSID, which was generated by an AP, was unique for each AP. It could be employed to identify which AP transmits the Beacon frames, so it connected the Beacon frames OWPT measurement with specific position information.

In the AP block, there were two timers working together. One timer represented the AP clock and its working frequency was 1 MHz. This timer generated the Beacon frames Timestamp and counted the Beacon Interval. The Beacon Interval was set to be 1 time unit in the simulation, which was 1,024 μs. After every Beacon Interval, another timer, which had a working frequency of 1 GHz, started to count the random delay Δt before transmitting a Beacon frame. A Random Integer Generator

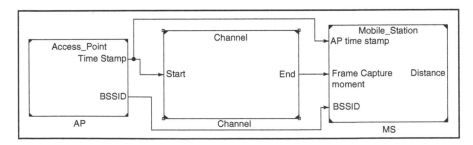

Fig. 6 AP-channel-MS model

generated random integer series. One integer in the random series represented the number of ns delay for one Beacon frame. The integer was in the range between 0 and 999. The Random Integer Generator generated a random series according to an Initial Seed. Different Initial Seeds led the generator to generate irrelevant random series.

After the Beacon frames transmitted through the Channel block, a propagation time was added to represent the transferring distance. The third block, MS, was responsible for analysing the received Beacon frames. The OWPT Measurement Time Resolution Improvement and synchronization algorithms were implemented here. The MS had its own receiver clock which works at 1 GHz frequency to record Beacon frames arrival time. When the MS received a Beacon frame, at first, the pseudo propagation time $\Delta T'$ was calculated. Then different Beacon frames' $\Delta T'$ in the same measurement group were processed by the algorithms described in Section 3 to improve the time measurement resolution or to synchronize AP and MS in OWPT measurements. According to the algorithm chosen, the output could be the synchronization bias between AP and MS or the most accurate OWPT measurement in one measurement group.

4.2   Simulation Results and Evaluation

Figure 7 was the simulation results of OWPT Measurements Time Resolution Improvement. Each line in Fig. 7 represented a measurement group with 1,000 OWPT measurements. The horizontal axis was logarithmic scale. It showed the number of beacon frames which have been processed. The vertical axis showed the decrease of the time delay Δt between the Timestamp and the broadcasting time during the simulation. Δt ranges from 0 to 999 ns.

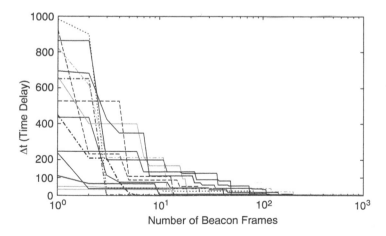

Fig. 7 Δt attenuation

There were 15 measurement groups displayed in Fig. 7. Each group employs a random integer series to generate various Δt for each beacon frame in that group. Different groups had irrelevant random series. There were eight groups which Δt started with more than 500 ns delay. The highest Δt was 983 ns. Utilizing the proposed algorithms, all the Δt were attenuated dramatically. After 10 beacon frames were received, the Δt in most of groups were under 200 ns. The biggest delay at this stage was just 212 ns, while the smallest Δt was 1 ns. When 100 beacon frames were received by the MS, the highest Δt was 39 ns, corresponding to a 11.7 m localization deviation. When the simulation finished, there were only three groups' Δt bigger than 0. These Δt were 5, 4 and 1 ns respectively. The biggest Δt led to a 1.5 m localization deviation. Hence in the simulation, the proposed algorithm successfully eliminated the time delay Δt in 80% of OWPT measurement groups. In these groups, the OWPT was accurately measured with time resolution of ns. In the worst situation of the remaining 20% OWPT measurement groups, the proposed algorithm could still achieve a 1.5 m localization deviation. These results simulated how well the OWPT Measurement Time Resolution Improvement could be in 1 s Beacon frames capturing.

Figure 8 illustrated four simulation results of OWPT Synchronization algorithm. The solid line in Fig. 8 was the synchronization bias τ which was expected to be measured when the calibration finished. As the synchronization algorithm was developed from the OWPT Measurement Time Resolution Improvement algorithm, the simulation results showed the similar measurement errors attenuation curves. But the process time for synchronization was much longer than the process time to eliminate the time delay Δt. The more beacon frames were received, the higher accuracy of the OWPT synchronization could be achieved. As shown in Fig. 8, after 60,000 Beacon frames were received in 1 min, the OWPT synchronization algorithm found all the synchronization bias τ with ns precision.

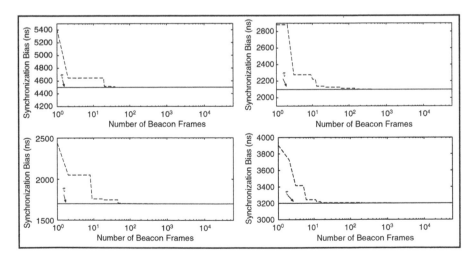

Fig. 8 Synchronization bias τ measurement

The simulation results confirm that the proposed algorithms can efficiently improve the OWPT measurement time resolution and synchronize AP and MS in OWPT measurements.

5 Conclusion

This paper presents algorithms to improve the OWPT resolution to ns and to highly synchronize AP and MS in OWPT measurements. Lower time resolution and synchronization used to be two obstacles which limited the application of OWPT. The proposed algorithms utilize a ns resolution MS clock and a μs AP clock to measure Beacon frames propagation time and synchronization bias between an AP and an MS. The simulation results showed that the proposed algorithms can accurately measure most of propagation time and synchronization bias with ns precision. These proposed algorithms provide the possibility to apply OWPT measurements on WLAN for indoor localization. In the next stage, the proposed algorithms will be implemented on 802.11 WLAN hardware and the algorithms will be verified in experiments. Multipath effects should also be explored in the future to further improve the OWPT measurement accuracy.

References

1. Kushki, A., Plataniotis, K.N., & Venetsanopoulos, A.N. (2007). Kernel-based positioning in wireless local area networks. *IEEE Transactions on Mobile Computing, 6*(6), 689–705.
2. Olivera, V.M., Plaza, J.M.C., & Serrano, O.S. (2006). WiFi localization methods for autonomous robots. *Robotica, 24*(4), 455–461.
3. Ciurana, M., Barcelo-Arroyo, F., & Izquierdo, F. (2007). *A ranging system with IEEE 802.11 data frames*. Radio and Wireless Symposium, 2007 IEEE, (pp. 133–136). Long Beach, CA, United States.
4. Gunther, A., & Hoene, C. (2005). *Measuring round trip times to determine the distance between WLAN nodes* (vol. 3462, pp. 768–779). Waterloo, Ontario, Canada.
5. Golden, S.A., & Bateman, S.S. (2007). Sensor measurements for Wi-Fi location with emphasis on time-of-arrival ranging. *IEEE Transactions on Mobile Computing, 6*(10), 1185–1198.
6. Gast, M. (2005). *802.11 wireless networks: The definitive guide*. Sebastopol, CA: O'Reilly Media.

Chapter 18
Residual Energy Based Clustering for Energy Efficient Wireless Sensor Networks

Noritaka Shigei, Hiroki Morishita, Hiromi Miyajima, and Michiharu Maeda

Abstract This chapter describes two types of clustering for wireless sensor networks (WSNs). The goal of these methods is to prolong the life time of sensor nodes powered by batteries. The first type is a centralized method, and the second type is a distributed method. These methods effectively take into account the residual energy of sensor nodes. As a result, they achieve better performances than conventional methods such as LEACH, HEED and ANTCLUST.

Keywords Wireless sensor network · Energy efficient · Clustering · Residual energy · Vector quantization

1 Introduction

The wireless sensor network (WSN) technology is a key component for ubiquitous computing. A WSN consists of a large number of sensor nodes as shown in Fig. 1. Each sensor node senses environmental conditions such as temperature, pressure and light and sends the sensed data to a base station (BS), which is a long way off in general. Since the sensor nodes are powered by limited power batteries, in order to prolong the life time of the network, low energy consumption is important for sensor nodes. In general, radio communication consumes the most amount of energy, which is proportional to the data size and proportional to the square or the fourth power of the distance. In order to reduce the energy consumption, a clustering and data aggregation approach has been extensively studied [1]. In this approach, sensor nodes are divided into clusters, and for each cluster, one representative node,

N. Shigei (✉), H. Morishita, and H. Miyajima
Graduate School of Science and Engineering, Kagoshima University, Kagoshima 890-0065, Japan
e-mail: shigei@eee.kagoshima-u.ac.jp; k4368746@kadai.jp; miya@eee.kagoshima-u.ac.jp

M. Maeda
Fukuoka Institute of Technology, 3-30-1 Wajiro-higashi, Fukuoka 811-0295, Japan
e-mail: maeda@fit.ac.jp

S.-I. Ao et al. (eds.), *Intelligent Automation and Computer Engineering*,
Lecture Notes in Electrical Engineering 52, DOI 10.1007/978-90-481-3517-2_18,
© Springer Science+Business Media B.V. 2010

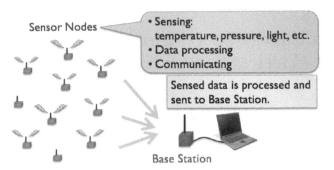

Fig. 1 The concept of wireless sensor network

which called cluster head (CH), aggregates all the data within the cluster and sends the data to BS. Since only CH nodes need long distance transmission, the other nodes save the energy consumption.

In order to manage effectively clusters and CHs, distributed clustering methods have been proposed such as LEACH, HEED, ACE and ANTCLUST [2–4, 8]. LEACH, which is the most popular method, guarantees that every nodes evenly become CHs but does not take into account battery level and the interrelationship among nodes [3]. HEED, ACE and ANTCLUST achieve better performance than LEACH by taking into account battery level, communication cost, node density, etc. In order to keep a dense and wide coverage area for the observation over the long term, node's residual energy has to be effectively reflected in clustering. Further, additional inter-node communications for determining clusters and CHs have to be small.

In this chapter, we propose two types of clustering methods based on node's residual energy, which have been originally proposed in [6] and which are generalized versions of the previous ones. The first type is a centralized method, and the second type is a distributed method. The centralized method employs vector quantization (VQ) [5] for effective clustering, and then determines representative nodes (CH: cluster head) based on a hybrid measurement of distance and residual energy. In the centralized method, BS determines clusters and CHs according to remaining energy level and node location. The second method determines clusters and CHs in a distributed autonomous fashion, and takes into account not only residual energy and but also node density. In the distributed method, clustering is performed by the interaction among proximity nodes, and the involved additional inter-node communication is also less. For both methods, effective values for importance factors, which indicate how much node's residual energy is weighted in clustering, are investigated by simulations. Further, the proposed methods with the effective values are compared with conventional methods LEACH, HEED and ANTCLUST. The effectiveness of the proposed methods are demonstrated by numerical simulation.

2 Wireless Sensor Network Model

This section describes the wireless sensor network (WSN) model considered in this chapter [2–4, 8]. The WSN model consists of N sensor nodes and one base station (BS) node as shown in Fig. 1. All sensor nodes are identical and are assumed to have the following functions and features: (1) sensing environmental factors such as temperature, pressure, and light, (2) data processing by low-power micro-controller, (3) radio communication, and (4) powered by a limited life battery. In this chapter, the remaining power of node's battery is also referred as residual energy. The BS node is assumed to have an unlimited power source, processing power, and storage capacity. The data sensed by sensor nodes are sent to the BS node over the radio, and a user can access the data via the BS node.

In this WSN application, the clock synchronization of sensor nodes is an important issue. Because the time at which a data was sensed is important, which requires low clock skew among all the sensor nodes. We assume that the low clock skew requirement is guaranteed by using a clock synchronization method [7].

The radio communication consumes more energy than the data processing on a sensor node. We assume the following energy consumption model for radio communication. The transmission of a k-bit message with transmission range d m consumes $E_T(k, d)$ of energy.

$$E_T(k, d) = \begin{cases} k(E_{\text{elec}} + \varepsilon_{\text{fs}}d^2) & \text{for } d \leq d_0 \\ k(E_{\text{elec}} + \varepsilon_{\text{mp}}d^4) & \text{for } d > d_0, \end{cases} \tag{1}$$

where E_{elec} is the electronics energy, and ε_{fs} and ε_{mp} are the amplifier energy factors for free space and multipath fading channel models, respectively. The reception of a k-bit message consumes $E_R(k)$ of energy.

$$E_R(k) = k \cdot E_{\text{elec}} \tag{2}$$

3 Clustering Approach for WSN

In order to save the energy consumption of WSN, a clustering approach for WSN as shown in Fig. 2 has been considered. In the approach, N sensor nodes are divided into clusters, and each cluster has a representative sensor node called cluster head (CH). Each non-CH sensor node sends the sensed data to the CH node in its own cluster, instead of to BS. Each CH node aggregates the received data into smaller size and sends it to BS. This approach has the following advantages: (1) non-CH sensor nodes can save the energy consumption because the nodes can avoid long-distance communication and have only to send data to its own CH being nearby and (2) the amount of data to be sent to BS can be reduced, which also saves the energy consumption.

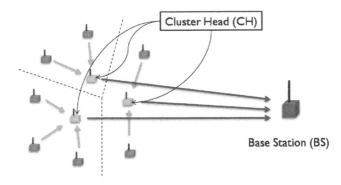

Fig. 2 The concept of the clustering approach for WSN

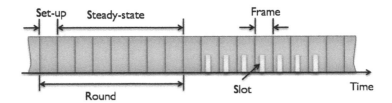

Fig. 3 The operating cycle in clustering methods

The operating cycle of clustering methods is shown in Fig. 3. Each round consists of consecutive frames. The first frame is for set-up, and the others are for steady-state. In the set-up frame, CH nodes and clusters are determined based on the used clustering algorithm, and each CH assigns a non-CH node to a slot in order to create time-division multiple-access (TDMA) schedule. In the steady-state frames, each non-CH node sends data to CH at the assigned slot in TDMA fashion, and CHs fuse (compress) the received data and send it to BS.

In order to decide CHs and clusters, clustering algorithms such as LEACH and HEED have been proposed [3, 8]. In LEACH, CHs are determined in a distributed autonomous fashion. At each round l, each node v independently decides to be a CH with probability $P_v(t)$ if the node v has not been a CH in the most recent $(l \bmod (N/k))$ rounds.

$$P_v(l) = \frac{k}{N - k\left(l \bmod \frac{N}{k}\right)}, \tag{3}$$

where k is the average number of CHs for each round. This means that each node becomes CH at least once every N/k rounds. However, LEACH does not take into account battery level and node distribution.

More effective clustering methods than LEACH have been proposed such as HEED, ACE and ANTCLUST [2, 4, 8]. However, they need additional inter-node communications for clustering. In the next two sections, we propose two types of clustering methods based on node's residual energy. The first method is a centralized method, and the second is a distributed method.

4   The Proposed Centralized Method

In this method, the BS node manages the clustering by utilizing a vector quantization (VQ) technique. The traditional VQ process approximates the distribution of the large set of vectors $X = \{x_1, \ldots, x_\nu\}$ by using a small set of vectors $W = \{w_1, \ldots, w_\kappa\}$, where $\nu \gg \kappa$ in general, $x \in X$ is called *input vector*, and $w \in W$ is called *weights* or *codebook vectors* [5]. In VQ, an input vector $x \in X$ is approximated by the nearest weight $w_{\iota(x,W)}$, where $||x - w_{\iota(x,W)}|| = \min_{w \in W} ||x - w||$. An adaptive VQ algorithm such as LBG and K-means trains the weight set so as to minimize the approximation error $E = \frac{1}{|X|} \sum_{x \in X} ||x - w_{\iota(x,W)}||^2$. Assuming that each $w \in W$ and each $x \in X$ are correspond to a CH node and a non-CH node, respectively. This assignment of CHs and clusters will minimize the total energy consumption of intra-cluster communication because the energy consumption of transmission is proportional to the squared distance as shown in Eq. 1.

As shown later, the direct application of VQ will not work because for the same X it always gives a same assignment, which brings specific nodes down quickly. In order to successfully apply VQ to the clustering of WSN, we introduce a hybrid metric of Euclid distance and residual energy level as follows:

$$d_H(x, w, e) = \frac{||x - w||}{e^\alpha}, \tag{4}$$

where e is the residual energy level of the node corresponding to x, and α is an importance factor of residual energy, which indicate how much node's residual energy is weighted.

We assume that the distance between a node and BS is informed by using data packets via CH. The algorithm is presented as follows:

Algorithm Centralized Clustering
Input:
$A = \{\text{All active nodes}\}$.
$X = \{\text{The coordinate of node } v \; x_v | v \in A\}$.
Step 1: Each active node $v \in A$ sends its coordinates x_v and its remaining battery level e_v to BS.
Step 2: BS performs the following VQ procedure.
(2-1) Initialize the weight set $W = \{w_1, \ldots, w_k\}$ by random numbers and $t \leftarrow 0$.
(2-2) Select randomly a node $v \in A$ with probability p_v.

$$p_v = \frac{e_v}{\sum_{u \in A} e_u} \tag{5}$$

(2-3) Find the nearest $w_k \in W$ to x_v, where $||x_v - w_k|| = \min_{w \in W} ||x_v - w||$.
(2-4) Update w_k as follows.

$$w_k \leftarrow w_k - \alpha(w_k - x_v). \tag{6}$$

(2-5) $t \leftarrow t + 1$. If $t = T_{\max}$ then go to Step 3. Otherwise go to (2-2).
Step 3: For each $k \in \{1, \ldots, K\}$, kth CH is assigned to the nearest node $v \in A$ to $w_k \in W$, where $d_H(x_v, w_k, e_v) = \min_{u \in A} d_H(x_u, w_k, e_u)$. Let $C = \{v \in A$ is CH.$\}$ be the set of CH nodes. Let $c(k) \in C$ denote the kth CH node.
Step 4: Each $v \in A \setminus C$ is assigned to kth cluster whose CH is the nearest CH from v, that is $||x_v - x_{c(k)}|| = \min_{j \in \{1,\ldots,K\}} ||x_v - x_{c(j)}||$.
Step 5: BS broadcasts the decided CH and cluster assignments.   □

In Step 3, the hybrid metric d_H is used for assignment of CH. The importance factor α is a key parameter in the algorithm. In order to find an effective value for α, we perform a simulation. In the simulation, N sensor nodes are randomly distributed in the square region of size 100×100 m and the base station is 75 m away from the center of a side as shown in Figs. 8 and 9. The simulation is performed for $N = 100$ and $N = 300$, and the parameters used in the simulation is summarized in Table 1. The simulation results are shown in Figs. 4 and 5. Six cases of $\alpha = 0.0 - 5.0$ and five cases of $\alpha = 1.0 - 5.0$ are examined for $N = 100$ and $N = 300$, respectively. Note that $\alpha = 0.0$ means node's residual energy is not taken into account at all. As α increases, the algorithm more emphasizes node's residual energy. The choice $\alpha = 0.0$ can keep alive some nodes for a very long term as shown in Fig. 4. However, a large proportion of nodes quickly go down, which means this choice cannot keep a dense and wide coverage area for the observation over the long term. For both $N = 100$ and $N = 300$, the choice $\alpha = 4.0$ achieves the longest term where more than 95% alive nodes are kept. Therefore $\alpha = 4.0$ is a reasonable choice.

Table 1 Conditions used in the simulations

For energy model	
d_0	75 m
E_{elec}	50 nJ/bit
E_{fusion}	5 nJ/bit
ε_{fs}	100 pJ/bit/m^2
ε_{mp}	1.3 fJ/bit/m^4
Initial battery level	0.5 J
Energy for data aggregation	5 nJ/bit/signal
For packet model	
Data packet size	800 bit
Broadcast packet size	200 bit
Packet header size	200 bit
For distributed method	
R_{inf}	20 m
R_{cnd}	55 m

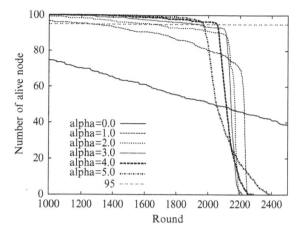

Fig. 4 The number of alive nodes versus round for different α's ($N = 100$)

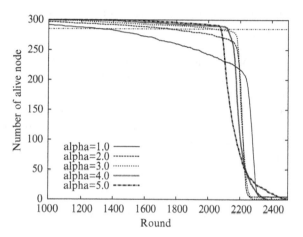

Fig. 5 The number of alive nodes versus round for different α's ($N = 300$)

5   The Proposed Distributed Method

The second proposed method is performed in a distributed autonomous fashion. In the method, the role of BS is just to receive sensed data from CHs. There exist some conventional distributed methods such as LEACH, HEED and ANTCLUST. The distinguished features of the proposed method are as follows: (1) aware of remaining battery level, (2) aware of node density, and (3) a small communication overhead in the clustering process. LEACH has a small communication overhead but is not aware of remaining battery level and node density. HEED and ANTCLUST are aware of remaining battery level and node density but their communication overheads are large.

Unlike our centralized version in the previous chapter, in our distributed version, each sensor node broadcasts its own existence within its proximity. Based on the proximity information and their own battery level, the CH nodes and the clusters are autonomously determined among sensor nodes. In the process, every nodes broadcast some messages at most twice within a small range of R_{inf} or R_{cnd} radius. The proposed algorithm consists of four phases which performed in a setup frame. In lth round, each node begins each phase $q \in \{1, 2, 3, 4\}$ at a specific time $(l-1)T_{rnd} + \sum_{j=1}^{q-1} T_j$, where $T_{rnd} = \sum_{j=1}^{4} T_j + T_{ss}$ and T_{ss} is the period of the stead state. The algorithm is presented as follows:

Algorithm Distributed Clustering
Phase 1: All active nodes broadcast their node IDs within R_{inf} meters radius. All nodes count how many IDs are received. Let m_v be the counted number for node v.
Phase 2: Each node $v \in A$ performs the procedure (**A**) in descending order of the following evaluation function

$$f(m_v, e_v) = m_v e_v^\beta, \tag{7}$$

where β is an importance factor of residual energy.

When each node $v \in A$ receives a candidacy for CH from other node, performs the procedure (**B**).
(**A**) If a node v has not received a candidacy for CH from other node, broadcasts its candidacy for CH within R_{cnd} meters radius.
(**B**) A node v accept the candidacy from other node.
Phase 3: The nodes that broadcast the candidacy in Phase 2 become CHs. The other nodes becomes non-CHs. The non-CH nodes send *intentions of participating* to the nearest candidate for CH.
Phase 4: Each CH node creates a TDMA schedule and send it to the non-CH nodes as the registration approval. □

Figure 6 shows examples of Phases 1 and 2 for $\beta = 4$. In Fig. 6a for Phase 1, all the active nodes broadcast their node IDs, and nodes 1, 2, 3 and 4 recognize one,

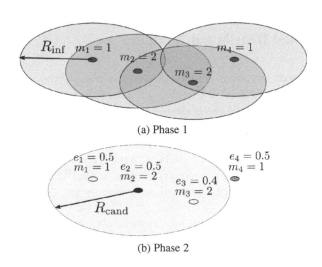

(a) Phase 1

(b) Phase 2

Fig. 6 Examples of phases 1 and 2 for $\beta = 4$

two, two and one neighboring nodes, respectively. In Fig. 6b for Phase 2, node 2 has the maximum $m_v \cdot e_v^4$ and broadcasts its candidacy for CH in the first place. As a result, nodes 1 and 3 give up their candidacy, and node 4 will broadcast its candidacy in time.

In the algorithm, the way to execute the phase 2 is not obvious. Let us explain the phase 2 in detail. The period of "phase 2" T_2 consists of the candidacy period T_{2c} and the post-margin T_{2m}, that is $T_2 = T_{2c} + T_{2m}$. The margin T_{2m} is inserted to prevent the effect of unavoidable clock skew and communication delay. In lth round, each node v broadcasts its candidacy for CH at time $t_{cnd}(l, m_v, e_v)$, where

$$t_{cnd}(l, m, e) = (l - 1)T_{rnd} + T_{2c}\left(1 - \frac{m \cdot e^\beta}{\gamma(l)}\right), \qquad (8)$$

and $\gamma(l)$ is a decreasing function with round l such that $\gamma(l) < \max_{v'} m'_v \cdot e'^\beta_v$. If $m_v \cdot e_v = \max_{v'} m'_v \cdot e'^\beta_v$ then $t_{cnd}(l, m_v, e_v) > t_{cnd}(l, m'_v, e'_v)$ for any $v' \neq v$, that is the first broadcast of candidacy is performed by the node v. If $m_v \cdot e_v^\beta = \min_{v'} m'_v \cdot e'^\beta_v$ then $t_{cnd}(l, m_v, e_v) < t_{cnd}(l, m'_v, e'_v)$ for any $v' \neq v$, that is the last broadcast of candidacy is performed by the node v. Further, if $m_v \cdot e_v > m'^\beta_v \cdot e'^\beta_v$ then the node v broadcasts earlier than the node v'.

In Phase 2, the evaluation function $f(m_v, e_v)$ with β is used for determining the order of candidacy. The importance factor β is a key parameter in the algorithm. In order to find an effective value for β, we perform a simulation same as in the previous section for $N = 100$ and 300. The simulation results are shown in Figs. 7 and 8. For both results, five cases of $\beta = 1.0$–5.0 are examined. As β increases, the algorithm more emphasizes node's residual energy. The simulation result shows that, a higher β provides a better performance and the performances for $\beta = 4.0$ and $\beta = 5.0$ seem to be almost same. Since the autonomous candidacy mechanism in the proposed method depends on the value of $m_v e_v^\beta$, a smaller β is preferable in terms of computational cost. Therefore, $\beta = 4.0$ is a reasonable choice.

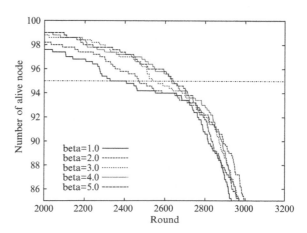

Fig. 7 The number of alive nodes versus round for different β's ($N = 100$)

Fig. 8 The number of alive
nodes versus round for
different β's ($N = 300$)

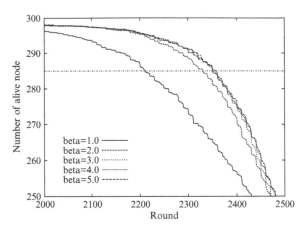

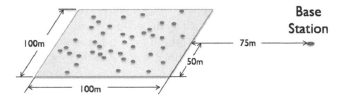

Fig. 9 The node arrangement in the simulation

6 Numerical Simulation

In this section, the effectiveness of the proposed methods is demonstrated by numerical simulation. The proposed methods are compared with the conventional methods LEACH, HEED and ANTCLUST. The simulation is performed for $N = 100, 300$ and 1,000 as in the previous sections. For the proposed methods, we use the parameter choices $\alpha = 4.0$ and $\beta = 4.0$, whose effectiveness is shown in the previous sections.

The simulation results for $N = 100, 300$ and 1,000 are shown in Figs. 10, 11 and 12, respectively. In the graphs, "Centralized" and "Distributed" are our proposed methods in Sections 4 and 5, respectively. For $N = 100$ and $N = 300$, our distributed method is the best and our centralized method is the second best. The difference between our proposed methods for $N = 300$ is small than for $N = 100$. For $N = 1,000$, our centralized method is the best and our distributed method is the second best. For any N, our proposed methods achieve better performance than the conventional methods LEACH, HEED and ANTCLUST. Our simulation result shows that our centralized and distributed methods are suitable for larger and smaller numbers of sensor nodes, respectively.

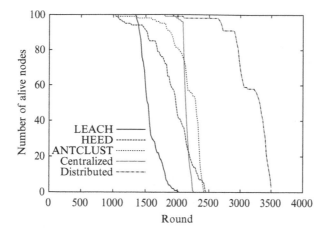

Fig. 10 The number of alive nodes versus round for $N = 100$

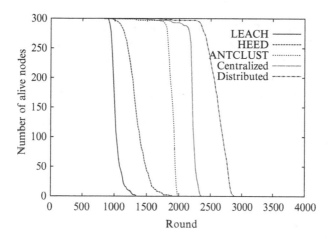

Fig. 11 The number of alive nodes versus round for $N = 300$

7 Conclusions

In this chapter, we described two types of clustering methods for WSNs. The first type, which is based on centralized management, employs vector quantization (VQ) for clustering and a hybrid metric of Euclid distance and residual energy. The second type, which is performed in a distributed autonomous fashion, takes into account remaining battery level and node density. Both methods involve importance factors α or β which indicates how much node's residual energy is weighted in the cluster head selection. The reasonable choices for α and β were shown by simulations, and the found ones are same as used in [6]. The effectiveness of the proposed methods were demonstrated in the numerical simulation. In the simulation, our proposed methods prolong the network lifetime longer than the conventional methods

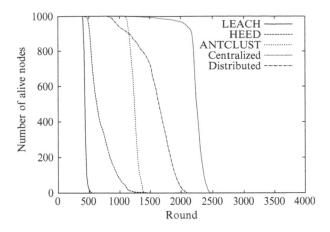

Fig. 12 The number of alive nodes versus round for $N = 1,000$

LEACH, HEED and ANTCLUST. Further, our simulation results show that the centralized method and the distributed method are suitable for larger and smaller sensor nodes, respectively. Our future works are theoretical analysis of the proposed methods, further improvement of prolonging performance, consideration on other models such as WSN models with solar cell, and evaluation on a WSN testbed.

Acknowledgments This work is supported by Grant-in-Aid for Scientific Research (C) (No. 20500070) of Ministry of Education, Culture, Sports, Science and Technology of Japan.

References

1. Abbasi, A.A., & Younis, M. (2007). A survey on clustering algorithms for wireless sensor networks. *Computer Communications, 30*, 2826–2841.
2. Chan, H., & Perrig, A. (2004). ACE: An emergent algorithm for highly uniform cluster formation. *Proceedings of the 1st Euro workshop sensor networks* (pp. 154–171).
3. Heinzelman, W.B., Chandrakasan, A.P., & Balakrishnan, H. (2002). An application-specific protocol architecture for wireless microsensor networks. *IEEE Transactions on Wireless Communications, 1*, 660–670.
4. Kamimura, J., Wakamiya, N., & Murata, M. (2006). A distributed clustering method for energy-efficient data gathering in sensor networks. *International Journal of Wireless and Mobile Computing, 1*, 113–120.
5. Linde, Y., Buzo, A., & Gray, R.M. (1980). An algorithm for vector quantizer design. *IEEE Transactions on Communications, 28*, 84–95.
6. Shigei, N., Miyajima, H., Morishita, H., & Maeda, M. (2009). Centralized and distributed clustering methods for energy efficient wireless sensor networks. *Proceedings of international multiconference of engineers and computer scientists, I*, 423–427.
7. Sundararaman, B., Buy, U., & Kshemkalyani, A.D. (2005). Clock synchronization for wireless sensor networks: a survey. *Ad Hoc Networks, 3*, 281–323.
8. Younis, O., & Fahmy, S. (2004). HEED: A hybrid, energy-efficient, distributed clustering approach for ad hoc sensor networks. *IEEE Transactions on Mobile Computing, 3*, 366–379.

Chapter 19
Fast Dissemination of Alarm Message Based on Multi-Channel Cut-through Rebroadcasting for Safe Driving

Pakornsiri Akkhara, Yuji Sekiya, and Yasushi Wakahara

Abstract Many applications in Vehicular Ad Hoc Networks (VANETs) are attracting a lot of research attention from academic community and industry, especially car industry, as a part of Intelligent Transport System (ITS). One important feature of the applications in VANETs is the ability to extend the line-of-sight of the drivers by the extensive use of on-board devices in order to improve the safety and efficiency of road traffic. However, due to mobility constraints and driver behaviors in VANETs, the broadcasting approaches used in Mobile Ad Hoc Networks (MANETs) cannot be properly applied to the applications in VANETs. Moreover, the conventional broadcasting methods for broadcasting an alarm message have a defect of long time required for the complete dissemination, which leads to the degradation in the safety of road traffic in case of emergency. We propose a new broadcasting method called Cut-Through Rebroadcasting (CTR) for alarm message dissemination scenarios based on the minimization of the number of rebroadcasting vehicles and the overlap rebroadcasting by making use of multiple-channels.

Keywords Alarm message · Broadcasting method · Multiple channels · Vehicular ad hoc network

1 Introduction

An interest has been increasing in the application of advanced information technology to transportation systems for providing improved comfort and additional safety in driving. Existing Intelligent Transport System (ITS) deployments, e.g. Advanced Cruise-Assist Highway System (AHS) [1], mainly rely on networks in the roadside infrastructure or Road-Vehicle Communication (RVC). While such systems provide substantial benefits, their deployment is very costly, which prevents

P. Akkhara (✉), Y. Sekiya, and Y. Wakahara
Department of Frontier Informatics, Graduate School of Frontier Sciences, The University of Tokyo, Information Technology Center 2–11–16 Yayoi Bunkyo-ku Tokyo Japan 113–8658
e-mail: pakornsiri@cnl.k.u-tokyo.ac.jp; sekiya@wide.ad.jp; wakahara@nc.u-tokyo.ac.jp

S.-I. Ao et al. (eds.), *Intelligent Automation and Computer Engineering*,
Lecture Notes in Electrical Engineering 52, DOI 10.1007/978-90-481-3517-2_19,
© Springer Science+Business Media B.V. 2010

them from reaching their full potential. Due to this problem, there is a trend of equipping vehicles with the communication technology allowing the vehicles to contact with other equipped vehicles in their vicinity, which is referred to as Inter-Vehicle Communication (IVC). IVC has two key advantages: low latency due to direct communication among vehicles and broader coverage beyond areas where roadside infrastructure equipments have been deployed.

The vehicles with such IVC capability form ad hoc networks called Vehicular Ad Hoc Networks (VANETs). Their specific characteristics allow the development of Comfort Application and Safety Application [2]. Although much effort is needed in order to make these applications reality, methods to disseminate various messages seem to be one of the most important challenges. In addition, the huge social and economical cost related to road accidents makes research of proactive safety services a task of primary importance in the ITS. A fundamental application for providing this safety service is the fast and reliable propagation of an alarm or warning message to the upcoming vehicles in case of hazardous driving situations such as accidents and dangerous road surface conditions. However, the existing broadcasting methods have some serious defects such as long delay for the complete propagation and high cost due to the use of wide frequency band. Because of these serious problems, they have not yet been actually put into wide commercial use.

In order to solve such problems, we propose a method that can reduce the broadcasting time required for the alarm or warning message propagation by utilizing multiple channels available e.g. in IEEE 802.11 standard as well as the Global Positioning System (GPS). This proposed method is called Cut-Through Rebroadcasting (CTR).

2   Alarm Message Broadcasting

As shown in Fig. 1, when a vehicle located in the middle of a road has an accident and recognizes itself as crashed by using some sensors that detect events like airbag ignition, this vehicle will start to broadcast an alarm message to propagate the information about its accident to nearby vehicles as shown by arrows. It will be possible for the drivers of other vehicles to take suitable actions to avoid the secondary accident by using this information. However, in order to guarantee safety, the following two factors have to be considered.

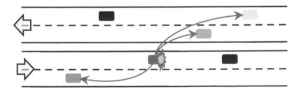

Fig. 1 Alarm message
broadcasting application

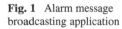

- Maximum allowed speed of the vehicle is about 100 km/hr (according to the country regulation). Consequently, the vehicle has very short period of time for communication with other vehicles encountered on the road.
- Human reaction time is 0.3 s, but 0.1 s will be used for the acquisition of the information by various sensors and 0.1 s for processing the information. Therefore, at most only 0.1 s is left for vehicle-to-vehicle communication [1]. Furthermore, if the acquiring and processing the information cannot be achieved in 0.2 s, the vehicle to vehicle communication has to be done in less than 0.1 s.

Based on these two factors, this alarm information is judged to have a very short useful lifetime. Thus, the information about the accident should reach the concerned vehicles with low delay and high reliability.

3 Related Works

Many broadcasting methods have been proposed, and Flooding seems to be the simplest among them. However, this method has some problems such as high collision or contention probability and high data redundancy because every vehicle receiving the message has an obligation to immediately rebroadcast the message to all of its neighbors. This can result in inefficiency in terms of radio resource usage, promptness of the message delivery and reliability, which has been referred to as Broadcast Storm Problem [2]. Consequently, a lot of broadcasting methods have been proposed in order to solve this problem and they can be taken as candidates for the alarm message broadcasting application. However, they have in practice serious problems from the viewpoint of the characteristics of VANETs as follows:

Probability Based Method [3, 4] In the probability based method, each vehicle decides to rebroadcast the message with some probability in order to decrease data redundancy and collision. Although the required average broadcasting time is rather short, this method still cannot entirely solve the redundancy problem. Moreover, its delivery ratio is generally rather low depending on the probability, which leads to the serious problem of low reliability.

Area Based Method In this method, each vehicle decides to rebroadcast the message by considering the additional coverage area of the transmission range achieved by the rebroadcasting. In the Distance Based Scheme [2, 5], when a vehicle receives the message, the vehicle will not rebroadcast the message if the distance between itself and its nearest neighbor vehicle which has previously rebroadcasted the same message is smaller than a predetermined threshold because the rebroadcasting is judged redundant. The Location Based Schemes [5–10] make more precise estimation of the additional coverage area by making use of the means to determine its own location, e.g. GPS.

An Area Based Method named Optimized Dissemination of Alarm Message (ODAM) [10] assigns the duty of message rebroadcasting to only furthest neighbor from the source vehicle in order to ensure the largest additional coverage area which has not yet been covered by the source vehicle. The intermediate vehicles that receive the message should not rebroadcast the message immediately. Instead, these vehicles must wait for some waiting time, whose duration length is inversely proportional to the distance between itself and the source vehicle. At the expiration of the waiting time, if a vehicle has not received the same message coming from another vehicle, it rebroadcasts the message. Although this method is considered efficient in terms of overhead cost and redundancy, it does not take into account a tight time delay constraint of the alarm message broadcasting application and thus its required time for the complete propagation of the message is rather long.

Cluster Based Method [11] In this method, all the related vehicles are structured into some clusters and the task for rebroadcasting the message is assigned to only the cluster head vehicle of each cluster. Although this method can work efficiently, the cost to create and maintain the cluster structure is rather high because of high speed move of the vehicles, which leads to the large traffic overload and long delay of the message propagation in general.

Topology Based Method [12, 13] Topology based methods are based on the complete knowledge of the network topology which is obtained by exchanging the control messages beforehand. Although this method is efficient in terms of redundancy and collision reduction, this method is not considered feasible in the VANETs because the high control traffic load is required just like cluster based method.

Cut-through Based Method According to the strict delay constraint of safety application, the approaches with effectively shortened forwarding latency e.g. cut-through forwarding method are required. The cut-through forwarding method has been used in the packet switch technology to allow frame (or packet) forwarding before the frame is entirely received [14]. Unfortunately, the cut-through forwarding method has not been studied for wireless networks until recently because, in general, forwarding latency was not the primary concern for the traffic in the wireless networks. However, this is not the case for the safety application. One of the broadcasting methods that utilize cut-through forwarding has been proposed by Shagdar [15]. In this broadcasting method, each vehicle that received the message has an obligation to rebroadcast the message. Thus, the Multiple Access Interference (MAI) increases in accordance with the number of simultaneously rebroadcasting vehicles. Moreover, the wide bandwidth is required for the proposed Code Division Multiple Access (CDMA).

4 Cut-Through Rebroadcasting for Alarm Message

It is to be noted that all the existing conventional broadcasting methods except the cut-through based method use only a single frequency channel and make no use of the rest channels that are actually available e.g. in IEEE 802.11 standard. However, in order to achieve the targets of safety application, we propose CTR method that utilizes multiple channels available e.g. in IEEE 802.11 standard as well as GPS function. In CTR, high priority to rebroadcast the message is given to some specific vehicles to avoid the interference problem and the multiple channels are utilized effectively to achieve overlap broadcasting.

4.1 The Characteristics and Assumptions for VANETs

We focus on the alarm message broadcasting in the highway scenario where there are a number of vehicles moving towards both directions of the highway with possibly multiple lanes. In this scenario, the alarm message will be destined to many or all of the vehicles located away from the accident vehicle (source vehicle) and in less than some predetermined coverage distance. In other words, the position information will be used as an attribute to limit the broadcasting process. It is assumed that the highway is rectilinear and that there are no obstacles for the radio wave propagation along the highway e.g. buildings on the road.

All the vehicles are assumed to be equipped with sensing, calculation, communication capabilities and GPS so that each vehicle can sense an accident, gather information about the accident, transmit the alarm message to the nearby vehicles, and determine its own position relatively to the other vehicles. Moreover, each vehicle is equipped with at least two half-duplex transceivers based on e.g. IEEE 802.11 standard and a dedicated channel is assigned to each transceiver. With this assignment, the vehicle can transmit a message on one channel and listen to and receive a different message on the other channel at the same time. Furthermore, all the antennas are assumed non-directional.

4.2 Targets to be Achieved

Efficiency of the alarm message broadcasting method can be measured in general by whether the following targets can be achieved or not.

- According to the aforementioned human reaction time, the time required for all the vehicles located in the predetermined coverage distance to receive the alarm message completely is shorter than 0.1 s.
- Since the alarm message is broadcasted in a multi-hop manner, the number of vehicles that newly receive the alarm message in each hop should be as large as possible and thus the number of rebroadcasting vehicles should be smallest.

4.3 Cut-Through Rebroadcasting

The basic idea is to give high priority to the furthest vehicle in the transmission range from the source vehicle to rebroadcast the alarm message after recognizing it from

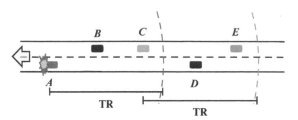

Fig. 2 Alarm message broadcasting scenario

its header. This priority control leads to the avoidance of collision of rebroadcasted alarm messages by the vehicles in the transmission range of the source vehicle by suppressing the rebroadcasting of vehicles with low priority and by making only the vehicle with high priority rebroadcast the message. Moreover, this method is characterized by utilizing cut-through-like forwarding approach or the overlap operation of alarm message transmission by some vehicles under the assumption that each vehicle is equipped with at least two transceivers and different channels are assigned to the transceivers in the individual hops to avoid the collision in broadcasting. For overlap broadcasting for more than 2 hops, at least three different channels are required for efficient transmission without interference.

In Fig. 2, A is assumed to have just had an accident, and B, C and D, E are assumed to be in the transmission range of the transceivers equipped on A and in that of the transceivers equipped on C respectively. After A recognizes an accident event based on the information received from various sensors, A acts as the source vehicle and starts to broadcast an alarm message to notify nearby vehicles including B and C of the accident. After recognizing that the received message is the alarm message, B and C calculate their own waiting times $T_{wait}(B)$ and $T_{wait}(C)$, respectively. The waiting time is used by each vehicle to make decision on whether it should be responsible for rebroadcasting the alarm message in the next hop or not. It should be remarked that the waiting time is longer for vehicles that are closer to the source vehicle. The details of waiting time calculation will be described later. When a waiting time of a vehicle expires and it has not received any alarm message from any other following vehicles, it starts to rebroadcast the alarm message in the following hop. On the other hand, if a vehicle has received an alarm message from any other following vehicles before the expiration of its waiting time, then it decides not to rebroadcast the alarm message. In Fig. 3, the furthest vehicle C will have priority to rebroadcast the alarm message and start to rebroadcast the alarm message just after the expiration of its waiting time by utilizing a channel which differs from the one used by the source vehicle A in order to avoid the interference of the messages and to reduce the broadcasting time.

Then, C becomes a source vehicle in the next hop to rebroadcast the alarm message and then in almost the same manner, only E will have high priority to rebroadcast the message. Such rebroadcasting will be repeated to cover all the vehicles in the predetermined coverage distance from the original source vehicle A.

Fig. 3 Basic idea of
Cut-Through Rebroadcasting

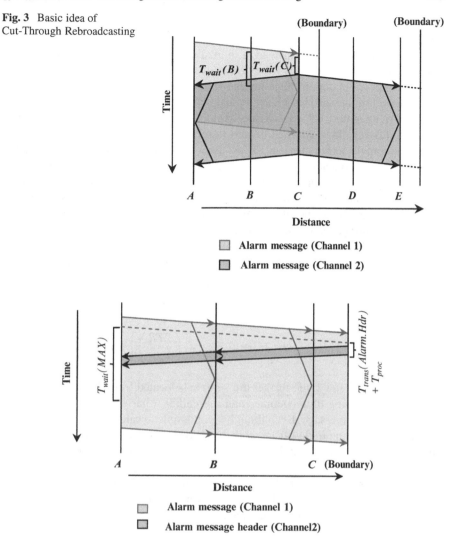

Alarm message (Channel 1)

Alarm message (Channel 2)

Alarm message (Channel 1)

Alarm message header (Channel2)

Fig. 4 Basic idea of waiting time calculation

4.4 Waiting Time Calculation

The waiting time calculation is based on the basic idea that the header of the alarm
message sent by the furthest vehicle, which should be responsible for rebroadcasting
the alarm message, should arrive at the vehicles located closer to the source vehicle
before the waiting time expiry of these vehicles. This basic idea can be elaborated
as follows.

In Fig. 4, after an intermediate vehicle, which is located in the transmission range
of A, recognizes that the message it has started to receive is the alarm message

broadcasted by A from its header, the vehicle should wait for some time to be notified whether there is a further vehicle which will responsible for rebroadcasting the alarm message or not instead of immediate rebroadcasting. This notification is achieved by the recognition of the header of the alarm message rebroadcasted by the further vehicle if any. By this approach, it becomes possible to avoid the collision of the alarm message broadcasting and achieve the largest additional coverage distance. An imaginary vehicle is assumed at the boundary of the transmission range of A, and this vehicle is assumed to start to rebroadcast the alarm message just after recognizing the alarm message. Thus, the waiting time of intermediate vehicles should be defined so that they wait long enough to receive the alarm message header rebroadcasted by this assumed vehicle. Furthermore, in the designing of the waiting time, the time required for the transmission, propagation and processing of header of the alarm message should be taken into account and the waiting time should become longer for an intermediate vehicle closer to the source vehicle.

In general, the waiting time for vehicle X located D_{SX} away from the source vehicle S can be represented by the following equations with a parameter Delta (Δ):

$$T_{wait} = \frac{(TR - D_{SX})}{TR} \times T_{wait}(MAX) \tag{1}$$

$$T_{wait}(MAX) = \left[T_{trans}(Alarm.Hdr) + T_{proc} + \left(\frac{2TR}{V_{prop}} \right) \right] \times (1 + \Delta) \tag{2}$$

By utilizing Eqs. 1 and 2, the further the vehicle is located from the source vehicle, the shorter its waiting time becomes and the earlier it has a chance to access the channel to send its message. Even though it is possible to calculate the waiting time by other methods, the trade-off between the broadcasting time and the number of rebroadcasting vehicles has to be taken into account. This trade-off is discussed in Section 5.

After various experiments, 0.0 is basically chosen as the value of Delta based on the consideration about the broadcasting time required to cover the coverage distance. Some discussion on this choice is also given in Section 5.

5 Evaluation Results

5.1 Evaluation Scenario

The average broadcasting time and the number of rebroadcasting vehicles of CTR, ODAM, and the pure flooding method are comparatively evaluated under NS-2 simulation environment in order to demonstrate the efficiency of CTR. ODAM is selected for this comparison because it is considered most efficient for the alarm message broadcasting application among the conventional broadcasting methods as described in Section 3.

Figure 5 depicts the simulation scenario of a straight highway with one lane where the distance between any two consecutive vehicles is randomly chosen from the values between two predetermined distances. In this scenario, the alarm message will be rebroadcasted in the multi-hop manner until it becomes possible to cover the predetermined coverage distance from the source vehicle. In addition, as mentioned in Section 4, each vehicle must be equipped with at least 2 transceivers assigned with different channels. In the simulation, the time required for a vehicle which is located furthest from the source vehicle within the coverage distance to completely receive the alarm message is evaluated as the broadcasting time for various transmission range values. The simulation parameters and their values are shown in Table 1.

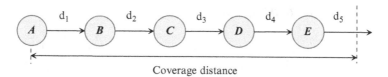

Coverage distance

Fig. 5 Simulation topology

Table 1 Simulation parameters and their values in NS-2 environment

Simulation parameters	Value
Data speed	1 Mbps
Radio propagation speed	3×10^8 m/s
MAC layer	CSMA/CA
Propagation model	Two-ray ground
Antenna type	Omni antenna
Processing time required for sending from application layer to MAC layer	0.075 ms
Processing time required for sending from MAC layer to application layer	0.025 ms
DCF Interframe Space (DIFS)	0.050 ms
Contention window	31 (default)
Slot time	0.020 ms
Alarm message size	500–2,500 bytes (default 1,425 bytes)
Alarm message header size	43 bytes
Transmission range	100–500 m (default 250 m)
Distance between two consecutive vehicles	20–40, 40–60, 60–80 m (default 20–40 m)
Speed of vehicle	20–27 m/s
Delta (Δ)	−1.0–15.0 (default 0.0)
No. of channels	1 channel (Flooding, ODAM); 2, 3 channels (CTR) (default 3 channels)
Coverage distance	1,000, 3,000 m (default 1,000 m)
Number of lanes	1–5 lanes (default 1 lane)
Lane width	3.5 m
No. of repetitions for simulation	100 times

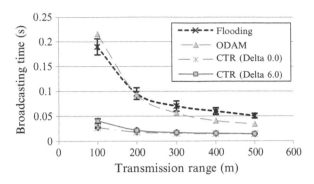

Fig. 6 Average broadcasting time required to cover the coverage distance

5.2 Average Broadcasting Time

Figure 6 shows the average broadcasting time of the above mentioned three methods for the case where the coverage distance is 1,000 m and the distance between two consecutive vehicles is randomly chosen from the values between 20–40 m. Error bars in Fig. 6 and the following figures show the 95% confidence interval of the results concerned. It is understood that CTR can achieve the broadcasting time shorter than the pure flooding method, by reducing the possibility of the collision in the alarm message rebroadcasting and giving high priority to rebroadcast the alarm message to the furthest vehicle in the transmission range from the source vehicle.

Compared with ODAM, CTR can achieve significantly shorter average broadcasting time by allowing the overlap in the alarm message broadcasting and rebroadcasting just after the expiration of the waiting time, while a vehicle in ODAM has to completely receive the alarm message and wait for the expiration of waiting time before they can make a decision on whether to rebroadcast the alarm message.

Because CTR starts to rebroadcast the alarm message after the expiration of waiting time, the average broadcasting time mainly consists of the transmission time of the alarm message header, its processing time, waiting time in each hop and the transmission time of the alarm message in the last hop, which are about 0.76 ms, less than 0.77 ms, less than 1.53 and 11.8 ms, respectively.

According to Fig. 6, CTR can achieve much shorter than 0.1 s average broadcasting time for every transmission range in the evaluation. The average broadcasting time of CTR decreases as the transmission range increases due to the decrease in the number of hops in the alarm message rebroadcasting to cover the coverage distance. Additionally, the average broadcasting time of CTR decreases as the value of Delta decreases due to the decrease in the value of $T_{wait}(MAX)$ used in the waiting time calculation.

5.3 Number of Rebroadcasting Vehicles

Figure 7 illustrates the total number of vehicles which were located in the coverage distance 1,000 m and rebroadcasted the alarm message by the above mentioned three

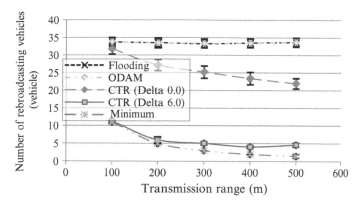

Fig. 7 Average number of rebroadcasting vehicles required to cover the coverage distance

methods and also the theoretically minimum value of the number of rebroadcasting
vehicles for reference. By giving priority control in the alarm message rebroadcasting, the number of rebroadcasting vehicles of CTR is significantly smaller than those
of ODAM and the pure flooding method which obliges every vehicle to rebroadcast
the alarm message. The reason why the number of rebroadcasting vehicles by CTR
where Delta is 6.0 is a little larger than the theoretically minimum value and is
smaller than the case where Delta is 0.0 is as follows.

According to $T_{wait}(MAX)$ in Eq. 2, because propagation time of a message
between two consecutive vehicles is negligibly small in comparison with other
required time, $T_{wait}(MAX)$ can be considered as a constant value. With the same average distance between two consecutive vehicles, the difference between the waiting
times of any two consecutive vehicles, which can be calculated by Eq. 1, decreases
as the transmission range increases. In addition, $T_{wait}(MAX)$ decreases as the value
of Delta decreases, resulting in decrease in the difference between the waiting time
of any two consecutive vehicles as well. Thus, the possibility that the leading vehicles will not receive the header of the alarm message from the rebroadcasting
vehicle before the expiration of their waiting time and start to rebroadcast the alarm
message will increase, resulting in larger number of rebroadcasting vehicles.

Although it is possible to increase the value of $T_{wait}(MAX)$, the average broadcasting time increases accordingly. Thus, the trade-off between the average
broadcasting time and the number of rebroadcasting vehicles has to be considered in the waiting time calculation.

According to the influence of the value of Delta on the average broadcasting time
and the number of rebroadcasting vehicles, choosing 0.0 as the value of Delta can
achieve the shortest broadcasting time when compared with other positive values of
Delta. In addition, the broadcasting time when 0.0 is chosen as the value of Delta
is not significantly different from other negative values of Delta but the smaller
number of rebroadcasting vehicles can be achieved.

It should be noted that CTR may be improved for reducing the number of rebroadcasting vehicles by the cancellation of rebroadcasting the alarm message after

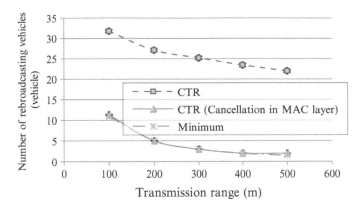

Fig. 8 Improvement of CTR by the cancellation of rebroadcasting the alarm message suspended by the operation like CSMA/CA

the reception of the header of the alarm message if the rebroadcasting has not yet actually started and has been suspended by the operation like CSMA/CA. However, the issues about layer violation have to be considered. According to Fig. 8, by allowing the cancellation of rebroadcasting the alarm message suspended by the operation like CSMA/CA in the MAC layer, CTR can achieve the optimum number of rebroadcasting vehicles and the minimum number of hops required to cover the whole coverage distance.

6   Conclusion and Further Researches

Reducing the broadcasting time of the alarm message will allow the drivers of the vehicles moving toward the accident place to have more time for a decision on the suitable action, resulting in more safety alarm message broadcasting application.

We have proposed a new broadcasting method in order to achieve such reduction in the broadcasting time by making use of multiple channels and GPS capability. This method is named Cut-Through Rebroadcasting (CTR). CTR can be characterized by the fact that the high priority to rebroadcast the alarm message is given to only the furthest vehicle within the transmission range. CTR can greatly reduce the broadcasting time mainly because of this priority control and the overlap rebroadcasting of the alarm messages by two or three vehicles. The resultant broadcasting time is well below the upper limit of 0.1 s even when the coverage distance is e.g. 3,000 m. Moreover, CTR is able to solve the Broadcast Storm Problem as well. In addition, CTR may be improved for reducing the number of rebroadcasting vehicles by the cancellation of rebroadcasting the alarm message after the reception of the header of the alarm message if the rebroadcasting has not yet actually started and has been suspended by the operation like CSMA/CA in case of collision. However, more study is required to confirm and evaluate this improvement in detail. Future study is also considered important to minimize the waiting time of vehicles in order to further decrease the broadcasting time.

We have an assumption of rectilinear road. If the road is not rectilinear and has a shape of e.g. curve, then some vehicles in the coverage distance may not be able to receive the alarm message according to so far proposed efficient broadcasting method. Thus, our future researches will focus on the approaches to cope with such a problem.

References

1. Advanced Cruise-Assist Highway System Research Association (AHSRA) (2008). Development of Road-to-Vehicle Communication Systems. Available: http://www.ahsra.or.jp/eng/c04e/index.htm.
2. Tseng, Y., Ni, S., Chen, Y., & Sheu, J. (Mar–May 2002). The broadcast storm problem in a mobile ad hoc network. *Wireless Networks, 8*(2/3), 153–167.
3. Williams, B., & Camp, T. (2002). Comparison of broadcasting techniques for mobile ad hoc networks. *The 3rd ACM International Symposium on Mobile ad hoc Networking and Computing*, pp. 194–205. Lausanne, Switzerland.
4. Alshaer, H., & Horlait, E. (2005). An optimized adaptive broadcast scheme for inter-vehicle communication. *The 61st IEEE Vehicular Technology Conference 2005*, VTC 2005-Spring, Stockholm.
5. Osafune, T., Lin, L., & Lenardi, M. (2006). Multi-Hop Vehicular Broadcast (MHVB). *The 6th International Conference on ITS Telecommunications 2006*, Chendu, China.
6. Briesemeister, L., & Hommel, G. (Aug 11, 2000). Role-based multicast in highly mobile but sparsely connected ad hoc networks. *First Annual Workshop on Mobile ad hoc Networking and Computing 2000* (MobiHOC. 2000) (pp. 45–50).
7. Sun, M., Feng, W., Lai, T., Yamada, K., Okada, H., & Fujimura, K. (2000). GPS-based message broadcasting for inter-vehicle communication. *International Conference on Parallel Processing* (pp. 279–286).
8. Fasolo, E., Furiato, R., & Zanella, A. (Sept 2005). Smart broadcast algorithm for inter-vehicular communications. *IWS 2005/WPMC 2005*, Aalborg, Denmark.
9. Fasolo, E., Zanella, A., & Zorzi, M. (2006). An effective broadcast scheme for alert message propagation in vehicular ad hoc networks. *IEEE International Conference on Communications 2006* (ICC '06). 11–15 June, Istanbul, Turkey.
10. Benslimane, A. (2004). Optimized dissemination of alarm message in vehicular ad-hoc networks (VANET). *The 7th IEEE International Conference HSNMC 2004*, pp. 655–666. Toulouse, France: Springer.
11. Little, T.D.C., & Agrawal, A. (Sept 13–16, 2005). An information propagation scheme for VANETS. The 8th international *IEEE Conference on Intelligent Transportation Systems*.
12. Zenella, A., Pierobon, G., & Merlin, S. (2004). On the limiting performance of broadcast algorithms over unidimensional ad-hoc radio networks. *WPMC'04*, Italy.
13. Wan, P.J., Alzoubi, K., & Frieder, O. (2002). Distributed construction of connected dominating set in wireless ad hoc networks. *IEEE INFOCOM'2002*. June 23–27, New York, USA.
14. Intel, Switches – What are forwarding modes and how do they work? Available: http://support.intel.com/support/express/switches/sb/cs-014410.htm.
15. Shagdar, O., Shirazi, M.N., Tang, S., Suzuki, R., & Obana, S. (2007–2010). Improving reliability of cut-through packet forwarding in CDMA vehicular network. IEICE Technical Report, NS2007–86.

Chapter 20
Lightweight Clustering Scheme for Disaster Relief Wireless Sensor Networks

Hiroshi Mineno, Yi Zheng, and Tadanori Mizuno

Abstract Most wireless sensor networks are driven with a battery. Lifetime maximization is thus very important when designing such networks. Clustering a network is an effective topology control scheme to enhance energy efficiency and scalability of large-scale wireless sensor networks. However, when using a wireless sensor network after a disaster, if sensor nodes cannot join any clusters, the detection area of the wireless sensor network is narrowed. Maximizing lifetime and minimizing the number of sensor nodes, that cannot join clusters is very important. In this work, we propose a lightweight clustering scheme for wireless sensor networks used in disaster relief. Simulation results show that the proposed clustering scheme is efficient and effective for maximizing the lifetime of a network and the number of communicable sensor nodes compared with other traditional clustering schemes.

Keywords Wireless sensor networks · Clustering · Lifetime · Disaster relief

1 Introduction

A rescue operation immediately following a disaster is extremely important for saving lives. Following the Hanshin-Awaji Earthquake, the survival rate after the second day was very low (Fig. 1) [1].

By analyzing many big earthquakes in history, survival rates were found to become extremely low after 72 h, so the first 72 h after an earthquake are critical. However, immediately after such a disaster, especially during the first day, people panic easily, and rescue operations are often inefficient. A wireless sensor network can be used to improve the efficiency of rescue operations in which the number of rescue workers is limited, because it can immediately detect a survivor by

H. Mineno (✉), Y. Zheng, and T. Mizuno
Department of Computer Science, Shizuoka University, 3-5-1 Johoku, Naka-ku,
Hamamatsu 432-8011, Japan
e-mail: mineno@inf.shizuoka.ac.jp; teiiz@mizulab.net; mizuno@inf.shizuoka.ac.jp

S.-I. Ao et al. (eds.), *Intelligent Automation and Computer Engineering*,
Lecture Notes in Electrical Engineering 52, DOI 10.1007/978-90-481-3517-2_20,
© Springer Science+Business Media B.V. 2010

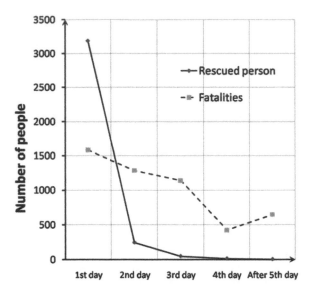

Fig. 1 Hanshin-Awaji earthquake survival rates versus rescue time

identifying conditions such as sounds, vibrations, and concentration of carbon dioxide without a cable network infrastructure. However, sensor nodes used in wireless sensor networks are generally battery-driven, thus limiting the lifetime of the network. The clustering protocol has been proposed as an effective way to reduce the energy consumption of a wireless sensor network. If a sensor node is not in the area of any cluster, it cannot send its sensing data to a cluster head, and an event that occurs in its sensing area cannot be reported. The area of detection of the entire wireless sensor network is related to the number of communicable sensor nodes, which is influenced by the position of the cluster head. Each sensor node lifetime is the time until its remaining energy becomes 0. The network lifetime is the time when the first sensor node in the network depletes its energy. In this paper, we propose the Lightweight Clustering Scheme (LiCS), which selects cluster heads to cover all sensor nodes using a lightweight algorithm without exact location information on sensor nodes, and we evaluate the performance via simulation.

2 Related Work

There have been several studies on reducing battery power consumption and extending network lifetime. Other methods have been proposed to maintain the detection capability of wireless sensor networks.

2.1   Sensor Node with Sleep Mode

Turning off sensor nodes can extend a wireless sensor network's lifetime. A MAC layer protocol named S-MAC [2] was proposed, which sets sensor nodes into sleep mode to reduce energy consumption by a trade-off with latency. S-MAC uses a static duty cycle with adaptive listening. In contrast, T-MAC [3], which uses an adaptive duty cycle, and D-MAC [4], which renews active periods adaptively, have been proven to be more effective than S-MAC via simulation. For rare-event detection, if the incidence of events is low and event duration is long, for example during a forest fire, the sampling interval of the sensor node is set to operate for a longer period of time. Then microcontrollers that control the sensor node and the communication modules switch to sleep mode, and power consumption can be controlled. However, a long sampling interval disables sensor nodes from detecting events immediately, so the detection delay is longer.

When two or more sensor nodes are arranged in the detection radius of an event such as a fire, a different sleep mode time is set for each sensor node, and sleep scheduling is optimized to minimize the detection delay [5].

Research on the sleep mode of the wireless sensor node has been performed, but little research has been done on applications used in rescue operations after disasters such as earthquakes when there are incidences of high frequency events with short durations. An event may not be detected by a technique that increases the sampling interval time, and for that reason, this technique is not practical for rescue operations.

2.2   Clustering-Based Protocol

In multihop wireless sensor networks, when an event takes place in a detection area, the sensor node that detected the event generates and sends a information to a SINK node, which then analyzes the information. When a sensor node cannot transmit this information to a SINK node in a single hop, the information must be sent to the parent node of the sensor node. To relay the information, the parent node consumes more energy during multihop. This makes the parent node's lifetime shorter than that of an end sensor node.

In clustering-based wireless sensor networks, where the whole network is divided into many clusters, a cluster member transmits the sensing information to a cluster head, which relays the information like a parent node in a multihop wireless sensor network. The operation in clustering protocols is to select a set of cluster heads from the nodes in the network.

Low-Energy Adaptive Clustering Hierarchy (LEACH) [6] is a commonly used clustering-based protocol. In LEACH, repeated selection of cluster heads is performed by using a probability, which means each sensor node takes the role of cluster heads in a random rotating shift. Normal LEACH is not based on exact location information of sensor nodes.

GPS-based LEACH (LEACH-C) was also proposed because electing cluster heads in an appropriate place will increase the lifetime of a wireless sensor network. LEACH assumes that all nodes can communicate directly with cluster heads, so the protocol is not suitable for large-scale applications in wireless sensor networks.

To enable the clustering protocol to handle a large-scale application, clusters with cluster communication ranges are proposed that only sensor nodes that are in the cluster communication range can become cluster members. Exact location information, for example, GPS, was used to select cluster heads, which can make all sensor nodes capable of cluster members [7, 8].

In addition, localization algorithms without exact location information are proposed. The weighted clustering algorithm (WCA) [9] elects a cluster head by using parameters such as distance and angle. The hybrid, energy-efficient, distributed clustering approach (HEED) [10] elects cluster heads by using parameters such as the number of neighbor nodes and the power consumption of the communication module. Furthermore, if the sensing data packets are locally correlated, then they can be compressed and fused into shorter data packets at the cluster heads for more energy-efficient transmission [11].

Researchers have shown that a clustering-based protocol is highly effective in a network simulation. The error of GPS is about 10 m, which is too large in disaster relief. In addition, the power consumption of the localization module and the increase in the sensor node's cost cannot be disregarded. On the other hand, parameter exchanges in localization algorithms increase the traffic overhead of the network and require a high performance SINK node such as a computer. However, preparing a lot of computers after a disaster is difficult. A lightweight clustering scheme with a simple algorithm is therefore needed for rescue operations.

3 Lightweight Clustering Scheme

3.1 Network and System Models

In this work, we consider a wireless sensor network to be similar to the network model used in other studies [6, 10]. We assume the following properties about wireless sensor networks.

- All sensor nodes are stationary and uniformly distributed in a field.
- There is only one SINK node in the wireless sensor network. The SINK node does not have an energy constraint. For example, the SINK node can be connected to a large-capacity battery or some sort of energy-charge system such as solar panels.
- The network topology has a two-layer structure (Fig. 2). The lower layer consists of one cluster head and many cluster members. The upper layer consists of may cluster heads and a SINK node. In our work, a cluster head cannot change its state to sleep mode, and an end sensor node can change its state to sleep mode

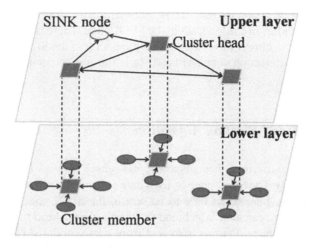

Fig. 2 Network and system model

Election phase	Data collection phase	Election phase	Data collection phase	Election phase

1 round

Fig. 3 A time interval called a "round"

when it does not need to transmit any sensing information. In the lower layer, an end sensor node transmits sensing information to a cluster head in a single hop. In the upper layer, cluster heads transmit signals to one another via mesh network topology. This enables the cluster heads to receive sensing information from an end sensor node and send it to a SINK node.

- Sensor nodes can control radio transmission power by using two levels. Cluster members only need to communicate with cluster heads when cluster members are in the communication range of cluster heads. However, cluster heads need to communicate with other cluster heads and the SINK node, which may be further than the cluster communication range.
- Cluster heads are elected by a time interval called a "round" (every several hours) as well as by other clustering protocols (Fig. 3). Some sensor nodes are elected to be cluster heads in one round, and when the next round starts, new cluster heads are elected and new clusters are constructed. Since a cluster head cannot change its state to sleep mode in this study, there is no need to have a strict time synchronization.
- Initially, each sensor node has the same energy capacity, but the energy consumption of each sensor is different. The batteries cannot be changed after sensor nodes are deployed.

- The data flow of the entire network is unidirectional from cluster members to cluster heads and from cluster heads to the SINK node. There is no data flow from the SINK node to cluster heads and cluster members, so the SINK node does not participate in the election of cluster heads or in the construction of clusters.

3.2 Cluster Head Election Algorithm

The target of this clustering scheme is a wireless sensor network that elects cluster heads autonomously. The SINK node is just a data collector, not a network construction coordinator. No parameters have to be sent to the SINK node, and the SINK node does not need to calculate which node will be a cluster head in the next round based only on a parameter that has been sent from sensor nodes. At the first round, cluster head election depends on the sensor node starting time. Figure 4 shows the flowchart of electing a cluster head from the second round. At the start of a round,

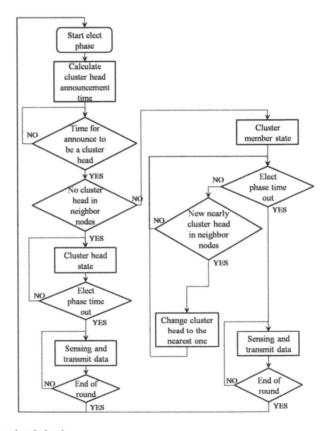

Fig. 4 Cluster head election

each sensor node calculates its cluster head announcement (CHA) time. The CHA time is decide by the amount of energy remaining in the sensor node.

$$Time_a = \frac{E_i - Er}{E_i} \times k, \tag{1}$$

where E_i is the initial energy, E_r is the remaining energy of a sensor node, and k is a coefficient constant. Equation 1 indicates that a sensor node with low remaining energy will have a later CHA time.

Sensor nodes that have an earlier CHA time become cluster heads, and broadcast their CHA message and end time of round. Sensor nodes check whether there are any cluster heads among their neighbor nodes before announcing they are cluster heads. If there are any cluster heads among their neighbor nodes, sensor nodes do not announce they are cluster heads, but they join the nearest cluster as a cluster member. A sensor node that has become a cluster member checks for new cluster heads until an election phase timeout. When a new cluster head is nearer (has a stronger RSSI) than the current cluster head appears, end sensor nodes change their cluster head to the new one. However, if there is no cluster head among its neighbor nodes, a sensor node announces that it is a cluster head. After a round, sensor nodes will start a new election phase and elect cluster heads of the wireless sensor network.

4 Performance Evaluation

4.1 Simulation Outline and Parameters

We evaluated the performance of LiCS via simulations. The wireless sensor node consisted of a microcontroller, communication module, sensor module, and battery module.

Energy consumption of the wireless sensor node is given by

$$E = E_T + E_R + E_S. \tag{2}$$

Here, E_T is the energy consumption of a transmission contains data processing and signal amplification. The energy consumption of data processing, E_{proc} depends on factors such as modulation, filtering, and encoding. E_R is the energy consumption of a reception that also contains data processing, and E_S is the energy consumption of a sensor module.

Cluster heads in LiCS must announce they are cluster heads, receive sensing data from members of their clusters, and relay the data to the SINK node. Of course, cluster heads also detect events around them. Therefore, the energy consumption of a cluster head is expressed by

$$E_{CH} = E_{Td} + E_{Rd} + E_{Ta} + E_S. \tag{3}$$

Here, E_{Td} and E_{Rd} are the energy consumption for transmitting and receiving sensing data, and E_{Ta} is the energy consumption for transmitting CHA message.

Cluster members in LiCS need to receive CHA message, detect nearby event, and transmit sensing data to the cluster head. Therefore, the energy consumption of a cluster member is given by

$$E_{CM} = E_{Ra} + E_{Td} + E_S. \tag{4}$$

The value E_{Ra} is the energy consumed to receive a CHA message from the cluster heads, and E_{Td} is the energy consumed to transmit sensing data.

The energy consumption of sensor nodes without a cluster head is expressed by

$$E_{CNM} = E_{Standby}. \tag{5}$$

Here, $E_{Standby}$ is the standby energy consumption. We used a transmit and receive model [6]. When l-bits of sensing data is transmitted a distance d, the energy consumption of the transmission is as follows

$$E_T = l \times (E_{proc} + \epsilon_{fs} \times d^2), \tag{6}$$

and the energy consumption to receive this sensing data is

$$E_R = l \times E_{proc}. \tag{7}$$

Table 1 lists system parameters used in our simulations.

We assume that 100 wireless sensor nodes are dispersed in a field with uniform distribution and that cluster heads can communicate with the SINK node in a sin-

Table 1 Simulation parameters		
	Number of sensor nodes	100
	Simulation field	50 × 50 m
		100 × 100 m
		150 × 150 m
	Position of SINK node	[0 m, 0 m]
	Transmission interval	1 s
	Length of sensing data (l_d)	500 bits
	Length of CHA message (l_a)	50 bits
	Initial energy	50 J
	k	2 s
	T_{ep}, time of election phase	2 s
	E_{proc}	10 pJ/bit/m^2
	ϵ_{fs}	50 nJ/bit
	E_S	200 nJ/s
	$E_{Standby}$	100 nJ/s
	1 round	360 s
	R_c, cluster communication range	30 m

gle hop. To evaluate the influence of different node densities, we set a simulation field with dimensions 50×50 m, 100×100 m, and 150×150 m and the position of the SINK node was [0 m, 0 m]. The sensing and transmission interval was 1 second, the length of sensing data (l_d) was 500 bits, and the length of a CHA message (l_a) was 50 bits. Energy consumption per second of a cluster head in the data collection phase is given by

$$E_{chc} = l_d \times (n + 1) \times (E_{proc} + \epsilon_{fs} \times d^2_{to\ SINK\ node}) + l_d \times n \times E_{proc}, \qquad (8)$$

and energy consumption per second of a cluster head in the cluster head election phase is

$$E_{chs} = l_a \times (E_{proc} + \epsilon_{fs} \times d^2_{R_c}). \qquad (9)$$

Energy consumption per second of a cluster member in the data collection phase is

$$E_{cmc} = l_d \times (E_{proc} + E_{proc} + \epsilon_{fs} \times d^2_{to\ cluster\ head}), \qquad (10)$$

and energy consumption per second of a cluster member in the cluster head election phase is

$$E_{cms} = l_a \times E_{proc}. \qquad (11)$$

Figure 5 is an illustration of the topology used in simulations. All cluster members are marked by solid circles, and cluster heads are marked by solid squares. The big white circle is the SINK node.

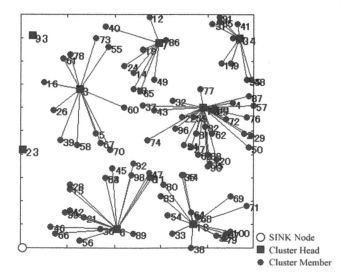

Fig. 5 Simulation topology

4.2 Simulation Results

To evaluate the performance of LiCS, we compared it with the performance of LEACH and HEED using a simulation. LEACH randomly chose a cluster head for the next round without using exact location information, and when there was no cluster communication range in LEACH, every node could communicate with any other node in the network. The expected number of cluster heads in LEACH is 5 [6].

Simulation results of the numbers of communicable nodes are shown in Figs. 6, 8 and 10. In the figures, we observe that the performance of LiCS was better than that of LEACH, which used a random selection algorithm. LiCS also shows the same drop in the number of communicable nodes as HEED. In LiCS and HEED, although the selection algorithms are different, a sensor node will announce it is a cluster head if there is no CH in its cluster communication range, and the average number of CH numbers will be almost the same, which means that LiCS has the same drop in the numbers of communicable nodes as HEED. We also observed that the low density network reduces the effectiveness of LEACH and also that of LiCS and HEED. In LEACH, the distance between cluster members and cluster heads was longer than that of the high-density point, which increases the energy consumption of sensor nodes. In LiCS and HEED, to maximize the number of nodes that can communicate, more cluster heads were needed than those at a high-density point, which makes the node lifetime short.

The results of the number of CHAs sent by sensor nodes in each round are shown in Figs. 7, 9 and 11. The horizontal axes are the rounds, and the vertical axes are the number of CHA messages. In the previous results, LiCS and Heed have the same lifetime. However, LiCS cover all sensor nodes with fewer CHs than HEED. Therefore, the transmission frequency of the CHA messages in the election phase decreases, too. A lighter selection algorithm resulted in LiCS having fewer CHA messages than HEED.

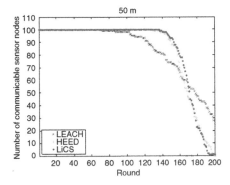

Fig. 6 Number of communicable nodes in 50 × 50 m field

Fig. 7 Number of CHA
messages in 50 × 50 m field

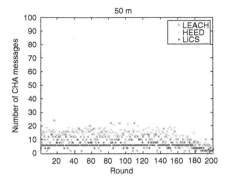

Fig. 8 Number of
communicable nodes
in 100 × 100 m field

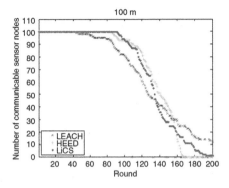

Fig. 9 Number of CHA
messages in 100 × 100 m
field

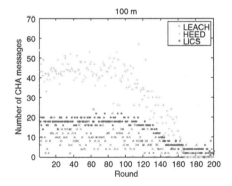

5   Conclusion

In this paper, we proposed LiCS for disaster relief, a simple clustering scheme to
increase the lifetime of wireless sensor networks and maximize the number of communicable sensor nodes. LiCS did not require exact location information of sensor
nodes, which calculated by using localization modules such as GPS or complex localization algorithms. Not requiring a localization module means that the cost of
sensor nodes was reduced, and no energy for a localization module was necessary.

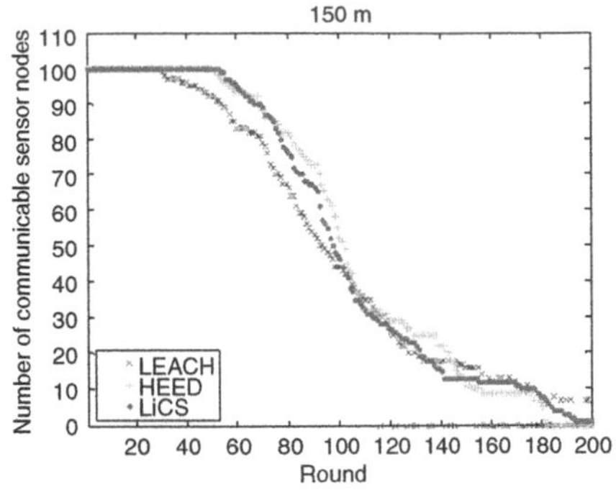

Fig. 10 Number of
communicable nodes
in 150 × 150 m field

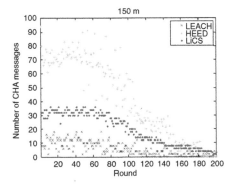

Fig. 11 Number of CHA
messages in 150 × 150 m
field

Having no complex localization algorithms means that the SINK node did not need
to join the cluster head election phase, which means that low-performance SINK
nodes can also be used in LiCS, and the overhead of parameters for clustering algorithms will not be transmitted in wireless sensor networks. In this study, we only
evaluated the performance of LiCS with a static clustering communication range. In
the future, we plan to evaluate the performance of LiCS with a dynamic communication range caused by a radio irregularity.

References

1. Police, H. (1996). Rescue report. *Police activity of Hanshin-Awaji earthquake* (pp. 71–75).
2. Ye, W., Heidemann, J., & Estrin, D. (2002). An energy-efficient MAC protocol for wireless
   sensor networks. *Proceedings INFOCOM, 3,* 1567–1576.
3. van Dam, T., & Langendoen, K. (2003). An adaptive energy-efficient mac protocol for wireless sensor networks. *SenSys.* Proceedings of the 1st international conference on Embedded
   networked sensor systems (2003), Los Angeles, California, USA, pp. 171–180.
4. Lu, G., Krishnamachari, B., & Raghavendra, C. (2004). An adaptive energy-efficient and low-
   latency MAC for data gathering in sensor networks. *IEEE WMAN.* Parallel and Distributed
   Processing Symposium, 2004. Proceedings. 18th International, 26–30 April 2004, Santa Fe,
   New Mexico, USA, 863–875.

5. Cao, Q., Abdelzaher, T., He, T., & Stankovic, J. (2005). Towards optimal sleep scheduling in sensor networks for rare-event detection. *IPSN*, 20–27. Proceedings of the 4th international symposium on Information processing in sensor networks table of contents, Los Angeles, California, USA, April 25–27.
6. Heinzelman, W.R., Chandrakasan, A., & Balakrishnan, H. (2002). An application-specific protocol architecture for wireless microsensor networks. *IEEE Transactions on Wireless Communications*, *1*(4), 660–667.
7. Bae, K., & Yoon, H. (2005). Autonomous clustering scheme for wireless sensor networks using coverage estimation-based self-pruning. *IIEICE Transactions on Communications*, *E88-B*(3), 973–980.
8. McLaughlan, B., & Akkaya, K. (2007). Coverage-based clustering of wireless sensor and actor netwoks. *IEEE international conference on pervasive services* (pp. 45–54).
9. Chatterjee, M., Das, S.K., & Turgut, D. (2002). WCA: a weighted clustering algorithm for mobile ad hoc networks. *Cluster Computing*, *5*, 301–322.
10. Younis, O., & Fahmy, S. (2004). HEED: a hybrid, energy-efficient, distributed clustering approach for ad-hoc sensor networks. *IEEE Transactions on Mobile Computing*, *3*(4), 660–669.
11. Gupta, H., Navda, V., Das, S.R., & Chowdhary, V. (2005). Efficient gathering of correlated data in sensor networks. *ACM international symposium on mobile ad hoc networking and computing* (pp. 402–413).

Chapter 21
On the Complexity of Some Map-Coloring Multi-player Games

Alessandro Cincotti

Abstract Col and Snort are two-player map-coloring games invented respectively by Colin Vout and Simon Norton where to establish who has a winning strategy on a general graph is a $\mathcal{PSPACE}$-complete problem. However, winning strategies can be found on specific graph instances, e.g., strings or trees. In multi-player games, because of the possibility to form alliances, cooperation between players is a key-factor to determine the winning coalition and, as a result, the complexity of three-player Col played on trees is $\mathcal{NP}$-complete and the complexity of n-player Snort played on bipartite graphs is $\mathcal{PSPACE}$-complete.

Keywords Complexity · Map-coloring game · Multi-player game

1 Introduction

Col and Snort are two map-coloring games invented respectively by Colin Vout and Simon Norton. Every instance of these games is defined as an undirected graph $G = (V, E)$ where every vertex is uncolored, black or white.

Two players, First and Second, play in turn and First starts the game. In the beginning, all the vertices are uncolored. First has to paint an uncolored vertex using the color black, and Second has to paint an uncolored vertex using the color white. There exists only one restriction: in the game of Col two adjacent vertices cannot be painted with the same color, but in the game of Snort two adjacent vertices cannot be painted with different colors. In normal play, the first player unable to paint an uncolored vertex is the loser.

The value of some Col and Snort positions and the description of some general rule for simplifying larger positions is presented in [1, 9]. Recently, Col has been solved on complete k-ary trees [8] and on a specific class of trees where all the

A. Cincotti (✉)
School of Information Science, Japan Advanced Institute of Science and Technology,
1-1 Asahidai, Nomi, Ishikawa 923-1292, Japan
e-mail: cincotti@jaist.ac.jp

S.-I. Ao et al. (eds.), *Intelligent Automation and Computer Engineering*,
Lecture Notes in Electrical Engineering 52, DOI 10.1007/978-90-481-3517-2_21,

internal nodes have at least two children and the depth of all the leaves is either even or odd [6]. Moreover, Col and Snort are proved to be two $\mathcal{PSPACE}$-complete problems on a general graph [13].

In this paper we show that in multi-player games, because of the possibility to form alliances, cooperation between players is a key-factor to determine the winning coalition and, as a result, three-player Col played on trees is $\mathcal{NP}$-complete [7] and n-player Snort played on bipartite graphs is $\mathcal{PSPACE}$-complete.

2 Three-Player Col

Three-player Col is a three-player version of Col. Every instance of this game is defined as an undirected graph $G = (V, E)$ where every vertex is labeled by an integer $j \in \{1, 2, 3\}$. Three players, First, Second and Third, take turns making legal moves in cyclic fashion (First, Second, Third, First, Second, Third, . . .) and First starts the game.

In the beginning, all the vertices are unlabeled. First has to label an unlabeled vertex with "1", Second has to label an unlabeled vertex with "2", and Third has to label an unlabeled vertex with "3". There exists only one restriction: two adjacent vertices cannot be labeled with the same number.

In normal play, if one of the players is unable to move, then he/she leaves the game and the remaining players continue in alternation until one of them cannot move. Then that player leaves the game and the remaining player is the winner. In other words, the last player to move wins.

Three-player games [2, 3], are difficult to analyze because of queer games [4, 12], i.e., games where no player has a winning strategy. Moreover, because of the possibility to form alliances, cooperation between players is a key-factor to determine the winning coalition. There exist at least two different ways to establish the winning coalition:

- **Weak Coalition Convention** If one of the player is unable to move, then he/she leaves the game but the other players of his/her coalition are still able to play. In other word, the coalition of the player that makes the last move wins.
- **Strong Coalition Convention** If one of the player is unable to move, then he/she and all the players of his/her coalition leave the game. In other word, a coalition wins if all its players are able to make a move until the end of the game.

In weak coalition convention, the fact that one of the players is not able to move does not affect the other players of the same coalition because the main goal is to be able to make the last move. Differently, in strong coalition convention, the fact that all the players are able to make a move until the end of the game becomes a crucial point.

In a previous work [5], we studied the complexity of three-player Hackenbush played on strings under weak coalition convention. In the next section, the complexity of three-player Col played on trees under strong coalition convention is investigated.

3 Three-Player Col Played on Trees Is $\mathcal{NP}$-Complete

We prove that to solve three-player Col played on trees under strong coalition convention is a $\mathcal{NP}$-complete problem.

We briefly recall the definition of Subset Sum Problem.

Definition 1. Let

$$\mathcal{U} = \{u_1, \ldots, u_n\}$$

be a set of natural numbers and K a given natural number. The problem is to determine if there exists $\mathcal{U}' \subseteq \mathcal{U}$ such that

$$\sum_{u_i \in \mathcal{U}'} u_i = K$$

This problem is known to be $\mathcal{NP}$-complete [10]. Starting from a general instance of Subset Sum Problem it is possible to create an instance of three-player Col on trees as shown in Fig. 1 where U is the sum of all the elements in $\mathcal{U}$, i.e.,

$$U = \sum_{u_i \in \mathcal{U}} u_i$$

Who has a winning strategy if Second and Third form a coalition and they play under strong coalition convention?

We observe that in the tree shown in Fig. 1a, First can make exactly $U + n$ moves. Moreover, in the trees shown in Fig. 1b and c, Second and Third can make

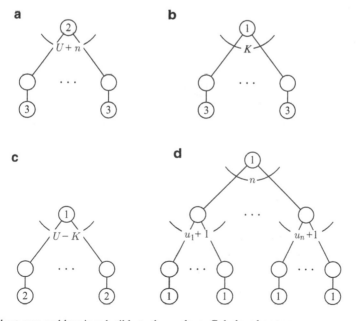

Fig. 1 Subset sum problem is reducible to three-player Col played on trees

respectively K and $U - K$ moves. Therefore, if Second and Third form a coalition, then two different scenarios are possible:

- If Second and Third are able to make respectively $U - K + n$ and $K + n$ moves in the tree shown in Fig. 1d, then First does not have a winning strategy.
- If Second and Third are not able to satisfy the previous condition, then First has a winning strategy.

The problem to determine if First has a winning strategy or not is strictly connected to Subset Sum Problem as shown in the following theorem.

Theorem 2. *Let G be a general instance of three-player Col played on trees. Then, to establish the outcome of G is a $\mathcal{NP}$-complete problem.*

Proof. The problem is clearly in $\mathcal{NP}$.

We show that it is possible to reduce every instance of Subset Sum Problem to G. Previously we have described how to construct the instance of three-player Col, therefore we just have to prove that Subset Sum Problem is solvable if and only if the coalition formed by Second and Third has a winning strategy, i.e., First does not have a winning strategy.

If $\mathcal{U}'$ is a solution of Subset Sum Problem, then the coalition formed by Second and Third has a trivial way to win the game. All the roots of the sub-trees corresponding to $u_i \in \mathcal{U}'$ will be labeled 2 and all the leaves of such sub-trees will be labeled 3. All the roots of the sub-trees corresponding to $u_i \notin \mathcal{U}'$ will be labeled 3 and all the leaves of such sub-trees will be labeled 2. In this way, Second and Third will make respectively $U - K + n$ and $K + n$ moves in the tree shown in Fig. 1d.

Conversely, if the coalition formed by Second and Third has a winning strategy, then Second and Third are able to make respectively $U - K + n$ and $K + n$ moves in the tree shown in Fig. 1d, i.e., all nodes must be labeled.

Because of the rules of the game, two adjacent vertices cannot be labeled with the same number, therefore there exist a set of sub-trees corresponding to $\mathcal{U}' \subset \mathcal{U}$ such that

$$\sum_{u_i \in \mathcal{U}'} u_i = K.$$

Therefore, the problem to establish the outcome of G is $\mathcal{NP}$-hard and $\mathcal{NP}$-complete. $\square$

4 N-Player Snort

N-player Snort is a n-player version of Snort. Every instance of this game is defined as an undirected graph $G = (V, E)$ where every vertex is labeled by an integer $j \in \{1, 2, \ldots, n\}$. N players take turns making legal moves in cyclic fashion (1st player, 2nd player,. . . ,nth player, 1th player, 2nd player, . . .) and 1th player starts the game.

In the beginning, all the vertices are unlabeled. 1st player has to label an unlabeled vertex with "1", 2nd player has to label an unlabeled vertex with "2", and so on. There exists only one restriction: two adjacent vertices cannot be labeled with two different numbers.

In normal play, when one of the n players is unable to move, then that player leaves the game, and the remaining $n - 1$ players continue playing in the same mutual order as before. The last remaining player is the winner.

In the next section, the complexity of n-player Snort played on bipartite graphs under weak coalition convention is investigated.

5 N-Player Snort Played on Bipartite Graphs Is $\mathcal{PSPACE}$-Complete

We prove that to solve n-player Snort played on bipartite graphs under weak coalition convention is a $\mathcal{PSPACE}$-complete problem.

The $\mathcal{PSPACE}$-complete problem of *Quantified Boolean Formulas* [11], QBF for short, can be reduced by a polynomial time reduction to n-player Snort played on bipartite graphs.

Let $\varphi \equiv \exists x_1 \forall x_2 \exists x_3 \cdots Q x_n \psi$ be an instance of QBF, where Q is $\exists$ for n odd and $\forall$ otherwise, and $\psi \equiv C_1 \wedge C_2 \wedge \cdots \wedge C_k$ is a quantifier-free Boolean formula in conjunctive normal form. If n is the number of variables and k is the number of clauses in ψ, then the instance of n-player Snort has $n + k + 1$ players and the bipartite graph $G = (V_1 \cup V_2, E)$ is organized as follows:

$$V_1 = \{v_1, v_2, \ldots, v_{n+1}, \overline{v}_1, \overline{v}_2, \ldots, \overline{v}_{n+1}, lc_1, lc_2, \ldots, lc_k\}$$

where the vertices lc_i with $1 \le i \le k$ are labeled with $n + 1 + i$,

$$V_2 = \{lv_1, lv_2, \ldots, lv_{n+1}, c_1, c_2, \ldots, c_k, d_1, d_2, \ldots, d_k\}$$

where the vertices lv_i with $1 \le i \le n + 1$ are labeled with i,

$$\begin{aligned}
E = \{ \ & (v_i, lv_i) : 1 \le i \le n + 1, \\
& (\overline{v}_i, lv_i) : 1 \le i \le n + 1, \\
& (lc_i, c_i) : 1 \le i \le k, \\
& (lc_i, d_i) : 1 \le i \le k, \\
& (v_i, c_j) : x_i \in C_j, 1 \le i \le n, 1 \le j \le k, \\
& (\overline{v}_i, c_j) : \overline{x}_i \in C_j, 1 \le i \le n, 1 \le j \le k \}.
\end{aligned}$$

Let us assume that:

- If n is even then the first coalition is formed by the 1st, 3rd,...,$(n-1)$th, and $(n+1)$th player. The second coalition is formed by the remaining $(n/2)+k$ players.
- If n is odd then the first coalition is formed by the 1th, 3rd,...,nth, and $(n+1)$th player. The second coalition is formed by the remaining $\lfloor n/2 \rfloor + k$ players.

An example is shown in Fig. 2 where

$$\varphi \equiv \exists x_1 \forall x_2 \exists x_3 \forall x_4 (C_1 \wedge C_2 \wedge C_3)$$

and

$$C_1 \equiv (x_1 \vee \overline{x}_2 \vee x_3)$$
$$C_2 \equiv (\overline{x}_1 \vee x_2 \vee x_4)$$
$$C_3 \equiv (\overline{x}_1 \vee x_3 \vee \overline{x}_4)$$

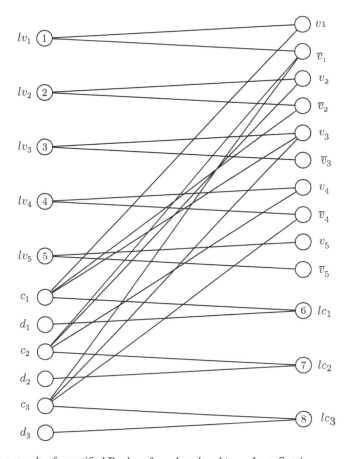

Fig. 2 An example of quantified Boolean formula reduced to n-player Snort

The problem to determine the winning coalition is strictly connected to the problem of QBF, as shown in the following theorem.

Theorem 3. *Let G be a general instance of n-player Snort played on bipartite graphs. Then, to establish if a given coalition has a winning strategy or not is a $\mathcal{PSPACE}$-complete problem.*

Proof. The problem is clearly in $\mathcal{PSPACE}$ because each strategy can be verified in polynomial time.

We show that it is possible to reduce every instance of QBF to G. Previously we have described how to construct the instance of n-player Snort, therefore we just have to prove that QBF is satisfiable if and only if the first coalition has a winning strategy.

If QBF is satisfiable, then there exists an assignment of x_i such that ψ is true with $i \in 1, 3, \ldots, m$ where

$$m = \begin{cases} n - 1 & \text{if } n \text{ is even} \\ n & \text{if } n \text{ is odd} \end{cases}$$

If x_i is true, then the ith player will label the vertex v_i with i. If x_i is false, then the ith player will label the vertex $\bar{v}_i$ with i. Because of the rules of the game, two adjacent vertices cannot be labeled with different numbers, therefore, no player can move into the vertices $c_1, c_2, \ldots, c_k$. Moreover, the $(n+1)$th player is always able to make two moves (labeling the vertices v_{n+1} and $\bar{v}_{n+1}$), therefore, the first coalition has a winning strategy.

Conversely, if the first coalition has a winning strategy, then the vertices $c_1, c_2, \ldots, c_k$ cannot be labeled. In other words, each clause has at least one true literal and QBF is satisfiable.

Therefore, to establish whether or not a coalition has a winning strategy in n-player Snort played on bipartite graphs is $\mathcal{PSPACE}$-hard and $\mathcal{PSPACE}$-complete.

Acknowledgments The author wishes to thank Mark G. Elwell for a careful reading of the manuscript.

References

1. Berlekamp, E.R., Conway, J.H., & Guy, R.K. (2001). *Winning way for your mathematical plays*. Natick, MA: AK Peters.
2. Cincotti, A. (2005). Three-player partizan games. *Theoretical Computer Science, 332*, 367–389.
3. Cincotti, A. (2007). Counting the number of three-player partizan cold games. In H.J. van den Herik, P. Ciancarini, & H.H.L.M. Donkers (Eds.), *Computers and games*, volume 4630 of *LNCS* (pp. 181–189). Springer, Germany.
4. Cincotti, A. (2008a). The game of cutblock. *INTEGERS: Electronic Journal of Combinatorial Number Theory, 8(G06)*, 1–12.

5. Cincotti, A. (2008b). Three-player hackenbush played on strings is $\mathcal{NP}$-complete. In S.I. Ao, O. Castillo, C. Douglas, D.D. Feng, & J. Lee (Eds.), *International multiconference of engineers and computer scientists 2008* (pp. 226–230). Newswood Limited, Hong Kong.
6. Cincotti, A. (2009a). Further results on the game of col. In *Proceedings of the 8th Annual Hawaii international conference on statistics, mathematics and related fields* (pp. 315–318).
7. Cincotti, A. (2009b). Three-player col played on trees is $\mathcal{NP}$-complete. In S.I. Ao, O. Castillo, C. Douglas, D.D. Feng, & J. Lee, (Eds.), *International multiconference of engineers and computer scientists 2009* (pp. 445–447). Newswood Limited, Hong Kong.
8. Cincotti, A., & Bossart, T. (2008). The game of col on complete k-ary trees. In C. Ardil (Ed.), *Proceedings of world academy of science, engineering and technology*, volume 30 (pp. 699–701).
9. Conway, J.H. (2001). *On numbers and games*. Natick, MA: AK Peters.
10. Garey, M.R., & Johnson, D.S. (1979). *Computers and intractability*. New York: Freeman.
11. Papadimitriou, C.H. (1994). *Computational complexity*. Reading, MA: Addison-Wesley.
12. Propp, J.G. (2000). Three-player impartial games. *Theoretical Computer Science*, *233*, 263–278.
13. Schaefer, T.J. (1978). On the complexity of some two-person perfect information games. *Journal of Computer Systems and Science*, *16*(2), 185–225.

Chapter 22
Methods to Hide Quantum Information

Gabriela Mogos

Abstract The revolution of the digital information determined deep changes in society and in our lives. The many advantages of the digital information generated as well new challenges and opportunities in the field of innovation. New devices have appeared beside the powerful software, like printers, scanners, digital cameras, MP3 players, etc. for the creation, handling and utilization of multimedia data. The Internet and the wireless network offered channels for exchanging information. The security, and the correct use of multimedia data, as well as the delivery of the multimedia content to different users constitute another very important subject. The solution to these problems does not contribute only to our better understanding of the speed with which technology develops, but also offers new exploration opportunities. All that follows addresses to the subject concerning hiding the data in multimedia files, and applications in multimedia and communication security. Taking into account the easiness with which digital data is edited or reproduced, the protection and the prevention of the unauthorized use of the multimedia data (audio, image, video, documents) become extremely important. One of the most frequently used electronic formats in which information can be found nowadays is the digital image.

Keywords Hiding information · Qutrits · Quantum secret key · Quantum digital signature

1 Introduction

The cryptography for images represents the process of alteration of the data from images so as different pieces of information could be inserted. Here is some normal text. Here is some normal text. Here is some normal text. The embedding procedures

G. Mogos (✉)
Computer Science Department, Al.I.Cuza University, General Berthelot, 16,
Room 201, Iasi, 700483, Romania
e-mail: gabi.mogos@gmail.com

S.-I. Ao et al. (eds.), *Intelligent Automation and Computer Engineering*,
Lecture Notes in Electrical Engineering 52, DOI 10.1007/978-90-481-3517-2_22,
© Springer Science+Business Media B.V. 2010

of quantum information which will be presented in this work are based on the studies realized by Zizzi [1] and Mütze [2]. According to the holographic principle, the pixel can be seen as an object delimited by a closed surface, being a carrier of information. In 1999, Zizzi in his work [1] demonstrates that in a classical holographic image a pixel is a system with two states, i.e., a classical bit of information. Later, in the year 2008, Ulrich Mütze went further with this demonstration for the case of the digital images, and his conclusion was that a pure state of a qubit can be associated to a pixel belonging to a color image. The pure state of a qubit can be represented by a dot on the surface of Bloch sphere, and the impure states, by dots situated inside the sphere. A quantum analogy of natural colors can be realized with the help of the systems which can be completely described in the terms of the three-orthogonal states, systems named qutrits. The color space RGB is defined as a Cartesian system of coordinates whose primary axes are red, green, blue, each axial domain being between zero and one (pure black $[0, 0, 0]$ and pure white $[1, 1, 1]$). The three coordinate axes in which the qutrit is represented are labeled $\{|1\rangle, |2\rangle, |3\rangle\}$. Replacing the three-dimensional system $\{|1\rangle, |2\rangle, |3\rangle\}$ with $\{|R\rangle, |G\rangle, |B\rangle\}$, we can write the state of such a qutrit as a linear combination of the projections on the three coordinate axes:

$$|\Psi\rangle = r|R\rangle + g|G\rangle + b|B\rangle \qquad (1)$$

There are three basic colors, red, green and blue, corresponding to:

$$|R\rangle = \begin{bmatrix} 1 \\ 0 \\ 0 \end{bmatrix} \quad |G\rangle = \begin{bmatrix} 0 \\ 1 \\ 0 \end{bmatrix} \quad |B\rangle = \begin{bmatrix} 0 \\ 0 \\ 1 \end{bmatrix} \qquad (2)$$

The color images made of such qutrits placed under the form of a matrix will be used as a support for hiding the information using different algorithms. This work has the purpose to place the presentation of two algorithms used for hiding the quantum information.

2 Quantum Data Hiding

The algorithm starts from the study developed by Grudka and Wójcik [3] according to whom a chosen state of a qubit can be reconstructed from a qutrit. Thus, starting from a state of a qutrit of the image, we presume that two qubits can be randomly reconstructed, with the states (Fig. 1):

$$|\Psi\rangle = \sin\phi|R\rangle + \cos\phi|G\rangle \qquad (3)$$

respectively

$$|\Phi\rangle = \sin\theta|B\rangle + \cos\theta|K\rangle \qquad (4)$$

Fig. 1 Two qubits can
be rebuilt

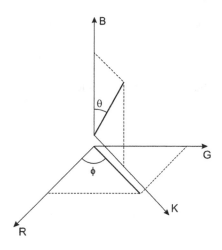

Table 1 Coding every bit
of the key in a quantum value

	0	1
Basis 1: $\{|R\rangle, |G\rangle\}$	$|R\rangle$	$|G\rangle$
Basis 2: $\{|B\rangle, |K\rangle\}$	$|B\rangle$	$|K\rangle$

The procedure of information encapsulation has the purpose to hide a message encrypted with a secret key within a digital image. The image is made of qutrits, and the secret, the key, and the message are made of qubits.

2.1 Hiding the Information

Alice wants to send a secret message to Bob, and she uses a digital image in which she will encapsulate it. In order to better protect the information, she will encrypt the message, and then she will introduce it in the image. The encryption of the message will be made using the encryption key obtained with the help of the protocol Bennett-Brassard [4], coding every bit of the key in a quantum value with the help of the two bases $\{|R\rangle, |G\rangle\}$ respectively $\{|B\rangle, |K\rangle\}$ (Table 1). The procedure of information encapsulation consists of two aspects: coding the information, and hiding the coded message in the image. The most important aspect which should be taken into account is the one referring to the keeping of the qutrit characteristics after the data were encapsulated, so as visually no difference should be perceived between the original image and the one containing the hidden data. As a first step, Alice will reconstruct one by one the states of two qubits $|\Psi\rangle$ respectively $|\Phi\rangle$. Then she will take one of the two qubits, and she will entangle it maximally with a qubit from the key, and one from the message, after which she will combine the entanglement resulted with the second qubit which was left unused. We presume that the state of the reconstituted qubit is:

$$|\Psi\rangle_i = \sin\phi|R\rangle + \cos\phi|G\rangle \tag{5}$$

The qubit belonging to the secret key has the state:

$$|\Psi\rangle_k = \sin\gamma|R\rangle + \cos\gamma|G\rangle \tag{6}$$

The qubit belonging to the secret message has the state:

$$|\Psi\rangle_m = \sin\delta|R\rangle + \cos\delta|G\rangle \tag{7}$$

The state of the maximally entanglement obtained among the three qubits is:

$$|\Psi\rangle_{mki} = \sin\phi\sin\gamma\sin\delta|RRR\rangle_{mki} + \cos\phi\cos\gamma\cos\delta|GGG\rangle_{mki} \tag{8}$$

We can write the state of the entanglement as it follows:

$$|\Psi\rangle_{mki} = \sin\chi|\mathbf{R}\rangle + \cos\chi|\mathbf{G}\rangle \tag{9}$$

where:

$$\sin\chi = \sin\phi\sin\gamma\sin\delta; \cos\chi = \cos\phi\cos\gamma\cos\delta \tag{10}$$

We will examine the way in which the "presence" of the qubits of the key and of the message could influence the characteristics of the qubit of the image, or, in other words, how much can the values of the angles ϕ, γ and δ, influence the value of the angle χ, so as the encapsulation should not be perceived visually. For this study we used the programme Absolute Color Picker, choosing a random combination R,G,B. The purpose is to determine by mathematical calculation the limits between which the angles θ and ϕ can be varied so as the shade of the color does not modify itself. For this, we varied an angle, maintaining the other constant. Starting from a combination of randomly chosen colors: $R = 0,231; G = 0,835; B = 0,125$ (the color green) the angles θ and ϕ were calculated. The following values were obtained by mathematical calculations: $\theta = 82°13'$ respectively $\phi = 74°31'$. Maintaining the angle ϕ constant, a maximum variation of $\pm 2'$ ($R = 0,234; G = 0,83; B = 0,126$) of the angle θ was obtained, and for the angle θ constant, the maximum variation of the angle ϕ is $\pm 1'$ ($R = 0,231; G = 0,835; B = 0,125$). Knowing the limits between which the two angles can vary, we can see if it is necessary to choose the expression of the states of the qubits of the key, respectively of the message, so as the encapsulation in the qubit of the image does not determine a change of the angle ϕ.

$$R = \sin\theta\cos\phi$$
$$G = \sin\theta\sin\phi \tag{11}$$
$$B = \cos\theta$$

Thus, if from the qutrit of the image a qubit with the state (3), is reconstituted, after the encapsulation, the angle χ of the state (9) obtained should not go beyond the admitted limits of variation of the angle ϕ. Therefore, the angle χ must meet the condition:

$$\phi - 1' \leq \chi \leq \phi \tag{12}$$

We have:

$$\sin(\phi - 1^{'}) \le \sin \chi \le \sin \phi$$
$$\cos(\phi - 1^{'}) \le \cos \chi \le \cos \phi \tag{13}$$

We study the first expression.

$$\sin(\phi - 1^{'}) \le \sin \phi \sin \gamma \sin \delta \le \sin \phi \tag{14}$$

Noting $z = \sin \gamma \sin \delta$, we get:

$$\sin(\phi - 1^{'}) \le \sin \phi \cdot z \le \sin \phi \tag{15}$$

which leads to:

$$0 \le z \le 1 \tag{16}$$

Replacing z, the equation (16) becomes:

$$0 \le \sin \gamma \sin \delta \le 1 \tag{17}$$

which is:

$$0 \le \sin \delta \le 1 \; for \; 0 \le \sin \gamma \le 1 \tag{18}$$

Taking into account the cosines function, we meet the condition (18) for: $\delta, \gamma \in [0, \frac{\pi}{2}]$.

We notice that the angles γ and δ can get any value; the presence of the encoded message in the image does not determine the change of the angle χ more than the maximum admitted. This means that the encapsulation does not change the characteristics of the qubit of the image. Taking into account that from a qutrit only a qubit with a certain state can be reconstituted, Alice will reconstitute from the same qutrit (1), each of the two qubits (3), (4) one after the other, in the bases $\{|R\rangle, |G\rangle\}$ respectively $\{|B\rangle, |K\rangle\}$, and she will keep them, and in one of them she will encapsulate the coded message. Then, she will recombine the states $|\Theta\rangle = \sin \chi |\mathbf{R}\rangle + \cos \chi |\mathbf{G}\rangle$ and $|\Phi\rangle = \sin \theta |B\rangle + \cos \theta |K\rangle$ getting a partially separable state. We notice that the state resulted will keep the initial shade, each of the states participating in its construction maintaining their characteristics. Due to the fact that the number of the qubits from the message is greater than the number of the qubits of the key, Alice will divide it in sub-messages with the number of qubits equal with the one of the key, in order to code the message. The states partially separable obtained through the combination of the states $|\Theta\rangle$ and $|\Phi\rangle$, will be placed on the surface of the image as an oriented graph, so as the message can be reconstituted correctly. These states will be the nodes of the graph, and the connection of the nodes can be made either by states entanglement, or by using controlled-phase gates [5]. For this study, the link between the nodes was realized by the states entanglement of the qubits from the message which must be sent. Only after the realization of this entanglement Alice will divide the message. The states entanglement is very important, because, according to Einstein-Podolski-Rosen, no matter how far away the quantum systems (the qubits) are, they still remain "connected". In the Fig. 2 it is the symbol

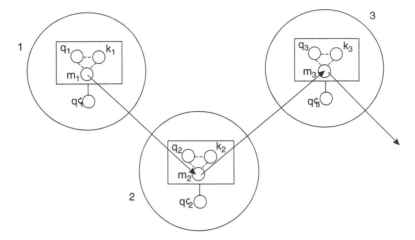

Fig. 2 The encrypted information will be placed on the surface of the image as an oriented graph

of a graph, oriented as it was obtained. Every node has a partially separable state, and it is made of the qubit of the message (m), the qubit of the key (k) and the two qubits reconstructed from a qutrit of the image (q, q'). The generic state of a node is the following:

$$|\Psi\rangle_n = |\Phi\rangle \otimes |\Psi_{mki}\rangle \tag{19}$$

2.2 Extracting the Information

Bob, after receiving the image with embedded data, must extract the information. For that, he first needs to know the position on the image where the first qubit of the message was hidden, the rest of the locations being determined due to the quantum connection of qubits. The coordinates of this position will be communicated by Alice. It is as important to know the base in which he should make the projection of the node state in order to obtain $|\Psi_{mki}\rangle$. The two of them previously establish that the base of the "reading" of the node state should be in the base in which the qubit of the key was coded.

When Bob will measure the node state using the base accordingly, the system will collapse toward the state $|\Psi_{mki}\rangle$. All he has to do is to extract the first qubit of the message and in a similar way all the other qubits of the hidden information.

For that, Bob must have the original image, with no hidden information, and he must reconstitute the qubit of the image. In a similar way, at the qubit reconstruction, he used the base in which the qubit of the key had been coded. After that, he will apply on them the transformations $C - NOT$ and Hadamard in order to obtain the Bell bases:

$$|\Psi_{\pm}\rangle_{ik} = \frac{1}{\sqrt{2}}(|RR\rangle_{ik} \pm |GG\rangle_{ik})$$

$$|\Phi_{\pm}\rangle_{ik} = \frac{1}{\sqrt{2}}(|RG\rangle_{ik} \pm |GR\rangle_{ik})$$

Depending on these bases, the state $|\Psi_{mki}\rangle$ can be written:

$$|\Psi\rangle_{mki} = \frac{1}{2}[|\Psi_+\rangle_{ik}(\sin\delta|R\rangle_m + \cos\delta|G\rangle_m) + |\Psi_-\rangle_{ik}(\sin\delta|R\rangle_m - \cos\delta|G\rangle_m)$$

$$+|\Phi_+\rangle_{ik}(\cos\delta|R\rangle_m + \sin\delta|G\rangle_m) + |\Phi_-\rangle_{ik}(-\cos\delta|R\rangle_m + \sin\delta|G\rangle_m)]$$

We presume that Bob realizes a projective measurement in one of the Bell bases, which is $|\Psi_+\rangle_{ik}$. The state $|\Psi\rangle_{mki}$ will collapse in the state $\sin\delta|R\rangle_m + \cos\delta|G\rangle_m$, which is in fact that of a qubit belonging to the information.

As the qubits are extracted, they are kept, and the order of message reading is eventually from the first to the last qubit extracted.

2.3   Security Analysis

In the case of quantum cryptography, the attack consists in finding techniques to intercept the key, to recover, modify or remove the message encapsulated, but the ways of realization differ from the classical case.

2.3.1   Intercepting the Encryption Key

The security of the distribution protocol of the quantum key was previously discussed, however we will restate that the presence of an intruder (Eve) can be easily discovered due to big quantum errors that can appear during the protocol (the error goes beyond an accepted limit). Even if Eve can attack the quantum channel and listen to the discussions between Alice and Bob from the classical channel, the result she obtains is not at all the expected one. Eve can attack the qubits in two steps. First, she can let them go through a device which is "stocking" their states, and then measure them. Though she is listening to the discussion between Alice and Bob and she can find the base that Bob used at the measurement, she will not gain anything by measuring the sample in that base, because she cannot know if the base used is also the correct one. Re-measuring is practically impossible, because the qubit deteriorates itself after the previous operation.

2.3.2   Attack Against the Message

In order to interfere in any way (recover or modify) with the hidden message, it is necessary to know the encryption key, but also the position of the first knot of the graph. Even if Eve will succeed to find out the position of the first knot on the image, the lack of the encryption key of the message does not help her to reconstitute it. The attempt to remove the first qutrit of the graph by extracting it and replacing it with a qutrit desired by the intruder can be easily discovered by Bob through the absence of the quantum connection between that knot and the others from the graph.

3   Quantum Digital Signature

The digital signature has the purpose to identify its titular, to attest the origin of the document, to associate one or several persons with the content of the document. The digital signature is a succession of bits obtained from the transformation of a message and of a secret information, known only by the sender. Any digital signature must be able to be checked, and the result can only be "true" or "false". The signature has the purpose to certify the originality of the data transmitted between different parties.

The following scheme presents the case in which the digital signature is made of n qutrits, and it relies on the quantum procedure of sharing the secret. In order to encapsulate the digital signature, Alice will place it on the surface of the image under the form of an oriented graph. The connection among the nodes of the graph is realized by the phenomenon of entanglement. The steps of this scheme are:

Generating and distributing the key.   Alice will define a secret key whose dimension will be chosen so as the number of qutrits forming it should be twice as big as those of the signature. Half of the qutrits of the key will be kept by Alice, and the other half will be given to John.

Generating GHZ states and the signature embedding.   Before sending, Alice will generate GHZ states among the qutrits of the signature, key, and image. The distribution of the encrypted signature will be realized as an oriented graph on the surface of the image.

Checking the authenticity.   Bob, using the public key, and helped by John, will determine the signature, which he will compare to the one possessed by Alice in order to check its belonging.

At the first and second step of the scheme, Alice will generate a secret key which she will share with John. The protocol of the quantum key distribution will be Bechmann-Pasquinucci and Peres [6], the number of the bits from its structure being even. Then, she will create an entanglement among the qutrits of the signature so as a quantum "connection" should form among them. The connection thus established is necessary for the realization of the oriented graph, and it has an important role to reconstitute the digital signature. In every node of the graph we will have an entanglement made of: the qutrit of the signature (Fig. 3), Alice's qutrit, John's qutrit, and the image qutrit.

At the last step, the sender (Bob) will want to check the authenticity of the signature. For this, he will use a public key, and helped by John, he will be able, in the end, to be in the possession of the signature, which he will then compare with the one in Alice's possession.

The procedure of signature encapsulation is realized in a similar way for every qutrit from its structure. Initially, Alice will realize a GHZ entanglement among her qutrit (A), John's (J) and the one of the signature (s), which he will then combine with the image qutrit (i), obtaining the state:

$$|\Psi\rangle_{AisJ} = |\Psi\rangle_i \otimes |\Psi\rangle_{AsJ}$$

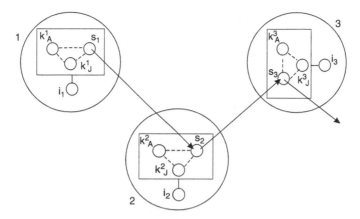

Fig. 3 The encrypted message will be placed on the surface of the image as an oriented graph

where:

$$|\Psi\rangle_s = \alpha|R\rangle_s + \beta|G\rangle_s + \gamma|B\rangle_s$$

with α, β, γ complex numbers satisfying the relation: $|\alpha|^2 + |\beta|^2 + |\gamma|^2 = 1$, it is a qutrit belonging to the signature, and the state of GHZ entanglement:

$$|\Psi\rangle_{AsJ} = \frac{1}{\sqrt{3}}(\alpha|RRR\rangle_{AsJ} + \beta|GGG\rangle_{AsJ} + \gamma|BBB\rangle_{AsJ})$$

When Bob wants to check a signature, he will need both Alice's and John's help. The first step in checking the authenticity is made by Alice, who will realize a projective measurement over the state $|\Psi\rangle_{AisJ}$ using the Bell bases obtained from the qutrit of the image and from the one in her possession. Thus, the $3^2 = 9$ possible states in which the system with the state $|\Psi\rangle_{sAJi}$ can collapse are the following:

$$\langle\Psi_{RR}^{Ai}|\Psi|\Psi_{RR}^{Ai}\rangle = \frac{1}{3}|\Psi_{RR}\rangle_{Ai}(\alpha|RR\rangle_{sJ} + \beta|GG\rangle_{sJ} + \gamma|BB\rangle_{sJ})$$

$$\langle\Psi_{RG}^{Ai}|\Psi|\Psi_{RG}^{Ai}\rangle = \frac{1}{3}|\Psi_{RG}\rangle_{Ai}(\alpha|GG\rangle_{sJ} + \beta|BB\rangle_{sJ} + \gamma|RR\rangle_{sJ})$$

$$\langle\Psi_{RB}^{Ai}|\Psi|\Psi_{RB}^{Ai}\rangle = \frac{1}{3}|\Psi_{RB}\rangle_{Ai}(\alpha|BB\rangle_{sJ} + \beta|RR\rangle_{sJ} + \gamma|GG\rangle_{sJ})$$

$$\langle\Psi_{GR}^{Ai}|\Psi|\Psi_{GR}^{Ai}\rangle = \frac{1}{3}|\Psi_{GR}\rangle_{Ai}(\alpha|RR\rangle_{sJ} + e^{\frac{4\pi i}{3}}\beta|GG\rangle_{sJ} + e^{\frac{2\pi i}{3}}\gamma|BB\rangle_{sJ})$$

$$\langle\Psi_{BR}^{Ai}|\Psi|\Psi_{BR}^{Ai}\rangle = \frac{1}{3}|\Psi_{BR}\rangle_{Ai}(\alpha|RR\rangle_{sJ} + e^{\frac{2\pi i}{3}}\beta|GG\rangle_{sJ} + e^{\frac{4\pi i}{3}}\gamma|BB\rangle_{sJ})$$

$$\langle\Psi_{GG}^{Ai}|\Psi|\Psi_{GG}^{Ai}\rangle = \frac{1}{3}|\Psi_{GG}\rangle_{Ai}(\alpha|GG\rangle_{sJ} + e^{\frac{4\pi i}{3}}\beta|BB\rangle_{sJ} + e^{\frac{2\pi i}{3}}\gamma|RR\rangle_{sJ})$$

$$\langle\Psi_{BG}^{Ai}|\Psi|\Psi_{BG}^{Ai}\rangle = \frac{1}{3}|\Psi_{BG}\rangle_{Ai}(\alpha|GG\rangle_{sJ} + e^{\frac{2\pi i}{3}}\beta|BB\rangle_{sJ} + e^{\frac{4\pi i}{3}}\gamma|RR\rangle_{sJ})$$

$$\langle\Psi_{GB}^{Ai}|\Psi|\Psi_{GB}^{Ai}\rangle = \frac{1}{3}|\Psi_{GB}\rangle_{Ai}(\alpha|BB\rangle_{sJ} + e^{\frac{4\pi i}{3}}\beta|RR\rangle_{sJ} + e^{\frac{2\pi i}{3}}\gamma|GG\rangle_{sJ})$$

$$\langle\Psi_{BB}^{Ai}|\Psi|\Psi_{BB}^{Ai}\rangle = \frac{1}{3}|\Psi_{BB}\rangle_{Ai}(\alpha|BB\rangle_{sJ} + e^{\frac{2\pi i}{3}}\beta|RR\rangle_{sJ} + e^{\frac{4\pi i}{3}}\gamma|GG\rangle_{sJ})$$

The Bell bases obtained from de pairs (A, i) necessary for the measurements of the graph nodes constitute the public key.

We presume that in the public key, the realization of the projection on Bell basis $|\Psi_{RR}\rangle_{Ai}$ is indicated. In this case, the qutrits s and J will collapse in the state:

$$|\Psi\rangle_{sJ} = \frac{1}{3}(\alpha|RR\rangle_{sJ} + \beta|GG\rangle_{sJ} + \gamma|BB\rangle_{sJ}) \tag{20}$$

In order to get the signature qutrit, Bob needs John's help. He will measure the state $|\Psi\rangle_{sJ}$ using one of the mutually unbiased bases [6] of the base $\{|R\rangle, |G\rangle, |B\rangle\}$. We presume that the basis chosen is:

$$
\begin{aligned}
|\varphi_R\rangle &= \frac{1}{\sqrt{3}}(|R\rangle + |G\rangle + |B\rangle) \\
|\varphi_G\rangle &= \frac{1}{\sqrt{3}}(|R\rangle + e^{2\pi i/3}|G\rangle + e^{4\pi i/3}|B\rangle) \\
|\varphi_B\rangle &= \frac{1}{\sqrt{3}}(|R\rangle + e^{4\pi i/3}|G\rangle + e^{2\pi i/3}|B\rangle)
\end{aligned}
\tag{21}
$$

Depending on the basis, $|\Psi\rangle_{sJ}$ can be expressed as it follows:

$$
\begin{aligned}
|\Psi\rangle_{sJ} = \frac{1}{3}\Bigg[&\frac{1}{\sqrt{3}}|\varphi_R\rangle_J(\alpha|R\rangle_s + \beta|G\rangle_s + \gamma|B\rangle_s) \\
&+\frac{1}{\sqrt{3}}|\varphi_G\rangle_J(\alpha|R\rangle_s + e^{-2\pi i/3}\beta|G\rangle_s + e^{-4\pi i/3}\gamma|B\rangle_s) \\
&+\frac{1}{\sqrt{3}}|\varphi_B\rangle_J(\alpha|R\rangle_s + e^{-4\pi i/3}\beta|G\rangle_s + e^{-2\pi i/3}\gamma|B\rangle_s)\Bigg]
\end{aligned}
$$

If the result of the measurement will be $|\varphi_R\rangle_J$ then the state $|\Psi\rangle_{sJ}$ will collapse towards $\alpha|R\rangle_s + \beta|G\rangle_s + \gamma|B\rangle_s$, which is the state of the signature qutrit.

If the result of John's measurement is $|\varphi_G\rangle_J$, then the state $|\Phi\rangle_{sJ}$ will collapse in $\alpha|R\rangle_s + e^{-2\pi i/3}\beta|G\rangle_s + e^{-4\pi i/3}\gamma|B\rangle_s$. Bob will reconstruct the signature qutrit applying the operator:

$$
O_1 = \begin{pmatrix} 1 & 0 & 0 \\ 0 & e^{2\pi i/3} & 0 \\ 0 & 0 & e^{4\pi i/3} \end{pmatrix}.
$$

If the result of John's measurement is $|\varphi_B\rangle_s$, then the state $|\Phi\rangle_{sJ}$ will collapse in $\alpha|R\rangle_s + e^{-4\pi i/3}\beta|G\rangle_s + e^{-2\pi i/3}\gamma|B\rangle_s)$. Bob will reconstruct the signature qutrit applying the operator:

$$
O_2 = \begin{pmatrix} 1 & 0 & 0 \\ 0 & e^{4\pi i/3} & 0 \\ 0 & 0 & e^{2\pi i/3} \end{pmatrix}.
$$

After Bob extracts all the qutrits, he will reconstruct the signature, and then he will check its belonging by direct confrontation with the one in Alice's possession. The signature consists of the succession of the results obtained after "reading" the first node of the graph, respectively the last node of the graph. If the result obtained by Bob coincides with the signature in Alice's possession, it means that the authentication succeeded.

John also measured the state $|\Psi\rangle_{sJ}$ using the other two bases mutually unbiased:

$$|\varphi'_R\rangle = \frac{1}{\sqrt{3}}(e^{2\pi i/3}|R\rangle + |G\rangle + |B\rangle)$$

$$|\varphi'_G\rangle = \frac{1}{\sqrt{3}}(|R\rangle + e^{2\pi i/3}|G\rangle + |B\rangle)$$

$$|\varphi'_B\rangle = \frac{1}{\sqrt{3}}(|\alpha\rangle + |\beta\rangle + e^{2\pi i/3}|\gamma\rangle)$$

respectively:

$$|\varphi''_R\rangle = \frac{1}{\sqrt{3}}(e^{4\pi i/3}|R\rangle + |G\rangle + |B\rangle)$$

$$|\varphi''_G\rangle = \frac{1}{\sqrt{3}}(|R\rangle + e^{4\pi i/3}|G\rangle + |B\rangle)$$

$$|\varphi''_B\rangle = \frac{1}{\sqrt{3}}(|R\rangle + |G\rangle + e^{4\pi i/3}|B\rangle),$$

Bob succeeds eventually to obtain the qutrit s by applying one of the operators O_1 respectively O_2.

3.1 Security Analysis

The security of the method of information encapsulation is quantified by the easiness with which an intruder could get it. The intruder could try to find the encryption key, and then to extract the signature and to replace it with his own signature.

It was proved [7] that the security of the protocol H. Bechmann - Pasquinucci and A. Peres of the quantum key distribution is much higher than Bennett-Brassard's. Therefore, the cryptographic key in Alice's and John's possession is a safe key.

However, the most unpleasant situation is when either Bob or John is not honest. In Bob's case, there is the possibility that he extracts the information and replaces it afterwards with false information, taking thus the image in his own possession. In this case, Alice can approach John, who will reconstitute the signature separately, demonstrating thus Bob's dishonesty. If during several processes Bob proves to be dishonest, he will be replaced.

In the case in which John proves to be dishonest, Bob will not obtain the original signature. This is easy for Alice to find out, because she also knows the part from the secret key in John's possession, and she would be able to prove his dishonesty.

In a similar way, if Alice does not recognize her own signature, John can prove her dishonesty by the secret key he possesses, and with which he can remake the original signature.

4 Conclusions

The encapsulation schemes of quantum information presented above are made of three phases: the initial phase of establishing the encryption key, the intermediary phase of preparing the information to be hidden, and the final phase of checking its belonging. No matter the scheme used to hide the information, the purpose must be reached, i.e., to keep the secrecy of the encapsulated data. The analysis of the security of each scheme is considered important in reaching the objective proposed. The studies regarding the security of the quantum key distribution for the protocols that use two and three-level quantum systems show that the presence of an intruder can be easily discovered, the secret key obtained through such a protocol being safe. The only problem is the possibility that an intruder might make copies of the qutrits of the image (he copies the general image), which he could stock in order to extract afterwards the information encapsulated. The use of the secret keys, of a trustful person, or of the sharing of the quantum secret could be some of the alternatives to protect the information encapsulated against the attacks.

References

1. Zizzi, P.A. (1999). Holography, quantum geometry and quantum information theory. *The 8th UK foundations of physics meeting*, 13–17 September, London, UK.
2. Mütze, U. (2008). *Quantum image dynamics – an entertainment application of separated quantum dynamics*. www.ma.utexas.edu/mp arc/c/08/08-199.pdf.
3. Grudka, A., & Wójcik, A. (2003). *How to encode the states of two non-entangled qubits in one qutrit*. quant-ph/0303168.
4. Bennett, C.H., & Brassard, G. (1984). *Proceedings of the IEEE international conference on computers, systems and signal processing*, IEEE, New York.
5. Kuchment, P. (2008). Quantum graphs: an introduction and a brief survey. *Analysis on graphs and its applications. Proceedings of the Symposium of the Pure Mathematics*, AMS, 2008.
6. Bechmann-Pasquinucci, H., & Peres, A. (2000). Quantum cryptography with 3-state systems. *Physical Review Letters, 80*, 3313.
7. Bru, D., & Macchiavello, C. (2002). Optimal eavesdropping in cryptography with three-dimensional quantum states, http://arxiv.org/abs/quant-ph/0106126v2.

Chapter 23
Analyses of UWB-IR in Statistical Models for MIMO Optimal Designs

Xu Huang and Dharmendra Sharma

Abstract The third generation partnership projects spatial channel model has been attracting great and wider interests from the researchers for a stochastic channel model for MIMO systems and multi-antenna-based multi-input multi-output (MIMO) communications as they become the next revolution in wireless data communications. MIMO has gone through the adoption curve for commercial wireless systems to the today's situation, all high throughput commercial standards, i.e. WiMax, Wi-Fi, cellular, etc., have adopted MIMO as part of the optional. This paper is to present our investigations of the behaviors of the MIMO Ultra-Wide-Band-Impulse Radio (UWB-IR) systems, which will contribute to optimal designs for the low-power high-speed data communication over unlicensed bandwidth spanning several GHz, such as IEEE 802.15 families. We have developed and analyzed three no coherent transceiver models without requiring any channel estimation procedure. The massive simulations are made based on the established models. Our investigations show that the Poisson distribution of the path arriving will affect the signal-noise ratio (SNR) and that for the Nakagami distributed multipath fading channel the "m" factor, together with receiver number, will impact on the SNR of the MIMO UWB-IR systems.

Keywords MIMO · WiMax · UWB-IR · Poisson distribution · Nakagami distribution

1 Introduction

The multiple-input multiple-output (MIMO) system called Vertical Bell Laboratories layered space-time (V-BLAST) is used for very high spectral efficiency [2]. Although a maximum likelihood (ML) detection scheme has the best detection performance among the existing detection schemes; the complexity of ML detection

X. Huang (✉) and D. Sharma
Faculty of Information Sciences and Engineering, University of Canberra, 2601, Australia
e-mail: Xu.Huang@canberra.edu.au; Dharmendra.Sharma@canberra.edu.au

S.-I. Ao et al. (eds.), *Intelligent Automation and Computer Engineering*,
Lecture Notes in Electrical Engineering 52, DOI 10.1007/978-90-481-3517-2_23,
© Springer Science+Business Media B.V. 2010

is excessively high. Multiple-input multiple-output (MIMO) technique brings a relevant increase not only in capacity but also in coverage, reliability, and spectral efficiency. It is recalled that multi-antenna-based MIMO communications first occurred in the mid-1990s when researchers at Bell Labs and Stanford were looking for ways to increase system throughput without increasing bandwidth. After that thousands of research papers have been written on the topic dealing with both physical layer and network layer ramifications of the technology. In MIMO case, the overall transmit channel is described as a matrix instead of a vector, and the spatial correlation properties of the channel matrix define the number of available parallel channels for data transmission. In fact all high throughput commercial standards, such as WiMax, Wi-Fi, cellular, etc., have adopted MIMO as part of the optional. The adoption of MIMO into military wireless communications systems has played important role as well as in the commercial arena.

Ultra Wide Band Impulse Radio (UWB-IR) is an emerging wireless technology, proposed for low power high speed data communication over unlicensed bandwidth. This technology has been drawing great attentions from the researchers [1–9]. Currently the transceiver architectures have been showing the tendency of extending this technology to next generation WLAN compliant operating scenarios. Therefore, exploiting both spatial and temporal diversity and combing the MIMO technology with the UWB-IR become inevitable, which becomes our motivation to this current paper.

It is well known that the design of a MIMO communication system depends on the degree of knowledge of the channel state information (CSI), which is normally very expensive. In this paper, as normal way did, it is based on the UWB-IR statistical channel models. We take the noncoherent transceiver [7, 13–17, 19] and focus on wireless three models, namely Gaussian, Nakagami and log-normal distribution channels. We, in particularly, extended the previous research results [11–15] to investigate how the Nakagami m factor impacts on the signal to noise ratio (S/N) in the statistical channel model. The idea of increasing the efficiency of a wireless communication system by applying multiple input and output antennas goes back to 1970. A.R. Kaye and D.A. George worked on a wireless communication system that tried to improve bandwidth efficiency with multiple input and output antennas. During the 1980s Jack Winters at Bell Laboratories published several papers on MIMO applications. Spatial Multiplexing using MIMO was first patented in 1993 by Arogyaswami Paulraj and Thomas Kailath (US Patent No 5,345,599). In 1998 Bell labs demonstrated Spatial Multiplexing in 1998. Today couple of companies, Beceem Communications, Samsung, Runcom Technologies, have developed MIMO based solutions for IEEE 802.16e WIMAX. Other companies like Broadcom and Intel have successfully applied MIMO-OFDM in IEEE802.11n, which is supposed to take over IEEE802.11g pretty soon. Anybody can buy a wireless n router and a wireless card and experience the benefits of MIMO. 4G will be implementing MIMO-OFDM or MIMO-CDMA. In this paper we will take a look at how MIMO works and in the following papers we will weigh the pros and cons of OFDM and CDMA.

The simulations under various conditions have been done in this paper, our simulations presented the following suggestions: (a) if we take so called single-cluster Poisson model [7], namely the random integer valued number, then the mean of this random integer valued number will impact on the MIMO S/N regardless which of three models (Gaussian, Nakagami and log-normal distribution channels); and (b) for the Nakagami distribution channel, as we expected that the "m" factor will impact on the MIMO S/N.

In the next section the MIMO UWB-IR statistical channels are to be investigated. Then, the models for those discussed statistical channel will be established in Section 3. In Section 4 simulations will be given for the investigated models in Section 3. In the final section the conclusions are presented.

To insert images in *Word*, position the cursor at the insertion point and either use Insert | Picture | From File or copy the image to the Windows clipboard and then Edit | Paste Special | Picture (with "Float over text" unchecked).

IMECS 2006 reserves the right to do the final formatting of your paper.

2 Statistical Models for MIMO UWB-IR

It is well known that multiple-input multiple-output technology can significantly improve wireless link performance. The performance enhancement, however, comes at the cost of higher radio frequency (RF) hardware complexity. For systems that have a large number of antennas in both transmit and receive sides, the increase may be rather formidable. Most of the previous studies on antenna selection in MIMO systems were concerned only with either the receiver antenna selection or the transmit antenna selection. The baseband point to point (P2P), shown in Fig. 1, is composed by N_t transmit and Nr receive antennas working on an UWB-IR MIMO channel.

At the signaling period T_s second the source of Fig. 1 generates an L-ary ($L \geq 2$) information symbol b, i.e. $b \in \{0, 1, \ldots, L - 1\}$. The multi-antenna transmitter

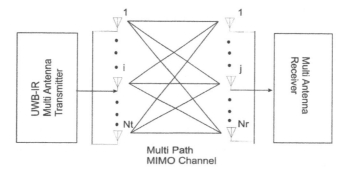

Fig. 1 The MIMO point to point UWB-IR system with N_t transmit and N_r receive antennas. The MIMO UWB-IR channel is affected by multipath fading that is described by $N_t \times N_r$ baseband impulse channel responses

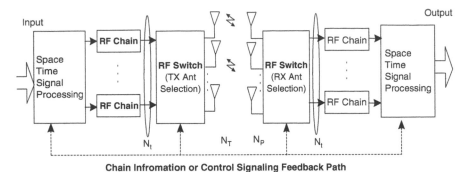

Fig. 2 A typical block diagram for transmit receive antenna frame in MIMO systems

maps b onto N_t M-ary baseband signals of time duration limited by T_s. It is noted that the USB baseband pulse is limited to pulse time, T_p, and repeated N_f times over each signaling period, T_s, here N_f is the number of frames and the time for frames of duration denoted by T_f. In order to avoid inter-frame interference (IFI), we must have $T_f > T_\mu$, where T_μ is the UWB channel delay spread time.

For the details for the Fig. 1, we may describe a typical block diagram as Fig. 2 as transmit and receive antenna frame in MIMO systems.

It is well known that we have single input single output (SISO) UWB-IR channel by IEEE 802.14. If we take the impulse channel responses in Fig. 1 as $h_{ji}(t)$, $0 \leq j \leq N_t$, $0 \leq j \leq N_r$, we may collect these impulse responses into the corresponding $(N_t \times N_r)$ matrix $\mathbf{H}(t)$ [2]. Therefore, as IEEE 802.15 recommend that each SISO impulse response $h_{ji}(t)$ in Fig. 1 is modeled as the superposition of several path clusters, with both inter-cluster and intra-cluster inter-arrival times being exponentially distributed.

Hence, we have [1, 7]:

$$h_{ji} = \sum_{n=0}^{V} h_n(j,i)\delta(t - \tau_n)$$

$$= \sum_{n=0}^{V} \beta_n(j,i)\alpha_n(j,i)\delta)t - \tau_n) \qquad (1)$$

here, $1 \leq i \leq N_t, 1 \leq j \leq N_r$

It is noted the integer valued number V of received paths over a signaling period T_s is a Poisson distributed random variable with mean value $E\{V\} = \lambda T_s$, where λ is rate in $(ns)^{-1}$, τ_n is the non-negative arrival time of the nth path, in ns. We use $h_n(j,i)$ for the nth path gain of SISO link going from the ith transmit antenna to the j-th receive one. The random variable (r.v.) $\beta_n(j,i) \in \{-1, 1\}$ and the non-negative r.v. $\alpha_n(j,i)$ are the corresponding phase and amplitude, respectively. As the previous references [1, 7, 11–14, 17] show that the statistic of the fading affecting rich-scattered medium-range quasi-LOS UWB-IR links may be well modeled by resorting to the Nakagami distribution, long-normal distributed channel amplitudes, $\alpha_n(j, i)$ may be suitable for less scattered LOS short-range indoor propagation environments and the log-normal distribution is recommended by IEEE 802.15 workgroups for WPAN and sensor applications [1, 2].

The central limit theorem [2, 18] underpin the fact that zero-mean Gaussian distributed channel coefficients well model highly scattered outdoor NLOS propagation environments.

For the space-time orthogonal PPM (OPPM) modulated the size M of the employed OPPM format equates LN_t and N_t columns of the l-th matrix codeword Φ_l are constituted by the N_t unit-vectors of R^M with index i ranging from $i = lN_t$ to $((l + 1)N_t) - 1$, i.e.

$$\Phi_l = [e(lN_t) \ldots e((l + 1)N_t - 1)], \quad 0 \le l \le L - 1 \tag{2}$$

Because of orthogonal and unitary we have:

$$\begin{aligned} \Phi_l^T \Phi_m &= 0, \qquad for\ l \ne m \\ \Phi_l^T \Phi_l &= I_{Nt}, \qquad for\ any\ l \end{aligned} \tag{3}$$

We also have the relation between Bit-Error-Probability $P_E^{(b)}$ and the corresponding Word Error Probability P_E [7] as shown below:

$$P_E^{(b)} = \left(\frac{L}{2(L-1)} \right) P_E \tag{4}$$

As the general equation we have the decision statistics set $\{z_l\}$ can be expressed by

$$\Phi_{ML} = \arg\max_{0 \le l \le L-1} \{z_l\} \tag{5}$$

Our next target is try to use the obtained analytically mathematical expresses to closely look at different statistic models for various communication cases. Now we can take z_l as different statistics for the three major models we discussed above, namely, (1) Nakagami distribution, (2) log-normal distribution, and (3) Gaussian distribution. The models for those channels will be investigated in the next section.

3   Channels Analyses with MIMO UWB-IR of Different Statistic Models

We first investigate how does the r.v. parameter, V, impact on the S/N in the above three different channels with statistic models?

For the Nakagami distribution multipath fading channel, we have [7]:

$$z_l = \sum_{n=0}^{V} \sum_{j=1}^{N_r} \sum_{i-1}^{N_t} \ln\{\cosh[\phi(y_j(n)^T e_i(l))]\} \tag{6}$$
$$where,\ l = 0\ldots, L - 1$$

and ϕ_n is defended as

$$\varphi_n = \beta e^{c\mu_n},\ n = 0, \ldots, V\ and\ c = \tfrac{1}{20}\ln(10),$$
$$\mu_n = E\{\alpha_n(j, i)\}.$$

Also as the Appendix of [7] mentioned, we have the word error probability (WEP):

$$P_E \le (L - 1) \left(\frac{4e^2\Gamma(2m)}{\Gamma(m)}\right)^{N_t N_r (V+1)} \cdot \prod_{n=0}^{V} \left[\frac{\left(1 + \frac{\sigma_n^2\beta^2}{m}\right)}{\left(1 + 2\frac{\sigma_n^2\beta^2}{m}\right)}\right]^{m N_t N_r} \tag{7}$$

Here G(.) is the Gamma function [10–13], it is noted that if r.v. V is large enough Eq. 6 can be simplified further format.

We now consider the situation of log-normal distributed multipath fading, i.e. the fading amplitudes $\{\alpha_n(j, i)\}$ is log-normal distribution with $m \ge 0.5$, we have [14]:

$$P_E \le (L - 1) \left(\frac{2}{\sqrt{\pi}}\right)^{(V+1)N_t N_r} \prod_{n=0}^{V}$$
$$\times \left[\int_{-\infty}^{+\infty} \exp\{-t^2 - \beta\phi_n e^{c\mu_n} \exp\{c\sqrt{2}\sigma_r t\}\}dt\right]^{N_t N_r} \tag{8}$$

Finally let's have a closer look at the Gaussian distribution, we have [14]:

$$P_E \le (L - 1) \prod_{n=0}^{V} \left[\frac{\left(1 + \sigma_n^2\beta^2\right)}{\left(1 + \frac{1}{2}\sigma_n^2\beta^2\right)^2}\right]^{N_t N_r / 2} \tag{9}$$

It is noted that the situation similar to Eq. 7 and that when r.v. V is larger than unit we can simplify Eq. 9. Now we have the major distributions with their analytic formats.

The following section we shall present a number of simulations under different conditions to explore the behaviors of MIMO UWB-IR of different statistic channels.

4 Simulations of MIMO UWB-IR of Different Channels with Statistic Models

Our target is, based above induced results, to investigate two situations that lead the optimal designs for MIMO UWB-IR communications, namely (a) because we don't want to have expensive channel state information (CSI), we take the "single cluster Poisson Model" for capturing the behavior of each $h_{ji}(t)$. Therefore, question occurs: how does the r.v. V impact on the S/N of the MIMO UWB-IR transceiver channels? (b) As Nakagami distribution is of important wireless communication distributions and the major parameter, m, will impact on the Nakagami distributions. Hence, the second question occurs: how does the factor m of the Nakagami distribution impact on the S/N of the MIMO UWB-IR transceiver channels?

In the following section, the massive simulations, based on above theories, are made for those two questions.

For the first question without loss generality we take simple case, $L = 2$, and the corresponding SISO impulse responses $\{h_{j,i}(t)\}$ in Eq. 1 and we have been generated according to the CM 6 UWB-IR channel model, i.e. IEEE 802.15.4 with $\lambda = 1.13(1/\text{ns})$, $T_\mu = 15.9(\text{ns})$, $\gamma = 9.3$ (ns), $N_f = 8$, and the spectral efficiency of 1/200 (bit/sec/Hz). The simulations first take $N_r = 1$ and then let $N_t = 1, 2, 3, 4$, namely investigating the MISO situations.

Under the above conditions, Figs. 3, and 4 show the $V = 5, 15$ with Nakagami distribution multipath channels. Here we have the parameters: Nakagami distribution multipath channel with $N_r = 1$ and $N_t = 1, 2, 3,$ and 4 the S/N is in "dB".

It is clearly to show by those figures that under the same statistic distribution the random variable V has impacted on the S/N under the same BER. For example, for the targeted BER, 10^{-5}, when the $N_t = 2$ there are 1.6 dB dropped and in general case it is obviously that with V increasing the S/N will significantly dropped (Figs. 3 and 4). It is noted that we did not change any Nakagami parameter such as "m," as

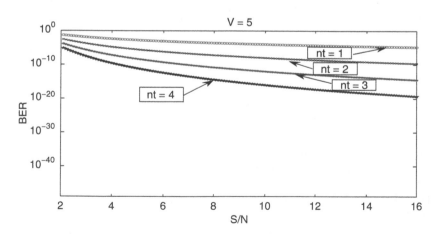

Fig. 3 Nakagami distribution multipath channel with $N_r = 1$ and $N_t = 1, 2, 3,$ and 4 the S/N is in "dB"

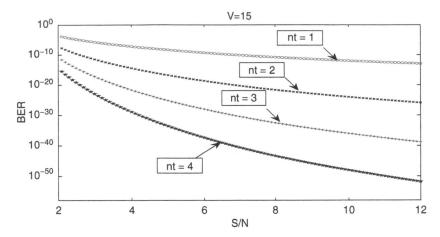

Fig. 4 Nakagami distribution multipath channel with $N_r = 1$ and $N_t = 1, 2, 3,$ and 4 the S/N is in "dB"

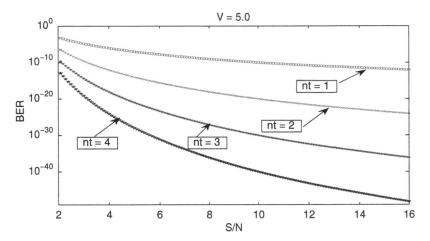

Fig. 5 Log-normal distribution multipath channel with $N_r = 1$ and $N_t = 1, 2, 3,$ and 4 the S/N is in "dB"

we are now focus on how the random variable V impact on the S/N for the MIMO communications. After that we are going to show how the Nakagami parameter m and random variable together impact on the S/N to the MIMO communications.

Figures 5 and 6 presented the almost similar situations as that in Figs. 3 and 4 except for the distribution changed from Nakagami distribution to log-normal distribution.

Here we have Log-normal distribution multipath channel with $N_r = 1$ and $N_t = 1, 2, 3,$ and 4 the S/N is in "dB" in Figs. 5 and 6. Again, we are focus on the in this particular distribution how the MIMO parameters (N_t) impact on the S/N. Even though the distributions are changed from the Nakagami- to log-normal-distributions, the simulation conclusions are highly similar, which can be evidenced

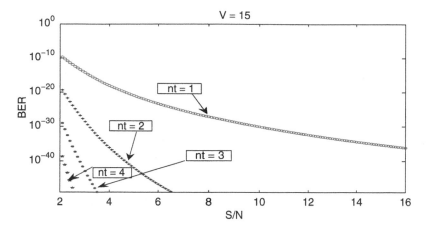

Fig. 6 Log-normal distribution multipath channel with $N_r = 1$ and $N_t = 1, 2, 3$, and 4 the S/N is in "dB"

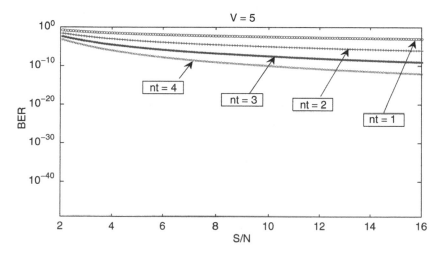

Fig. 7 Gaussian distribution multipath channel with $N_r = 1$ and $N_t = 1, 2, 3$, and 4 the S/N is in "dB"

by the observations from Figs. 5 and 6. However, it is noted that under the similar conditions log-normal distribution will cause more S/N drops if we compare the simulation results obtained from Fig. 2 with that from Fig. 4. This is not surprised as the samples increased those two distributions approach the common nature.

Let's have a look at Figs. 6 and 7, which show the other different distributions. The distribution becomes zero mean Gaussian distribution, which models highly-scattered outdoor NLOS propagation environments.

Here we have Gaussian distribution multipath channel with $N_r = 1$ and $N_t = 1$, 2, 3, and 4 the S/N is in "dB" in Figs. 7 and 8. It is indeed, as we observed, the more drops under the same conditions.

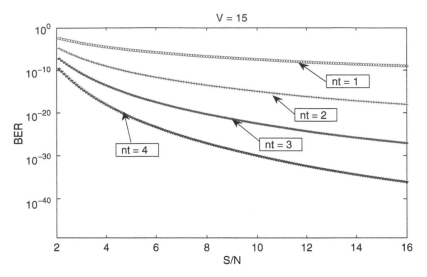

Fig. 8 Gaussian distribution multipath channel with $N_r = 1$ and $N_t = 1, 2, 3,$ and 4 the S/N is in "dB"

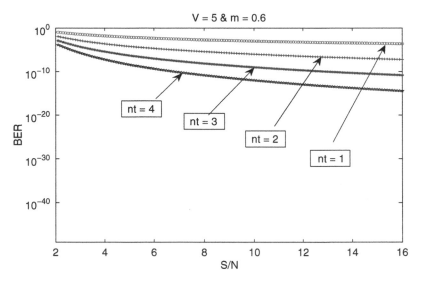

Fig. 9 Nakagami distribution with m $= 0.6$ V $= 5$ the rest parameters are the same as that in previous figures

The observations from Figs. 7 and 8 presented that as the communications is modeling for the out door the environmental situations are completed in terms of noises the impacts on the SNR would be stronger as expected.

In order to investigate how does the m factor affect the Nakagami distribution as the above second question described, we have taken the factor $m = 0.6$ and 0.9 and the random variable $V = 5$ and 15 and the simulations are shown in Figs. 9–12.

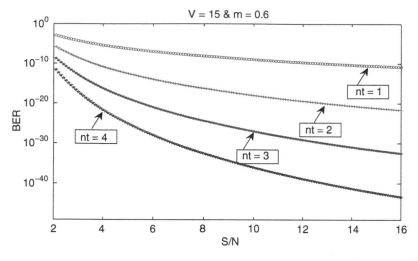

Fig. 10 Nakagami distribution with $m = 0.6\ V = 15$ the rest parameters are the same as that in previous figures

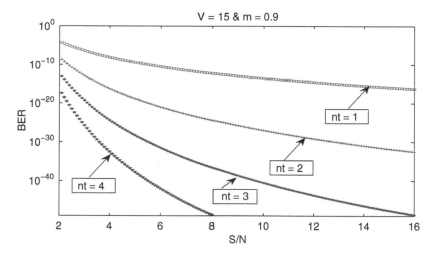

Fig. 11 Nakagami distribution with $m = 0.9\ V = 5$ the rest parameters are the same as that in previous figures

For example, for the targeted 10^{-5} when $N_t = 2$ under the same conditions except for $m = 0.6$ and $m = 0.9$ the former S/N dropped 1.9 dB in comparison with later (referring Figs. 9 and 11). Also from Figs. 8 and 10 for $N_t = 3$, at the targeted 10^{-5}, we have S/N dropped about 2 dB from $m = 0.6$ to $m = 0.9$ with the same r.v. V values.

Figures 13–15 show the same BER vs. S/N with the comparable parameters but for receiver number, $N_r = 1$, 2 and 3. From those simulations we can observe

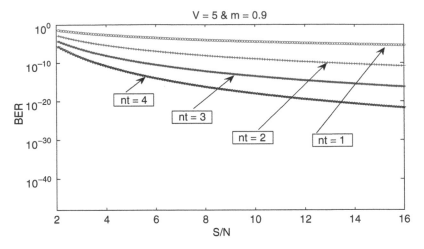

Fig. 12 Nakagami distribution with $m = 0.9\, V = 15$ the rest parameters are the same as that in previous figures

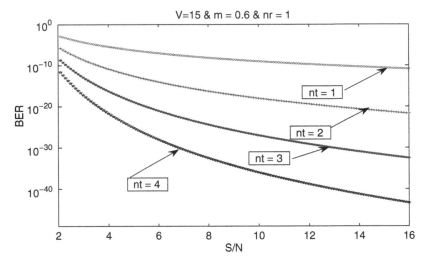

Fig. 13 BER vs. S/N with the same condition as mentioned above and $N_r = 1$, $m = 0.6$, $V = 15$ for the Nakagami distributions

that as the receiver number increasing, for the Nakagami distribution multipath communication channels, the S/N will drop because of this model (Nakagami distribution) focus on the case that the communication channel approaches to quasi-LOS, which is now deviating from the assumptions when the N_r becomes lager.

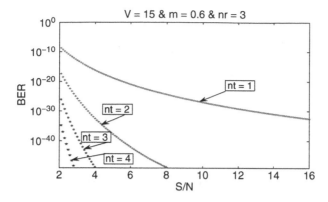

Fig. 14 BER vs. S/N with the same condition as mentioned above and $N_r = 2$, $m = 0.6$ and $V = 15$ Nakagami distributions

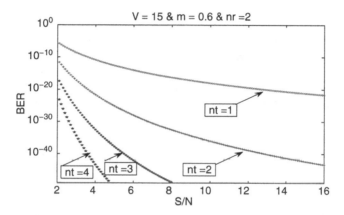

Fig. 15 BER vs. S/N with the same condition as mentioned above and $N_r = 3$, $m = 0.6$ and $V = 15$ Nakagami distributions

5 Conclusions

It is well known that multiple-input multiple-output technology can significantly improve wireless link performance. It was a groundbreaking development pioneered by Jack Winters of Bell Laboratories in his 1984 article [20], since then, many academics and engineers have made significant contribution to the understanding of MIMO systems. In 1996, radically novel approaches were invented to increase signaling efficiency over MIMO channels, Gregory G. Raleigh and V.K. Jones wrote a paper [21] arguing that multi-path channels can have a multiplicative capacity effect if the multi-path-signal propagation in used in an appropriate communications structure. In the same year, Foschini [22] introduced the BLAST concept in his paper. BLAST is one of the most widely examined techniques today. In 1999, the Shannon capacity of an isotropic fading MIMO channel was calculated by I. Emre Telatar

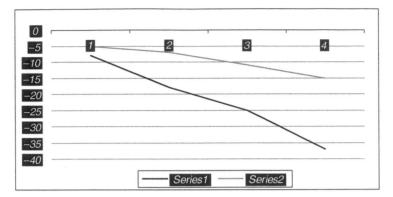

Fig. 16 Comparison with $V = 5$ and $V = 15$ when S/N $= 6\,$dB in Nakagami distribution without changing all other parameters

[23]. He stated that the channel capacity increases with the number of antennas and is proportional to the minimum number of transmit or receive antennas. This basic information theoretic result drew widespread attention to MIMO communications. Also in 1999, Gigabiy Wireless Inc. and Stanford University successfully held the first outdoor prototype demonstration. Iospan Wireless Inc. produced the first commercial product in 2002. As one of the first, a 4×4 MIMO academic test-bed started operation at University of Alberta in 2003 [24].

In this paper, in order to have optimal designs for wireless MIMO UWB-IR transceiver multipath communication channels, in particularly, under the condition of that there is no expensive CSI we have established statistic models for three major situations in MIMO UWB-IR communications. They are Nakagami distribution, log-normal distribution, and Gaussian distribution. The random variable changes produce impacts on the output signal-to-noise ratio are shown in Fig. 16 for Nakagami distribution. The pink one (series 2) in the Fig. 16 is for the random variable, $V = 5$ and the blue one (series 1) is for the random variable, $V = 15$. It is clear that the larger random variable will make larger changes of the signal to noise ratio (in dB).

Our paper focuses on (a) how does the random variable V affect MIMO UWB-IR multipath communication channels? (b) If we stick with general LOS case, Nakagami fading channel, how does the major "m" factor affect the MIMO UWB-IR communication channels? Our have presented massive simulations, based on theoretically investing, which show the answers for above questions we concerned. The simulation results also offer better information for the optimal designs for MIMO UWB-IR transceiver multipath communication channels. Finally we also investigate how the receiver number affects the MIMO UWB-IR S/N. All those results will give the optimal designs for MIMO UWB-IR transceiver multipath communication channels.

References

1. Reed, J.H. (2005). *An introduction to ultra wideband communication systems*. Englewood Cliff, NJ: Prentice Hall.
2. Molishch, A.F., Cassion, D., et al. (2006). A comprehensive standardized model for ultra-wideband propagation channels. *IEEE Transactions on Antennas and Propagation, 54*(11), 3151–3166.
3. Ohtsuki, T. Space-time Trellis coding for UWB-IR. *Vehicular Technology Conference, VTC 2004*, Spring 2004 IEEE 59th, *2*, 1054–1058.
4. Ezaki, T., & Ohtsuki, T. (2005). Rake performance for UWB-IR system with SISO and MISO. *IEICE Transactions on Communications, E88-B*(10), 4112–4116.
5. Barton, R.J., & Rao, D. (2008). Performance capabilities of long-range UWB-IR TDOA localization systems. *EURASIP Journal on Advances in Signal Processing, 2008*(81), 1–17.
6. Lee, S.-H., Kim, N.-S., Kang, H.-J., Kim, S.-G. (2006). Performance improvement of intelligent UWB-IR communication system in multipath channel. *ICIC 2006* (pp. 1103–1108). Berlin, Heidelberg: Springer.
7. Reed, J.H., Baccarelli, E., Biagi, M., Pelizzoni, C., Cordeschi, N. (Jan 2008). Optimal MIMO UWB-IR transceiver for Nakagami-fading and Poisson-Arrivals. *Journal of Communications, 3*(1), 27–40.
8. Promwong, S., Hanitach, W., Takada, J.-I., Koon, P.S., Tangtisanon, P. (Oct–Dec 2003). Measurement and analysis of UWB-IR antenna performance for WPANs. *Thammasat International Journal of Science and Technology, 8*(4), 56–62.
9. Duenas, S.R., Dno, X., Yamac, S., Lsmail, M., Zheng, L.-R. (2006). CMOS UWB IR noncoherent receiver for RF-ID applications. *Applications, Circuits and Systems, IEEE*, 2006 IEEE North-East Workshop on Circuits and Systems. Gatineau, Que., Canada, 18–21 June 2006, 213–216.
10. Proakis, J. (2001). *Digital communications* (4th ed.). New York: McGraw-Hill.
11. Huang X., & Madoc, A.C. (2005). Image multi-noise removal via Levy process analysis. *Lecture Notes in Computer Science* (vol. 3684, pp. 22–25). Berlin, Heideberg: Springer.
12. Huang, X. (June 23–26, 2008). Noise removal for image in Nakagami fading channels by Wavelet-based Bayesian Estimator. *IEEE International Conference on Multimedia & Expo 2008* (pp. 21–24). Germany.
13. Huang, X., & Sharma, D. (11–13 Feb 2009). MIMO UWB-IR noncoherent transceiver with Poisson wireless models. *IEEE International Symposium on Wireless and Pervasive Computing* (pp. 11–16). Melbourne, Australia. ISBN: 978-1-4244-2966-0.
14. Huang, X., & Sharma, D. (Mar 2009). Behaviours of MIMO UWB-IR transceiver with statistical models. *International Multi-Conference of Engineers and Computer Sciences 2009* (Proc. pp. 440). (IMECS 2009), Hong Kong.
15. Huang, X., & Sharma, D. (18–19 Mar 2009). Behaviours of MIMO UWB-IR transceiver with statistical models. *International Multi-Conference of Engineers and Computer Sciences 2009* (pp. 440). (IMECS 2009), Hong Kong.
16. Baccarelli, E., Biagi, M., Pelizzoni, C., Cordeschi, N. (2007) Non-coherent transceivers for multipath-affected MIMO UWB-IR communications. http://infocom.uniroma1.it/~pelcris/tech.1.pdf
17. Nagesh Polu, V.V.S., Colpits, B.G., Peterson, B.R. (2006). Symbol-wavelength MMSE gain in a multi-antenna UWB system. *IEEE Proceedings of the 4th Annual Communication Networks ans Service Research Conference* (CNSR'06), (pp. 1–5).
18. Ezaki, T., & Ohtsuki, T. (2004). Performance evaluation of space hopping ultra wideband impulse radio (SH-UWB-IR) system. *IEEE International Conference on Communication* (ICC'04), *6*, 3591–3595.
19. Wei, Y.R., & Wang, M.Z. (Feb 2007). Efficient capacity-based joint transmit and receive antenna selection schemes in MIMO systems. *IEICE Transactions on Communication, E90-B*(2), 372–376.

20. Winters, J. (Aug 1984). Optimum combining in digital mobile radio with cochannel inter-ference. *Special Issue on Mobile Radio Communications IEEE Journal on Selected Areas in Communications. IEEE Transactions on Vehicular Technology,* 2(4), 528–539.
21. Releigh, G.G., & Jones, V.K. (May 1999). Multivariate modulation and coding for wireless communication. *IEEE Journal of Selected B Areas in Communication,* 17(5), 851–866.
22. Gerard, J.F. (Oct 1996). Layer space-time architecture for wireless communication in a fading environment when using multi-element antennas. *Bell Laboratory Technical Journal,* 1(2), 41–59.
23. Emre Telatar I. (Nov 1999). Capacity of multi-antenna Gaussian channels. *European Transactions on Telecommunications,* 10, 585–595.
24. Bolcskei, H. (13 Oct 2005). *MIMO Systems.* ETH Zurich: Communication technology laboratory.

Chapter 24
Intruder Recognition Security System Using an Improved Recurrent Motion Image Framework

Chuan Ern Wong and Teong Joo Ong

Abstract In this paper, we present an extension to the Recurrent Motion Image (RMI) motion-based object recognition framework for use in the development of an automated video surveillance Intruder Recognition Security (IRS) System. We extended the original object classes of RMI to include *four-legged animal* (such as dog and cat), and various enhancements are made to the object detection and classification algorithms for better object segmentation, error tolerance and recognition accuracy. Under the new framework, object blobs obtained from background subtraction of scenes are tracked using region correspondence. In turn, we calculate the RMI signatures based on the silhouettes of the object blobs for proper classification. The framework functions as the core of the IRS System to provide intruder recognition function and to reduce nuisance alarms since the system is capable of differentiating different category of objects in the surveillance area. A recognition rate of approximately 98% (40 out of 41 moving objects in the experiments were correctly classified) was achieve in our tests based on several real world 320 × 240 resolution color image sequences captured with a low-end digital camera, and also on the PETS 2001 dataset. Thus, indicating the applicability of the new RMI framework to minimize nuisance alarms in an IRS System.

Keywords Intruder recognition · Moving object recognition · Recurrent motion image · Surveillance security

1 Introduction

Conventional security sensors are commonly installed at residences and workplaces to detect intrusion or motion at various entry points [1]. Nuisance alarm is a frequent occurrence for such system in that the users may eventually grow accustomed

C.E. Wong (✉) and T.J. Ong
Faculty of Information and Communication Technology, Universiti Tunku Abdul Rahman,
Petaling Jaya, 46200 Selangor, Malaysia
e-mail: wongce1@mail2.utar.edu.my; ongtj@utar.edu.my

S.-I. Ao et al. (eds.), *Intelligent Automation and Computer Engineering*,
Lecture Notes in Electrical Engineering 52, DOI 10.1007/978-90-481-3517-2_24,
© Springer Science+Business Media B.V. 2010

to ignoring or turning off the alarms rather than investigating the causes of them. To circumvent this problem, some systems [2, 3] measure the size of the object before generating an intrusion alert. For example, systems with pet-immunity function distinguish between human and animal by measuring the object size to reduce nuisance alarms caused by the pets. However, size discrimination is not the best solution since we may still keep human-sized pets, such as a large dog, in the premises [4].

The need for more sophisticated security systems, such as a video surveillance system with computer monitor outputs, is increasing to provide better security, area surveillance and nuisance alarms prevention. However, a video surveillance system without proper intruder detection and recognition functions has to rely on security personnel to manually inspect and detect intrusions. Such a manned system is insufficient to address the needs of creating a more secured environment because it is highly susceptible to human negligence and errors.

Due to the aforementioned reasons, an automated Intruder Recognition Security (IRS) System that detects and tracks moving objects in a surveillance area, and classify them into various predefined categories (*human*, *four-legged animal* and *vehicle*) is invaluable to the users since specific intrusion alerts and action plans can be customized easily based on the category of the objects that trespasses the surveillance area. Deployment of an IRS System can greatly reduce time and human resources wasted on handling nuisance alarms and active monitoring of the terminals. In addition, it reduces many of the human error factors that commonly plague manned systems.

2 Overview of the IRS System

The proposed IRS System consists of two main modules: Intruder Recognition Framework (IRF), and Control Center (CC) (see Fig. 1). Input image sequence from the surveillance area is processed by the IRF which forms the core of the IRS System. The IRF detects, tracks and classifies intruders within the surveillance area and its outputs are used by the CC to trigger the appropriate intrusion alerts setup by the user. The alerts are delivered via the CC, to acknowledge users upon an intrusion event.

Log of all intrusion events, organized by time and intruder category, is also accessible from the CC. In addition, this system allows users to select the object classes that are prohibited in the surveillance area and vice versa to reduce nuisance alarms (see Sections 3 and 4).

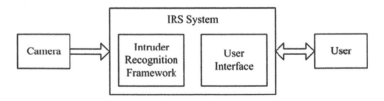

Fig. 1 Overall structure and connections of IRS system

3 Intruder Recognition Framework

Moving object recognition has been an active research area for computer vision and pattern analysis applications. It is devoted to detect and track moving objects in a surveillance area, and classify the objects of interest into various predefined categories. Many approaches, such as [5–8], have been presented for such purposes.

The RMI method [5] is one of the few approaches that produce high recognition rate while remaining computationally and space efficient. A specific feature vector called RMI was proposed to estimate the recurrent motion behavior of moving objects. Different kinds of object have different motion behaviors, yielding different RMIs. Consequently, a moving object can be classified as a *single person*, *group of persons* or *vehicle* based on its RMI. The areas of RMI demonstrating high motion recurrence are used to determine the object class. For example, the RMI of a walking human shows high recurrence near the hands and legs, whereas the RMI of a moving vehicle shows no motion recurrence.

Experiments conducted in [5] indicate that this approach yields correct classification in about 97% of all tested samples. However, the shadow removal algorithm in the original framework suffered an error rate of 30% due to segmentation failure. The segmentation algorithm failed to divide cast shadows and self shadows in different regions. Besides, the framework has only been tested on a small set of object classes (*human* and *vehicle*). Also, the error tolerance of the original algorithm is low since it is unable to accommodate slight deviations in the RMI data. For instance, a person who walks with hands in the pockets will not be recognized as a *human* because the resultant RMI does not exhibit significant hand movements. The person is categorized as *other object* since the RMI does not match with any of the predefined classes. Such limitations, in essence, confined the recognition range and accuracy of the framework.

Due to the advantages of [5] over other recognition approaches, the RMI approach is adopted for use in the IRF. Several refinements are introduced to the original RMI framework before incorporating it in the IRS System to improve its recognition accuracy and robustness:

1. Improved shadow removal by using a better segmentation algorithm (see Section 3.1)
2. Extended classification list by including *four-legged animal* (such as dog and cat) as a new object class (see Section 3.3)
3. Improved error tolerance by modifying the existing classification algorithm, especially for recognition of a *human* (see Section 3.3)

In general, the intruder recognition framework is divided into three phases called detection, tracking and classification. The algorithms and enhancements for each phase are detailed in the subsequent sections.

3.1 Detection

The original RMI framework [5] uses a mixture of K Gaussian distributions [9] to perform background subtraction, followed by connected components labeling [10] to segment the foreground pixels into regions. A combination of color segmentation using K-means approximation of the EM algorithm and gradient direction [5] is applied to locate and remove shadows. Their experimental results indicated that the shadow removal process failed in about 30% of the frames that contain significant shadows. The errors were caused by failure to divide cast shadows and self shadows in different regions. To improve the framework's shadow removal and object segmentation functions, we implemented a different detection algorithm and multiple levels of noise filtering are applied at preprocessing for better moving object segmentation (see Fig. 2).

The detection algorithm starts with background subtraction which is carried out by computing an L-inf distance image [11] in the Red-Green-Blue (RGB) color space. Foreground points are extracted by applying a low threshold (±0.08 in our

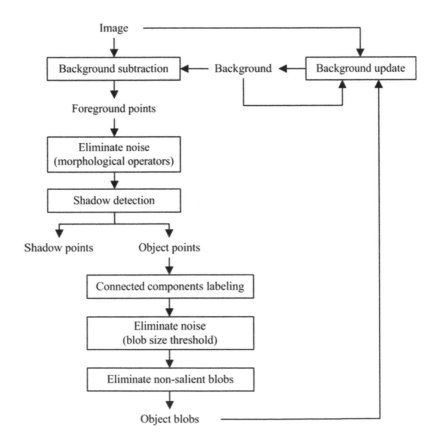

Fig. 2 Object detection algorithm

experiments) to the L-inf distance image. These foreground points will go through the first denoise layer where morphological opening [12] operation is performed. Subsequently, shadow points are located by transforming the image pixel values to the Hue-Saturation-Value (HSV) color space for better accuracy [13] and removed from the foreground points. The resultant object points will go through connected components labeling to obtain the foreground blobs. Blob analysis is carried out to filter noise clutters using a blob size threshold. Lastly, a high threshold (± 0.4 in our experiments) is applied to the L-inf distance image to select points with large difference from the background. Blobs consisting of at least one of these salient points are validated whereas the others are removed as non-salient blobs.

In addition, we adopted a simple median background update to increase the framework's adaptability to long term illumination changes. As denoted in Eq. 1, the background is updated using object-level reasoning which is proven to be more reliable and less sensitive to noise [11] in contrast to point-level selectivity. For pixels that are associated to the detected moving objects or shadows in frame t, the background model $B^{t+\Delta t}$ adopts the pixel values of current background B^t. On the other hand, the values of background model $B^{t+\Delta t}$ for pixels that do not belong to any of the moving objects or shadows in frame t, are obtained from the statistical model $B_S{}^{t+\Delta t}$ which performs median function to a set of elements as shown in Eq. 2. This set consists of current image I^t, n previous image frames sub-sampled at a rate of one every Δt, and current background B^t with its weight ω_b.

$$B^{t+\Delta t}(p) = \begin{cases} B^t(p) & if \quad p \in \{object \cup shadow\}^t \\ B_S^{t+\Delta t}(p) & otherwise \end{cases} \tag{1}$$

$$B_S^{t+\Delta t}(p) = median\left[\left\{I^t(p), I^{t-\Delta t}(p), \dots, I^{t-n\Delta t}(p)\right\} \cup \omega_b\left\{B^t(p)\right\}\right] \tag{2}$$

3.2 Tracking

Blobs obtained from the detection phase are tracked using region correspondence [14]. Various parameters and descriptors such as centroid, bounding box, size, velocity and change in size of each blob, are extracted from the blobs. Correspondences between regions in the previous and current frames are established using the minimum cost criteria [5] to update the status of each object over the frames.

As shown in Fig. 3, there might be non-corresponding regions in the previous and current frames. Since object exit or occlusion events may be associated to some of the regions in the previous frame, they must be examined based on the following rules:

1. If a region's predicted position exceeds the frame boundary, the corresponding object is determined to have exited the surveillance area; otherwise, object occlusion may have happened.

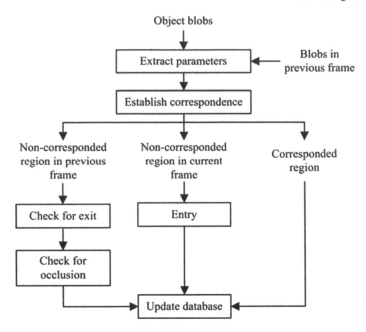

Fig. 3 Object tracking algorithm

2. If an object's bounding box overlaps the bounding box of another region Q in the current frame, Q is marked as an occluded region, and all of the non-corresponded regions in previous frame overlapping Q are, thus, marked as occluding each other.
3. Lastly, non-corresponded region in the current frame is set to be an object entry.

3.3 Classification

The classification phase categorizes each of the moving objects detected and tracked in the previous phases into the following object classes: (1) *single person*, (2) *group of persons*, (3) *vehicle*, and (4) *four-legged animal*. Recurrent motion (represented by RMI) which is denoted as repetitive changes in the shape of object is the essential feature that differentiates the object classes. As shown in Eqs. 3 and 4, RMI is generated by determining the areas of a moving object's silhouette that shows repetitive changes. In the equations, S_a is a binary silhouette for object a at frame t, whereas DS_a is a binary image that indicates the areas of motion for object a in between frame t and $t-1$. RMI_a which is the RMI for object a calculated over T frames, has high values at pixels where motion occurred repetitively and low values at pixels where little or no motion occurred.

$$DS_a(x, y, t) = S_a(x, y, t-1) \oplus S_a(x, y, t) \tag{3}$$

$$RMI_a = \sum_{k=0}^{T} DS_a(x, y, t-k) \tag{4}$$

Subsequently, the RMI is partitioned into N equal-sized blocks in order to compute the average recurrence for each block. Blocks with average recurrence value greater than the threshold τ_{RMI} are set to 1 (white) and the rest are set to 0 (black). Hence, white blocks indicate object regions with high motion recurrence, whereas, black blocks indicate the regions with insignificant or no motion recurrence. These blocks serve as cues for the classifier.

Based on previous research [5, 15], an object is classified as *human* (*single person* or *group of persons*) when white blocks are present at the middle and bottom sections of the partitioned RMI (which correspond to recurrent motion at the hands and legs regions). However, in our experiments, we discovered that this rule is only applicable for most of the common cases when the human subjects demonstrate periodic motion at both the hands and legs while walking. It is insufficient to account for cases when the humans are walking with hands in the pockets, at the back, or lifting or carrying things as they walk. Humans in these special cases will be misclassified as *other object* if the aforementioned rule is applied. Figure 4 illustrate situations when it is inappropriate to rely on hands movement as a cue to classify an object as *human* because no periodic motion is observed at the hands of the person.

To overcome the problem, we extended the classification rule of *human* into two sets:

1. Generic case, where a walking human demonstrates high motion recurrence at the hands and legs, as shown in Fig. 5
2. Special case, where a walking human demonstrates high motion recurrence at the legs only, such as the walking humans in Fig. 4

The generic cases can be handled by the classification rule proposed in [5], whereas special cases are handled by noting that the RMIs in Fig. 4 demonstrated high motion recurrence near the legs region, as evident from high concentration of white blocks at the bottom section of the partitioned RMI. Therefore, to account for the special cases when the rule for generic cases failed, the algorithm should search for region with significant recurrent motion. If white blocks are detected at the

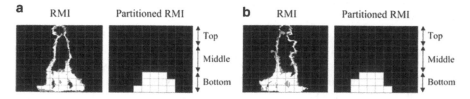

Fig. 4 RMI of a walking human with (**a**) hands in pockets, (**b**) hands carrying things

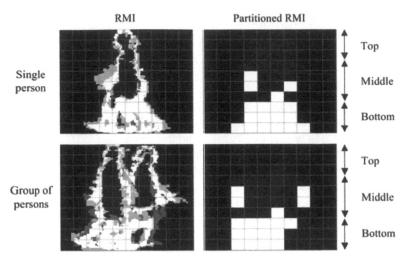

Fig. 5 RMI of a *single person* and a *group of persons*

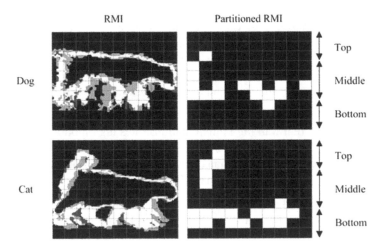

Fig. 6 RMI of a dog and a cat

bottom section of the partitioned RMI which correspond to the legs region, the object is classified as *human*, since recurrent motion at the legs is always seen from all human RMIs.

Analogously, derivation of the criteria for classifying a moving object as a *four-legged animal* is based on the observations of the recurrent motion behavior of dogs and cats. As shown in Fig. 6, the RMIs of dog and cat reminisces each other, whereby their legs and tail demonstrated periodic motion. White blocks tend to occur in all top, middle and bottom sections of the resultant partitioned RMIs. However, as for dogs and cats without tail, the white blocks are observable only at the middle and bottom sections. Moreover, dogs and cats exhibiting similar silhouette to the Dachshund, may cause white blocks to occur only at the middle section due to their smaller height to length ratio. On the other hand, dogs and cats exhibiting bigger height to length ratio such as the Irish Setter and Alaskan Malamute, may cause white blocks to occur only at the bottom section. Based on these observations, we noted that the location of white blocks in the partitioned RMI for dogs and cats may differ on their size and the presence of a tail.

The classification rule of *four-legged animal* can be derived based on the pattern exhibited by the black area within the RMI of an object where the object demonstrated no recurrent motion. As observed in the RMIs of *human* and *four-legged animal*, the two object classes generally do not show any recurrent motion at the main part of the body where the backbone is located. Hence the black area within the human RMI has a vertically aligned major axis, whereas the black area within the RMI of a dog or a cat has a horizontal major axis. This alignment of major axis serves as the cue for differentiation between *human* and *four-legged animal*.

Therefore, if the algorithm detects white blocks at the middle and bottom sections of a partitioned RMI, the black area within the respective RMI is extracted and examined for classification based on the following rules:

1. If the black area has a vertical major axis, the corresponding object is classified as a *human*.
2. Otherwise, it is classified as a *four-legged animal*.

For special cases of human subject where white blocks are detected at the bottom section (legs region) only, the same major axis alignment rule is applied to confirm the object class.

Subsequently, a moving object that is classified as a *human* will be further categorized as a *single person* or a *group of persons* based on any of the following rules:

1. Multiple peak points in a silhouette indicate more than one headcount, therefore representing a *group of persons*, for instance there are two peak points in the group of persons in Fig. 5 since there are two headcounts.
2. Normalized area of recurrence response at the top section of RMI for a *group of persons* is greater than that for a *single person*, due to presence of multiple heads.

Lastly, if there are no white blocks in a partitioned RMI, which indicates no recurrent motion, the corresponding object is classified as a *vehicle*, as shown in Fig. 7.

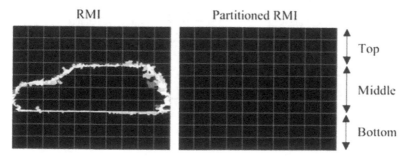

Fig. 7 RMI of a vehicle

An object that does not fall into any of the predefined categories (*single person, group of persons, vehicle* and *four-legged animal*) will, instead, be classified as *other object.*

4 Control Center

The CC is developed to facilitate user interactions with the intruder recognition framework (see Fig. 8). It fetches image sequences along with user-defined thresholds as inputs to the intruder recognition framework, and outputs intrusion records and alerts based on results from the recognition framework. In addition, CC allows users to authorize access permission for various predefined object classes in the surveillance area.

For instance, if *human* and *vehicle* are selected as the prohibited object classes, the alarm will be triggered upon entry of a human or vehicle. However, a four-legged animal entering the surveillance area will not trigger the alarm. This feature allows the system to ignore uninterested object classes (an "immunity" function) to minimize nuisance alarm.

A screenshot of the CC module is shown in Fig. 9. Before running the system, the prohibited object classes are selected from the "Object to Trigger Alarm" panel while the algorithm thresholds can be altered based on conditions of the images on the screen. "Indoor" is the preset threshold setting which is suitable for common indoor scenes, whereas "Outdoor" is meant for common outdoor scenes. In most cases, "Indoor" and "Outdoor" preset settings are sufficient for producing fine results. However, users may alter the threshold values to suit different task environments. Upon an intrusion event, visual and audio intrusion alerts will be triggered and delivered through computer screen and speakers to notify the users. Furthermore, the system will also update the list of intrusion records so that the user can export them into a text file for future references.

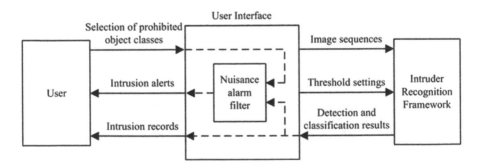

Fig. 8 General functionalities of user interface

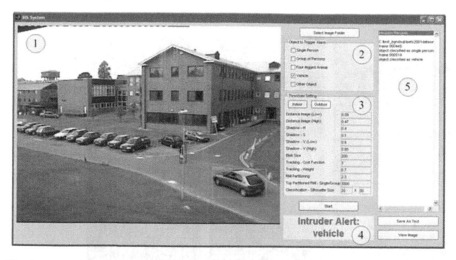

Fig. 9 Screenshot of the user interface program with an intrusion alert, where the numbered items are, as follows: (1) image frame, (2) "Object to Trigger Alarm" panel, (3) "Threshold Setting" panel, (4) intrusion alert, and (5) intrusion records

5 Results

The IRS System is implemented in Matlab [16] and executed on a 1.5 GHz Core 2 Duo CPU using several image sequences captured with a low-end digital camera (Olympus FE-280) at various housing areas. The image sequences consist of scenes with a variety of single persons, groups of persons, vehicles, and four-legged animals (dogs and cats). The frames are 320×240 pixels in size and sampled at a rate of 8 frames per second. Figure 10 shows several instances of moving object detected and classified by this framework. In our experiments, the RMI of a moving object was generated for 1 s duration after the object has completely entered the scene, and the partitioned RMI was computed with a threshold (τ_{RMI}) of 2. The framework took an average of 4 s (starting from the complete entrance of a moving object in the scene) to process the image sequence to produce the classification result. All of the moving objects tested were correctly classified into the predefined categories (see Table 1).

In addition to the image sequences mentioned above, we have also tested the IRS System with the PETS 2001 dataset from the Second IEEE International Workshop on Performance Evaluation of Tracking and Surveillance [17]. The dataset consists of several vehicles, single persons, and groups of persons from a wide surveillance area in 768×576 frame resolution. Figure 11 illustrates one of the scenes from the dataset where an occlusion between a vehicle and a single person, indicated by the red rectangles, was also successfully handled by the IRS. As listed in Table 2, all of the moving objects in the PETS 2001 dataset were properly classified, except for a vehicle that was misclassified as *four-legged animal* due to the size of the

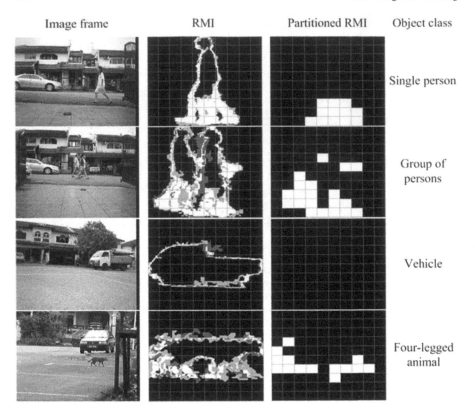

Image frame	RMI	Partitioned RMI	Object class

Fig. 10 Examples of moving object detected and classified

Table 1 Classification results for our datasets

Object class	Number of samples tested	Number of samples correctly classified
Single person	10	10
Group of persons	5	5
Vehicle	6	6
Four-legged animal	9	9

vehicle silhouette in the JPEG image sequence. The misclassified vehicle is small in size and the image sequence is noisy which caused improper segmentation of the silhouette that led to an inaccurate RMI.

While the IRS System is able to classify intruders into predefined categories properly, its effectiveness in reducing nuisance alarm is validated by the experiments. Immunity to certain object classes can be activated by stating which object classes are prohibited and which ones are consented in a surveillance area, by checking the boxes under "Object to Trigger Alarm" panel (see Fig. 9).

a   b

Fig. 11 Screenshots of PETS 2001 dataset (**a**) during occlusion, (**b**) after the occlusion

Table 2 Classification results for PETS 2001 dataset		

Object class	Number of samples tested	Number of samples correctly classified
Single person	5	5
Group of persons	2	2
Vehicle	4	3

6 Conclusion

The IRS correctly classified a total of 40 out of 41 objects from our experiments, thus indicating its backward compatibility and successful integration of the new classification list (*single person, group of persons, vehicle* and *four-legged animal*) in the new RMI framework. In addition, the modified detection and classification algorithms enhanced the recognition accuracy and practicability of this framework, where better classification, shadow handling, and adaptivity to long term illumination changes are observed. However, the misclassified sample in our experiments, due to weak segmentation for objects in noisy image sequences, indicates that smoothing and image enhancing routines may be applied to the rough silhouettes obtained from noisy images to reduce misclassifications.

The CC provides complete control over all algorithm thresholds to the users when the preset settings are inadequate for certain surveillance areas. The setup process requires experiences, and it may take some efforts to fine-tune the thresholds. Future works in this area should focus on automatic thresholding or tuning of the IRS system based on various conditions of the task environment. In addition, more classification categories can be incorporated into the recognition framework to extend the capabilities of the framework.

Lastly, this IRS System reduces nuisance alarms by allowing users to specify access authorization based on different object classes via the CC. It can be realized as a real-time system for various security applications in the future with the appropriate code translation (such as, C++).

References

1. Home Technology Store, "HardWired Security." [Online], from http://www.home-technology-store.com/home-security/wired.aspx.
2. Arco Infocomm Inc., "RK410PT PIR motion detector with PET immunity." [Online], from http://www.acesuppliers.com/company/information_34790.html.
3. Napco Security Technologies Inc., "C100 SERIES." [Online], from http://www. napcosecurity.com/dual.html
4. Garcia, M.L. (2006). *Vulnerability assessment of physical protection systems* (pp. 8). Burlington: Elsevier Butterworth-Heinemann.
5. Javed, O., & Shah, M. (2002). Tracking and object classification for automated surveillance. *Proceedings of the 7th European Conference on Computer Vision-Part IV*, pp. 343–357.
6. Toth, D., & Aach, T. (2003). Detection and recognition of moving objects using statistical motion detection and Fourier descriptors. *Proceedings of the 12th International Conference on Image Analysis and Processing*, pp. 430–435.
7. Hayfron-Acquah, J.B., Nixon, M.S., & Carter, J.N. (2001). Recognising human and animal movement by symmetry. *Proceedings of the IEEE International Conference on Image Processing*, pp. 290–293.
8. Bogomolov, Y., Dror, G., Lapchev, S., Rivlin, E., & Rudzsky, M. (2003). Classification of moving targets based on motion and appearance. *Proceedings of the British Machine Vision Conference*, pp. 429–438.
9. Stauffer, C., & Grimson, W.E.L. (2000). Learning patterns of activity using real-time tracking. *IEEE Transactions on Pattern Analysis and Machine Intelligence, 22*(8), 747–757.
10. Gonzalez, R.C., Woods, R.E., & Eddins, S.L. (2004). *Digital image processing using Matlab* (pp. 359). Upper Saddle River, NJ: Prentice Hall.
11. Cucchiara, R., Grana, C., Piccardi, M., & Prati, A. (2003). Detecting moving objects, ghosts and shadows in video streams. *IEEE Transactions on Pattern Analysis and Machine Intelligence, 25*(10), 1337–1342.
12. Gonzalez, R.C., Woods, R.E., & Eddins, S.L. (2004). *Digital image processing using Matlab* (pp. 347). Upper Saddle River, NJ: Prentice Hall.
13. Cucchiara, R., Grana, C., Piccardi, M., Prati, A., & Sirotti, S. (2001). Improving shadow suppression in moving object detection with HSV color information. *Proceedings of the IEEE International Conference on Intelligent Transportation Systems*, pp. 334–339.
14. Jahne, B., & HauBecker, H. (2000). *Computer vision and applications: A guide for students and practitioners* (pp. 379). San Diego: Academic.
15. Wong, C.E., & Ong, T.J. (2009). A new RMI framework for outdoor objects recognition. *Proceedings of the International Conference on Advanced Computer Control*, pp. 555–559.
16. The MathWorks, Inc., "MATLAB – The Language of Technical Computing." [Online], from http://www.mathworks.com/products/matlab/.
17. The University of Reading, UK, "PETS2001 Datasets." [Online], from http://ftp.pets.rdg.ac.uk/PETS2001/.

Chapter 25
Automatic Recognition of Sign Language Images

J. Ravikiran, Kavi Mahesh, Suhas Mahishi, R. Dheeraj, S. Sudheender, and Nitin V. Pujari

Abstract The objective of the research presented in this chapter is to enable communication between people with hearing impairment and those with visual impairment. Computer recognition of sign language snapshots is one of the most challenging research problems in this area. This chapter presents an efficient and fast algorithm for identification of the number of fingers opened in a gesture representing an alphabet of the American Sign Language. Finger detection is accomplished based on the concept of boundary tracing and finger tip detection. A significant feature of the solution is that it does not require the hand to be perfectly aligned to the camera or use any special markers or input gloves.

Keywords Boundary tracing · Computer access for disabled · Finger detection · Image processing · Sign language recognition

1 Introduction

The long-term goal of our research is to enable communication between visually impaired (i.e., blind) people on the one hand and hearing and speech impaired (i.e., deaf and dumb) people on the other. Since the former cannot see but can speak and the latter use sign language but cannot hear, there is currently no means of communication between such people who are unfortunately in significantly large numbers in a country such as India.

Our project aims to bridge this gap by introducing an inexpensive computer in the communication path so that the sign language can be automatically captured, recognized and translated to speech for the benefit of blind people. In the other direction, speech must be analyzed and converted to either sign or textual display on the screen for the benefit of the hearing impaired (Fig. 1).

J. Ravikiran (✉), K. Mahesh, S. Mahishi, R. Dheeraj, S. Sudheender, and N.V. Pujari
Department of Computer Science and Engineering, PES Institute of Technology, Ring Road, Banashankari 3rd Stage, Bangalore-560085, India
e-mail: ravikiran.j.127@gmail.com; drkavimahesh@gmail.com; suhasmahishi@gmail.com; rdheera@gmail.com; sudheender.s@gmail.com; nitin.v.pujari@pes.edu

S.-I. Ao et al. (eds.), *Intelligent Automation and Computer Engineering*,
Lecture Notes in Electrical Engineering 52, DOI 10.1007/978-90-481-3517-2_25,
© Springer Science+Business Media B.V. 2010

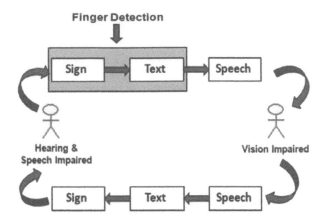

Fig. 1 Possible communication paths between hearing impaired and vision impaired

As seen in the above figure, enabling communication between the vision impaired and the hearing impaired is accomplished in two phases in either direction. In order to enable the hearing impaired to interact with the blind, we need to convert hand gestures representing specific signs into text which is then fed to a text-to-speech software engine that reads it aloud for the benefit of the blind. A similar reverse process occurs when a vision impaired person wishes to communicate with a hearing and speech impaired person. In this case, initially, the speech commands from the blind are to be mapped into text using a speech recognition engine. This is further converted into familiar visual signs that are recognized by the hearing impaired when displayed on the computer screen.

The algorithm [13] presented in this chapter concerns the first phase where a hearing impaired person wants to communicate with a blind person. Our approach does not require the use of any special hardware device such as sensors or colored gloves. We have been working with the added constraints of minimal calibration of the system between different users. Further, our solution also does not require the users to be computer literate or even literate.

The solution comprises a special-purpose image processing algorithm that we have developed to recognize signs from the American Sign Language with high accuracy. As noted above, a significant contribution of this result is that it does not require the person making the signs to use any artificial gloves or markers.

2 Related Work

There have been many previous attempts at sign language recognition that extracted certain features of the hand for finger detection. Some common features extracted include hand silhouettes [2, 3], contours [4], key points distributed along the hand (fingertips, joints) [5–8, 11, 12], and distance-transformed images [9].

Currently, the only technology that satisfies the advanced requirements of hand-based input for human-computer interaction is glove-based sensing. This technology, however, has several drawbacks including that it hinders the ease and naturalness with which the user can interact with the computer-controlled environment and it requires lengthy calibration and setup procedures. In contrast, computer vision has the potential to provide more natural, non-contact solutions. As a result, there have been considerable research efforts to use the hand as an input device through computer vision.

There have also been attempts where finger detection has been accomplished via color segmentation and contour extraction [8, 10]. But these techniques require fine-tuning every time the system switches to a new user as the color complexion varies from person to person. Contour extraction needs initial guesses (after appropriate thresholding and morphological operation) and the possible human profile region is then found out according to the projection of the image using the detected region boundary as the initial evolving curve.

In view of the limitations posed by the schemes discussed above there is a need to devise an efficient and robust technique for finger detection. The next section discusses our solution to this problem.

3 Proposed Methodology

In our research, a number of approaches were proposed to solve the given problem before the boundary tracing approach was adopted. The initial approaches and their shortcomings are discussed below.

The first approach proposed was a template matching technique. It involved storing a collection of images of all the gestures representing the alphabets of the American Sign Language that would serve as a standard for a pixel-by-pixel comparison. The alphabet that had the highest percentage of a match would be chosen as the recognized output. The shortcomings of this method were that the entire matching operation required an enormous amount of processing which affected the performance of the system and the system was dependent on the skin complexion of the user. Hence, this approach was discarded.

The second approach proposed was a color segmentation technique. It involved extraction of the portion of the image containing a gesture (the palm and the fingers) which would then be subjected to further processing to identify the gesture. This approach was excessively dependent on the environment in which the symbols were being shown (such as background and illumination) and hence the accuracy of the technique was low. As a result, this technique too was discarded.

The third technique involved dividing the input image into pre-defined grids. The grids would then be scanned for the presence of fingers by using colour based segmentation. The results thus obtained would then be used to identify the gesture shown by the user. The technique depended entirely on the orientation of the hand.

Fig. 2 Finger detection flowchart

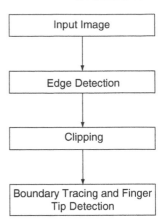

This shortcoming resulted in the occurrence of a number of false detections and a decrease in the accuracy and performance of the system. Hence, this method was discarded as well.

Finally we designed the boundary tracing approach which is a robust and efficient technique for finger detection. This method has three main phases of processing viz., Edge Detection, Clipping and Boundary Tracing.

The first phase employs Canny edge operator [1] and produces an edge detected image reducing the number of pixels to be processed at runtime. The next phase clips the undesirable portions of the edge-detected image for further processing. The final phase traces the boundary of the image and in the process detects finger tips which aid in finger detection (Fig. 2).

3.1 Canny Edge Detection

Edge detection is a phenomenon of identifying points in a digital image at which the image brightness changes sharply or, more formally, has discontinuities. The Canny algorithm [1] uses an optimal edge detector based on a set of criteria which include finding the most edges by minimizing the error rate, marking edges as closely as possible to the actual edges to maximize localization, and marking edges only once when a single edge exists for minimal response.

The first stage involves smoothing the image by convolving with a Gaussian filter. This is followed by finding the gradient of the image by feeding the smoothed image through a convolution operation with the derivative of the Gaussian in both the vertical and horizontal directions (see the equations below).

$$G(x, y) = \frac{1}{2\pi\sigma^2} e^{\frac{x^2 - y^2}{2\sigma^2}} \tag{1}$$

$$\frac{\partial G\,(x,y)}{\partial x}\alpha\,xe^{-\frac{x^2-y^2}{2\sigma^2}} \quad \frac{\partial G\,(x,y)}{\partial y}\alpha\,ye^{-\frac{x^2-y^2}{2\sigma^2}} \tag{2}$$

$$G = \sqrt{Gx^2 + Gy^2} \tag{3}$$

$$\theta = \arctan\left[\frac{Gy}{Gx}\right] \tag{4}$$

The non-maximal suppression stage finds the local maxima in the direction of the gradient, and suppresses all others, minimizing false edges. The local maximum is found by comparing the pixel with its neighbors along the direction of the gradient. This helps to maintain the single-pixel thin edges before the final thresholding stage.

Instead of using a single static threshold value for the entire image, the Canny algorithm introduced hysteresis thresholding, which has some adaptivity to the local content of the image. There are two threshold levels, t_h, high and t_l, low where $t_h > t_l$. Pixel values above the t_h value are immediately classified as edges. By tracing the edge contour, neighboring pixels with gradient magnitude values less than t_h can still be marked as edges as long as they are above t_l. While the results are desirable, the hysteresis stage slows the overall algorithm down considerably.

The performance of the Canny algorithm depends heavily on the adjustable parameters, σ, which is the standard deviation for the Gaussian filter and the threshold values, t_h and t_l. σ also controls the size of the Gaussian filter.

The bigger the value for σ, the larger is the size of the Gaussian filter. This implies more blurring which is necessary for noisy images as well as detecting larger edges. Smaller values of σ imply a smaller Gaussian filter which limits the amount of blurring, maintaining finer edges in the image. The user can tailor the algorithm by adjusting these parameters to adapt to different environments with different noise levels. The threshold values and the standard deviation for the Gaussian filter are specified as 4.5, 4.7 and 1.9 for the input source and background environment used in our system (Fig. 3).

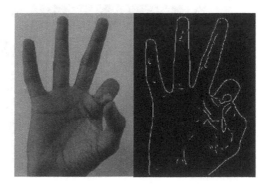

Fig. 3 Image of a hand gesture before and after edge detection

3.2 Clipping

Clipping is used to cut a section through the data (image) currently being rendered. The image contents that pertain to only the area of interest are retained after clipping. The edge-detected image contains portions which are unnecessary for further analysis. Hence we eliminate them by adopting two techniques as discussed below.

The first technique examines pixels from the bottommost 'y level' and at each level checks to see if there are three or more consecutive white pixels. If this condition is satisfied we mark this y-level as "y1" (Fig. 4).

The second technique exploits the fact that most of the edge detected images of hand gestures have the wrist portion (and below) which has a constant width. When it approaches the palm and the region of the hand above it, this difference increases drastically. We make use of this fact and find the y-level where this event occurs and mark this y-level as "y2" (Fig. 5).

Now we choose the maximum of (y1, y2) as the clipping y-level. All the pixels below this y-level are cleared by overwriting them with a black pixel.

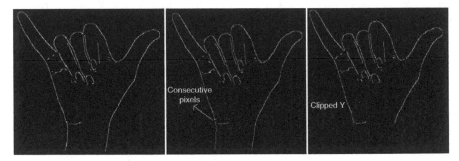

Fig. 4 Steps of Clipping Technique 1 based on finding consecutive white pixels from the bottommost 'y level'

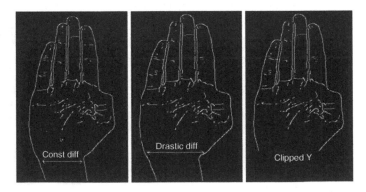

Fig. 5 Steps of Clipping Technique 2 based on drastic difference above wrist portion

Variable	Meaning
minx	Minimum x-coordinate of the set of white pixels in the clipped image.
maxx	Maximum x-coordinate of the set of white pixels in the clipped image.
miny	Minimum y-coordinate of the set of white pixels in the clipped image.
maxy	Maximum y-coordinate of the set of white pixels in the clipped image.
dx	Difference between minx and maxx.
dy	Difference between miny and maxy.
UD	Up/Down flag which indicates upward or downward direction trace.

Fig. 6 Variables used in boundary tracing

3.3   Boundary Tracing

This phase of the algorithm is the heart of processing. It consists of the following steps: (i) identifying the optimal y-level, (ii) identifying the initial trace direction, (iii) tracing with appropriate switch of direction, (iv) rejoining the trace on encountering breaks, and (v) finger tip detection. In the explanation of the above steps, the following are assumed (Figs. 6 and 7):

3.3.1   Identifying the Optimal y-Level

This step involves identifying the y-level to start the trace of the image indicated as step 1 in the above figure. By experimenting with different y-levels for various image samples, we fixed the optimal y-level as 30–35% of dy from the top of the edge-detected, clipped image. Hence the starting pixel for trace is the first white pixel found as a result of scanning from *minx* to *maxx* on the optimal y-level. This is a vital part of the algorithm because we will be eliminating a lot of pixels below the optimal y-level which are insignificant for finger detection, which helps by improving the efficiency of the algorithm.

3.3.2   Identifying the Initial Trace Direction

From the initial trace point, we proceed towards miny without changing the current x-coordinate, until there is no pixel to proceed. Then we examine the neighboring white pixels and set the flag to "left" or "right" appropriately. This flag serves as the initial trace direction, shown as step 2.

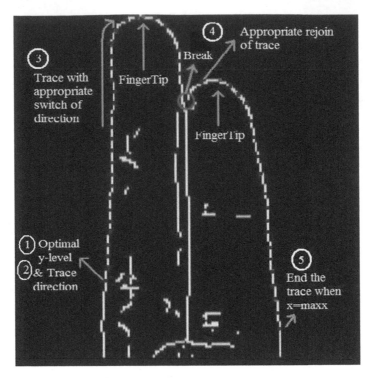

Fig. 7 Steps in boundary tracing

3.3.3   Tracing with Appropriate Switch of Direction

After identifying the trace direction, the system proceeds by tracing pixels. For every five pixels traced, we write the fifth pixel with blue color. Also whenever we find no pixels in the current direction, we check if there is a pixel in the other direction. If present, we toggle the direction of trace, else it is a break. This is indicated as step 3.

3.3.4   Rejoining the Trace on Encountering Breaks

Breaks may be encountered while tracing upwards or downwards along the boundary of a finger. They are handled separately based on the flag "UD" which indicates whether we are travelling up ($+1$) or down (-1).

Varying the x-coordinate from current value to *maxx*, we scan from the current y-level towards the upper or lower boundary of the image based on the value of "UD" for a white pixel. If a white pixel is found, we then start the trace again from this pixel re-initializing the count to zero. This is indicated as step 4.

3.3.5 Finger Tip Detection

Whenever the flag "UD" switches from $+1$ to -1, it indicates the change in trace direction from up to down. This event signifies the detection of a finger tip and hence we write this pixel with red. After finding a finger tip there are two techniques to find the starting point of the next finger.

In the first technique we trace downwards from the finger tip position to the optimal y-level, then from that position we increment the x-coordinate until we find a white pixel, this serves as the starting point for processing the next finger.

The second technique is employed when the fingers are adjoined. In this technique we check if a white pixel exists towards the right/left of the current downward trace pixel, if found, this serves as the starting point of processing next finger.

The algorithm iterates through the above process until the x-coordinate reaches *maxx*. The count of the finger tips at this stage is the count of the number of fingers that were open. This is indicated as step 5.

4 From Fingers to Signs

We now discuss how we map finger detection to letters in the sign language. The count of fingers that are open and the co-ordinates of detected finger tips are obtained after boundary tracing has been done with the help of the above algorithm. Using this data, we identify each symbol and the corresponding alphabets (precisely B, D, F, I, K, L, U, V and W, see Fig. 8). The techniques used to identify each symbol are discussed below.

If the count of fingers that are open is equal to four, then the symbol is B. If the count of fingers that are open is equal to three then symbol identified would be

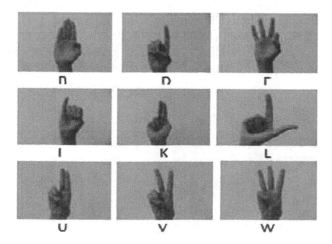

Fig. 8 American sign language gestures with fingers open

either W or F. Between them, if the finger tips identified are in the increasing order of height, then the symbol is F else the symbol is W.

If the count of fingers open is equal to two, then the symbol is either V or K or U. Between them, if the difference between the finger tips located is greater than the critical value (i.e., > 100 pixels), then the symbol is V, else it is K or U. Between these, if the thumb is detected near the open fingers then, the symbol is K, else it is U.

When the count of fingers that are open is one then the symbol is one among D, I or L. The co-ordinates of the finger tips are now used to identify each symbol uniquely. If the x co-ordinate of the finger tip is near the minimum x-coordinate then the symbol is identified as I. To differentiate between D and L, a critical value, the difference between the x co-ordinate value of the finger tip and maximum x-coordinate, is defined. If the critical value is greater than hundred then the symbol is L else the symbol is identified as D.

5 Implementation and Results

In this section we describe the accuracy of our algorithm. The application has been implemented in Java 1.6 (Update 7) using the ImageIO libraries. The application has been tested on a Pentium IV running at 3.00 GHz. The images have been captured using a 6 Mega Pixel Canon PowerShot S3 IS.

The captured images are of resolution 640 × 480. For the performance evaluation of the finger detection, the system has been tested multiple times on samples authored by a set of five different users.

The first figure shows a subset of American Sign Language gestures which have fingers open. The second figure shows the performance evaluation results. These results are plotted on a graph, where the y-axis represents the number of tests and the x-axis represents the various gestures of the American Sign Language corresponding to alphabets. The columns are paired for each gesture: the first column is the number of times the fingers are correctly identified in the gesture; the second column is the total number of times that the test on the gesture has been carried out. As can be seen in Fig. 9, the finger recognition works accurately for 95% of the cases.

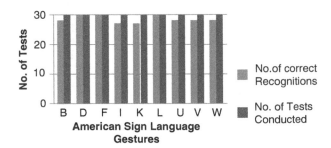

Fig. 9 System performance evaluation

6 Conclusion and Future Work

A boundary-trace based finger detection technique was presented wherein cusp detection analysis is done to locate the finger tip. This algorithm is a simple, efficient and robust method to locate finger tips and enables us to identify a class of hand gestures belonging to the American Sign Language that have fingers open. The accuracy obtained in this work is sufficient for the purposes of converting sign language to text and speech, since a dictionary can be used to correct spelling errors up to 5% from our gesture recognition algorithm.

In future work, sensor based contour analysis can be employed for differentiating the fingers between open fingers and closed fingers. This will give more flexibility to interpret the gestures. Furthermore, hand detection method using texture and shape information can be used to maximize the accuracy of detection in cluttered backgrounds.

More importantly, we need to develop algorithms to cover other signs in the American Sign Language that have all the fingers closed. An even bigger challenge will be to recognize signs that involve motion (i.e., where various parts of the hand move in specific ways). In our future work, we also plan to complete other modules of the overall solution needed to enable communication between blind and deaf people. In particular, we will focus on translating the recognized sequences of signs to continuous text (i.e., words and sentences) and then to render the text in speech that can be heard by blind people.

Acknowledgements The authors would like to acknowledge the contribution of Bharadvaj J., Ganesh S., Ravindra K., Vinod D. towards the development of one part of the solution presented in this chapter.

References

1. Canny, J. (Nov 1986). A computational approach to edge detection. *IEEE Transactions on Pattern Analysis and Machine Intelligence, 8,* 679–714.
2. Shimada, N. et al. (1998). Hand gesture estimation and model refinement using monocular camera – Ambiguity limitation by inequality constraints. In *Proceedings of the 3rd Conference on Face and Gesture Recognition,* pp. 268–273.
3. Sonka, M., Hlavac, V., & Boyle, R. (1999). *Image processing, analysis, and machine vision.* Pacific Grove, CA: Brooks/Cole Publishing Company.
4. Starner, T., & Pentland, A. (21–23 Nov 1995) Real-time American sign language recognition from video using hidden Markov models. In *Proceedings of International Symposium on Computer Vision,* pp. 265–270.
5. Lee, J., & Kunii, T. (Sept 1995) Model-based analysis of hand posture. *IEEE Computer Graphics Applications, 15*(5), 77–86.
6. Fillbrandt, H., Akyol, S., & Kraiss, K.F. (Oct 2003). Extraction of 3D hand shape and posture from image sequences for sign language recognition. *IEEE International Workshop on Analysis and Modeling of Faces and Gestures, 17,* 181–186.
7. Starner, T., Pentland, A., & Weaver, J. (Dec 1998) Real-time American sign language recognition using desk and wearable computer based video. *IEEE Transactions on Pattern Analysis and Machine Intelligence, 20*(12), 1371–1375.

8. Imagawa, K., Lu, S., & Igi, S. (14–16 Apr 1998) Color-based hands tracking system for sign language recognition. In *Proceedings of the Third IEEE International Conference on Automatic Face and Gesture Recognition*, pp. 462–467.
9. Teh, C.H., & Chin, R.T. (Jul 1988) On image analysis by the methods of moments. *IEEE Transactions on Pattern Analysis and Machine Intelligence, 10*(4), 496–513.
10. Kang, S.K., Nam, M.Y., & Rhee, P.K. (Aug 2008) Color based hand and finger detection technology for user interaction. *International Conference on Convergence and Hybrid Information Technology.*
11. Shirai, Y., Tanibata, N., & Shimada, N. (2002). Extraction of hand features for recognition of sign language words. VI'2002, *Computer-Controlled Mechanical Systems*. Graduate School of Engineering, Osaka University.
12. Hamada, Y., Shimada, N., & Shirai, Y. (May 2004) Hand shape estimation under complex backgrounds for sign language recognition. In *Proceedings of the 6th International Conference on Automatic Face and Gesture Recognition*, pp. 589–594.
13. Ravikiran, J., Mahesh, K., Mahishi, S., Dheeraj, R., Sudheender, S., & Pujari, N.V. (2009). Finger detection for sign language recognition, ICCS2009. In *Proceedings of the International Conference on Computer Science, International Multi-Conference of Engineers and Computer Scientists – IMECS 2009*, March 2009, Hong Kong.

Chapter 26
Algorithm Using Expanded LZ Compression Scheme for Compressing Tree Structured Data

Yuko Itokawa, Koichiro Katoh, Tomoyuki Uchida, and Takayoshi Shoudai

Abstract Due to the rapid growth of information technologies, the use of electronic data such as XML/HTML documents, which are a form of tree structured data, has been rapidly increasing. We have developed an algorithm for effectively compressing tree structured data and one for decompressing a compressed tree that are based on the Lempel–Ziv compression scheme. Next, we have implemented both compression and decompression algorithms by applying our algorithms for the XMill compressor and XDemill decompressor presented by Liefke and Suciu. Then, testing using synthetic large ordered trees and real-world tree structured data demonstrated the effectiveness and efficiency of our algorithms.

Keywords Tree structured data · Lemplel–Ziv compression scheme · XMill · XDemill

Y. Itokawa (✉)
Department of Kansei Design, Hiroshima International University, 555-36, Kurose Gakuendai, Higashi Hiroshima, Hiroshima, 739-2695, Japan
e-mail: y-itoka@he.hirokoku-u.ac.jp

K. Katoh
Enterprise Server Division Department I Server Development, Hitachi Ltd.,1 Horiyamashita, Hadano, Kanagawa, 259-1392, Japan
e-mail: kohichiroh.katoh.gr@hitachi.com

T. Uchida
Faculty of Information Sciences, Hiroshima City University, 3-4-1, Ozuka-Higashi, Asa-Minami-Ku, Hiroshima, 731-3194, Japan
e-mail: uchida@hiroshima-cu.ac.jp

T. Shoudai
Department of Informatics, Kyushu University, 744 Motooka, Nishi-ku, Fukuoka 819-0395, Japan
e-mail: shoudai@inf.kyushu-u.ac.jp

S.-I. Ao et al. (eds.), *Intelligent Automation and Computer Engineering*,
Lecture Notes in Electrical Engineering 52, DOI 10.1007/978-90-481-3517-2_26,
© Springer Science+Business Media B.V. 2010

1 Introduction

Due to the rapid growth of information technologies, the use of electronic data such as XML/HTML documents has been rapidly increasing. Since such data have no rigid structure but rather a tree structure, they are called *tree structured data*. By using a variant of Object Exchange Model introduced by [1], we can be represented by a rooted tree without vertex label but with edge labels. A rooted tree with internal vertices that have ordered children is called an *ordered tree*. Part of an example tree of structured XML data *Sample_html* and ordered tree T, which represents *Sample_html*, are shown in Fig. 1. The number on the left of each vertex in T denotes the ordering on its siblings. The edge labels in T are either a tag, such as "`<table>`" and "`<tr>`", or the text written in the PCDATA field in T, such as "`Text 1-A`" and "`Text 1-B`". A sequence of ordered trees can be treated as a single ordered tree whose root has the roots of all ordered trees in the sequence as children.

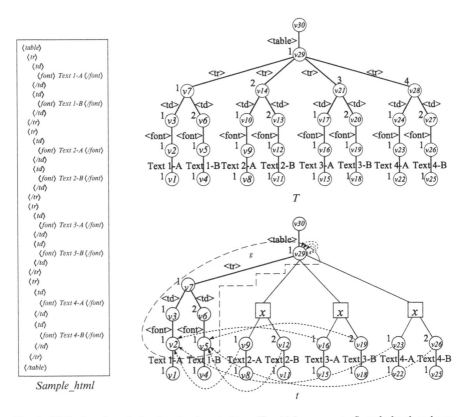

Fig. 1 XML data *Sample_html*, ordered rooted tree T, which represents *Sample_html*, and term tree t. A variable is represented by a box with lines to its elements. The *letter* in the box represents the variable label. The *number on the left* of each vertex denotes the ordering of its siblings

We have developed an algorithm for efficiently compression an ordered tree without loss of information that is based on a Lempel–Ziv compression scheme for ordered trees. In a Lempel–Ziv compression scheme for strings such as LZSS [10], a previously seen text is used as a dictionary, and phrases in the input text are replaced with pointers into the dictionary to achieve compression. In our version of the Lempel–Ziv compression scheme for ordered trees, the first occurring tree, f, in the postorder traversal of ordered tree T is used as a dictionary, and the subgraphs in T that are isomorphic to f are replaced by variables with pointers into the dictionary to achieve compression. We represent the compression of an ordered tree T by a pair of a term tree t and a substitution θ such that T is obtained by applying θ to t, where a term tree and a substitution over term trees are introduced by [8]. An example term tree is shown in Fig. 1; also shown is substitution $\{x := [g, (v29, v2, v5)]\}$ as an example of a dictionary, where g is a subgraph given in the figure. The variables in a term tree are represented by squares with lines to their elements. The letter in each square represents the variable label.

We also developed an algorithm for decompressing a compressed tree. It works by applying a substitution to the tree pattern. For example, by replacing all variables having the label x in t in Fig. 1 with ordered trees isomorphic to g given in Fig. 1, we can obtain ordered tree T in Fig. 1 from term tree t and substitution $\{x := [g, (v29, v2, v5)]\}$.

Several compressors and decompressors for tree structured data have been developed, including XMill and XDemill [7], LZCS [2], XGrind [12] and XMLPPM [3]. The compressor, XMill, has an architecture that leverages existing compression algorithms and tools for application to XML data. XMill obtains the tree structure by parsing the given XML data with respect to the XML tags and attributes and then separates it from the data, which consists of a sequence of data items (strings) representing element contents and attribute values. The sequence describing the tree structure is compressed by using a compressor that works over strings, such as zip. We have implemented compression algorithm for applying our version of Lempel–Ziv compression scheme to the XMill compressor. In comparison, our compressor can directly treat a tree structure of given XML data and produce a compression over ordered trees. It can be used as a compressor in the XMill structure container. The compressor, LZCS, developed by [2], obtains the tree structure by replacing frequently repeated subtrees by using backward references to their first occurrence. A subtree having vertex v as a root consists of v and all of its descendants. In contrast, our compression algorithm replaces frequently repeated connected subgraphs with references to their first occurrence. We evaluated the performance of compression and decompression algorithms by implementing them on a PC and applying them to synthetic ordered trees that were randomly generated and to XML files, which are real-world data.

The Subdue system, developed by [4], compresses electronic data having a graph structure. The algorithm, developed by [5], compresses geometric data without loss of graph structures. [13] presented a grammar-based lossless compression algorithm for ordered tree T. It achieves a grammar-based compression of T by identifying the frequent subtrees in T and then repeatedly replacing the isomorphic ones with a variable having the same label until the frequent isomorphic subtrees are exhausted.

This paper is organized as follows. In Section 2, we formally define a term tree and its substitution, which leads us to Lempel–Ziv compression for an ordered tree without loss of information. In Section 3, we give our Lempel–Ziv compression scheme for ordered trees by regarding a substitution as a dictionary. In Section 4, we present compression and decompression algorithms for ordered trees, by applying our Lempel–Ziv compression scheme to the XMill structure container. In Section 5, we present experimental results of applying our algorithm to both synthetic large trees and XML data which are real-world data.

2 Ordered Term Trees and Substitutions

For set or list D, the number of elements in D is denoted by $|D|$.

Let $T = (V_T, E_T)$ be an ordered tree with vertex set V_T and edge set E_T. Let $\ell \geq 1$ be an integer. A list $h = (u_0, u_1, \ldots, u_\ell)$ of vertices in V_T is called a *variable* if $u_1, \ldots, u_\ell$ is a sequence of consecutive children of u_0, i.e., u_0 is the parent of $u_1, \ldots, u_\ell$ and u_{j+1} is the next sibling of u_j for j $(1 \leq j < \ell)$. Two variables, $h = (u_0, u_1, \ldots, u_\ell)$ and $h' = (u'_0, u'_1, \ldots, u'_{\ell'})$, are said to be *disjoint* if $\{u_1, \ldots, u_\ell\} \cap \{u'_1, \ldots, u'_{\ell'}\} = \emptyset$. Let H_T be a set of pairwise disjoint variables of $T = (V_T, E_T)$. An *ordered term tree* on T and H_T is triplet $t = (V_t, E_t, H_t)$, where $V_t = V_T$, $E_t = E_T - \bigcup_{h=(u_0, u_1, \ldots, u_\ell) \in H_T} \{(u_0, u_i) \in E_T \mid 1 \leq i \leq \ell\}$, and $H_t = H_T$. Because T and H_T are easily found from $t = (V_t, E_t, H_t)$, we do not write T and H_T explicitly. Hereinafter, we call an ordered term tree simply a *term tree*. Term tree $t = (V_t, E_t, H_t)$ is called a *ground term tree* if $H_t = \emptyset$. Let Λ and X be finite alphabets such that $\Lambda \cap X = \emptyset$, whose elements are called *edge labels* and *variable labels*, respectively. Every variable label $x \in X$ has a nonnegative integer $rank(x)$. Every variable h has a variable label x such that $rank(x) = |h|$. A *term tree over* $\langle \Lambda, X \rangle$ is a term tree t such that all edges and variables in t are labeled with elements in Λ and X, respectively. If Λ and X are clear from the context, we often omit them. We use ordered tree terminology of ordered trees for term trees, for example, *parent, child,* and *leaf*.

For vertices u, u', u'' of term tree t, we write $u' <^t_u u''$ if u' and u'' are children of u and u' is smaller than u'' in the order of the children. Term tree $t = (V_t, E_t, H_t)$ *is isomorphic to* term tree $g = (V_g, E_g, H_g)$, denoted by $t \equiv g$, if there is a bijection $\pi : V_t \to V_g$ such that for $v_0, v_1, \ldots, \in V_t$, (1) $(v_1, v_2) \in E_t$ if and only if $(\pi(v_1), \pi(v_2)) \in E_g$, (2) (v_1, v_2) in E_t and $(\pi(v_1), \pi(v_2))$ in E_g have the same edge label, (3) $v_1 <^t_{v_0} v_2$ if and only if $\pi(v_1) <^g_{\pi(v_0)} \pi(v_2)$, (4) $(v_0, \ldots, v_\ell) \in H_t$ if and only if $(\pi(v_0), \ldots, \pi(v_\ell)) \in H_g$, and (5) variables $(v_0, \ldots, v_\ell) \in H_t$ and $(v'_0, \ldots, v'_\ell) \in H_t$ have the same variable label if and only if $(\pi(v_0), \ldots, \pi(v_\ell)) \in H_g$ and $(\pi(v'_0), \ldots, \pi(v'_\ell)) \in H_g$ have the same variable label. Such a bijection π is called an *isomorphism from t to g*.

Let $t = (V_t, E_t, H_t)$ be a term tree. Triplet $f = (V_f, E_f, H_f)$ is called a *term subtree* of t if f is a term tree such that $V_f \subseteq V_t$, $E_f \subseteq E_t$, and $H_f \subseteq H_t$. If f is a ground term tree, we call f simply a *subtree* of t. For two term subtrees $f = (V_f, E_f, H_f)$ and $g = (V_g, E_g, H_g)$ of t, we say that f and g are *overlap* in t if $((E_f \cap E_g) \cup (H_f \cap H_g)) \neq \emptyset$, $V_f \not\subseteq V_g$ and $V_g \not\subseteq V_f$. Let f and g be term trees over $\langle \Lambda, X \rangle$ having at least two vertices. Let $h = (v_0, v_1, \ldots, v_\ell)$ $(\ell \geq 1)$ be a variable in f and $\sigma = (u_0, u_1, \ldots, u_\ell)$ a list of $\ell + 1$ distinct vertices in g such that u_0 is the root of g and $u_1, \ldots, u_\ell$ are leaves of g. The pair $[g, \sigma]$ is called an $(\ell + 1)$-*hypertree over* $\langle \Lambda, X \rangle$. Hereinafter, we often omit $\langle \Lambda, X \rangle$. For $(\ell + 1)$-hypertrees $[g, (u_0, \ldots, u_\ell)]$ and $[f, (w_0, \ldots, w_\ell)]$, we denoted $[g, (u_0, \ldots, u_\ell)] \equiv [f, (w_0, \ldots, w_\ell)]$, if there is an isomorphism π from g to f such that for each i $(0 \leq i \leq \ell)$, $w_i = \pi(u_i)$. For variable h, term tree g and list σ of distinct vertices of g, the form $h \leftarrow [g, \sigma]$ is called a *variable replacement* for h. A new term tree, $f' = f\{h \leftarrow [g, \sigma]\}$, is obtained by applying $h \leftarrow [g, \sigma]$ to f in the following way. For the variable $h = (v_0, \ldots, v_\ell)$ in f, we attach g to f by removing h from f and identifying $v_0, \ldots, v_\ell$ of f with $u_0, \ldots, u_\ell$ of g in this order. We define a new ordering, $<_v^{f'}$, on every vertex v in f' so that, for vertex v in f' with at least two children (v' and v''), (1) if $v, v', v'' \in V_g$ and $v' <_v^g v''$ then $v' <_v^{f'} v''$, (2) if $v, v', v'' \in V_f$ and $v' <_v^f v''$ then $v' <_v^{f'} v''$, (3) if $v = v_0(=u_0)$, $v' \in V_f - \{v_1, \ldots, v_\ell\}$, $v'' \in V_g$, and $v' <_v^f v_1$ then $v' <_v^{f'} v''$, and (4) if $v = v_0(=u_0)$, $v' \in V_f - \{v_1, \ldots, v_\ell\}$, $v'' \in V_g$, and $v_\ell <_v^f v'$ then $v'' <_v^{f'} v'$.

Let x be a variable label and $[g, \sigma]$ an rank(x)-hypertree. Then, the form $x := [g, \sigma]$ is called a *binding* for x. A finite collection of bindings $\theta = \{x_1 := [g_1, \sigma_1], \ldots, x_n := [g_n, \sigma_n]\}$ is called a *substitution* if the x_i's are mutually distinct variable labels in X and no variable in g_i has a variable label in $\{x_1, \ldots, x_n\}$. For variable label x in X and term tree t, let $H_t(x)$ be the set of all variables in H_t with labels x. For term tree t and substitution $\theta = \{x_1 := [g_1, \sigma_1], \ldots, x_n := [g_n, \sigma_n]\}$, a new term tree, $t\theta$, is obtained from t by applying all variable replacements in $\bigcup_{i=1}^n \{e \leftarrow [g_i, \sigma_i] \mid e \in H_t(x_i)\}$ to t.

3 LZ Compression Scheme for Tree Structured Data

We expanded the LZ77 (Lempel–Ziv 1977) compression algorithm [10, 15], which works over strings, into a one for ordered trees and used it in both our compressor and decompressor. In the LZ77 compression scheme, a previously seen text is used as a dictionary, and phrases in the input text are replaced with pointers into the dictionary to achieve compression. Using the framework described in the previous section, we developed a Lempel–Ziv compression scheme for ordered trees by regarding a substitution as a dictionary.

Let T be a tree and t a term tree. If there is a substitution θ for t such that $T \equiv t\theta$ and the sum of the description lengths of t and θ is less than that of T, the pair (t, θ) gives a compression of T by using θ as a dictionary. Then, we define a Lempel–Ziv compression for tree structure data. Let T be a tree. A pair (t, θ) of term tree

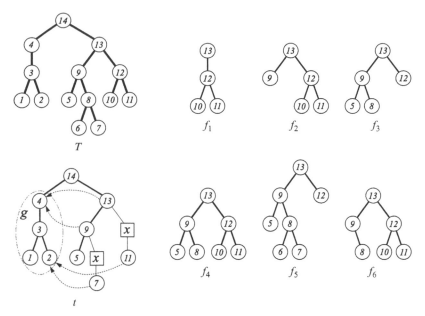

Fig. 2 Tree T is input tree-structured data, and f_1, f_2, f_3, f_4, f_5, and f_6 are subtrees of T; $(t, \{x := [g, (4, 2)]\})$ is an LZ compression of T. Tree f_1 and the list $(4, 2)$ in the binding are indicated by the *rough circle* connecting subtrees of t and the four pointers from 9 to 4, 7 to 2, 13 to 4, and 11 to 2

t and substitution θ is called a *Lempel–Ziv (LZ, for short) compression of* T if the following conditions hold.

(1) $T \equiv t\theta$.
(2) For each binding $x := [g, \sigma]$, t has a term subtree isomorphic to g.
(3) The sum of description lengths of t and θ is less than the description length of T.

In Fig. 2, we show an example of LZ compression. Hereinafter, for tree T and its LZ compression (t, θ), we denote by $\|T\|$ and $\|(t, \theta)\|$ the description lengths of T and (t, θ), respectively. If the subtraction of $\|(t, \theta)\|$ from $\|T\|$, i.e., $\|t\| - \|(t, \theta)\|$ is greater than 0, we call the subtraction by the *gain* of LZ compression (t, θ) from tree T. In next section, we will give concrete examples for the description lengths of a tree and an LZ compression.

An algorithm *Compressing_Ordered_Tree* for a given tree T and two integers *Min* and *Max* ($0 < Min < Max$) is shown in Fig. 3. It outputs an LZ compression (t, θ) of T. Since the algorithm *Compressing_Ordered_Tree* visits each vertex in the postorder traversal, we can see that the algorithm is based on Lempel–Zip compression scheme. For vertex v in a given tree, the function *Make_Term_Subtree* in the *Compressing_Ordered_Tree* algorithm outputs the set of hypertrees $[f, \sigma]$ satisfying conditions (1)–(6) below. Let $f = (V_f, E_f, H_f)$ and $\sigma = (u_1, \ldots, u_n)$.

Algorithm Compressing_Ordered_Tree
Input: Tree $T = (V_T, E_T)$, integers Min and Max $(0 < Min < Max)$
Output: LZ compression (t, θ) of T

$\quad t := T, \theta := \emptyset, \delta := \emptyset;$
$\quad$ for each vertex v in T, in postorder {
$\quad\quad S := Make\_Term\_Subtree\,(t, v, Min, Max);$
$\quad\quad$ if $\theta = \emptyset$ and $\delta = \emptyset$ then $\delta := S;$
$\quad\quad$ else {
$\quad\quad\quad$ for each hypertree $[f, \sigma] \in S$ in decreasing order of gains.
$\quad\quad\quad$ {
$\quad\quad\quad\quad S := S - \{[f, \sigma]\};$
$\quad\quad\quad\quad$ if there exists a binding $x := [f', \sigma']$ in θ with $[f', \sigma'] \equiv [f, \sigma],$
$\quad\quad\quad\quad$ then {
$\quad\quad\quad\quad\quad$ replace f in t with a new variable labeled with $x;$
$\quad\quad\quad\quad\quad S := S - \{[g, \sigma] \in S \mid f \text{ and } g \text{ are overlap }\};$
$\quad\quad\quad\quad$ }
$\quad\quad\quad\quad$ else {
$\quad\quad\quad\quad\quad$ if there exists a hypertree$[f', \sigma']$ in δ with $[f', \sigma'] \equiv [f, \sigma]$
$\quad\quad\quad\quad\quad$ then {
$\quad\quad\quad\quad\quad\quad$ replace f in t with a new variable having a new variable label $y;$
$\quad\quad\quad\quad\quad\quad \theta := \theta \cup \{y := [f', \sigma'] \};$
$\quad\quad\quad\quad\quad\quad \delta := \delta - \{[f', \sigma']\} - \{[g, \sigma] \in \delta \mid f \text{ and } g \text{ are overlap }\};$
$\quad\quad\quad\quad\quad\quad S := S - \{[g, \sigma] \in S \mid f \text{ and } g \text{ are overlap }\};$
$\quad\quad\quad\quad\quad$ }
$\quad\quad\quad\quad\quad$ else $\delta := \delta \cup \{[f, \sigma] \};$
$\quad\quad\quad\quad$ }
$\quad\quad\quad$ }
$\quad\quad$ }
$\quad$ }
$\quad$ return the pair (t, θ).

Fig. 3 Algorithm *Compressing_Ordered_Tree*

(1) f is a subtree of t, and v is the rightmost child of the root of f.
(2) $Min \le |V_f| \le Max$.
(3) For vertex $u \in V_f$, if there exists a vertex w in $V_T - V_f$ such that w is adjacent to u in T, u is the root of f or a leaf of f.
(4) Vertex u_1 in σ is the root of f.
(5) If $n > 2$, all $u_2, \ldots, u_n$ are adjacent to vertices in $V_T - V_f$.
(6) If $n = 2$, either u_2 is adjacent to a vertex in $V_T - V_f$ or u_2 is the rightmost leaf of f.

By appropriately setting integers Min and Max, we can bound the maximum number of hypertrees output by *Make_Term_Subtree*. Since the number of hypertrees in S affects the time complexity of *Compressing_Ordered_Tree*, Min and Max must be appropriately set by users for tree structured data. For example, let T be the tree in Fig. 2 and f_1, f_2, f_3, f_4, f_5, and f_6 be the subtrees of T. All the subtrees satisfy the conditions (1) and (2) for vertex 12, $Min = 4$, and $Max = 7$. However, f_6 does not satisfy condition (3) for the same parameters while the others do. Let us consider hypertrees $[f_1, (13, 10)]$ and $[f_1, (13, 11)]$. Since 11 is the rightmost leaf of f_1 and 10 is not, $[f_1, (13, 11)]$ is constructed by *Make_Term_Subtree*$(T, 12, 4, 7)$ and $[f_1, (13, 10)]$ is not. Hence, it outputs the set $\{[f_1, (13, 10, 11)], [f_2, (13, 9, 10, 11)], [f_3, (13, 5, 8, 12)], [f_4, (13, 5, 8, 10, 11)], [f_5, (13, 5, 6, 7, 12)]\}$.

Compressing_Ordered_Tree repeats the following processes. Let g be a term subtree generated by *Make_Term_Subtree* for vertex $v \in V_t$. If there is a vertex v' that was visited before v in postorder such that *Make_Term_Subtree* for v' outputs a term subtree g' isomorphic to g, we revise the term tree t by replacing g with a new variable having a new variable label, x, and add a new binding, $x := [g', \sigma]$, to θ. We introduce a new pointer from g into the first occurring subtree $g' = (V'_g, E'_g)$ in $t = (V_t, E_t, H_t)$ such that g' is isomorphic to g, and, for any vertex $v \in V'_g$, v is not adjacent to any vertex in $V_t - V'_g$ in t if v is neither the root of g' nor a leaf of g'. Such a pointer consists of a list of pointers into the root of g' and into some of the leaves of g'.

Our algorithm for decompressing an LZ compression (t, θ) works by repeatedly applying the following variable replacement processes to $t = (V_t, E_t, H_t)$ until t has no other variable.

(1) Choose the variable h in t that appears first in the postorder traversal.
(2) Find a binding $x := [g, \sigma]$ such that x is the variable label of h.
(3) Apply the variable replacement $h \leftarrow [g, \sigma]$ to t.

Since our decompression algorithm appears variable replacement for each variable that appears in the postorder traversal, we can see that our decompression algorithm is also based on Lempel–Ziv compression scheme (Fig. 4).

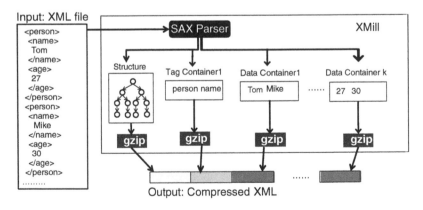

Fig. 4 Illustration of XMill

4 Application of Our Lempel–Ziv Compression Scheme to XMill Compressor

In 1999, Liefke and Suciu developed a user-configurable XML compressor, XMill [7]. XMill applies the following three principles to compress XML data.

- The *structure* consisting of XML tags and attributes, is compressed separately from the Data.
- Data items are grouped into *containers*, and each container is compressed separately. In order to compress XML data efficiently, PCDATAs with parents that have the same tag label are stored in the same container.
- XMill applies specialized compressors to different containers.

By applying our Lempel–Ziv compression scheme to structure container in XMill, we develop an XML compressor for XML data. In the same way as XMill, the structure of an XML file is encoded in preorder as follows. PCDATA is assigned 1, all end-tags are assigned positive integers greater than 1, while all start-tags are replaced with the token "/". For XML file T, we denote by code(T) the resultant coding of T. We define a *description length* of T, denoted by $\|T\|$, as the length of code(T), For example, for *Sample_html* in Fig. 1, we assigned `</table> = 2`, `</tr> = 3`, `</td> = 4`, and `</font> = 5`, and obtained the following sequences, which represents T.

```
/////1 5 4 ///1 5 4 3 ////1 5 4 ///1 5 4 3 ////1 5 4
///1 5 4 3 ////1 5 4 ///1 5 4 3 2
```

We can see that $\|T\| = 86$.

Let (t, θ) be LZ compression for XML data T. In our XML compressor for T, we use structure container for t and container for θ instead of structure container for tree T of XMill. First of all, we show container expression for θ below. We define the representation of a binding in θ in order to store it into a dictionary effectively. We assume that for each binding $x := [g, \sigma]$ in θ, t has a term subtree g' such that $g \equiv g'$. Let π be an isomorphism from g to g'. Let u_0 be the root of g, and let $u_1, \ldots, u_n$ be all leaves of g, which are listed in left-to-right order. The *corresponding list of g in t*, denoted by CL_t^g, is the list $(\pi(u_0), \pi(u_1), \ldots, \pi(u_n))$. To exploit a substitution as a dictionary in a Lempel–Ziv compression scheme, for each binding $x := [g, \sigma]$ in θ, the CL_t^g are stored in a dictionary instead of term tree g. The *corresponding dictionary of θ in t*, denoted by CD_t^θ, is the list obtained by sorting all elements of $\{CL_t^g \mid x := [g, \sigma] \in \theta\}$ lexicographically. For binding $x := [g, \sigma]$ in θ, the *port index* of σ in t, denoted by PI_t^σ, is a list $(k_2, k_3, \ldots, k_{|\sigma|})$ of $|\sigma| - 1$ distinct integers from 1 to n such that, for any i ($2 \leq i \leq |\sigma|$), $CL_t^g[k_i] = \pi(\sigma[i])$. We suppose that corresponding list CL_t^g is the ℓ_g-th element in CD_t^θ ($1 \leq \ell_g \leq |CD_t^\theta|$). The pair (ℓ_g, PI_t^σ) is called the *corresponding variable label* of binding $x := [g, \sigma]$, and denoted by $CVL(x := [g, \sigma])$. We note that, if σ is equal to the list $(u_0, u_1, ..., u_n)$, $CVL(x := [g, \sigma])$ can be given only by number ℓ_g. Hereinafter, we identify term tree g in binding $x := [g, \sigma]$ with term subtree g' of t such that $g \equiv g'$. For example, let T be the tree and t the term tree shown in Fig. 2. Let $\theta = \{x := [g,(4,2)]\}$ be a substitution, where g is a term subtree of t. Obviously, $T \equiv t\theta$ holds. This means that the corresponding list of g in t is $CL_t^g = (4,1,2)$. The corresponding dictionary of θ is the list $((4,1,2))$. The port index of σ in t is $PI_t^{(4,2)} = (2)$. And the corresponding variable label of binding $x := [g,(4,2)]$ is $(1,(2))$.

Next, we show structure container expression for term tree below. Let $t = (V_t, E_t, H_t)$ be a term tree obtained from XML file $T = (V_T, E_T)$ and variable set H_T, i.e., $V_t = V_T$, $E_t = E_T - \bigcup_{(u_0,u_1,\ldots,u_\ell) \in H_T} \{(u_0,u_i) \in E_T \mid 1 \le i \le \ell\}$, and $H_t = H_T$. In a similar way as XML files, a term tree $t = (V_t, E_t, H_t)$ is encoded by applying the following replacements to code(T). For each variable $(u_0, u_1, , \ldots, u_n)$ in H_T, we replace the coding, "/", of start-tag for the edge (u_0, u_1) with "(/", coding of end-tag for each edge $(u_0, u_i)(1 \le i \le n-1)$ with new token "$", and coding of end-tag for (u_0, u_n) with the string obtained by concatenating "$)" and coding of $CVL(\alpha)$. In the same way as tree, we denote the resultant coding by code(t), and define a *description length* of a term tree t as the length of code(t). For an LZ compression (t, θ), we define a *description length* of (t, θ), denoted by $\|(t, \theta)\|$, as $\|t\| + \sum_{x:=[g,\sigma] \in \theta} c \times |CL_t^g| + \sum_{x:=[g,\sigma] \in \theta'} c \times |\sigma|$, where $\theta' = \{x := [f, \delta] \in \theta \mid CL_t^f \ne \delta\}$, where c is a constant. For example, The following string is coding which represents LZ compression $(t, \{x := [g, (v29, v2, v5)]\})$ of T, where T, t, and g are shown in Fig. 1.

```
/////1 5 4 ///1 5 4 3 (/1 $ /1 $)6 (/1 $ /1 $)6
(/1 $ /1 $)6 2
```

where we assigned `</table>` = 2, `</tr>` = 3, `</td>` = 4, and `</font>` = 5, and $CVL(x := [g, (v29, v2, v5)]) = 6$.

Since $CL_t^g = (v29, v2, v5)$, we can see that $\|(t, \{x := [g, (v29, v2, v5)]\})\| = 62 + c \times 3$. If $c \le 7$, $\|(t, \{x := [g, (v29, v2, v5)]\})\|$ is less than $\|T\| = 86$.

5 Implementation and Experimental Results

We evaluated our compression and decompression algorithms by implementing them on a PC with a 3.4-GHz CPU (XEON) and 2-GB main memory.

Let T be a tree having n_T vertices and t be a term tree, which has n_t vertices, giving the LZ compression of T. Then, we define the *reduction rate* of t for T as $(n_t/n_T) \times 100$. Moreover, for a file p, size(p) denotes file size of p. Then, for a file p and its compressed file p', we define the *compression rate* of p' for p as $(size(p')/size(p)) \times 100$. We then used a data generator to randomly produce a synthetic large tree in which the degree of each vertex was less than or equal to 4 and the number of edge labels is less than or equal to 3. For each $N \in \{20,000, 40,000, 60,000, 80,000, 100,000\}$, we generated a set, $D(N)$, by using the data generator. We applied our algorithms to sets $D(20,000)$, $D(40,000)$, $D(60,000)$, $D(80,000)$, and $D(100,000)$. The settings were $Min = 4$ and $Max = 7$, and the temporary dictionary could store 10,000 candidate subtrees.

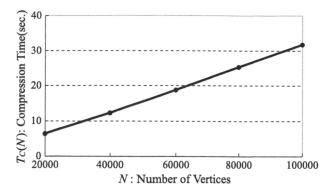

Fig. 5 Compression time vs number of vertices

Let $T_C(N)$ and $T_D(N)$ be the average running times of the compressions and decompressions, respectively, for one tree in $D(N)$. The running times depended on the capacity of the temporary dictionary and the settings of *Min* and *Max*. As shown in Fig. 5, $T_C(N)$ increased almost linearly w.r.t. N. A tree with 100,000 vertices was compressed in about 30 seconds. The reduction rate of each tree was about 80% of its original size. For the randomly generated trees, over 50 elements of the dictionary were generated. For subtree f in randomly generated tree T, the number of isomorphic subtrees to f in T is small, in general. Hence, we couldn't get high reduction ratios for randomly generated trees.

We applied our decompression algorithm to the compressions for $D(20,000)$, $D(40,000)$, $D(60,000)$, $D(80,000)$, and $D(100,000)$. $T_D(N)$ was substantially less than $T_C(N)$. For example, the average $T_D(N)$ for D(100,000) was about 1 second, and all the other $T_D(N)$ were less than 1 second. This shows that our Lempel–Ziv compression scheme has the same characteristics as a Lempel–Ziv compression scheme, such as LZSS over strings. Moreover, these results show that our algorithms are robust for large amounts of data.

We also applied our compression algorithm to HTML files as an example of real-world data. We used three sets of HTML files containing many of the same tree structures and one set of HTML files containing few of the same tree structure. We created a huge tree of structured data by connecting several smaller trees of structured data, which represent HTML files, and applied our algorithm to it. Table 1 shows the compression ratios for HTML file sizes and the running times for compression for each set. The running times for our compression algorithm were lower those for XMill. These experimental results show higher compression ratios than those obtained using XMill. Similar to the results for random data sets, the decompression time was quite short. For example, the "toyota" compressed set was decompressed in about 1 second. The other three sets were decompressed in less than a second.

Moreover, we applied our compression algorithm to XML files as other real-world data. We used nine XML files (Table 2), which can be downloaded from

Table 1 Experimental results for HTML files

HTML file set	#Vertices	Compression ratio(%)		Run-time for compression(s)		Run-time for decompression(s)	
		Ours	XMill	Ours	XMill	Ours	XMill
Nissan[*1]	5,142	2.84	16.08	5.69	1.86	1.87	1.76
Honda[*2]	7,369	2.56	15.61	7.32	1.84	2.10	1.90
Toyota[*3]	5,295	3.08	17.08	5.97	1.88	1.90	1.78
Various files[*4]	10,980	11.94	15.63	10.67	2.56	2.52	2.33

[*1] http://www.nissan.co.jp/CARLINEUP/
[*2] http://www.honda.co.jp/auto-lineup/
[*3] http://toyota.jp/Showroom/carlineup/index.html
[*4] Mixed data from official sites of European football teams.

Table 2 Real-World data used in experiments: all files can be downloaded from the site, http://snp.ims.u-tokyo.ac.jp/XML/

No.	File name
1	JSNP-Exp_Method.xml
2	JSNP-PCR_Region_Y.xml
3	JST_snp.chrY.xml
4	JSNP-NoneSnp-PCR_Region_Y.xml
5	JSNP-NoneSnp-PCR_Region_18.xml
6	JSNP-NoneSnp-PRC_Region_13.xml
7	JSNP-PCR_Region_13.xml
8	JSNP-Screened_SNP_21.xml
9	JST_gene.chr1.xml

Table 3 Experimental results for XML files

Original			Compressed		Reduction	Compression	
No.	Size	#Vertices	#Vertices	#Variables	Rate (%)	Rate (%)	Run-time (s)
1	58,748	1,616	852	300	52.7	2.715	0.531
2	200,927	4460	2,327	673	52.2	12.971	6.437
3	531,896	15,619	8,939	1,815	57.2	4.855	26.812
4	794,263	17,878	8,220	4,016	46.0	11.071	28.875
5	4,018,341	93,221	42,233	19,657	45.3	10.682	25.578
6	4,494,650	105,440	44,636	24,838	42.3	10.341	28.875
7	5,612,990	133,701	53,697	30,297	40.2	9.921	54.5
8	10,886,595	350,069	171,057	45,596	48.9	3.078	208.312
9	13,230,267	461,170	124,840	53,702	27.1	5.564	122.141

the site, http://snp.ims.u-tokyo.ac.jp/XML/. The depth of each XML file is from 7 to 9. Table 3 shows the performance of our algorithm, i.e., reduction rate for vertices of structures of XML files, compression ratios for XML file sizes, and running times for compression. We can easily see that the running time of our compression algorithm became slow according to the number of vertices of input XML file. We can see that the number of vertices of input XML file could be reduced to the half and our compression algorithm has high performance w.r.t. compression, regardless of the size of input XML file. Table 4 shows the comparison of our

Table 4 Experimental results of comparing original XMill with our algorithm

No.	Compression rate (%)			Run-time (s)	
	zip	XMill	Ours	XMill	Ours
1	3.103	**2.267**	2.715	0.031	0.531
2	13.507	**12.551**	12.971	0.062	6.437
3	6.906	**4.746**	4.855	0.219	26.812
4	12.176	**11.048**	11.071	0.312	28.875
5	12.426	10.710	**10.682**	4.406	25.578
6	12.112	10.363	**10.341**	5.39	28.875
7	11.724	9.926	**9.921**	7.313	54.5
8	5.205	3.144	**3.078**	10.172	208.312
9	7.564	5.627	**5.564**	24.203	122.141

compression algorithm with other compressors, zip and XMill. In all experiments, zip was the fastest in three compressors. Since our compression algorithm compressed the structure of input XML file, our compression algorithm was the slowest. However, for almost files, our compression algorithm gave better compression rates than zip. Moreover, for almost larger files, our compression algorithm was the best w.r.t. compression ratio.

From these above experiments for synthetic ordered trees that were randomly generated and to XML files, which are real-world data, we could see that our compression algorithm has advantage for compressing XML files whose depths are high and whose maximum degrees are small.

6   Conclusion

We have developed efficient compression and decompression algorithms for ordered trees that are based on a Lempel–Ziv compression scheme. They are an improved version of the XMill compressor and XDemill decompressor introduced by [7]. Evaluation of their performance by applying them to synthetic large ordered trees and real-world tree structured data demonstrated their effectiveness and efficiency.

In the fields of data mining and knowledge discovery, data mining and data compression techniques are closely related. For example, compressed pattern matching methods, which find all occurrences of a pattern in a compressed text without decompression, have been developed in ([6,9]). We plan to develop a pattern matching algorithm for ordered term trees without decompression that determines whether there is a substitution δ such that $t\theta \equiv s\delta$ or not when an LZ compression (t, θ) and a term tree s are given. Moreover, Suzuki et al. [11] have considered polynomial time inductively inferable from positive data, by giving a polynomial time algorithm for solving the minimal language problem for term trees. Yamasaki et al. [14] have introduced a block preserving graph pattern, which is an extension of a term tree, and have considered a polynomial time inductively inferable from positive data. Then, they have developed incremental graph mining algorithms for enumerating all frequent block preserving graph patterns with respect to a given finite set of connected outerplanar graphs. From these machine learning results, we plan to develop efficient graph mining algorithms for extracting interesting features from compressed graph structured data.

References

1. Abiteboul, S., Buneman, P., & Suciu, D. (2000). *Data on the Web: from relations to semistructured data and XML*. Morgan Kaufmann. San Francisco, CA, USA.
2. Adiego, J., Navarro, G., & de la Fuente, P. (2004). Lempel-ziv compression of structured text. *Proceedings of the IEEE data compression conference (DCC 2004)* (pp. 112–121).
3. Cheney, J. (2001). Compressing xml with multiplexed hierarchical ppm models. *Proceedings of the IEEE data compression conference (DCC 2001)* (pp. 163–172).
4. Cook, D.J., & Holder, L.B. (2000). Graph-based data mining. *IEEE Intelligent Systems, 15*(2), 32–41.
5. Itokawa, Y., Uchida, T., Shoudai, T., Miyahara, T., & Nakamura, Y. (2003). Finding frequent subgraphs from graph structured data with geometric information and its application to lossless compression. *Proceedings of the 7th Pacific–Asia conference on advances in knowledge discovery and data mining (PAKDD-2003)*, Springer, LNAI 2637 (pp. 582–594).
6. Kida, T., Matsumoto, T., Shibata, Y., Takeda, M., Shinohara, A., & Arikawa, S. (2003). Collage system: a unifying framework for compressed pattern matching. *Theoretical Computer Science, 1*(298), 253–272.
7. Liefke, H., & Suciu, D. (2000). Xmill: an efficient compressor for xml data. *Proceedings of the 2000 ACM SIGMOD international conference on management of data* (pp. 153–164).
8. Matsumoto, S., Hayashi, Y., & Shoudai, T. (1997). Polynomial time inductive inference of regular term tree languages from positive data. *Proceedings of the 8th workshop on algorithmic learning theory (ALT-97)*, Springer, LNAI 1316 (pp. 212–227).
9. Sakamoto, H. (2005). A fully linear-time approximation algorithm for grammar-based compression. *Journal of Discrete Algorithms, 3*(2–4), 416–430.
10. Storer, J.A., & Szymanski, T.G. (1982). Data compression via textual substitution. *Journal of the ACM, 29*(4), 928–951.
11. Suzuki, Y., Shoudai, T., Uchida, T., & Miyahara, T. (2006). Ordered term tree languages which are polynomial time inductively inferable from positive data. *Theoretical Computer Science, 350*, 63–90.
12. Tolani, P.M., & Haritsa, J.R. (2002). Xgrind: a query-friendly xml compressor. *Proceedings of the 18th international conference on data engineering* (pp. 225–234).
13. Yamagata, K., Uchida, T., Shoudai, T., & Nakamura, Y. (2003). An effective grammar-based compression algorithm for tree structured data. *Proceedings of the 13th international conference on inductive logic programming (ILP-03)*, Springer, LNAI 2835 (pp. 383–400).
14. Yasamaki, H., Sasaki, Y., Shoudai, T., Uchida, T., & Suzuki, Y. (2009). Learning block-preserving graph patterns and its application to data mining. *Machine Learning, 76*, 137–173.
15. Ziv, J., & Lempel, A. (1977). A universal algorithm for sequential data compression. *IEEE Transactions on Information Theory, IT-23*(3), 337–343.

Chapter 27
Using Finite Automata Approach for Searching Approximate Seeds of Strings

Ondřej Guth and Bořivoj Melichar

Abstract Seed is a type of a regularity of strings. A restricted approximate seed w of string T is a factor of T such that w covers a superstring of T under some distance rule. In this paper, the problem of searching of *all restricted seeds with the smallest Hamming distance* is studied and a polynomial time and space algorithm for solving the problem is presented. It searches for all restricted approximate seeds of a string with given limited approximation using Hamming distance and it computes the smallest distance for each found seed. The solution is based on a finite (suffix) automata approach that provides a straightforward way to design algorithms to many problems in stringology. Therefore, it is shown that the set of problems solvable using finite automata includes the one studied in this paper.

Keywords Hamming distance · Approximate seed · Suffix automaton · Stringology

1 Introduction

Searching regularities of strings is used in a wide area of applications like molecular biology, computer-assisted music analysis, or data compression. By regularities, repeated strings are meant. Examples of regularities include repetitions, borders, periods, covers, and seeds.

The algorithm for computing all exact seeds of a string was introduced in [4]. The first algorithm for searching all seeds using finite automata was introduced in [5].

Finding exact regularities is not always sufficient and thus some kind of approximation is used. An algorithm for searching approximate periods, covers, and seeds under Hamming, Levenshtein (also called edit), and weighted Levenshtein distance was presented in [2]. The algorithm for computing approximate seeds was originally

O. Guth (✉) and B. Melichar
Faculty of Electrical Engineering, Czech Technical University in Prague, Karlovo náměstí 13, 12135 Praha 2, Czech Republic
e-mail: gutho1@fel.cvut.cz; melichar@fel.cvut.cz

S.-I. Ao et al. (eds.), *Intelligent Automation and Computer Engineering*,
Lecture Notes in Electrical Engineering 52, DOI 10.1007/978-90-481-3517-2_27,
© Springer Science+Business Media B.V. 2010

introduced in [1]. An algorithm for searching all approximate covers with the smallest Hamming distance based on finite automata was presented in [3].

Finite automata provide common formalism for many algorithms in the area of text processing (stringology). Finite automaton as a data structure may be easily implemented. Therefore, using it as a base for similar approach to many algorithms is not only theoretical problem, as it may make development of software related with above mentioned areas easier, faster, and cost-reduced.

2 Preliminaries

Some notions and notations used in this paper are commonly used, thus they are not defined here. Exact definitions of such notions and notations could be found in [3].

Having string $T = a_1 a_2 \cdots a_{|T|}$, *reversed string* T is denoted by T^R and it is equal to $T^R = a_{|T|} a_{|T|-1} \cdots a_1$. An *effective alphabet* of a string T is denoted by A_T and consists just of symbols contained in T.

Suppose $p, s, u, w, x, p', s', T \in A^*$. T is a *superstring* of w if $T = pws$. p' is an *approximate prefix* of T with maximum Hamming distance k if $T = pu$ and $D_H(p, p') \leq k$. Set of all approximate prefixes of T with maximum Hamming distance k is denoted by $Pref^k(T)$. An *approximate suffix* and set of all approximate suffixes of T with respect to k is defined by analogy and is denoted by $Suff^k(T)$. The set of all factors of string T is denoted by $Fact(T)$.

We say that string w *occurs* in string T if $w \in Fact(T)$. Factor w *occurs at position* (end-position) i in T if $\forall j \in \{1, \ldots, |w|\} : w[j] = T[i - |w| + j]$. An *end-set* is a set of all i such that w occurs at position i in T. String w occurs approximately with maximum Hamming distance k at position i in string T (or w has approximate occurrence at position i in T) if there exists factor x that occurs at position i in T and $D_H(x, w) \leq k$.

String w is a *restricted approximate seed* of string T with maximum Hamming distance k if w is a factor of T and w is a restricted approximate cover of some superstring of T with maximum Hamming distance k.

The *smallest Hamming distance* of a restricted approximate seed w of string T is the smallest possible integer l_m such that w is a restricted approximate seed of T with maximum Hamming distance l_m.

Deterministic finite automaton is denoted by DFA, nondeterministic one is denoted by NFA. DFA is *partial* if there may exist a pair of state q_i and symbol a such that $\delta(q_i, a)$ is undefined. In this paper, only partial DFA are considered. A deterministic *trie* is a DFA that may be represented as a tree, i.e., for each state q^j, there exists at most one q^i such that for any $a \in A$, $\delta(q^i, a) = q^j$. An *extended transition function* denoted by δ^* is for a DFA defined for $a \in A, u \in A^*$ in this way:

$$\delta^*(q, \varepsilon) = q, \ \delta^*(q, ua) = \delta(\delta^*(q, u), a)$$

An extended transition function of an NFA is defined as:

$$\delta^*(q, \varepsilon) = \{q\}, \ \delta^*(q, au) = \bigcup_{q_i \in \delta(q,a)} \delta^*(q_i, u)$$

A *left language* of a state q of a DFA is set of all strings w for that holds $\delta^*(q_0, w) = q$ for initial state q_0. Left language of a state q of a trie contains one string, denoted by *factor(q)*.

A nondeterministic Hamming suffix automaton for string T and maximum Hamming distance k is denoted by $M_{SN}^k(T)$.

A deterministic suffix automaton for string T and maximum Hamming distance k, denoted by $M_{SD}^k(T)$, is a DFA that accepts all strings from $Suff^k(T)$ (see Fig. 9). A *depth* of a state q of the automaton is equal to length of the longest string w such that $\delta^*(q_0, w) = q$ for initial state q_0. A *d-subset* of a state of a DFA M_D is a set of corresponding states of an NFA M_N when M_D was created using subset construction from M_N. Each *element* of the d-subset of M_D corresponds to a state of M_N. An element of a d-subset is denoted by r_i, where the subscript i means an index (order) of the element r_i within the d-subset. A *level* of a state of an NFA or of an element of a d-subset corresponds to an approximation (i.e., number of errors in Hamming distance). In figures, states of nondeterministic automata and elements of d-subsets of deterministic automata are denoted by their depths and levels, e.g., $3''$ means state or element with depth 3 and level 2.

3 Problem Formulation

(All restricted seeds with the smallest Hamming distance.) Given string T and maximum Hamming distance k, find all restricted approximate seeds of T with respect to k and compute their smallest distances.

The algorithm for searching exact seeds in generalised strings [5] works for (non-generalised) strings as well, because string is special case of generalised string. It is based on the following idea. First, nondeterministic suffix automaton $M_{SN}(T)$ for string T is constructed (see Fig. 7 for an example). Equivalent deterministic automaton $M_{SD}(T) = (Q, A_T, \delta, q_0, F)$ is computed using subset construction (see Fig. 9 for an example). One of conditions for any factor to be a seed of string T is its length. Seed w must be long enough to cover "central part" of string T (i.e., the part of T between the leftmost and the rightmost position of w within T), the uncovered suffix of T must be covered by a prefix of w and the uncovered prefix of T must be covered by a suffix of w. All sufficiently long factors are then checked whether they cover the uncovered suffix, and prefix of T, respectively. If $M_{SD}(T)$ accepts some prefix of factor w then w covers uncovered suffix of T. If a suffix automaton $M_{SD}(T^R)$ for reversed string T accepts some prefix of reversed factor w then w covers uncovered prefix of T. When w satisfies all the conditions, w is a seed of T.

Computation of the smallest Hamming distance of a cover (presented in [3]) is based on the following idea: when the maximum approximation of the first and the last position of cover w in T is $l_{\min}$, for its smallest distance l_m holds $l_m \geq l_{\min}$, because cover is an approximate prefix and suffix of T and thus it cannot cover T without its first and last position. When cover w of T has positions with approximation at most l, for its smallest distance l_m clearly holds $l_m \leq l$. When the positions of w with the maximum approximation equal to l are no longer considered (the first and the last position must be still considered) and w is still cover of T, then for l_m holds $l_m \leq l - 1$. l_m is decremented till w still covers T. This may be used with modifications for computation of seeds.

The algorithm for searching exact seeds from [5] uses two phases: first, deterministic suffix automaton is constructed and then d-subsets are analysed and seeds are computed. This means that complete automaton or at least all the d-subsets to be analysed need to be stored in memory at a time. By contrast, the algorithm for searching covers from [3] uses merge of the phases, each d-subset is analysed just after its construction. A depth-first search like algorithm is used and the states that are no longer needed are removed. For approximate seeds searching, there is also no need to store all elements of d-subsets of the automaton at a time.

4 Problem Solution

Some properties are common for exact and approximate seeds with Hamming distance. Therefore, the algorithm presented in [5] is used as a base of algorithm for the problem studied in this paper, using some (but not all) techniques for searching covers in [3] for further improvements.

Every approximate restricted seed of string T is necessarily an exact factor of T with other possible approximate occurrences. Suffix automaton constructed for string T and maximum Hamming distance k has extended transitions defined for all factors of T with respect to k. When deterministic suffix automaton $M_{SD}^k(T)$ is constructed using subset construction from nondeterministic one $M_{SN}^k(T)$, each element of any d-subset of any state of $M_{SD}^k(T)$ contains information not only about position (depth within $M_{SN}^k(T)$) but also about approximation (level within $M_{SN}^k(T)$). Therefore, it may be easily determined whether string from left language of any state of $M_{SD}^k(T)$ is an exact factor of string T.

Note. For string T, string v such that v is not a factor of T, and any string u being a superstring of v holds:

$$\forall T, u, v \in A^*, v \in Fact(u) : v \notin Fact(T) \Rightarrow u \notin Fact(T)$$

Lemma 1. *For DFA* $M_{SD}^k(T) = (Q_D, A_T, \delta_D, q_0^D, F_D)$ *created using subset construction from* $M_{SN}^k(T)$ *and for its state* $q_i \in Q_D$ *with* d-subset $d(q_i)$ *such that*

$\forall r \in d(q_i) : level(r) > 0$ holds that any successor of q_i cannot contain element r^j such that $level(r^j) = 0$.

Corollary 2. *As only exact factor of T may be a restricted seed of T, there is no need to construct any state of $M_{SD}^k(T)$ having only non-zero-level elements in its d-subset, as such state contains no exact factor of T in its left language. Therefore, when such state is created during construction, it may be removed and any of its successors need not be constructed. Such deterministic suffix automaton containing only states having at least one zero-level element in its d-subset is denoted by $\tilde{M}_{SD}^k(T)$.*

Note. Special type of deterministic suffix automaton, suffix trie, is considered in this paper. Construction of the trie and left language extraction is simpler than for general suffix automaton. As left language of any state of the trie contains exactly one string, extraction of left language of any state takes linear time with respect to length of the string (e.g., using inverted transition function). See Fig. 9 for example.

The relationship between length and positions of any seed (presented in [5]) holds also for approximate positions with Hamming distance, as the distance is defined for strings of equal lengths only.

Note. When searching for covers with Hamming distance ([3]), it is possible to remove all states q of deterministic suffix trie that do not represent prefix, i.e., such q that $|factor(q)| < depth(r_1)$, where $d(q) = r_1, \ldots, r_{|d(q)|}$. Similar property between the first position and length of a factor is used for searching seeds: $|factor(q)| \leq \frac{1}{2} \cdot depth(r_1)$. Unlike in computing covers, this condition cannot be used for removing states q and their successors.

Example 3. Let us consider suffix trie for string $T = bbbbbaaabb$ and maximum Hamming distance $k = 2$. Factor aaa cannot be a seed of T as its first approximate position within T is 6. Factor $aaabb$ is a seed of T with respect to k. It is obvious that for states q_1, q_2 of the trie, where $factor(q_1) = aaa$ and $factor(q_2) = aaabb$, holds: q_2 is a successor of a state that is a successor of q_1. Therefore, q_1 must not be removed to be able to find $aaabb$.

For computation of the smallest distance l_m of each seed, the idea used for searching covers ([3]) may be used for searching seeds. Unlike searching covers, any position may be removed, including the first and the last, thus the only lower bound of l_m is 0. Determination whether continue to decrement l is for seeds more complex than for covers, as computation of covering of central part of T is not sufficient condition for seeds. For factor w of T, not only positions and their approximation need to be considered, but also distance of the uncovered prefix, suffix, of T, and some suffix, prefix, of w, respectively. See algorithm at Fig. 5 for further information.

Example 4. Let us have string $T = bbbbbaaa$ and maximum Hamming distance $k = 2$. One seed of T with respect to k is $bbba$. It may be seed of T with Hamming distance 2, because its positions in T are 4, 5, 6, 7, and 8 with maximum approxi-

Fig. 1 Possible covering
of string *bbbbbaaa* with
string *bbba* and Hamming
distance 2 from Example 4

$$
\begin{array}{r}
bbbbbaaa \\
bbba \\
bbba \\
bbba \\
bbba \\
bbba
\end{array}
$$

Fig. 2 Possible covering
of a superstring of *bbbbbaaa*
with *bbba* and Hamming
distance 1 from Example 4

$$
\begin{array}{r}
bbbbbaaa \\
bbba \\
bbba \\
bbba \\
bbba \\
b
\end{array}
$$

mation 2 (see Figs. 1 and 9). When the position 8 with approximation 2 is removed, *bbba* is still seed of T with positions 4, 5, 6, and 7, all with approximation at most 1 (Fig. 2).

The deterministic suffix trie is needed not only to determine positions of each factor w of T, but also for checking whether w is able to cover uncovered prefix and suffix of T (see algorithm at Fig. 6). Thus, the trie must be able to accept strings of length at least $|w| - 1$. Therefore, the depth-first search with removing states used in [3] cannot be used. By contrast, only elements of d-subset $d(q)$ may be removed after construction of all successors of q, transitions must be preserved. Thus, breadth-first search in the automaton is used (see algorithm at Fig. 3 and usage of queues L, L^R). As no state of trie $\tilde{M}^k_{SD}(T)$ is removed and the last element of each d-subset is preserved, it is possible to recognise all approximate suffixes of T of length at least $|w| - 1$ and their distance.

Like an exact seed, an approximate one must also cover the uncovered prefix and suffix of string T (i.e., some prefix of seed w must be an approximate suffix of T and some suffix of w must be an approximate prefix of T). Similar technique as for exact seeds ([5]) is used (algorithm at Fig. 6), but with tries $\tilde{M}^k_{SD}(T)$ and $\tilde{M}^k_{SD}(T^R)$. When some suffix of seed w of T (i.e., some prefix of reversed w) is accepted by $\tilde{M}^k_{SD}(T^R)$, i.e., $w = factor(q)$ for some final state q of $\tilde{M}^k_{SD}(T^R)$, w covers the uncovered prefix of T with approximation equal to level of the last element $r_{|d(q)|}$ of $d(q)$. Similarly for a prefix of w, the uncovered suffix of T and $\tilde{M}^k_{SD}(T)$.

For complete solution of the problem see algorithm at Fig. 3.

Example 5. Let us compute set of all seeds with maximum Hamming distance $k = 2$ for string $T = bbbbbaaa$. Nondeterministic suffix automata $M^k_{SN}(T)$ (see Fig. 7) and $M^k_{SN}(T^R)$ (see Fig. 8) are constructed. Next, subset construction of deterministic suffix trie $M^k_{SD}(T)$ from $M^k_{SN}(T)$ starts state-by-state (see transition diagram of $M^k_{SD}(T)$ with all states, that need to be constructed, at Fig. 9), the same

Set-of-seeds(string T, maximum Hamming distance k)
$hseeds^k(T) := \emptyset$
construct $M^k_{SN}(T) = (Q_N, A_T, \delta_N, q_0^N, F_N)$
construct $M^k_{SN}(T^R) = (Q^R_N, A_T, \delta^R_N, q_0^{NR}, F^R_N)$
create state q_0^D of trie $M^k_{SD}(T) = (Q_D, A_T, \delta_D, q_0^D, F_D)$
create state q_0^{DR} of trie $M^k_{SD}(T^R) = (Q^R_D, A_T, \delta^R_D, q_0^{DR}, F^R_D)$
define $factor(q_0^D) = \varepsilon$, $depth(q_0^D) = 0$, $depth(q_0^{DR}) = 0$
create L, L^R new empty queues of states
enqueue(L, q_0^D), enqueue(L^R, q_0^{DR})
while(L^R is not empty)   /*construct complete $\tilde{M}^k_{SD}(T^R)$*/
  $q^{tR} := $ dequeue(L^R)
  for all ($a \in A_T$)
    compute new state q^{uR} – successor of q^{tR} for a (Figure 4)
    discard all elements of $d(q^{tR})$ but the last one
              /*all successors of $d(q^{tR})$ have just been computed*/
  end for
end while
while (L is not empty)   /*compute $\tilde{M}^k_{SD}(T)$ and seeds*/
  $q^t := $ dequeue(L)
  for all ($a \in A_T$)
    compute new state q^u – successor of q^t for a (Figure 4)
    if(exists $r \in d(q^u)$ where $level(r) = 0$)
    /*only state q^u that is part of $M^k_{SD}(T)$ is further processed*/
      define $w = factor(q^u) = factor(q^t).a$
      if(w is seed of T – positions from $d(q^u)$ (Figure 6))
        compute the smallest distance l_m of w (Figure 5)
        if($|w| > k$ or $l_m < |w|$)
                  /*all strings of length less or equal to l_m are seeds*/
          $hseeds^k(T) := hseeds^k(T) \cup \{(w, l_m)\}$
        end if
      end if
    end if
  end for
  discard all elements of $d(q^t)$ but the last one
end while
return $hseeds^k(T)$

Fig. 3 Algorithm for computing the set of all restricted approximate seeds of string T with maximum Hamming distance k

Compute-state-of-trie(**NFA** $(Q_N, A_T, \delta_N, q_0^N, F_N)$, **state of trie**
$q^t \in Q_D$, **symbol** $a \in A_T$, **queue** L **of states**)
    create new state q^u
    define $depth(q^u) = depth(q^t) + 1$
    for all($r^i \in d(q^t)$ (in order as stored in $d(q^t)$))
      append all $r^j \in \delta_N(r^i, a)$ to $d(q^u)$ in order by $depth(r^j)$
    end for
    if (exists $r \in d(q^u)$ where $level(r) = 0$)
      $Q_D := Q_D \cup \{q^u\}$
      enqueue(L, q^u)
      if ($r^u_{|d(q^u)|} \in F_N, d(q^u) = r^u_1, \ldots r^u_{|d(q^u)|}$)
        $F_D := F_D \cup \{q^u\}$
      end if
    end if
    return $\tilde{M}$

Fig. 4 Algorithm for computing a successor of a state of deterministic suffix trie $\tilde{M} = (Q_D, A_T, \delta_D, q_0^D, F_D)$

Smallest-distance(d–subset $d(q) = r_1, r_2, \ldots, r_{|d(q)|}$ **represent-
ing seed** w **of** T)
    $t := d(q)$
    $l_{max} := \max_{r \in t}\{level(r)\}$
    $l := l_{max}$
    repeat
      for all($r \in t : level(r) = l$)
        remove r from t
      end for
      $l := l - 1$
    until(w is seed of T using positions given by t and l (Fig. 6))
    $l_m := l + 1$
    return l_m

Fig. 5 Algorithm for computing the smallest distance of a seed of string T

is done with trie $M_{SD}^k(T^R)$ from $M_{SN}^k(T^R)$. Some states may have only elements with non-zero level in its d-subset (for example $7''8'$). Such states are removed and their successors are not constructed as strings from their left languages (e.g., *aaaa*) are not factors of T (follows by Corollary of Lemma 1).

All other states need to be checked whether their left languages contain some seeds. For example, state with d-subset $6''7'8$ contains string *aaa* in its left language. The string occurs approximately at positions 6, 7 in T and exactly at position 8 in T, thus it cannot be seed of T, as its leftmost occurrence within T ends at position 6 and its length is 3 (i.e., the occurrence starts at position 4), so any proper suffix of *aaa* cannot cover the uncovered prefix (positions 1 to 3) of T.

Other example is state with d-subset $4'5'67'8''$, which contains factor *bbba* of string T in its left language. This factor covers T with Hamming distance 2, and therefore it is seed of T (see Fig. 1). When all positions with the maximum distance

Is-seed(**constructed parts of DFA** $M_{SD}^k(T) = (Q, A_T, \delta, q_0, F)$,
$M_{SD}^k(T^R) = (Q^R, A_T, \delta_R, q_0^R, F^R)$, d–**subset** $t = r_1, r_2, \ldots, r_{|t|}$ **for**
$q \in Q$, $w \in factor(q)$, **maximum Hamming distance** l)
   $if\,((\text{for all } i = 2, 3, \ldots, |t| : r_i - r_{i-1} \le |w|)$
    $and\ (\exists p \in Pref^0(w), |p| \ge |T| - r_{|t|} :$
      $\delta^*(q_0, p) = q^1, q^1 \in F \wedge level(r_{|d(q^1)|}^{q^1}) \le l)$
    $and\ (\exists s \in Suff^0(w), |s| \ge r_1 - |w| :$
      $\delta_R^*(q_0^R, s) = q^2, q^2 \in F^R \wedge level(r_{|d(q^2)|}^{q^2}) \le l))$
   $return\ true$
  $else$
   $return\ false$
  $end\ if$

Fig. 6 Algorithm for determining whether string w is a seed of string T with maximum Hamming distance l

Fig. 7 Transition diagram of the nondeterministic suffix automaton $M_{SN}^k(T)$ for string $T = bbbbbaaa$ and maximum Hamming distance $k = 2$ from Example 5

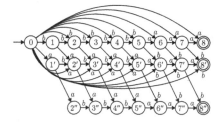

Fig. 8 Transition diagram of the nondeterministic suffix automaton $M_{SN}^k(T^R)$ for reversed string T, i.e., $T^R = aaabbbbb$, and maximum Hamming distance $k = 2$ from Example 5

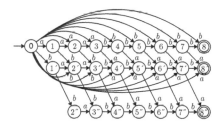

(i.e., 8) are not considered, $bbba$ is still seed of T, as proper prefix b of $bbba$ covers uncovered suffix a of T with Hamming distance 1 (see Fig. 2). See resulting table of all seeds and their distances in Table 1.

5 Time and Space Complexities

Note. As parts of automata $M_{SD}^k(T)$ and $M_{SD}^k(T^R)$ are constructed the same way in algorithm at Fig. 3, the time and space complexities of their construction are the same.

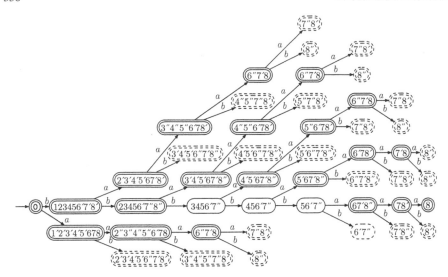

Fig. 9 Transition diagram of the constructed part of the suffix trie $M_{SD}^k(T)$ for string $T = bbbbbaaa$ and maximum Hamming distance $k = 2$ from Example 5; dashed states are removed as their left languages do not contain exact factor of T and thus they are not states of trie $\tilde{M}_{SD}^k(T)$

Table 1 All seeds of string $T = bbbbbaaa$ with maximum Hamming distance $k = 2$ and their smallest distances l_m; p is used prefix of a seed, s is used suffix (computed by algorithm at Fig. 6); see Example 5

Seed	d-Subset	l_m	Occurrences	p	s
ba	2'3'4'5'67'8'	1	2,3,4,5,6,7,8	ε	ε
baa	3''4''5''6'78'	2	3,4,5,6,7,8	ε	ε
bba	3'4'5'67'8''	1	3,4,5,6,7	b	ε
bbb	3456'7''	2	3,4,5,6,7	b	ε
baaa	6''7'8	2	6,7,8	ε	aa
bbaa	4''5''6'78'	2	4,5,6,7,8	ε	aa
bbba	4'5'67'8''	1	4,5,6,7	b	ε
bbbb	456'7''	2	4,5,6,7	b	ε
bbaaa	6''7'8	2	6,7,8	ε	a
bbbaa	5''6'78'	1	6,7,8	ε	a
bbbba	5'67'8''	1	5,6,7	b	ε
bbbbb	56'7''	2	5,6,7	b	ε
bbbaaa	6''7'8	1	7,8	ε	a
bbbbaa	6'78'	1	6,7,8	ε	ε
bbbbba	67'8''	1	6,7	b	ε
bbbbaaa	7'8	1	7,8	ε	ε
bbbbbaa	78'	1	7,8	ε	ε
bbbbbaaa	8	0	8	ε	ε

Lemma 6. *Left languages of states of deterministic suffix trie with set of states Q_D are distinct, i.e.,*

$$q_1, q_2 \in Q_D; q_1 \neq q_2 \Rightarrow factor(q_1) \neq factor(q_2)$$

Definition 7. Let us consider string T and maximum Hamming distance k. When a factor w approximately occurs e-times in T with respect to k, we say that *number of approximate repetitions of w in T with respect to k*, denoted by $R_w^k(T)$, is $e - 1$. Then *number of approximate repetitions of all factors of T with respect to k*, denoted by $R^k(T)$, is defined as

$$R^k(T) = \sum_{w \in Fact(T)} R_w^k(T)$$

Lemma 8. *Number of states of deterministic suffix automaton $\tilde{M}_{SD}^k(T)$ for string T and maximum Hamming distance k is*

$$\frac{1}{2} \cdot (|T|^2 + |T|) - R^k(T) + 1$$

Note. As restricted approximate seeds of string T are exact factors of T, it is meaningful to consider effective alphabet A_T only and $|A_T| \leq |T|$ always holds (recall that effective alphabet A_T consists only of symbols that occur in T). It is also meaningless to consider high k, because every factor of T having length less or equal to k is always approximate seed of T. Thus $k \leq |T|$ always holds. Usually k and $|A_T|$ are independent of $|T|$.

Lemma 9. *Number of states of deterministic suffix automaton $M_{SD}^k(T)$ constructed for string T and maximum Hamming distance k using algorithm at Fig. 3 is at most*

$$|A_T| \cdot \left(\frac{1}{2} \cdot (|T|^2 + |T|) - R^k(T) \right) + 1$$

Lemma 10. *For every d-subset of automaton $\tilde{M}_{SD}^k(T)$ constructed for string T and maximum Hamming distance k by algorithm at Fig. 3 holds that there are no two elements having the same depth.*

Lemma 11. *Number of elements of all d-subsets of deterministic suffix automaton $\tilde{M}_{SD}^k(T)$ constructed for string T and maximum Hamming distance k is not greater than*

$$\frac{1}{2} \cdot (|T|^3 + |T|^2) - |T| \cdot R^k(T) + 1$$

Lemma 12. *Number of elements of all d-subsets of deterministic suffix automaton $M_{SD}^k(T)$ constructed for string T over effective alphabet A_T and maximum Hamming distance k using algorithm at Fig. 3 is not greater than*

$$|A_T| \cdot \left(\frac{1}{2} \cdot (|T|^3 + |T|^2) - |T| \cdot R^k(T) \right) + 1$$

Lemma 13. *Time complexity of the check whether d-subset $d(q)$ of a state q of deterministic suffix automaton $\tilde{M}_{SD}^k(T)$ represents a seed $w = factor(q)$ of string T (algorithm at Fig. 6) is at most*

$$2 \cdot |d(q)| + 2 \cdot |w| - 2$$

that is $\mathcal{O}(|T|)$.

Lemma 14. *Having string* T, *its seed* w *and given maximum Hamming distance* k, *time complexity of the computation of the smallest distance of seed* $w = factor(q)$ *of* T *(algorithm at Fig. 5) is at most*

$$|d(q)| + k \cdot (3 \cdot |d(q)| + 2 \cdot |w| - 2)$$

that is $\mathcal{O}(k \cdot |T|)$.

Note. Number of all seeds is $\mathcal{O}(|T|^2)$ (like number of factors). Thus, the sum of their lengths is $\mathcal{O}(|T|^3)$, denoted by $|hseeds^k(T)|$.

Theorem 15. *Time complexity of computation of all restricted approximate seeds with their smallest distance for string* T *over effective alphabet* A_T *with maximum Hamming distance* k *(algorithm at Fig. 3) is*

$$\mathcal{O}(k \cdot |A_T| \cdot |T|^3)$$

Proof. Construction of nondeterministic suffix automaton $M_{SN}^k(T) = (Q_N, A_T, \delta_N, q_0^N, F_N)$ for string T and maximum Hamming distance k takes $\mathcal{O}(k \cdot |A_T| \cdot |T|)$ time. For each state of deterministic suffix automaton $\tilde{M}_{SD}^k(T)$ (Lemma 8) and for each symbol of A_T, new d-subset is constructed. As each element of any d-subset may be constructed in constant time (just using already known transition function δ_N) and the elements are naturally ordered (no need to sort – proven in [3]), all d-subsets are constructed in at most

$$|A_T| \cdot \left(\frac{1}{2} \cdot (|T|^3 + |T|^2) - |T| \cdot R^k(T) \right) + 1$$

time. Each d-subset is checked whether it contains element with zero level in linear time. The left language extraction of state takes linear time and thus by Lemma 13 and 5 the theorem holds.

Lemma 16. *During construction of deterministic suffix trie* $\tilde{M}_{SD}^k(T)$ *for string* T *(algorithm at Fig. 3), there are at most* $\mathcal{O}(|T|^2)$ *elements of* d-*subsets stored in memory at a time.*

Theorem 17. *Space complexity of computation of all seeds of string* T *with maximum Hamming distance* k *is*

$$\mathcal{O}(|T|^2 + |hseeds^k(T)|)$$

Proof. Space complexity of construction of nondeterministic suffix automaton $M_{SN}^k(T)$ is $\mathcal{O}(k \cdot |A| \cdot |T|)$ (proven in [3]). By Lemma 16, number of elements of

d-subsets stored in memory at a time is $\mathcal{O}(|T|^2)$, as no more elements of d-subsets than those in queue L plus $\mathcal{O}(|T|)$ new are in memory at a time. By Lemma 8, number of states of deterministic suffix automaton $\tilde{M}^k_{SD}(T)$ is $\mathcal{O}(|T|^2)$ (they all are stored in memory with one element each). As the automaton $\tilde{M}^k_{SD}(T)$ is constructed as trie, number of transitions is also $\mathcal{O}(|T|^2)$. The space complexity also depends on size of result, which is $|hseeds^k(T)|$.

6   Experimental Results

The Algorithm was implemented in C++ using STL and compiled by GNU C++ 3.4.6 with O3 parameter. The dataset used to test the algorithm is the nucleotide sequence of *Saccharomyces cerevisiae* chromosome IV (the dataset could be downloaded from http://www.genome.jp/). The string T consists of the first $|T|$ characters of the chromosome.

The set of tests was run on an AMD Athlon (1400 MHz) system, with 1.2 GB of RAM, under Gentoo Linux operating system (see Fig. 10).

Note. In comparison to experimental results shown in [2], the algorithm presented in this paper runs a bit faster for the same data, even on a slightly slower computer (1.3 s in [2] for text length 100 vs. maximum 0.7 s for text length 113 – see Fig. 10).

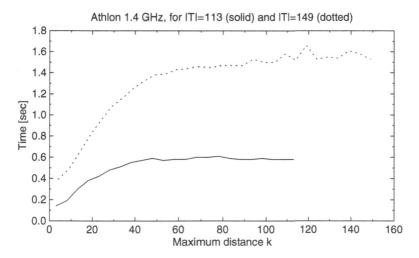

Fig. 10 Time consumption of the experimental run on the Athlon with respect to the maximum distance (see Section 6)

7 Conclusion

In this paper, we have shown that an algorithm design based on determinisation of a suffix automaton is appropriate for computation of all restricted approximate seeds with the smallest Hamming distance. The presented algorithm is straightforward, easy to understand and to implement and its theoretical and experimental time requirements are comparable to the existing approach [1].

For the future work, we would like to extend the algorithm for searching seeds to other distances and to utilise similar approach for searching other types of regularities.

Acknowledgements This research was supported by the Czech Technical University in Prague as grant No. CTU0803113 and as grant No. CTU0915313, by the Ministry of Education, Youth and Sports of the Czech Republic under research program MSM 6840770014, and by the Czech Science Foundation as project No. 201/06/1039 and as project No. 201/09/0807.

References

1. Christodoulakis, M., Iliopoulos, C.S., Park, K.S., & Sim, J.S. (2005). Approximate seeds of strings. *Journal of Automata, Languages and Combinatorics*, *10*(5/6), 609–626.
2. Christodoulakis, M., Iliopoulos, C.S., Park, K.S., & Sim, J.S. (2005). Implementing approximate regularities. *Mathematical and Computer Modelling*, *42*, 855–866.
3. Guth, O., Melichar, B., & Balík, M. (2008). Searching all approximate covers and their distance using finite automata. *Information technologies – applications and theory* (pp. 21–26). Košice: Univerzita P. J. Šafárika.   http://ftp.informatik.rwth-aachen.de/Publications/CEUR-WS/Vol-414/paper4.pdf.
4. Iliopoulos, C.S., Moore, D., & Park, K.S. (1993). Covering a string. *CPM '93: Proceedings of the 4th annual symposium on combinatorial pattern matching* (pp. 54–62). London, UK: Springer.
5. Voráček, M., & Melichar, B. (2006). Computing seeds in generalized strings. *Proceedings of workshop 2006* (pp. 138–139). Prague: Czech Technical University.

Chapter 28
Speech Recognizer with Dynamic Alternative Path Search and Its Performance Evaluation

Tsuyoshi Morimoto and Shin-Ya Takahashi

Abstract For a middle-size (around 1,000 words) vocabulary speech recognition, a Finite State Automaton (FSA) language model is widely used. However, defining a FSA model with sufficient coverage and consistency requires much human effort. We already proposed a method to automatically construct a FSA language model from learning corpus by use of FSA DP matching algorithm. Experiment results show that this model attains quite high recognition correct rate for closed data, but only low rate for open data. This is mainly because a necessary path does not appear in a generated FSA. To cope with this problem, we propose a new search algorithm that allows to jump dynamically to an alternative path when speech recognition of some words seems to fail. We report experiment results and discuss the effectiveness of the algorithm.

Keywords FSA language model · Automatic construction of a language model · Dynamic alternative path search · Speech recognition performance

1 Introduction

For a large vocabulary speech recognition system, like a dictation system, a statistical language model such as bi-gram or tri-gram is widely used. This kind of models is generated from a large amount of learning texts (corpus), such as newspaper articles covering several years. Here, if the size of corpus is not large enough, reliability of statistical information calculated from the corpus decreases and then so the efficiency of the generated statistical language model (this is called as *a sparseness problem*). Therefore, for a middle-size (around 1,000 words) vocabulary speech recognition system, such as on travel conversation, a Finite State Automaton (FSA)

T. Morimoto (✉) and S.-Y. Takahashi
Electronics and Computer Science Department, Fukuoka University, 8-19-1 Nanakuma, Jonan-ku, Fukuoka 814-0180, Japan
e-mail: morimoto@tlsun.tl.fukuoka-u-ac.jp; takahasi@tlsun.tl.fukuoka-u-ac.jp

S.-I. Ao et al. (eds.), *Intelligent Automation and Computer Engineering*,
Lecture Notes in Electrical Engineering 52, DOI 10.1007/978-90-481-3517-2_28,
© Springer Science+Business Media B.V. 2010

language model is usually used. However, defining a FSA model by hands with sufficient coverage and consistency requires much human effort, even for a middle-sized vocabulary.

On generating a FSA language model automatically from learning data, several methods have been proposed. Note that constructing an acyclic FSA from given data is a simple problem; one can construct a TRIE tree in which common prefix is shared, and then minimize the tree by merging equivalent states. This method, however, is computationally quite expensive. Several method have been proposed to improve the efficiency [1, 2], but they are still at a basic research stage and have not applied yet to practical applications. Meanwhile, other kind of approaches have been proposed [3–5] aiming at applying to actual speech recognition language modeling. However, since common key technique used in them for improving efficiency is to use stochastic features, a sparseness problem mentioned above arises again when corpus size is not large enough.

We already proposed a new method to generate a FSA language model by use of FSA DP (Dynamic Programming) matching method [6]. Experiment results shows that this model can attain high recognition correct rate for closed data, but only low rate for open data. This is mainly because a necessary path does not appear in a generated FSA for an open data. To cope with this problem, we propose, in this paper, a new search algorithm that allows to jump dynamically to an alternative path when speech recognition of some words seems to fail.

In the following section, we briefly explain the method on constructing a FSA language model from a corpus, and describe basic performance of a generated FSA model. In Section 3, our new search algorithm is proposed, and effectiveness of the algorithm is discussed. In paper [6], we have proposed another performance improving approach; *a post-filtering*. After completion of speech recognition, acceptability of a recognized result is evaluated and is judged whether the result can be accepted or not. In Section 4, an overall result by combining the new search method and the post-filtering is also reported.

2   Overview of Constructing an FSA Language Model

As mentioned above, we already proposed a method to generate a FSA language model by use of DP matching. Here, we briefly describe the method and also about basic performance of generated FSA models.

2.1   Constructing an FSA Language Model

(a) Before executing DP matching, sentences in a corpus are divided into some number of groups (clusters) based on distances between them. This is because to avoid unnecessary execution of DP matching between dissimilar sentences.

There are several ways to calculate a distance between two sentences, but we adopted a following equation for simplicity.

$$d\left(S_x, S_y\right) = 1 - \frac{Num\left(w \mid w \in S_x \cap S_y\right)}{Num\left(w \mid w \in S_x \cup S_y\right)} \qquad (1)$$

where

$d\left(S_x, S_y\right)$: distance between sentence S_x and S_y

$Num\left(w\right)$: number of words such as w

(b) For each cluster, one randomly chosen sentence (a target) is converted to a simple one path FSA [1] and is put on a y-axis. Next, another sentence (a reference) is picked up, converted to a one path FSA as well, and is put on an x-axis. Next, two FSA are aligned by DP matching (hereafter, we call this DP matching method as a FSA-DP matching: *FDP*). As a result of *FDP*, relationships between words such as '*equal*', '*substitution*', '*deletion*', or '*insertion*' are obtained. According to these relationships, nodes of a reference FSA are merged to a target FSA (see Fig. 1).

A target FSA constructed in this way is used as a target FSA in the next *FDP*, but it is no longer one-path FSA. Therefore, nodes in the target FSA are sorted topologically according to a distance from a start node, and put on a y-axis. Next *FDP* is executed along with paths defined by a target FSA as shown in Fig. 2.

A global distance at a point (x, y) is calculated according to Eq. 2

$$gd(x, y) = \min \begin{pmatrix} gd(x-1,\ y) \\ gd(x-1,\ \Delta y) \\ gd(x,\ \Delta y) \end{pmatrix} + ld(x,\ y) \qquad (2)$$

where

$gd(x, y)$: a global distance at a point (x, y)
$ld(x, y)$: a local distance between word w_x and word w_y,

and calculated as follows.

$$ld(x,\ y) = \begin{pmatrix} 0.0 & (w_x = w_y) \\ 1.0 & (w_x \neq w_y) \end{pmatrix}$$

: a location on a y-axis of the previous word.

An each sentence in a cluster is taken out as a reference in turn and the above procedure is repeatedly applied to, and thus the target FSA is incrementally extended.

(c) When the procedure has finished for all clusters, the same number of FSA models with the number of clusters are obtained. Finally, these FSA models are combined into one FSA so that they share a common ST (start) node and an ED (end) node.

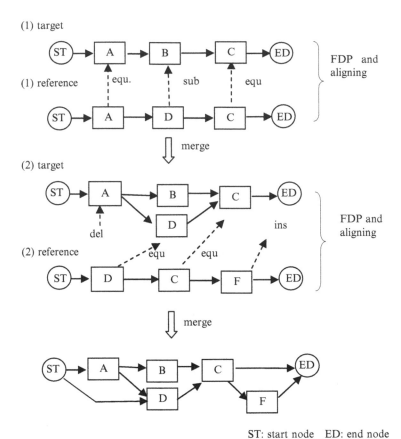

(1) target

(1) reference

merge

(2) target

(2) reference

FDP and aligning

FDP and aligning

del

ins

ST: start node   ED: end node

Fig. 1 FDP and merging

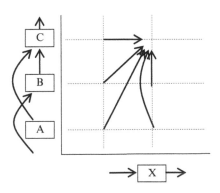

Fig. 2 DP matching path

2.2   Basic Performance of a Generated FSA

2.2.1   Corpus and Speech Recognizer

We conducted several speech recognition experiments on generated FSA language models. As a corpus, Japanese travel conversation sentences have been collected from four Japanese-English travel conversation textbooks. Here, sentences being too colloquial or fragmental were removed. As number of collected sentences is a slightly small (about 950 sentences), then several new sentences (about 50 sentences) modified little from the originally collected ones were added so that number of sentences becomes 1,000. Features of the corpus are listed in Table 1, and some example sentences appearing in the corpus is shown in Table 2.

In speech recognition experiments, HVite [7] is used as a decorder, and a speaker-independent tri-phone HMM of Julius [8] is used as a Japanese acoustic model. As a test data for speech recognition experiments, 60 utterances spoken by three male speakers are used.

2.2.2   Experiment for Closed Data

All 1,000 sentences were used as a learning data. We have constructed several FSA language models for different number of clusters, but the results of speech recognition experiments among them did not differ much. Therefore, we only show the result for a 70 clusters model (Table 3). In the table, an experiment result using bi-gram language model is also shown for a reference. From this table, we can see that the FSA language model constructed by our method has very small branching factors and then attains high speech recognition rate. Especially, a sentence %correct is 20 points or more higher than that of a bi-gram model.

Table 1 Corpus

Vocabulary	1,254 words
Sentences	1,000 sentences
Average words per sentence	8.87 words

Table 2 Example sentences

1	*Bizyutsu-kan meguri-no tsua-wa ari-mase-n-ka?* (Is there a tour of art museum?)
2	*Okurimono-you-ni housou-site morae-masu-ka?* (Can I gift wrapped?)
3	*Koko-de tabako-wo sutte-mo ii-desu-ka?* (May I smoke here?)

Table 3 Speech recognition result (for closed data)

Num. of clusters	Ave. branching factor	Word %correct	Sentence %correct
70	1.40	98.9	91.7
Bi-gram	5.88 (perplexity)	93.0	70.0

2.2.3 Experiment for Open Data

We took out 60 sentences as test data from the corpus, and constructed FSA models from the rest 940 sentences. Here, the amount of leaning sentences are absolutely insufficient, simple elimination of test data will cause deletion of several words from a learning corpus. To avoid this problem as far as possible, noun words appeared in 940 sentences were replaced by appropriate some semantic classes, and then FSA models were constructed by *FDP*. In *FDP*, equation of calculating local distance is modified as next equation.

$$ld(x, \ y) = \begin{cases} 0.0 \ (w_x = w_y) \\ 0.5 \ (w_x \neq w_y, sem_x = sem_y) \\ 1.0 \ ((w_x \neq w_y, (sem_x \neq sem_y \ or \ w_x \neq noun \ )) \end{cases} \tag{3}$$

When $ld(x, \ y)$ is 0.5, a node denoting that semantic class is generated in a *FDP* stage. For converting from a word w_x to a semantic class, we used *Bunrui-Goi-Hyo* (*BGH*) [9] as a thesaurus. In *BGH*, each word is categorized in five levels. We use two decimal places as a semantic code.

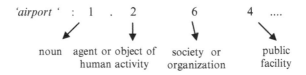

Bunch of semantic levels:

'airport' : 1 . 2   6   4 ....

noun   agent or object of   society or   public
        human activity   organization   facility

After completion of *FDP*, a FSA containing some semantic classes is obtained. In the next step, these nodes are expanded to the all of nouns which belong to that semantic class, and then the final FSA (a word level FSA) is generated.

Experiment result is shown in Table 4. In the table, a result only for 70 clusters is shown. Compared with the result in Table 3, we see that speech recognition correct rates, especially sentence %correct, drop very much, though they are still better than that of bi-gram. This is because some words disappeared in the final FSA model an/or even some paths did so because the size of the training data was fairly small. We investigated the reasons for recognition errors, and the result is shown in Table 5. From this table, we see that many of sentence recognition errors are due to deletion of corresponding paths.

Table 4 Speech recognition result (for open data)

Num. of clusters	Ave. branching factor	Word %correct	Sentence %correct
70	1.68	79.4	28.3
Bi-gram[1]	6.36 (perplexity)	58.4	3.3

[1] When constructing a bi-gram language model, 60 sentences are not eliminated but back-off (Witten-Bell) smoothing method is applied.

Table 5 Details of sentence recognition error

Due to	Num. of sentences
Deletion of words	4 (23.5%)
Deletion of paths	13 (76.5%)
Total	17 (100%)

3   Dynamic Alternative Path Search

3.1   Dynamic Alternative Path Search

As described above, main reason for large decrease of sentence %correct is due to deletion of a necessary path. In most of current speech recognition system using a FSA language model, search is executed along with paths defined in the model and never strays from them Therefore, if a necessary path breaks on the way, recognition thereafter will never succeed. To cope with this problem, the following dynamic path search algorithm was incorporated into the speech recognizer's decorder (see Fig. 3).

- When recognition of each word is finished, the recognition score of the word is examined whether the recognition result seems to be correct or incorrect. If seemed to be incorrect, words that are the same with the one prior word on the different paths are searched for.
- If found, recognition information (recognition score and path information) of that prior word was copied and attached to those words, and searches are restarted from these words.
- Search on the original path is also continued to avoid misjudgment on incorrectness of recognition result.

Here, a worrisome problem is that how we should detect a recognition error on a certain word, because an acoustic score is not always reliable. For the moment, we use the following criteria for simplicity.

- For each recognition path, a *score per frame* (*spf*) of that path is maintained.
- When recognition for a certain word is finished, a *spf* of the word is calculated.
- When the following inequality holds, the recognition of that word is regarded to be wrong, otherwise to be correct. Here, β is decided experimentally (in the following experiment, it was set to be 1.4).

$$(a\ spf\ \text{of a word})\ \leqq\ \beta \times (a\ spf\ \text{of a path}) \qquad (4)$$

Here, log-likelihood is used as a score, in the same way in HVite [7]. Now, hereafter, we will call this dynamic alternative path search algorithm as *DAPS*.

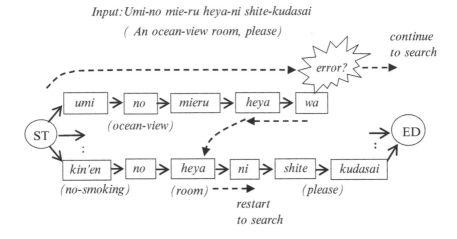

Fig. 3 Dynamic alternative path search

3.2 Experiment and Evaluation

We evaluated the effectiveness of *DAPS* by conducting speech recognition experiment. In the experiment, not only for complete open data, but data with intermediate open rates were also tested: that is, some of test data are bring back to the learning data. Here, we define *open rate* of test data as follows.

$$\text{open rate} = \frac{\text{number of test sentences not included in learning sentences}}{\text{number of test sentences}} \times 100$$

Note that "open rate is 0%" means that test data is completely closed, while "open rate is 100%" means that test data is completely open. The result is shown in Fig. 4. From this figure, performance of *DAPS* is worse than that of the original system for a closed data. This might happen because 'detection of word recognition error' made mistake. However, as open rate increase, performance of *DAPS* becomes better than that of the original system. For 100% open rate, performance, especially sentence %correct, of *DAPS* is considerably better than that of the original system; sentence %correct of the former is 40.0%, while that of the latter is 28.3% (11.7 points up).

The reason of improvement in sentence %correct might be that common sentence final expressions are frequently used in spoken Japanese. Even if recognition failure occurs at some word in an utterance, an alternative path that can follow might exist.

3.3 *Combining with a Post Filter*

In paper [6], we have proposed another approach to improve performance, that is a *post-filter*. As mentioned above, some words or paths necessary for correct recognition do not appear in a FSA for open data. Therefore, speech recognition performance would depend on a distance of a test text from a learning corpus; as a distance becomes large, recognition performance drops. Note here that we do not know a correct word string of a test text. However, as seen in Fig. 4, word %correct is still high even for a complete open data, it can be expected that the same tendency still remains for a distance between a recognition result and a learning corpus. From such backgraound, we introduced an additional mechanism, a post-filter. This filter decides whether a recognized result can be accepted or should be rejected according to a distance; a result is accepted if a distance is smaller than a certain threshold, but is *rejected* if a distance is larger than it (see Fig. 5).

Here, a distance is calculated according to the next equation.

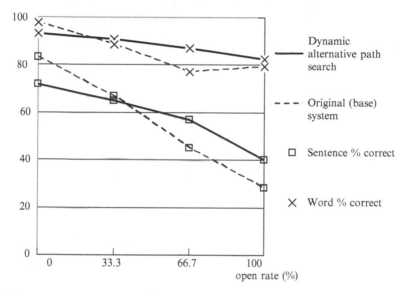

Fig. 4 Recognition performance for unseen data

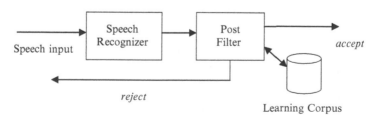

Fig. 5 Post-filter

Table 6 Effectiveness of
a post-filter

	The decision is	
Decision	right(%)	wrong(%)
Accept (71.7%)	53.5	46.5
Reject (18.3%)	95.5	4.5

$$D(S_x, C) = \min d(S_x, S_y | S_y \in C)$$
where
$d(S_x, S_y):$ the same with in Eq. (1)
$D(S_x, C):$ distance between sentence S_x and a corpus C

We evaluated the effectiveness of this mechanism combined with *DAPS*. The distance threshold was set to be 0.5. The result for 100% open data is shown in Table 6.

In the table, *accept-right* means that the recognition result is accepted and the decision is right (the recognition result is correct), *reject-right* means, to the contrary, that the result is rejected and the recognition result is actually erroneous, and so on. From this table, we see that, 71.7% of the results were accepted and 53.5% of the accepted results were actually correct. In other words, by combining with the post-filter, sentence %correct can furthermore be risen up to 53.5% (this is 13.5 points up from *DAPS*).

4 Conclusion

We developed a new method to construct a FSA language model automatically. With this model, however, speech recognition sometimes fails because a necessary path will break on the way. To cope with this problem, we propose a new search algorithm that allows to jump dynamically to an alternative path when speech recognition of some words seems to fail. Experiment for 100% open data, sentence %correct is improved remarkably (from 28.3% to 40.0%). This shows the effectiveness of this algorithm.

An FSA language model is quite convenient for a middle-sized vocabulary speech recognition. However, it has a serious problem that speech recognition fails even when a recognition path slightly stray from a path defined in a model. The proposed method could be thought to be a kind of a back-off method for a FSA language model to cope with this problem. From this point of view, there will be also other methods as following.

- When forward recognition fails, backward recognition is carried out, and two recognized fragments are tried to be glued.
- When recognition fails somewhere after a pause, all possible paths that follow a pause will be tried to be recognized.

We also reported the performance evaluation for the system in which *DAPS* and the post filter were combined. This post filter decides whether a recognized result can be accepted or should be rejected according to a distance between a recognized result and a learning corpus. The combined system accepts 71.1% of the recognized results, and sentence %correct for them is 53.5%; it is totally improved from 28.3% of the original system to 53.5% (25.1 points up).

References

1. Lang, K.J., Pearlmutter, B.A., & Price, R. (1998). Results of the Abbadingo One DFA Learning Competition and a New Evidence Driven State Merging Algorithm. *Proceedings of the International Colloquium Grammatical Inference*, pp. 1–12.
2. Lucas, S.M., & Reynolds, T.J. (Jul 2006). Learning deterministic finite automata with a smart state labeling evolutionary algorithm., *IEEE Transactions on Pattern Analysis and Machine Intelligence, 27*(7).
3. Kermorvant, C., de la Hinguera, C., & Dupont, P. (2004). Learning typed automata from automatically labeled data, *Journal Électronique d'Intelligence Artificielle, 6*(45).
4. Hu, J., Turin, W., & Brown, M.K. (1996). Language Modeling with Stochastic Automata. *Proceedings of ICSLP-1996.*
5. Riccardi, G., Pieraccini, R., & Boccieri, E. (1996). Stochastic automata for language modeling. *Computer Speech and Language, 10*(4), 265–293.
6. Morimoto, T., & Takahashi, S. (2008). Automatic Construction of FSA Language Model for Speech Recognition by FSA DP-Matching. In O. Castillo et al. (Eds.), *Trends in intelligent systems and computer engineering* (pp. 515–524). Springer.
7. Young, S., et al. (1999). The HTK Book (for Ver. 3.0)". http://htk.eng.cam.ac.uk/.
8. Kawahara, T., Lee, A., Takeda, K., Itou, K., & Shikano, K. (2004). Recent Progress of Open-Source LVCSR Engine Julius and Japanese Model Repository – Software of Continuous Speech Recognition Consortium. *Proceedings of ICSLP-2004.* (http://julius.sourceforge.jp/en/julius.html).
9. National Language Research Institute (1994). Bunrui-Goi-Hyo (Word List by Semantic Principles). Syuei-Shuppan, (in Japanese).

Chapter 29
Text Mining Decision Elements from Meeting Transcripts

Caroline Chibelushi and Mike Thelwall

Abstract The frequent but unfortunate need to rework software development projects may often be caused by inappropriate decision making. The first step in addressing this issue is to explore decision making processes and to extract the tangible elements of decision making within meetings. This chapter explores the hypothesis that text mining techniques can be used to extract the elements of decision making from software development project meetings and can ultimately be used as a facility to develop a decision management system. Theories of discourse, lexical chaining and cohesion are presented and used as the basis for the analysis of meeting transcripts. Information retrieval and data mining methods are also used. To assess the performance of the algorithm the C99 and TextTiling algorithms are used as comparators. The evaluation results show that our method is able to identify and extract the needs and actions of decision making with a high recall of 85–95% at a precision of 54–68%.

Keywords Decision making process · Rework · Software development projects · Text mining [1]

1 Introduction

Many software development projects have been unsuccessful partially as a result of communication failures between decision makers and incorrect or inappropriate decision making [1]. Rework involves altering, revising or restarting certain project activities because the previous work was incorrect, incomplete or inconsistent. It is a major cost factor in system development projects and accounts for over 50% of additional effort and substantial costs, particularly for large projects [2, 3]. This

C. Chibelushi (✉) and M. Thelwall
School of Computing and IT, University of Wolverhampton, City Campus – South,
Wulfruna Street, Wolverhampton, WV1 1SB, UK
e-mail: C.Chibelushi@wlv.ac.uk; M.Thelwall@wlv.ac.uk

S.-I. Ao et al. (eds.), *Intelligent Automation and Computer Engineering*,
Lecture Notes in Electrical Engineering 52, DOI 10.1007/978-90-481-3517-2_29,
© Springer Science+Business Media B.V. 2010

research considers decision management systems as a means of controlling rework and failure in system developments projects.

Rework can be necessitated by incorrect or inappropriate decision making. However, decisions could be 'hidden' and difficult to identify. Evidence of their existence in meetings is related to the issues discussed, the meeting participants' needs, the actions taken to satisfy the participants' needs, and the agents to perform the chosen actions. Such detailed information is often not found in meeting minutes, which may result in unmanaged decisions. This research argues that by investigating recorded and transcribed meeting conversations, it is possible to identify and extract evidence of meeting decision making processes, referred to herein as the elements of decision making. By understanding the relationship between these elements, managers will be able to better detect incorrect decisions and communication failures, and to understand their impact on the necessity for rework.

Many software development projects become successful after iterative processes [4], e.g., specification, design, coding and testing processes. However, in order to correctly perform these iterative processes, decision makers have to communicate and relate their decisions, understand the impact of each decision made on each iterative process and the risks involved in making each decision. For example, a decision maker can decide on *design issues* without relating them to the decisions made on the *testing issues*, and without keeping detailed and accurate documentation. In addition, software development project meetings may contain decisions that lack an action, or decisions that contain an action but lack an agent to perform that action. An example of a catastrophic failure that occurred as a result of the above mentioned factors is the 4th June 1996 explosion of the Ariane 5 rocket whilst lifting-off on its maiden flight [5, 6]. Rework also occurs in many organisational systems, especially in the public sector [7]. For example, rework has recently occurred in the major projects shown in Table 1.

Another factor that may cause software development complications is *incorrect analyses* made by the requirements engineers. In the case of Ariane 5, for example, this factor led to *incorrect decisions* to re-use the Ariane 4 alignment code that was not needed by Ariane 5 after its lift-off. The code remained operational in Ariane 5 without satisfying any specific requirement. The Ariane 5 IRS (Inertial Reference System) shut down 37 s after launch as because some Ariane 4 values overflowed in a computation that was not required by Ariane 5. This would have been prevented if there was communication between decision makers and engineers, as well as detailed and accurate documentation [8] on decisions made before launching Ariane 4 that could have been accessed by the integrators and users. If such documentation existed it would have been easier to be aware of the problems that may arise when adequate validation, verification, testing and review were not conducted on the system.

Despite the costliness of rework being well recognised, there has been little research focused on the analysis of meeting transcripts with the intention to reduce rework and failure in software development projects. Instead, the issue has been addressed through technical tools and techniques such as software language specifications or requirements management tools. More precisely, the problem of rework

Table 1 Recent software failures in the UK public sector

Software failure	Responsible software company	Failure date	Description	Reference
Educational maintenance allowance system	Liberta	Sept 2008	Students were unable to receive their study grants due to failure of an applications system and telephone helpline	www.kablenet.com
Standard Assessment Tasks (SATs) marking systems	ETS Europe	May 2008	School Children's exam results were delayed due to Technical and Operational problems associated with the systems and the management of information	www.bbc.co.uk
British Airways Terminal 5 failure	British Airports Authority (BAA) and British Airways (BA)	May 2008	A disastrous launch of terminal 5 was in part due to lack of software testing	www.computing.co.uk
A system to ensure that patient records are available electronically to all clinicians National Health Service (NHS)	iSoft	August 2006 – ongoing	A £12.7bn NHS project was supposed to be implemented in 2004. But it is 4 years delayed as iSoft has financial difficulties	www.guradian.co.uk www.telegraph.co.uk www.computerweekly. co.uk
Passport Office computer failure	Siemens	1999	The failure that led to a backlog of more than 500,000 unprocessed passport applications, costing the tax payer approximately £12 m.	http://news.bbc.co.uk
National Insurance Recording System II (NIRS II) failed	Andersen Consulting	September 1998	Allowance and some incapacity benefits are being paid "blind" after the National Insurance Recording System II (NRS II0) failed, erasing personal records	http://news.bbc.co.uk/1/ ji/uk/168277.stm

has been addressed through changing requirements. In addition, most tools support the process of carrying out rework, but do not help eliminate rework caused by inappropriate decision making. The first step in solving this problem is to explore decision making processes and propose a model for the tangible elements of decision making.

2 The Corpus

The research project used 17 transcripts recorded from three diverse meeting environments: industrial, organizational and educational, each involving a multi-party conversation with an accurate and unedited record of the meetings, corresponding speakers and no pre-set agendas. The meeting transcripts varied in size, ranging from 2,479 to 25,670 words, posed various complexities due to their informal style, their lack of structure, their argumentative nature, and the usage of common colloquial words. The transcripts contain incomplete sentences, sentences related to social chatting, interruptions, and references by participants made to visual contexts. The total corpus with a total of 247,238 words was used to identify an appropriate model and to illustrate the algorithm proposed for the analysis of transcripts. However, in this paper, only a transcript with a total of 25,670 words was analysed. This was the longest and most argumentative meeting, hence its selection.

3 The Ania Model

There have been various views on the number of distinct processes followed in organisational decision making, ranging from three to five main phases [9–12]. These have been developed by many other researchers resulting in different iterative models [13–16].

Another 3 models of decision making which are commonly used in many organisations and share similar phases to the above iterative models include the rational [15, 17], garbage can [18], and political model [19]. The three models represent current thinking with regard to formal organisational decision making and hence are preferred by many organisations.

The existing models need to be simplified as some (e.g. the rational model) are impossible to implement because they require comprehensive knowledge of every facet of the problem [20]. However, the models are iterative and share some of the same decision making components [21].

Although it is obvious that the aims of the above models are to assist decision makers to make appropriate decisions, these models do not explicitly identify and relate the key features of a decision. The obvious requirement to fully understand decision making or discriminate between different ways of managing decisions and subsequently rework is to identify the necessary features of decision making within the decision management domain; this has led this research to propose the ANIA model for decision making process.

The ANIA model is developed using the key shared components of the above mentioned models. ANIA represents *Agents, Needs, Issues and Actions* within decision making processes. In order to be able to develop the ANIA model, the research introduces the concept of *'decision making elements'* which are generated from the decision making processes.

The model perceives meetings as activities conducted because one or more *'agents'* have *'needs'* for a particular *issue* or a *topic*, and each need can be fulfilled by one or more *'actions'*. The *agents, needs, issues* and *actions* are referred to in this research as the elements of decision making.

Each element can be part of one or more decisions. The identification of these elements and the recording of them in the decision management system will enable decision makers to identify the actions, the information that led to them, the reason why the actions were taken and the result of the actions. Decision makers will thus be able to understand the decisions made by other decision makers, and why they were made.

4 Text Analysis Theories Pertinent to the Analysis of Transcripts

Grosz and Sidner [22] have proposed a theory for discourse structure to understand and determine the relationships between sets of words uttered by different speakers across dialogue turns. Any attempt to automate the process of identifying Grosz and Sidner's discourse structure requires a method of identifying linguistic segments in text. In order to achieve this, Morris et al. [23] extended Grosz and Sidner's theory to implement a lexical chaining technique.

Discourse is made up of functional words and content words. Examples of functional words are: *'the', 'is', 'a', and 'for'*. These words are likely to be used in the text about any subject. Examples of content words are: *'software', 'application',* and *'date'*. These are mostly represented in the text as nouns. Researchers such as Hasan [24], Hearst [25], and Reynar [26] have observed that nouns and noun phrases, sometimes called content phrases, are mostly used in human language to convey the information in a text. Lexical cohesion is a result of identifying and relating content phrases that contribute to the continuity of lexical meaning, hence identifying issues of conversation in meetings. For this reason, finding text structure involves finding units of text (content phrases) that are about the same thing [23]. When these units are semantically clustered, they are referred to as *lexical chains*.

Morris and Hirst [23] hypothesised that in order to be able to capture discourse structure, they needed to divide the text into cohesive segments. They employed Halliday and Hasan's [24] theory to analyse the cohesiveness of a text segment. The first step in Morris and Hirst's [23] algorithm was to link sequences of related words from a document to form lexical chains. They believed that each segment would be represented by the span of a lexical chain in the text.

Halliday and Hasan [24] identify five (not always distinct) cohesive relations that contribute coherence to a document: conjunction, substitution, reference, ellipsis and lexical cohesion [27]. Lexical cohesion is a linguistic device for investigating the discourse structure of texts, and lexical chains have been found to be an adequate means of exposing this structure. In the usual case where a discourse contains a set of related terms, Morris et al. [23] claim that cohesion is a useful sign of coherence in a text, especially since the identification of coherence itself is not computationally feasible at present. Stairmand [28] supports this claim by emphasizing that although cohesion fails to account for grammatical structure (i.e. readability) in contrast to coherence, cohesion can still account for the organisation of meaning in a text, and so, by implication, its presence corresponds to some form of structure in the text. This research uses lexical cohesion (as in Wordnet) to develop lexical chains.

5 Text Mining

Text mining allows the extraction of significant features from unstructured textual data such a meeting transcript. This research adapts the CRISP-DM (**CR**oss-Industry **S**tandard **P**rocess for **D**ata **M**ining) methodology for text mining by presenting a new methodology called the **Dec**ision **M**anagement using Text Mining (DecM-Text Mining). This methodology consists of three main phases as shown in Fig. 1.

The three stages of DecM-Text mining are as follows.

5.1 Pre-processing

This stage prepares the textual data to allow for more precise results and faster processing. It includes three steps: (i) *transcript cleaning* involves removing signs and characters which can be irrelevant for text mining, (ii) *text normalisation* involves tokenization (dividing transcripts into utterances), case folding (removing differences between capital and lower case words), identification of compound words and removal of stop words, and (iii) part-of-speech tagging is done using the

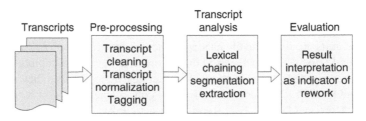

Fig. 1 DecM-Text mining

'Sense' as for stoplist (transcript 1)	'Sense' as for keeplist (transcript 2)
Ali: I am saying this, er, in a *sense* that, I am not trying to be funny but, er, it is impossible to do this by myself.	Abdul: ....to be able to do that you should apply word *sense* disambiguation methods

Fig. 2 A comparison of stoplists and keeplists

online tool Wmatrix [29]. This research pre-processing differs from others [27, 30]. For example, the task of generating stop words in some applications have used previously generated stop lists and do not regard tagging as a necessary task in this phase. Hearst [25] demonstrated that the use of a simple stop list is enough to remove the common words and to give good results. DecM-Text mining uses a content based approach, and a word is regarded as redundant if its potential contextual contribution to the transcript is small. For example the word *'sense'* can be classified as a stop list word in one transcript and as a keep list word in another as shown in Fig. 2.

For this reason, DecM-Text mining was unable to use previously generated stop lists as the use of such a list may jeopardise its output. Instead, DecM-Text mining pre-processing provides individual document content-dependent stoplists.

5.2   Transcript Analysis

The DecM-Text mining transcript analysis phase comprises three main tasks: lexical chaining, segmentation, and extraction.

5.2.1   Lexical Chaining

Lexical cohesion is used as a linguistic device for investigating the discourse structure of texts, and lexical chains have been found to be an adequate means of exposing this structure. Morris et al. [23] define *lexical chain* as a term used to identify sequences of cohesive words in a text, where lexical cohesive relationships between words are established using an auxiliary knowledge source such as a dictionary or a thesaurus.

Lexical chaining implements feature clustering, i.e. grouping together related features in a document or across document collections. In contrast to generic techniques like hierarchical clustering [31–33], k-means [34] and two dimensional clustering in Self Organizing Maps [35, 36], lexical chaining is less complex and takes into account the context of the document. In lexical chaining, words are clustered depending on the semantic relationships (synonym, hypernym, meronym etc.) between them, and hence the analysis is contextual.

5.2.2 Segmentation

Previous text segmentation techniques have focused on statistically based and linguistically driven methods [37, 38] Most statistical approaches involve probability distributions [39], machine learning techniques such as neural networks [40], support vector machines [41] and Bayesian networks [42] Linguistic text segmentation approaches utilize [24]'s lexical cohesion theory, term repetition to detect topic changes [30, 43, 44] n-gram word or phrases [45], or word frequency [44, 46]. Also, lexical chains [27, 30], or prosodic clues are often used to identify topic changes [44, 47]. However most lexical cohesion-based segmentation approaches use lexical repetition as a form of cohesion and ignore the other types of lexical cohesion such as synonym, hypernymy, hyponymy, meronymy [48]. Researchers such as [49] have also combined decision trees with spoken text linguistic features. The above segmentation methods were developed for applications that use written and structured texts, and not transcribed text such as the one used in this research. Based on complexities posed by transcripts (discussed in Section 2), this research developed a method to divide transcripts by their topical boundaries. The method includes temporary segmentation based on the cosine similarity measure, the application of lexical chains (which involves content) to identify topic boundaries, and final boundary identification to refine further the new segments by searching for speech cue phrases to confirm or cast doubt upon the topic boundaries.

5.2.3 Extraction

The aim of the DecM-Text mining Extractor (DTE) is to extract topics or issues, agents, actions, and needs, from meeting transcripts. To extract topics, Katz's theory [50] is implemented. In this, the extraction of agents, actions, and needs relies on linguistic pattern recognition methods. The extraction process depends on the appearance of elements related to decision making within a segment.

Topics boundaries are identified as discussed in the previous section. These boundaries are used to limit the area in which the related elements can appear. That is, the related elements of decision making can appear only within a segment and not outside the segment. Linguistic patterns identify the agent, needs and action patterns surrounding the main topic. These are the output of the DecM-Text mining.

5.2.4 Expected Applications

The DecM-Text mining output is fed into a decision management tool as shown in Fig. 3. The tool allows searching, browsing and linking meeting outcomes through the various elements associated with the decisions made in various meetings. Each element contains information describing the element type, the other related elements, the sub-elements, and the date when the element was recorded.

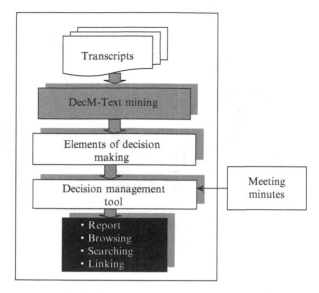

Fig. 3 The decision management tool

The decision management tool is designed to handle meeting minutes from which similar elements can be extracted using keywords.

The decision management tool aims to relate decision making elements from DecM-Text mining and identify elements that need other elements to become complete. For example, if there is a need, issue and an action but not an agent to fulfil that particular action, the decision management tool aim to remind/alert participants to provide the remaining element. Also it will provide an alert for any delay or changes in requirements.

6   Evaluation

Evaluation of the system was performed in two stages; segmentation and extraction. Segmentation was evaluated by comparing the **DecM-T**ext Mining **T**ranscript **S**egmentation (DTTS) against the two standard methods: Textilling and C99. Three types of evaluation metrics were used, P_k [46], P_k', and *WindowDiff* [51]. Each metric measures values ranging from 0 to 1 inclusive. These measures aim to calculate the average amount of error from each approach, and so the approach with the lowest score in each metric is the best performing algorithm. The results are shown as shown in Fig. 4.

Figure 4 show that DTTS has considerably fewer errors than TextTiling and C99. The difference between the P_k and P_k' values of the TextTiling and C99 approaches is relatively large. The P_k' values in C99 and TextTiling are approximately twice as large as P_k, which means that these approaches suffer from false positives.

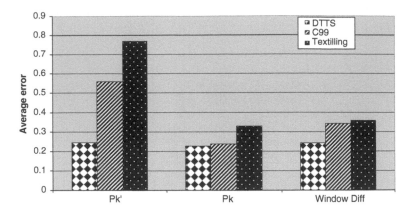

Fig. 4 DTTS, C99 and Textilling segmentation accuracy

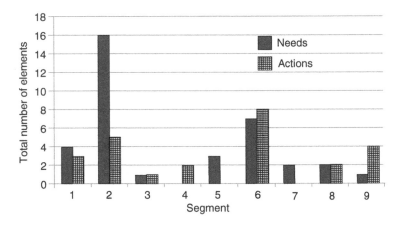

Fig. 5 The extracted elements

Since DTTS topic boundaries are relatively accurate, this implies that the discussed issues are also separated accurately. Similarly, because each utterance has the name of the corresponding speaker, when the utterance with a need or action is extracted, it will automatically have corresponding speaker (agent) associated with it. For this reason, this research focused on the evaluation of the needs and actions.

Figure 5 shows the 'needs' and 'actions' which were extracted for this particular transcript.

The extracted needs and actions were evaluated using precision and recall as shown in Fig. 6.

The goal of the DTE approach is to extract as many needs and actions as possible from meeting transcripts. The linguistic patterns used include actions or needs with repetition, faulty tagged phrases and interruption – incomplete actions, or needs and actions which are not portraying any meaning without some cognitive knowledge. As a consequence, precision suffers due to extracting needs or actions which are not

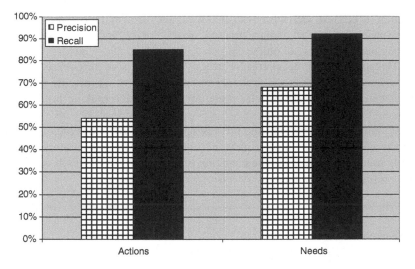

Fig. 6 Recall and precision of the extracted 'actions' and 'needs' from the corpus

useful. However, in the extraction of needs, the system's performance is satisfying because the recall of the transcript measured 100%. Apart from interruption and repetition, there was little irrelevant material. Most *needs* linguistic patterns do not have a wide variation in their appearances. Test results used to calculate recall and precision were identified by a group of 6 Natural language processing researchers.

7   Conclusions

This chapter describes a new text mining approach to extract key information regarding decisions from software development meeting transcripts. The approach combines aspects from many disciplines: decision theories, linguistics, natural language processing, information retrieval, text mining and statistics. These are incorporated in a text mining approach called **Dec**ision **M**anagement using Text mining, or DecM-Text mining. This was applied to software development project meeting transcripts to identify and extract the relevant elements of the decision making process. DecM-Text mining achieves this by providing the contextual analysis that is necessary to understand thematic aspects within software development project meeting conversations. The approach is an adaptation of the CRISP-DM methodology for mining unstructured text by combining algorithms from different fields.

The results of both the segmentation and extraction tasks were obtained and evaluated. Three metrics were applied to two different lexical cohesion based segmentation algorithms: C99 [52] and TextTiling [25]. The results show that DTTS outperforms the C99 and TextTiling algorithms. The results of the Dec-Text mining extraction (DTE) task were also evaluated using precision and recall measures. The average measure for recall on the extracted 'needs' is 92%, and its precision is 68%, while the measure for recall on the extracted 'actions' is 85%, while its precision is 54%. This suggests that the approach correctly detects most of the cohesive relationships necessary to identify the topic regions and shifts within meeting transcripts, and has extracted the relevant related elements of decision making found in each region.

Text mining is still evolving and this research has contributed some insights into methodologies and algorithms. This is the first study to use text mining to analyse meeting transcripts, to extract the elements of decision making which can be used as a basis for the development of a decision management system.

References

1. Rayson, P., Sharp, B., Alderson, A., Cartmell, J., Chibelushi, C., Clarke, R., Dix, A., Onditi, V., Quek, A., Ramduny, D., Salter, A., Shah, H., Sommerville, I., & Windridge, P. (2003). Tracker: a framework to support reducing rework through decision management. Presented at *Proceedings of 5th International Conference on Enterprise Information Systems ICEIS2003*, Angers, France.
2. Boehm, B., & Basil, V.R. (2001). Software reduction defects top 10 list. *IEEE Computer, 34*, 135–137.
3. Jones, T. (1986). *Software Productivity*. McGraw-Hill, OH, USA.
4. Melo, W., Briand, L., & Basili, V. (1995). *Measuring the impact of reuse on quality and productivity in object-oriented systems*, Department of Computer Science, University of Maryland, College Park, MD, Technical Report CS-TR-3395.
5. ESA (1996). Ariane 5 – Flight 501 Failure. European Space Agency.
6. Nuseibeh, B. (1997). Ariane 5: Who dunnit? *IEEE Software, 14*, 15–16.
7. Cross, S. (2002). *Message from the Director*, Software Engineering Institute, Annual Report 2002.
8. Goodman, J. (2002). Lessons learned from flights of "Off the shelf. aviation navigation units on the space shuttle," presented at Joint Navigation Conference, Orlando, Florida.
9. Simon, H. (1960). *The new science of management decision*. New York: Harper & Row.
10. Janis, I.L. (1968). *Stages in the decision making process*. In Theories of Cognitive Consistency: A Sourcebook, ed. Robert P. Abelson et al. Chicago: Rand McNally, pp. 577–588.
11. Schrenk, L.P. (1969). Aiding the decision maker – a decision process model, *Ergonomics* (pp. 543–557).
12. Witte, E. (1972). Field research on complex decision-making processes – the phase theorem. *International studies of management and organization* (pp. 156–182).
13. Henrikke, B. (1998). *Life cycle assessment and decision making-theories and practices*. GÖteborg: Chalmers University of Technology.
14. Adair, J. (1985). *Effective decision-making*. London: Pan Books.
15. Fulop, L., Linstead, S., Lilley, S., & Clarke, R.J. (1999). *Decision making in organizations, in management: a critical text* (pp. 462–494). Australia: Macmillan Education.
16. Pescatore, C., & Vári, A. (2004). Stepwise approach to decision making for long-term radioactive waste management, The NEA Radioactive Waste Management Committee NEA No. 4429.
17. Mintzberg, H., & Westley, F. (2001). Decision making: it's not what you think. *MIT Sloan Management Review, 42*, 89–93.
18. March, J.G. (1984). *Theories of choice and making decisions*. In Lane, F.A. (Ed.), Current Issues in Public Administration, St Martin's Press, New York, USA

19. Hoy, W., & Miskel, C. (1991). *Educational administration: theory, research, and practice* (4th edn.). New York: McGraw-Hill.
20. Chen, J. (2002). *A comparative study on decision making approach in a dilemma between taiwanese and italian business, in graduate institute of management sciences.* Taiwan: Tamkang University.
21. Harrison, F. (1999). *The managerial decision-making process.* USA: Houghton Mifllin Company.
22. Grosz, B.J., & Sidner, C.L. (1986). Attention, intentions, and the structure of discourse. *Computational Linguistics, 12*, 175–204.
23. Morris, J., & Hirst. G. (1991). Lexical cohesion computed. *Computational Linguistics, 17*, 21–48.
24. Halliday, M., & Hasan, R. (1976). *Cohesion in English.* London: Longman.
25. Hearst, M. (1997). TextTilling: segmenting text into multi-paragraph subtopic passages. *Computational Linguistics, 23*, 33–64.
26. Reynar, J. (1998). *Topic segmentation: algorithms and applications, in computer and information science.* College Park, Maryland: University of Pennsylvania.
27. Stokes, N. (2004). *Applications of lexical cohesion analysis in the topic detection and tracking domain.* Department of Computer Science: University College Dublin.
28. Stairmand, A.M. (1996). *A computational analysis of lexical cohesion with applications in information retrieval.* Department of language Engineering: University of Manchester Institute of Science and Technology.
29. Rayson, P. (2001). Wmatrix: a web-based corpus processing environment. Presented at *ICAME 2001 Conference.* Belgium: Université Catholique de Louvain.
30. Hearst, M. (1994). Multi-paragraph segmentation of expository text, presented at *Proceedings of the 32th Annual Meeting of the Association for Computational Linguistics.* Las Cruces, New Mexico.
31. Mandreoli, F., Martoglia, R., & Tiberio, P. (2005). Text clustering as a mining task, in text mining and its applications to intelligence. In A. Zanasi (Ed.), *CRM and knowledge management.* GB: WIT, pp. 75–108.
32. Manning, C., & Schütze, H. (2003). *Foundations of natural language processing* (6 edn.). London: MIT.
33. Meij, J., & Bosch, A. (2002). Text mining techniques. In J. Meij (Ed.), *Dealing with the data flood* (pp. 746–753). The Hague, Netherland: STT/Beweton.
34. Steinbach, M., Karypis, G., & Kumar, V. (2000). A comparison of document clustering techniques, presented at *Proceedings of the KDD Workshop on Text Mining.* Boston, MA, USA.
35. Lagus, K. (2000). *Text mining with the WEBSOM, in computer science and engineering* (pp. 58). Helsinki: Helsinki University.
36. Kohonen, T. (1989). *Self organization and associative memory.* Berlin, Heidelberg, New York: Springer.
37. Chibelushi, C., & Sharp, B. (2008). Tracker text segmentation approach: integrating complex lexical and conversation cue features, presented at *Proceedings of the 5th International Natural Language Processing and Cognitive Science Workshop.* Bacelona, Spain.
38. Manning, C. (1998). *Rethinking text segmentation models: an information extraction case study,* University of Sydney, Sydney, Australia.
39. Beeferman, D., Berger, A., & Laffety, J. (1999). Statistical models for text segmentation, machine learning, *Special Issue on Natural Language Processing, 34*, 177–210.
40. Bilan, Z., & Nakagawa, M. (2005). Segmentation of on-line handwritten japanese text of arbitrary line direction by a neural network for improving text recognition presented at *Proceedings of the Eighth International Conference on Document Analysis and Recognition.*
41. J. Renjie, Q. Feihu, L. Xu, & Wu G. (2006) Detecting and segmenting text from natural scenes with 2-stage classification presented at *Proceedings of the Sixth International Conference on Intelligent Systems Design and Applications* (ISDA'06)
42. Senda, S., & Yamada, K. (2001). A maximum-likelihood approach to segmentation-based recognition of unconstrained handwriting text, presented at *Proceedings of the Sixth International Conference on Document Analysis and Recognition.*

43. Youmans, G. (1991). *A new tool for discourse analysis: the vocabulary management profile, in languages, 67*(4), 763–789.
44. Reynar, J. (1999). Statistical models for topic segmentation, presented at *Proceedings of the Association for Computational Linguistics*. College Park, USA.
45. Kan, M., Klavans, J.L., & McKeown, K.R. (1998). Linear segmentation and segment relevance, presented at *Proceedings of the Sixth Workshop on Very Large Corpora*.
46. Beeferman, D., Berger, A., & Laffety, J. (1997). Text segmentation using exponantial models, presented at *Proceedings of EMNLP-2*. Madrid, Spain.
47. Levow, G. (2004). Prosodic cues to discourse segment boundaries in human-computer dialogue, presented at *Proceedings of the 5th SIGdial Workshop on Discourse and Dialogue*, USA.
48. Chibelushi, C. (2008). *Text mining for meeting transcripts analysis to support decision management*. In Faculty of Computing Engineering and Technology Stafford: Staffordshire University.
49. Passoneau, R., & Litman, D. (1997). Discourse segmentation by human and automated means. *Computational Linguistics, 23*, 103–139.
50. Katz, S.M. (1996). Distribution of context words and phrases in text and language modeling. *Natural language Engineering, 2*, 15–59.
51. Pevzner, L., & Hearst, M. (2002). Evaluation metric for text segmentation. *Computational Linguistics, 28*, 19–36.
52. Choi, F.Y. (2000). Advances in domain independent linear text segmentation presented at *Proceedings of the 1st Meeting of the North American Chapter of the Association for Computational Linguistics (ANLP-NAACL-00)*, Seattle.

Chapter 30
Double SVMSBagging: A Subsampling Approach to SVM Ensemble

Faisal Zaman and Hideo Hirose

Abstract In ensemble methods, pooling the decisions of multiple unstable classifiers often lead to improvements in the generalization performance substantially in many applications. We propose here a new ensemble method, Double SVMSBagging, which is a variant of double bagging. In this method we have used subsampling in order to make the out-of-bag samples larger and trained support vector machine as the additional classifier on these out-of-bag samples. The underlying base classifier is the decision tree. We have used radial basis function kernel, expecting that the new classifier can perform efficiently in both linear and non-linear feature space. We have studied the performance of the proposed ensemble method in several benchmark datasets with different subsampling rate (SSR). We have applied the proposed method in partial discharge classification of the gas insulated switchgear (GIS). We compare the performance of double SVMsbagging with other well-known classifier ensemble methods in condition diagnosis; the double SVMsbagging performed better than other ensemble method in this case. We applied the double SVMsbagging in 15 UCI benchmark datasets and compare its accuracy with other ensemble methods, e.g., Bagging, Adaboost, Random Forest and Rotation Forest. The performance of this method with optimum SSR generate significantly lower prediction error than Rotation Forest and Adaboost for most of the datasets.

Keywords Support vector machine · Double bagging · Subsampling rate · CART · Partial discharge analysis

1 Introduction

Support Vector Machines (SVM) are learning systems that use a hypothesis space of a linear functions in a high dimensional feature space, trained with a learning algorithm from optimization theory that implements a learning bias derived from

F. Zaman (✉) and H. Hirose
Department of Systems Design and Informatics, Kyushu Institute of Technology, Kyushu, Japan
e-mail: zaman@ume98.ces.kyutech.ac.jp; hirose@ces.kyutech.ac.jp

S.-I. Ao et al. (eds.), *Intelligent Automation and Computer Engineering*,
Lecture Notes in Electrical Engineering 52, DOI 10.1007/978-90-481-3517-2_30,
© Springer Science+Business Media B.V. 2010

statistical learning theory [9]. This high-dimensional feature space is non-linearly related to the input space. The SVM learns a separating hyperplane to maximize the margin and to produce a good generalization ability [7]. Recent theoretical research work has solved the existing difficulties of using the SVM in practical applications [19, 24]. The capability of SVM to have competitive generalization error than other classification methods and ensemble methods have also been checked [12, 22].

In this paper, we have used SVM as the additional classifier model in the ensemble method, double bagging [17]. In double bagging an additional classifier model is built on the out-of-bag samples and then this model is trained on both the inbag samples and test set to extract additional predictors for both in building the ensemble and testing it in the test set. As the SVM is a maximum margin classifier, which construct optimum separating hyperplane between the classes (for binary classification), we intend to use it in the first phase of the ensemble learning to attain the class *posteriori probabilities* consist of discriminative information between the classes and then integrate these as the additional predictors to construct the decision tree ensemble in the second phase. These posteriori probabilities are also used in the testing the decision tree ensembles. This procedure ensures a possibility of maximum separation of the classes and henceforth increases the prediction accuracy of the decision tree ensemble in discriminating the classes.

In this paper, we focused to endeavor Double SVMSBagging in classifying the type of partial discharge (PD) patterns in a model gas insulated switch gear (GIS) as a typical electric power apparatus. For condition monitoring purposes, it is considered to be important to identify the type of defects when monitoring discharge activities inside an insulation system. In the paper [15] authors first proposed to use the decision tree as a classification tool for diagnosing because it provides the tangible if-then rule, and thus we may have a possibility to connect the physical phenomena to the observed signals. In [16] authors used several ensemble methods in classifying the defect patterns in the electric power apparatuses. In [20] authors applied a SVM ensemble for fault diagnosis, based on the genetic algorithm (GA). They used the GA in order to find more accurate and diverse ensemble.

The paper is organized as follows. In Section 2, we have introduced the SVM. In Section 3, the Double SVMSBagging ensemble method is discussed. Here we have put together motivations behind constructing Double SVMSBagging. In Section 4 we have described the properties of the GIS datasets we have used in the experiments, this also include the description of the benchmark datasets used. Section 5 includes the experimental setup and discussion of the results of the paper. We have compared the performance of the new ensemble method with other ensemble methods, such as the bagging, the Adaboost.M1, random Forest and rotation Forest in benchmark datasets. We also compare its performance with double bagging with SVM [31]. For PD classification we have compared its performance with bagging, the Adaboost.M1, the logitboost and the double bagging (with subbagging) with LDA, k-NN and SVM. In Section 6, the conclusion of the study is stated.

2 Support Vector Machine (SVM)

The SVM models were originally defined for the classification of linearly separable classes of objects. It is easy to find a line that separates them perfectly. For any particular set of two-class objects, an SVM finds the unique hyperplane having the maximum margin (usually denoted as δ). A special characteristic of the SVM is that the solution to a classification problem is represented by the support vectors that determine the maximum margin hyperplane. This is regarded as an approximate implementation of the structural risk minimization (SRM) principle, which endows with good generalization performances independent of underlying distributions [19]. The SVMs algorithms are based on parametric families of separating hyperplanes of different Vapnik-Chervonenkis dimensions (VC dimensions). The SVMs can effectively and efficiently find the optimal VC dimension and an optimal hyperplane of that dimension simultaneously to minimize the upper bound of the expected risk. Usually the classification decision function in the linearly separable problem is represented by

$$f_{w,b} = sign(w \cdot x + b).$$

Thus, to find a hyperplane with minimum VC dimension, we need to minimize the norm of the canonical hyperplane $||w||$. So the distance between two hyperplanes, for example H_1 and H_2 is,

$$\delta = \frac{2}{||w||}.$$

where the hyperplanes H_1 defines the border with class $f_{w,b} = +1$ objects and hyperplane H_2 defines the border with class $f_{w,b} = -1$ objects, respectively. Consequently, minimizing the norm of the canonical hyperplane $||w||$ is equivalent to maximizing the margin δ between H_1 and H_2. The purpose of implementing SRM for constructing an optimal hyperplane is to find an optimal separating hyperplane that can separate the two classes of training data with maximum margin. Hence the optimal hyperplane separating the training data of two separable classes is the hyperplane that satisfies,

$$Minimize : F(w) = \frac{1}{2} w^T w, \ y_i (w \cdot x_i + b) \geq 1.$$

This is a convex, quadratic programming (QP), problem with linear inequality constraints.

In SVM for multi-class classification, mostly voting schemes such as one-against-one and one-against-all are used. In the one-against-one classification method (also called pairwise classification), $\binom{k}{2}$ classifiers are constructed where each one is trained on data from two classes. Prediction is done by voting where each classifier gives a prediction and the class which is most frequently predicted wins ("Max Wins"). In the one-against-all method k binary SVM classifiers are trained, where k is the number of classes, each trained to separate one class from the rest. The classifiers are then combined by comparing their decision values on a test data instance and labeling it according to the classifier with the highest decision value.

2.1 Advantage of SVM Over Other Classifiers in Data Based Condition Diagnosis

During the last years, Neural Network (NN) based models like multilayer perceptrons (MLP), radial basis function (RBF) networks or self organising maps (SOM) in application to the data-based fault diagnosis is widely studied in [23, 29]. With NN models it is possible to estimate a nonlinear function without requiring a mathematical description of how the output functionally depends on the input; NNs learn from examples. The most commonly mentioned advantages of NNs are their ability to model any non-linear system, the ability to learn, the highly parallel structure and the ability to deal with inconsistent or noisy data. SVM gives refreshing views on conventional pattern recognition and classification systems. It has several benefits compared to statistical classifiers or MLPs, e.g.

1. The most important benefit is its efficiency in high dimensional classification problems, where statistical classifiers often fail.
2. The other benefit of SVM compared to statistical classifiers is its general applicability to nonlinear problems. MLPs or RBF networks can also be applied in nonlinear problems, but SVM outperforms them when considering the globality of solution.
3. Training of the SVM results in a global solution for the problem under study, whereas MLPs and RBF networks may have many local minima leading to not a reliable solution.

2.2 Designing and Tuning of SVM in the Experiments

We have used the C-SVM in our paper. This name originates from the fact that the complexity of the C-SVM solely depends on the cost parameter C. Design of SVM for a classification task consists of two tasks: choosing the kernel function and setting a value for the parameter C. The parameter C is also called an error penalty, because it deals with the trade-off between maximum margin and the classification error during training. A high error penalty will force the SVM training to avoid classification errors. It is clear that with high error penalty, the optimizer gives a boundary that classifies all the training points correctly. In this paper we have used the R package e1071 [8] to implement the SVM. In the SVMs the optimization is done by SMO [24], which takes advantage of the special structure of the SVM quadratic problem (QP). The selection of kernel function has also influence on the decision boundary. When using polynomial kernel function, the order of the polynomial needs to be chosen, and when using RBF the spread (kernel width) γ, needs to be decided. In our experiments we have used grid search method to select the optimum combination of the parameters. In this search method the 10-fold cross-validation is used to search for the models with lowest prediction error. In our paper for multi-class classification we have used one–against–one rule. In all of our experiments we have used the *posteriori class probabilities* as output instead of *class labels* of SVM as the additional predictors. This is done by an improved implementation [21] of Platt's *a posteriori probabilities* [25].

$$Prob(y = 1|f) = \frac{1}{1 + e^{(Af+B)}}$$

where a sigmoid function is fitted to the decision values f of the binary SVM classifiers, A and B being estimated by minimizing the negative log-likelihood function. We extended the class probabilities to the multi-class case, combining all binary classifiers class probability output as proposed in [30].

3 Double SVMSBagging: Double Subbagging with SVM

The underlying idea of double bagging is in the spirit of Breiman [6], "Instead of reducing the dimensionality, the number of possible predictors available to the classification trees is enlarged and the procedure is stabilized by bootstrap aggregation." In this algorithm a *linear* classifier model LDA is constructed for each bootstrap sample using an additional set of observations: the out-of-bag sample. The prediction of this classifier is computed for the observations in the bootstrap sample and is used as additional predictors for a classification tree. The trees implicitly select the most informative predictors. The procedure is repeated sufficiently enough and a new observation is classified by averaging the predictions of the multiple trees. So we see that performance of the double bagging solely depends on two factors: (1) the classes of the dataset are linearly separable so that the additional predictors are informative (or discriminative), (2) the size of the out-of-bag samples as to construct LDA model: the underlying covariance matrix should be invertible. However, to handle real world classification problems, the base classifier should have more flexibility.

We know that SVC (support vector classifier) are maximum margin classifier, i.e., the support vectors construct the separating hyperplane with the maximal margin between the classes (for example in 2-class problem), it has an extra advantage regarding automatic model selection in the sense that both the optimal number and locations of the support vectors are automatically obtained during training [28]. So in the double bagging the use of SVM will ensure that the additional predictors (the class posteriori probabilities) extracted after training the SVM models on the inbag samples, will consist of optimum discriminative (maximal margin) information of the classes. Henceforth it will facilitate the base decision tree learn on the combined training sample (i.e., the bootstrap samples and the class posteriori probabilities) allow for more flexible and accurate split of the data. So it is evident theoretically that use of SVM in the double bagging therefore will have an improved performance.

As the success of the double bagging mostly lies on the classifier model build on the out-of-bag samples, to ensure large out-of-bag samples we also used subsamples instead of the bootstrap samples. We denote this as, "Double SVMSbagging." This modification ensures that the learning samples for the additional classifier model always contain large amount of observations of the training sample. This will be expedient in decreasing the probability of the additional classifiers to overfit the out-of-bag samples and also will increase the learning ability of the additional predictors. In the algorithm (Fig. 1), we see that its important to choose an optimum

---

Input:

- L: Training set
- X: the predictors in the training dataset
- B: number of classifier in the ensemble
- $\{w_1,...,w_c\}$: the set of class labels
- r: optimum subsampling rate
- x : a data point to be classified

Output: w: Class label for x.

Procedure ***Double SVMSBagging ()***

1 For b = 1,...,B

   (a) $L^{(b)} \leftarrow$ Subsample of size r from L. Let $X^{(b)}$ denote the matrix of predictors $X_1^{(b)}, \ldots, X_N^{(b)}$ from $L^{(b)}$.

   (b) $SVM^{(b)} \leftarrow$ An SVM model using the out-of-bag sample $L^{-(b)}$.

   (c) $CP^{(b)} \leftarrow$ A matrix with the columns are the class probability of the classes, after training $SVM^{(b)}$ on $L^{(b)}$

   (d) $C_{comb}^{(b)} \leftarrow (L^{(b)} \cup CP^{(b)})$: Construct the combined classifier

   (e) $TCP^{(b)} \leftarrow x$'s class posteriori probablity generated by $SVM^{(b)}$.

   (f) $c_{bj}(x, TCP^{(b)}) \leftarrow$ The probability assigned by the classifier $C_{comb}^{(b)}$ that x comes from the class w_j.

2 **EndFor**

3 Calculate the confidence for each class w_j, by the "average" combination rule:

$$\mu_j(x) = \frac{1}{B} \sum_{b=1}^{B} c_{bj}((x, TCP^{(b)})), j = 1, \ldots, c.$$

4 $w \leftarrow$ Class label with the largest confidence.

5 **Return** w

Fig. 1 Generic Framework of Double SVMSbagging Algorithm

SSR. In our experiments we have tried to find out the optimum SSR, by creating Double SVMSbagging with different SSR and choosing the one with lowest cross-validation error.

In Fig. 1, we have given the pseudocode of Double SVMSbagging algorithm. We see from Fig. 1 that in the first phase of training step SVMs are constructed using the out-of-bag samples, then to get additional predictors, these SVMs are trained on the subsamples to get the class posteriori probabilities (CP^b). In the second phase an ensemble of decision tree(DT^b) is built using these CP^bs and the subsamples (L^b). The SVMs are also trained in the test set to enlarge the size of the test set by the test posteriori class probabilities (TCP^b). Then these TCPs are included with test set as the additional predictors.

4 Data

The main objective of this work is threefold; first, get the optimum SSR for Double SVMSBagging, second, examine the performance of Double SVMSBagging in condition diagnosis and compare its prediction accuracy with other ensemble methods; third, investigate its classification performance in real world datasets along with other well-known ensemble methods.

The datasets used for condition diagnosis are GIS datasets which are transformed version of the electromagnetic signals measured by the sensors in the electric power substations. We assume three classes for possible abnormal conditions in the GIS; (1) the metal which is attached to the high voltage side conductor (abbreviated as "HV"), (2) the metal which is attached to the earth side tank (abbreviated as "TK"), and (3) the metal is freely movable (abbreviated as "FR"). The numbers of the observed samples are, 150, 377, 126, for HV, TK, FR. Here the first dataset consist of MLE (maximum likelihood estimates) of four parameters (two parameters for phase 0–180 and two parameters for phase 180–360) of the GND and two parameters for the single phase of the Weibull fitted to the observed PD patterns, and these are used as feature variables. For our second experiment we randomly selected 15 datasets from the UCI Machine Learning Repository [1]. The characteristics of the datasets are showed in Table 1.

5 Experimental Setup and Discussion of Results

In this paper we have conducted three experiments. In the first experiment we have tried to check the performance of DSVMsbagging with different subsampling ratio (SSR). From this experiment we got a rough idea about optimum SSR (OPSSR), for which Double SVMSBagging shows best generalization performance. In the second experiment we have applied Double SVMSBagging (with OPSSR) in the two GIS datasets. To compare the efficacy of the proposed double subbagging via SVM ensemble we have applied three different ensemble methods, bagging [3],

Table 1 Description
of the 15 Data used
in this paper

Dataset	Objects	Classes	Features
Boston housing	506	3	13
DNA	3, 186	3	180
E. coli	336	8	7
German-credit	1, 000	2	20
Glass	214	7	9
Cleveland-heart	297	5	13
Ionosphere	351	2	34
Iris	150	3	4
Liver-disorder	345	2	6
Pima-diabetes	768	2	8
Sonar	208	2	60
Vehicle	846	4	18
Vote	435	2	16
Wiscinson-breast	699	2	9
Zoo	101	7	16

Adaboost.M1 [10] and logitboost [11], with the double bagging (subbagging also)
with LDA, 5-NN and 10-NN classifier models in the GIS datasets. In the third
experiment we have checked the performance of the new ensemble method in 15
UCI repository datasets. We have compared its performance with Bagging, Ad-
aboost.M1, Random Forest [5] and Rotation Forest [26]. In all the experiments for
each dataset, we extracted the optimum parameters of the SVM using 10-fold cross-
validation and then use those parameters to construct the SVC to be used in each
out-of-bag sample.

5.1 Experiment to Get the Optimum SSR

In this experiment we have used six SSR, 0.20, 0.30, 0.50, 0.65, 0.75 and 0.80. This
implies that all the subsamples will consist SSR% of training set. In Fig. 2 we have
plotted the ten-fold cross-validation error of DSVMSBagging with different SSR.
It's quite surprising that DSVMSBagging with the SSR [0.20, 0.30] yields lower
error than SSR higher than 0.5. It is apparent from the Fig. 2 that with higher SSR
the performance of DSVMSBagging is deteriorating and the optimum region for
choosing SSR is between [0.20–0.50]. This is a very important finding to improve
the performance DSVMSbagging. In our next two experiments we will use only
optimum values of SSR.

5.2 Experiment with GIS Dataset

In this experiment, we have used CART [2] in bagging, double bagging and
Adaboost.M1 and decision stump (DS) [18] in Adaboost.M1 and logitboost as the

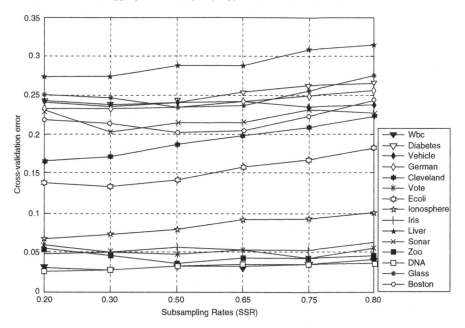

Fig. 2 Performance of Double SVMSBagging with different SSR in benchmark datasets

Table 2 Misclassification error of all the ensemble methods for GND and Weibull fitted GIS dataset

Classifiers	Abbreviations	GND	Weibull
Single decision tree	CART	0.08638	0.08191
Bagged CART	BCART	0.04407	0.03891
Double bagging with LDA	DBLDA	0.03798	0.03730
Double bagging with 5-NN	DB5NN	0.03889	0.03316
Double bagging with 10-NN	DB10NN	0.04610	0.04315
Double subbagging with LDA	DSBLDA	0.04086	0.04097
Double subbagging with 5-NN	DSB5NN	0.03824	0.03439
Double subbagging with 10-NN	DSB10NN	0.04314	0.04271
Double bagging with RBF kernel SVM	DBRBFSV	0.03198	0.03442
Double subbagging	DSVMSBagging	**0.02721**	**0.03269**
Adaboost.M1 CART	ADACART	0.05238	0.04717
Adaboost.M1 decision stump with 3-node	ADADS3	0.04671	0.04518
Logit boosted decision stump	LOGITDS	0.07221	0.03492

base classifier. We used here 2-node decision stump in case of logitboost and 3-node decision stump in case of Adaboost.M1. The results are shown in Table 2. We have used 10-fold cross-validation to estimate the misclassification error of the ensemble methods. We repeat this 5 times and report the average misclassification error of the 5 repetitions. In [15, 31] the accuracy of the bagging and double bagging ensemble was better with the ensemble size $B = 100$, for that the ensemble size for bagging and double bagging ensemble is 100 in this experiment; in case of Adaboost.M1 and logitboost we have used iterations $M = 100$. We have reported in Table 2, the lowest test errors of the classifiers. The best result is printed in bold. In the first column of Table 2 we have given the name of the ensemble methods and in the second column we have given the abbreviations we have used for the ensemble methods. For example for a Bagged CART ensemble we have used the acronym, "BCART."

In Table 2 we see that for the GND fitted dataset, the performance of BCART is (misclassification error 4.4%) better than single CART and Adaboost.M1 and logitboost. We see that 3-node DS has the highest prediction accuracy among the boosted algorithms. Among the results of DB5NN, DB10NN, DBLDA, DSB5NN, DSB10NN and DSBLDA we see that DBLDA has the highest accuracy than the other classifier (accuracy 96.02%) although DB5NN has 95.11% accuracy. We see here that the accuracy has increased (or misclassification error is decreased) than the best acquired by BCART (accuracy 95.6%). From the results we see that the DSVMSBagging has the lowest misclassification error 2.7% among all the classifiers here. For the Weibull fitted GIS dataset we see that the double bagging (also subbagging) with 5-NN has the second highest accuracy (96.84% and 95.61%). The DSVMSBagging performed best for this data also (an error rate of 3.2%). The performance of LOGITDS is satisfactory in this dataset, it is the fourth best performer (error 3.49%) in this dataset.

5.3   Experiment with the UCI Dataset

In this section we describe our findings of the comparative experiment with our new ensemble creation technique and several other ensemble creation techniques of CART (Bagging, Adaboost, Random Forest and Rotation Forest). For each dataset the optimum SSR is used to create the double SVMSbagging ensemble, which was reported in Fig. 2. For each data set and ensemble method, five 10-fold cross-validations were performed. The average accuracies and the standard deviations are reported. In this experiment for each data set, we used stratified ten-fold cross-validation method. A stratified n-fold cross-validation breaks the data set into n disjoint subsets each with a class distribution approximating that of the original data set. For each of the n folds, an ensemble is trained using $n - 1$ of the subsets, and evaluated on the held out subset. As this creates n non-overlapping test sets, it allows for statistical comparisons between approaches to be made.

To clarify how well double SVMsbagging performed in the benchmark datasets, we have plotted errors of double SVMsbagging and errors of double bagging with RBF SVM [31] and errors of best ensembles (ensemble having lowest error among bagging, Adaboost, random Forest and rotation Forest), see Fig. 3. If double SVMsbagging was better than the double bagging with RBF SVM for each data set, all the points on the graph (subplot a) would lie above the solid diagonal line which marks the equivalent scores. As all of the points in subplot a lie above the diagonal line, the figure demonstrates the higher accuracy of double SVMsbagging over

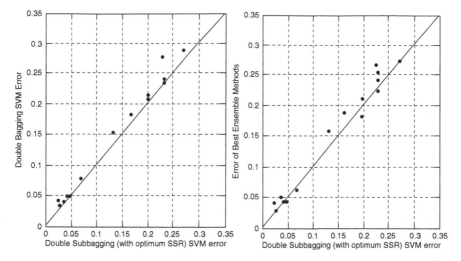

Fig. 3 Comparison of Double SVMSBagging with Double Bagging with SVM and other ensembles in benchmark datasets

double bagging with RBF SVM. In subplot b we also see that most of the points lie above the solid diagonal line, which show the better generalization ability of double SVMsbagging over other ensemble methods.

6   Conclusions

CART searches for partitions in the multivariate samples space, which may be seen as higher-order interactions or homogeneous subgroups defined by some combination binary splits of the predictors. SVM has advantage over other classifiers in (a) non-linear feature space, (b) dimensionality of the feature space and (c) generalization ability. In addition to these SVC construct the optimum separating hyperplane which maximize the margin between the classes (in binary classification). To build an ensemble of classifier with better generalization performance we combine these two methods.

A new subsampling approach to construct an SVM ensemble method has been proposed in this study, being a variant of another ensemble method named double bagging, where the SVM is used to construct additional classifier models using an independent sample than the training sample (the out-of-bag sample) to enhance the generalization performance of the ensemble method. Then, these additional predictors are combined with the CART to build the ensemble. This ensemble method relies on the selecting an optimum subsampling ratio, which enlarge the size of the out-of-bag samples and intensify the learning ability of the additional SVM models built on these.

The new method is used classify PD patterns in insulation system in order to model a better diagnosis system for the electric power apparatus. The proposed method outperformed other ensemble methods such as bagging, Adaboost.M1, logitboost and double bagging with LDA and k-NN ($k = 5$ and 10), in the experiments. The new ensemble method is tested on several UCI datasets and its performance is consistently better than popular ensemble methods like bagging, boosting and random Forest. Its performance is also competitive with the recent ensemble most successful ensemble method rotation Forest.

References

1. Blake. C., & Merz, C. (1999). UCI repository of machine learning databases, http://www. ics.uci. edu/mlearn/MLRepository.html.
2. Breiman, L., Friedman, J., Olshen, R., & Stone, C. (1984). *Classification and regression trees*. Belmont, CA: Wadsworth.
3. Breiman, L. (1996). Bagging predictors. *Machine Learning, 24*(2), 123–140.
4. Breiman, L. (1998). Arcing classifiers. *Annals of Statistics, 26*(3), 801–849.
5. Breiman, L. (2001). Random forests. *Machine Learning, 45*(1), 5–32.
6. Breiman, L. (2001). Statistical modeling: the two cultures. *Statistical Science, 16*(3), 199–231 (with discussion).
7. Burges, C. (1998). A tutorial on support vector machines for pattern recognition, *Data Mining and Knowledge Discovery, 2*, 121–167.
8. Chang, C., & Lin, C. (2001). LIBSVM: a library for support vector machines, software available at http://www.csie.ntu.edu.tw/ cjlin/libsvm.
9. Cortes, C., & Vapnik, V. (1995). support-vector networks, *Machine Learning, 20*, 273–297.
10. Freund Y., & Schapire, R. (1996). Experiments with a new boosting algorithm. *Proceedings of the thirteenth international conference machine learning* (pp. 148–156). San Francisco, MA: Morgan Kaufmann.
11. Friedman, J., Hastie, T., & Tibshirani, R. (2000). Additive logistic regression: a statistical view of boosting. *Annals of Statistics, 28*, 337–407(with discussion).
12. Gestel, T., Suykens, J., Baesens, B., Viaene, S., Vanthienen, J., Dedene, G., Moor, B., & Vandewalle, J. (2001). Bench marking least squares support vector machine classifiers, *Machine Learning, 54*(1), 5–32.
13. Gulski, E. (1995). Digital analysis of partial discharges. *IEEE Transactions on Dielectrics and Electrical Insulation, 2*(5), 822–837.
14. Hirose, H., Matsuda, S., & H., Hikita. (2006). Electrical insulation diagnosing using a new statistical classification method. *Proceedings of the 8th international conference on properties and applications of dielectric materials (ICPADM2006)* (pp. 698–701).
15. Hirose, H., Hikita, M., Ohtsuka, S., Tsuru, S., & Ichimaru, J. , (2008). Diagnosing the electric power apparatuses using the decision tree method. *IEEE Transactions on Dielectrics and Electrical Insulation, 15*(5), 1252–1261.
16. Hirose, H., Zaman, F., Tsuru, K., Tsuboi, T., & Okabe, S. (2008). Diagnosis accuracy in electric power apparatuses conditions using the classification methods. *IEICE Technical Report, 108*(243), 39–44.
17. Hothorn, T., & Lausen, B. (2003). Double-bagging: combining classifiers by bootstrap aggregation, *Pattern Recognition, 36*(6), 1303–1309.
18. Iba, W., & Langley, R. (1992). Induction of one-level decision trees. *Proceedings of the nineteenth international conference on machine learning*, Aberdeen, Scotland.
19. Joachims, T. (1999). Making large-scale support vector machine learning practical. *Advances in kernel methods: support vector machines* (pp. 169–184). Cambridge, MA: MIT Press.

20. Li, Y., Cal, Y., Yin, R., & Xu, X. (2005). Fault diagnosis based on support vector machine ensemble. *Proceedings of the 2005 international conference on machine Learning. Cybernet*, 6, 3309–3314.
21. Lin, H., Lin, C., & Wen, R. (2007). A Note on Platt's Probabilistic outputs for support vector machines, *Machine Learning, 68*(3), 267–276.
22. Meyer, D., Leisch, F., & Hornik, K. (2003). The support vector machine under test. *Neuro-computing*, 55, 169–186.
23. Patton, R., Lopez-Toribio, C., & Uppal, F. (1999). Artificial intelligence approaches to fault diagnosis, condition monitoring. *IEE Colloquium on Machinery, External Structures and Health (Ref. No. 1999/034)*, 5/1–5/18.
24. Platt, J. (1999). Fast training of support vector machines using sequential minimal optimization. In B. Scholkopf, C.J.C. Burges, & A.J. Smola (Eds.), *Advances Kernel methods – support vector learning* (pp. 185–208). Cambridge, MA: MIT.
25. Platt, J. (2000). Probabilistic outputs for support vector machines and comparison to regularized likelihood methods. In A.J. Smola, P. Bartlett, B. Scholkopf, & D. Schuurmans, (Eds.), *Advances in large margin classifiers* (pp. 61–74). Cambridge, MA: MIT Press.
26. Rodríguez, J., Kuncheva, L., & Alonso. C. (2006). Rotation forest: a new classifier ensemble method. *PAMI, 28*(10), 1619–1630.
27. Scholkopf, B., Smola, A., & Muller, K. (1998). Nonlinear component analysis as a kernel eigenvalue problem. *Neural Computation, 10*(5), 1299–1319.
28. Scholkopf, B., Burges, C., & Smola, A. (1999). Introduction to support vector learning. In B. Scholkopf, C.J.C. Burges, & A.J. Smola (Eds.), *Advances in kernel methods: support vector learning* (pp. 1–15). Cambridge, MA: MIT Press.
29. Sorsa, T. (1995). Neural network approach to fault diagnosis. Doctoral Thesis, Tampere University of Technology Publications 153.
30. Wu, T., Lin, C., & Weng, R. (2004). Probability estimates for multi-class classification by pairwise coupling. *Journal of Machine Learning and Research, 5*(Aug.), 975–1005.
31. Zaman, F., & Hirose, H. (2009). A new double bagging via the support vector machine with application to the condition diagnosis for the electric power apparatus. *International Conference on Data Mining and Applications (ICDMA'09)* (pp. 654–660).

Chapter 31
Clustering of Expressed Sequence Tag Using Global and Local Features: A Performance Study

Keng-Hoong Ng, Somnuk Phon-Amnuaisuk, and Chin-Kuan Ho

Abstract Clustering of expressed sequence tag (EST) plays an important role in gene analysis. Alignment-based sequence comparison is commonly used to measure the similarity between sequences, and recently some of the alignment-free comparisons have been introduced. In this paper, we evaluate the role of global and local features extracted from the alignment free approaches i.e., the compression-based method and the generalized relative entropy method. The evaluation is done from the perspective of EST clustering quality. Our evaluation shows that the local feature of EST yields much better clustering quality compared to the global feature. Furthermore, we verified our best clustering result achieved in the experiments with another EST clustering algorithm, *wcd*, and it shows that our performance is comparable to the later.

Keywords Sequence clustering · Expressed sequence tag · Alignment-free · Similarity measure · Grammar-based distance · Generalized relative entropy

1 Introduction

Expressed sequence tags (ESTs) were introduced in the early 1990s and they represent a significant advancement in modern biology [1]. This high-throughput technology provides the continuous flow of EST data that forms one of the richest resources for discoveries in genetics. An Expressed sequence tag is a tiny portion of an entire gene; it is produced by one-shot sequencing of a cDNA clone [2]. The cDNA clone is produced from an mRNA library. ESTs are easy to produce and they are valuable resources for different kinds of gene analysis e.g. gene identification, analysis of gene expression and gene structure analysis.

K.-H. Ng (✉), S. Phon-Amnuaisuk, and C.-K. Ho
Faculty of Information Technology, Multimedia University, Cyberjaya, 63100,
Selangor Darul Ehsan, Malaysia
e-mail: khng@mmu.edu.my; somnuk.amnuaisuk@mmu.edu.my; ckho@mmu.edu.my

S.-I. Ao et al. (eds.), *Intelligent Automation and Computer Engineering*,
Lecture Notes in Electrical Engineering 52, DOI 10.1007/978-90-481-3517-2_31,
© Springer Science+Business Media B.V. 2010

The characteristics of EST are low quality, high redundancy and short sequences. Therefore, unprocessed ESTs will not give any important information on gene analysis [3]. Clustering is usually the first step in EST data mining. It is a process of grouping ESTs that originate from the same gene. The goal is to construct gene indices, where all expressed data are partitioned into index classes such that expressed data are put into the same index class if and only they represent the same gene [4]. The ESTs in one cluster can be assembled to generate one or more consensus sequences [5]. Publicly available databases such as the Unigene of Genbank (http://www.ncbi.nlm.gov/unigene) and the Institute of Genome Research (http://www.tigr.org) accumulate and store the clustered EST data for gene research.

Methods employ in sequence clustering are commonly based on sequence comparison by alignment, which assumes conservation of contiguity between homologous segments. This alignment approach generates a similarity score, and this score can be calculated using BLAST [6] or FASTA [7]. TIGR uses this method for EST clustering, where it identifies all sequence overlaps using BLAST and FASTA [8]. Unigene [9] is another established player that uses the alignment method, where sequences are compared with the Smith–Waterman algorithm.

Although the alignment method gives satisfactory solutions, it is unfeasible to use it for long sequences because the computational load escalates as a power function of the sequence length [10]. The second drawback is the approach only considers local mutations of the genome, therefore it is not suitable to measure events and mutations that involve longer segments of genomic sequences [11]. For this reason alignment free sequence comparison has been recently introduced. Some of the non-alignment based EST clustering algorithms that are currently available are *d2_cluster* [12], *wcd* [13], *Xsact* [14], and *ESTmapper* [15]. Each of these algorithms is further discussed in the next section.

The aim of this paper is to study the effectiveness of global and local features in expressed sequence tag, in the clustering perspective. This study is important because we need to find out how well each of them contributes to the clustering process in EST. This information obtained from this study enables us to narrow down the research scope in EST clustering. The remainder of this paper is organized as follows. Section 2 reviews some related work in sequence clustering. The proposed method is presented in Section 3, while Section 4 discusses and compares the clustering results. Section 5 concludes the paper.

2 Related Work

In this section we survey and focus on some recent approaches used to define alignment-free distance measures of sequences, where the distances can later be used to perform sequence clustering. The alignment-free sequence comparison methods are based on counting the word frequencies, information theory and also data compression technique. This follows by reviewing several common clustering methods used in DNA sequences and the recent approaches used in EST clustering.

2.1   Methods Based on Word Frequencies

These methods transform a sequence into an object on which the analytical tools available in Linear Algebra and Statistical Theory can be applied. It starts with the mapping of sequences to vectors defined by the *k-tuple* counts. The obtained vectors represent the original sequence with the fixed word length k. The basic idea for this sequence comparison is that similar sequences should share common words, and then it can be quantified by many techniques. Blaisdell [16] is the pioneer who published sequence comparison report based on *k-tuple* counts, where the difference between two sequences was quantified by the Euclidean distance calculated between their word frequencies. For each word length k (or resolution), the Euclidean distance between sequence P and Q is defined as:

$$D_L(P, Q) = \sum_{i=1}^{j} \left(c_{L,i}^P - c_{L,i}^Q \right)^2 \tag{1}$$

The $c^P{}_L$ and $c^Q{}_L$ represent the word counts for the sequences and j is the number of possible *k-tuple* for the resolution k. For instance, the maximum number of *k-tuple* for word length $= 3$ is 64 (4^3). Even though this approach is alignment-free, but it is still length dependent in the sense that sequence comparisons are made for a fixed word-length. In fact, it can be recognized as local alignments between identical segments of sequences [11]. In a research work done by Peyzner [17], the distance based on word frequencies is regarded as a filtration method for sequence alignment algorithms. It increases the efficiency of the latter because it eliminates low similarity sequences which will directly reduce the input to the dynamic programming algorithm for sequence alignment.

Once this distance measure is established in sequence comparison, several methods originate from *k-tuple* frequencies are also quickly proposed. Petrilli [18] proposed a protein classification model which is based on di-peptide frequencies. It calculates the linear correlation coefficient between two sequences, from *k-tuple* frequencies and uses the conventional Pearson formalism. Mahalanobis distance is also introduced in sequence comparison, where it takes into account the data covariance structure.

2.2   Methods Based on Information Theory

In this method, the distance between two sequences is measured based on the *k-tuple* vectors and an information theory based metric is used to quantify the dissimilarity between them. In a research done by Wu et al. [19], the Kullback–Leibler discrepancy is proposed and it is computed from the *k-tuple* frequencies between two sequences P and Q. The equation of KL discrepancy is

$$D_k^{KL}(P, Q) = \sum_{i=1}^{n} f_{k,i}^P \times \log_2\left(f_{k,i}^P \Big/ f_{k,i}^Q\right) \qquad (2)$$

where $f^P{}_{k,i}$ is the *k-tuple* frequency of sequence P, integer n is the number of possible *k-tuples* with resolution k. The paper concludes that the KL discrepancy is preferred over the Mahalanobis distance and standard Euclidean distance in terms of computational efficiency, but it is not a good performing metric compared to the latter in the aspect of selectivity and sensitivity.

2.3 Methods Based on Compression Technique

The method is based on the basic idea that the more two sequences are similar, the more succinctly one sequence can be described given the other. It means that two sequences are considered close if one sequence is significantly compressible given the information contained in the other sequence. Similarities between sequences can be computed based on the well-known Lempel–Ziv parsing algorithm. Ziv and Merhav [20] introduce a measure of relative entropy between two sequences and it is a variant of the Lempel–Ziv parsing algorithm. Given two sequences x and y,

$$ZM(y|x) = (w_1, w_2, \dots, w_m, w_{m+1}, \dots, w_n) \qquad (3)$$

where $y = w_1, w_2, \dots w_n$, the block w_m is the longest prefix of $w_m w_{m+1} \dots w_n$ which occurs as factor in x. If such a prefix is different from the empty word and w_m is the first character of $w_m w_{m+1} \dots w_n$, otherwise. The integer n is the complexity of y relative to x. The idea is that the number of elements in $ZM(y|x)$ will be smaller if x and y are more similar.

Otu and Sayood [21] also introduce a distance measure based on the LZ parsing. Given two sequences P and Q, consider the sequence PQ and its exhaustive history. The number of component needed to build Q when appended to P is $c(PQ)-c(P)$, where $c(PQ)$ and $c(P)$ denote the number of components in the exhaustive history of sequence PQ and P. The number will be less than or equal to $c(Q)$, and it is dependent on the degree of similarity between P and Q. The closer between the two sequences, the fewer steps will be used to build Q in the production process of PQ. The paper shows that the algorithm constructed consistent phylogenies successfully.

2.4 Clustering Approaches in DNA Sequences

The agglomerative hierarchical clustering [22] is a popular merge-based clustering algorithm, it operates in a bottom up manner. In this algorithm, each sequence begins in its own cluster and it performs a series of merging of clusters. The merging process is based on the measure of similarity between clusters, where clusters that

Algorithm HierarchicalClustering(*S*)
Input: a dataset *S* with *N* sequences {S_1, S_2,..., S_N } and distance {*d*} matrix between sequences
Output: a hierarchy of clusters
1: for *a* ← 1 to *N*
2:    let C_a = {S_a} // start with *N* clusters
3: repeat step 4 – 6 until all all sequences are clustered into a single cluster of size *N*
4:    Find a pair of non-merged clusters, C_a and C_b where *d*(C_a, C_b) is the minimum
5:    Merge C_a and C_b
6:    Label C_a and C_b as merged
7: return the clusters' hierarchy

Fig. 1 The steps of hierarchical clustering algorithm are shown in pseudocode

are similar to each other are merged. This process is repeated until all clusters are merged into a single cluster [23]. Figure 1 shows the details of this clustering algorithm in pseudocode, the distance *d* is the similarity measure between sequences.

Another clustering algorithm to be highlighted is the famous *k*-means algorithm [24], which is a partitioning method. It starts with partitioning sequences in a dataset randomly, follows by re-computing the cluster centroids and reassigns each sequence to its nearest cluster repeatedly. The iteration terminates when there is no more sequence reassignment occurs.

The self-organizing map (SOM) is a clustering algorithm based on artificial neural network [25]. It is defined as a mapping of multi-dimensional input data spaces onto a map, typically a two-dimensional map. Neurons in the map are associated with reference vectors. As a result, the SOM can produce a visualization of cluster structures in the dataset. It can group sequences with similar features and also visualize their relationships with other groups that formed on the map. The drawback of the both algorithms compare to hierarchical clustering algorithm is the lack of providing hierarchical relationships between the input spaces.

Recent works on clustering include graph based clustering [26] where it can be naturally cast as a graph optimization problem, and ant-based clustering [27] that treats one gene as a node, every edge is associated with a certain level of pheromone intensity. The co-expression level between two genes determines the pheromone intensity of the edge. Then minimum spanning tree (MST) algorithm is used to break the linkages in order to generate clusters.

2.5 Recent Approaches in EST Clustering

The *d2_cluster* [12] is a well-known EST clustering method. It is a non-alignment based scoring method. This method begins with every sequence in a singleton cluster, and the clusters will be merged based on a series of similarity comparisons. It performs clustering according to the minimal linkage or transitive closure rules. The criterion for joining clusters is based on the percentage of matched words within a window size. The clustering process finishes after n (number of sequences) iterations of merging. The *wcd* [13] improves the performance of the above clustering algorithm, by reducing the number and cost of initializations. It uses zig–zag approach for window-to-window comparison. It enables the use of previous information to compute the initial score for a window and the computation speed can be improved with this approach. The details of this algorithm can be referred in the original paper.

Another clustering algorithm that is alignment free is *Xsact* [14]. It uses a suffix array, a lexicographically ordered array of all suffixes of the EST sequences. Radix sort is used to generate the suffix array, and then it will be used to find pairs of ESTs with long common substrings. *Xsact* calculates a score by finding the longest set of consistent matching substrings between each pair of EST. The clustering starts with the highest scoring pairs, where EST pairs above a certain similarity score are merged into a single cluster hierarchically. Clusters are then split according to the clustering threshold. The performance of this algorithm in terms of clustering quality is comparable to *d2_cluster* and alignment-based clustering, but it requires higher memory for the suffix construction.

Clustering algorithm such as *ESTmapper* [15] reads genome sequence and converts it into an eager WOTD (write only, top down) suffix tree. Each EST is mapped using the generated suffix tree, where it finds the long common substrings with the genome. The algorithm examines the list of common substrings and locations, and then combines substrings into a single gapped matching region if two common substrings are adjacent when mapped onto the genome. The longest matching region is used to determine the mapping of all ESTs to a location in the genome. ESTs are clustered if their sequences overlap or at nearby location in the genome. *ESTmapper* is efficient since ESTs can be compared to a suffix tree in linear time but its drawback is the consumption of large amount of memory.

3 Proposed Method

We are motivated by the problems encountered in EST clustering and the alignment-free similarity distance measures proposed in some research papers we highlighted in the above section. Hence, we propose a method to compare and evaluate the performance of derived global feature and local feature of EST in terms of clustering quality. To our best knowledge, there has never been any published work on this so far.

In our case, global feature can be defined as characteristics generalized from an entire EST sequence where the first subsequence is related to other subsequences and vice versa. On the other hand, local feature is mainly based on subsequence or location in a sequence and they are independent from each other. It indicates that the positions of subsequences in an EST sequence are not taking care of in the local feature. In this paper, we use grammar-based distance that based on LZ compression to represent the global feature of EST, while local feature is focused on the generalized relative entropy.

3.1 Dataset

The downloaded dataset contains 850 EST sequences from the Unigene database in the National Center for Biotechnology Information (NCBI) website. The dataset has ESTs from the cardiac muscle of heart organ in the organism named *Meleagris gallopavo* (turkey). These EST sequences have been grouped into 11 clusters by the Unigene and therefore it is a reliable source to be used for experiment. The dataset is pre-processed with *RepeatMasker* [28] to remove repeats in EST sequences. This process is important as the clustering quality will be affected in case they are not removed and masked. The sequences in the dataset are parsed to the grammar-based distance method and generalized relative entropy method for further processing.

3.2 Grammar-Based Sequence Distance

In this distance measure, it uses the fact that sequences share commonalities in their sequence structure if they have similar biological properties. This distance measure has been used to perform multiple sequence alignment in proteins and promising result is claimed by the authors [29]. Figure 2 gives an overview of the calculation of the grammar-based distance. It starts with the creation of LZ dictionaries for each EST sequence. Initially, the dictionary (G_p) for sequence P is empty; a fragment $f^1 = s_p(1)$ is set to the first residue of the corresponding sequence and it is visible to the algorithm. At ith iteration of the process, if fragment f^i is not reproducible from $s_p(1, \ldots, i-1)$, then f^i will be added to the dictionary $G^i{}_p = G^{i-1}{}_p + f^i$, and the fragment is reset. On the contrary, if the current dictionary contains enough rules to produce the current fragment, i.e. $G^i{}_p = G^{i-1}{}_p$, then it will not be added to the dictionary. The process continues until the visible sequence is equal to the entire sequence. For example, the dictionary for sequence $P = AACGTACC$ is $\{A, AC, G, T, ACC\}$.

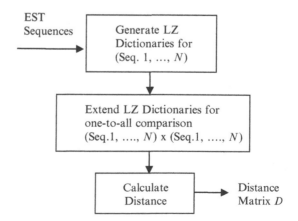

Fig. 2 Steps involved in the calculation of grammar-based distance

Each sequence is compared with all other sequences in the next step to generate the $N \times N$ size dictionaries. In this case, consider the comparison of sequence P and R. First, let the dictionary $G^1_{p,r} = G_p$, a fragment $f^1 = s_r(1)$ is set to the first residue of the sequence R, and the visible sequence is all rules in the dictionary of P. The algorithm operates as mentioned above. When it is complete, the new dictionary size will be smaller for sequences with higher similarity. The final step is the calculation of the distance using the dictionary sizes. The distance measure is based on one of the five suggested methods in the paper written by Otu and Sayood [21]

$$d_{p,r} = \frac{H_{p,r} - H_{p,p} + H_{r,p} - H_{r,r}}{\frac{1}{2}(H_{p,r} + H_{r,p})} \qquad (4)$$

where $p, r \in \{1, \ldots, N\}$ are the two sequences being compared, and H denotes the dictionary size of a sequence. The matrix distance D is generated from the calculation. This method compresses and builds the dictionary of an EST sequence based on the parsing of entire sequence string. Therefore, this method produces global feature for EST sequences.

3.3 Distance Based on Generalized Relative Entropy

This algorithm is one of the statistical distance measures used in protein or nucleotide sequences. Relative entropy has been explored as similarity measures such as *KLD* (Kullback–Leibler discrepancy) and *SimMM* (Similarity of Markov Models) to compare biological sequences. The drawback of *KLD* is when some entries of vectors are equal to 0 or 1, it becomes unsuitable. We adopt the generalized relative entropy described by Wang and Tang [30] in their work as the distance measure for EST sequences. It is denoted by *gre.d* and the following shows the calculation of the *gre.d* distance between sequence P and Q.

$$gre.d(P, Q) = \sum_{i=1}^{n} f^P(w_{k,i}) \times \log_2 \left[\frac{2 \times f^P(w_{k,i})}{f^P(w_{k,i}) + f^Q(w_{k,i})} \right] \qquad (5)$$

The $f^P(w_{k,i})$ and $f^Q(w_{k,i})$ are the k-word frequencies of sequence P and Q. The *gre.d* can deal with all kinds of k-word frequencies, including 0 and 1. We use this distance measure on several word sizes (k), ranging from 5 to 7 and a distance matrix will be generated for each of them. We use the average *gre.d* distance between two sequences because of the generated distances are not symmetric, i.e. $gre.d(P, Q) \neq gre.d(Q, P)$. This approach is based on the statistical measures of word frequencies in sequences and therefore it is regarded as local feature of EST sequences.

3.4   Evaluation of Clustering Quality in EST

Visualization is a powerful method for profiling clusters. By plotting the distance matrix into a 2D image [31], clusters can be seen in the image if there are a group of sequences with smaller distance among them. The objective of the clusters visualization is to evaluate the distance measures at coarse level, where we can find out from the plotted image whether the generated distance can give good clustering result or not. We decided to use the agglomerative hierarchical clustering algorithm that we mentioned in Section 2 to perform the EST clustering due to it runs in bottom-up approach and it can reveal the relationships among clusters in hierarchical manner.

The clustering quality is then measured with the non-weighted version of F-measure [22]. This measure is an external quality measure where it let us evaluate how well the clustering technique works by comparing the clustering result with the known classes, in our case it is the Unigene clusters. The F-measure combines recall (R) and precision (P) in the following formula:

$$F = \frac{2^* R^* P}{R + P} \tag{6}$$

4   Results and Discussion

In this paper, our goal is to evaluate the significance of the global and local features in EST sequences, in the perspective of clustering quality. We use a benchmark dataset that contains 11 clusters to assess the effectiveness of the proposed features as distance measures. Initially, we visualize and compare the two methods based on the images plotted from the generated distance matrices. We further investigate their roles play in EST clustering by performing hierarchical clustering, and the clustering quality of each method is shown in F-measure value. At the same time, we also use one of the latest EST clustering tools, *wcd* [13], which has been improved based on the d^2 distance function (it is computed based on Eq. 1), to perform clustering on the same dataset. The output of *wcd* is then compared with our best clustering result in terms of the number of clusters and their sizes.

4.1   Initial Evaluation of Features via Visualization

We perform an initial evaluation of both methods based on the plotted images. Figure 3 shows the grammar-based distance between sequences while Fig. 4 displays the generalized relative entropy (*gre.d*) distance with word size set to 5 and 7. The comparison of the images indicates that the latter distance measure performs

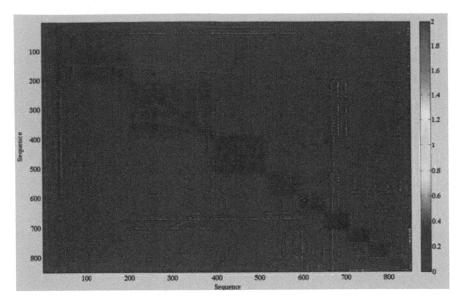

Fig. 3 Distance matrix calculated from the grammar-based method

better than the former, it is because we can see the 11 physically formed clusters in the images especially for k-words equal to 5 and 7.

A square or rectangle shape object in the image represents one cluster. These objects are formed due to the distance between EST sequences are small, which are shown in colour i.e. blue indicates small distance (<0.3). Furthermore, they also display larger distance with sequences from all other clusters. When comparing the images from Fig. 4, we can claim that the *gre.d* with word size 7 will give better clustering result compares to word size 5. It is because the former not only exhibits small distance among sequences in a cluster, but it also shows larger inter-cluster distance (red colour indicates distance between 0.8 – 1.0) compares to the latter with inter-cluster distance between 0.4 – 0.6 (light green).

The initial evaluation via visualization implies that the *gre.d* method will outperform the grammar-based method at the clustering stage. Furthermore, it also gives us a hint that the clustering quality in *gre.d* with larger word size will be higher compares to the smaller word size. It is because we find out two common things in larger word size i.e. (i) smaller intra-cluster distance, and (ii) larger inter-cluster distance.

4.2 Evaluation with Hierarchical Clustering Algorithm

The visualization results are verified by the hierarchical clustering algorithm, and then their outputs are evaluated using the F-measure method. Table 1 shows

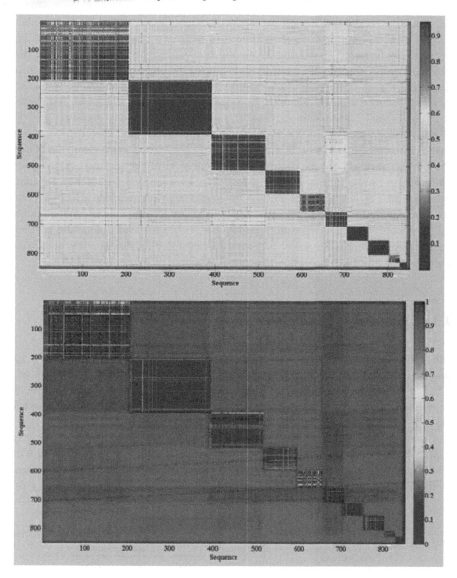

Fig. 4 Distance matrix generated with generalized relative entropy (*gre.d*), with k-word size set to 5 (*top*) and 7 (*down*) respectively

Table 1 Evaluation of grammar-based method and generalized relative entropy method with F-measure

Method	F-measure value
Grammar-based	0.1127
Generalized relative entropy (*gre.d*) with	
k-word $= 5$	0.5650
k-word $= 6$	0.8650
k-word $= 7$	0.8202

the clustering results in F-measure value, it is confirmed that the *gre.d* method outperforms the grammar-based method. The former obtains 0.8650 and 0.8202 respectively for word size 6 and 7. We did not extend the word size further due to the constraint of computational load. From the result, we can say that the local feature (*gre.d*) in ESTs plays a more important role towards the clustering quality as compares to the global feature (grammar-based method). The *gre.d* with word size 6 gives the best result among all others in terms of clustering quality.

We further investigate the reasons for the poor performance of the grammar-based distance measure in ESTs. Basically, the ESTs are sequenced from the cDNA library and they are not the complete representation of the parental cDNA [32]. Their length can be varying from sequence to sequence even though they originate from the same cDNA clone. As a result, the variance in length might affect the compression outcome since the EST sequence with larger length tends to produce richer LZ dictionary. Thus, this measure produces unreliable distance among the EST sequences. Another possible reason is the start position for the parsing, where different start positions for the same sequence give other versions of dictionaries.

4.3 Comparison with the wcd Clustering Algorithm

Other than verifying the clustering output with the Unigene, we also compare our best result with the *wcd* EST clustering tool in terms of number of clusters and size. This algorithm has been briefly introduced in Section 2. In this comparison, all default parameters of the *wcd* are used. Table 2 shows the clustering result between the *wcd* and our method.

Based on the result, the *wcd* produces 11 clusters with the cluster sizes range from 6 to 204. This clustering result is then compared with our *gre.d* method with word size = 6. It shows that the performance of our method is closed to the *wcd* method, where most of the clusters produced by both methods are generally the

Table 2 Comparative study of cluster number and size between the *wcd* method and the generalized relative entropy method

Method	wcd	Generalized relative entropy
Size for each cluster	204	203
	186	185
	121	121
	115	115
	59	59
	49	49
	48	65
	25	25
	20	20
	17	2
	6	6
Number of clusters	11	11

same, except for the two clusters with different sizes. The clustering results are similar and it might because of the distance functions of both methods are based on the word composition of sequences. The distance measure that we use is *gre.d*, and it is calculated based on word frequencies (Eq. 5). The generated distances among EST sequences are then used to perform clustering with the hierarchical clustering algorithm. The *wcd* method is also based on the word occurrences, the key difference is the word occurrence is not measured through the entire sequence, but it is based on a fixed window size in the sequence.

5   Conclusion

In this paper, we have presented a method to evaluate the significance of global and local features in ESTs based on alignment-free distance measures i.e. grammar-based method and generalized relative entropy method. We conclude that the local feature extracted from the generalized relative entropy method outperforms the global feature derived from the former method in terms of clustering quality. Furthermore, a comparison between the best result achieved in our experiment with the *wcd* clustering algorithm has been conducted, and it shows comparable output in terms of cluster number and size. In future work we will continue to enhance the EST clustering quality by exploring more alignment-free techniques and clustering algorithms.

References

1. Ptitsyn, A., & Hide, W. (2005). CLU: A new algorithm for EST clustering. *BMC Bioinformatics, 6.* doi:10.1186/1471-2105-6-S2-S3.
2. Malde, K., Coward, E., & Jonassen, I. (2005). A graph based algorithm for generating EST consensus sequences. *Bioinformatics, 21*(8), 1371–1375.
3. Hide, W., Miller, R., Ptitsyn, A., Kelso, J., Gopallakrishnan, C., & Christoffels, A. (1999). EST clustering tutorial. SANBI.
4. Burke, J.P., Wang, H., Hide, W., & Davison, D. (1998). Alternative gene form discovery and candidate gene selection from gene indexing projects. *Genome Research, 8*, 276–290.
5. Haas, S.A., Beissbarth, T., Ribals, E., Krause A., & Vingron, M. (2000). GeneNest: Automated generation and visualization of gene indices. *Trends Genetics, 16*, 521–523.
6. Altschul, S., Gish, W., Miller, W., Myers, E., & Lipman, D. (1990). A basic local alignment search tool. *Journal of Molecular Biology, 215*, 403–410.
7. Lipman, D.J., & Pearson, W.R. (1988). Improved tools for biological sequence comparison. *Proceedings of the National Academy of Sciences of the United States of America, 85*(8), 2444–2488.
8. Sutton, G., White, O., Adams, M.D., & Kerlavage, A.R. (1995). TIGR assembler: A new tool for assembling large shotgun sequencing projects. *Genome Science Technology, 1*, 9–18.
9. Boguski, M.S., & Schuler, G.D. (1995). Establishing a human transcript map. *National Genetics, 10*, 369–371.
10. Vinga, S., & Almeida, J. (2003). Alignment-free sequence comparison – a review. *Bioinformatics, 19*(4), 513–523.

11. Mantaci, S., Restivo, A., & Sciortino, M. (2008). Distance measures for biological sequences: Some recent approaches. *International Journal of Approximate Reason, 47*, 109–124.

12. Burke, J., Davison, D., & Hide, W. (1999). d2_cluster: A validated method for clustering EST and full length cDNA sequences. *Genome Research, 9*, 1135–1142.

13. Hazelhurst, S. (2008). Algorithms for clustering expressed sequence tag: The wcd tool. *South African Computer Journal, 40*, 51–62.

14. Malde, K., Coward, E., & Jonassen, I. (2003). Fast sequence clustering using a suffix array algorithm. *Bioinformatics, 19*(10), 1221–1226.

15. Wu, X., Lee, W.J., Gupta, D., & Tseng, C.W. (2005). ESTmapper: Efficiently clustering EST sequences using genome maps. *Proceedings of the 19th IEEE International Parallel and Distributed Processing Symposium*, 196a. doi:10.1109/IPDPS:2005.204.

16. Blaisdell, B.E. (1986). A measure of the similarity of sets of sequences not requiring sequence alignment. *Proceedings of the National Academy of Sciences of the United States of America, 83*, 5155–5159.

17. Pevzner, P.A. (1992). Statistical distance between texts and filtration methods in sequence comparison. *Computer Applications in the Biosciences, 8*, 121–127.

18. Petrilli, P. (1993). Classification of protein sequences by their dipeptide composition. *Computer Applications in the Bioscience, 9*, 205–209.

19. Wu, T.J., Hsieh, Y.C., & Li, L.A. (2001). Statistical measures of DNA sequence dissimilarity under Markov chain models of base composition, *Biometrics, 57*, 441–448.

20. Ziv, J., & Merhav, N. (1993). A measure of relative entropy between individual sequences with application to universal classification. *IEEE Transactions on Information Theory, 39*(4), 1270–1279.

21. Otu, H.H., & Sayood, K. (2003). A new sequence distance measure for phylogenetic tree construction. *Bioinformatics, 19*(16), 2122–2130.

22. Dong, G., & Pei, J. (2007). Classification, clustering, features and distances of sequence Data. *Sequence Data Mining, 33*, Springer US, 47–65. doi:10.1007/978-0-387-69937-0.

23. Ma, C.H., Chan, C.C., Yao, X., & Chiu, K.Y. (2006). An evolutionary clustering algorithm for gene expression microarray data analysis. *IEEE Transactions on Evolutionary Computation, 10*, 296–314.

24. Handl, J., Knowles, J., & Dorigo, M. (2003). Ant-based clustering: A comparative study of its relative performance with respect to k-means, average link and 1d-som. Technical Report TR/IRIDIA/2003-24, IRIDIA, http://dbkgroup.org/handl/TR-IRIDIA-2003-24.pdf.

25. Tamayo, P., Slonim, D., Mesiov, J., Zhu, Q., Kitareewan, S., Dmitrovsky, E., Lander, E.S., & Golub, T.R. (1999). Interpreting patterns of gene expression with self-organizing maps: Methods and application to hematopoietic differentiation. *Proceedings of the National Academy of Sciences of the United States of America, 96*(6), 2907–2912.

26. Xu, Y., Olman, V., & Xu, D. (2002). Clustering gene expression data using a graph-theoretic approach: an application of minimum spanning trees. *Bioinformatics, 18*(4), 536–545.

27. Zhou, D., He, Y., Kwoh, C.K., & Wang, H. (2007). Ant-MST: An ant-based minimum spanning tree for gene expression data clustering. *LNBI, 4774*, 198–205.

28. Smit, A.F.A., Hubley, R., & Green, P. (2004). RepeatMasker Open-3.0, 2004, http://www.repeatmasker.org.

29. Russell, D.J., Otu, H.H., & Sayood, K. (2008). Grammar-based distance in progressive multiple sequence alignment. *BMC Bioinformatics, 9*, 306. doi:10.1186/1471-2105-9-306.

30. Tai, Q., & Wang, T. (2008). Comparison study on k-word statistical measures for protein: From sequence to sequence space. *BMC Bioinformatics, 9*, 394. doi:10.1186/1471-2105-9-394.

31. Hathaway, R.J. & Bezdek, J.C. (2003). Visual cluster validity for prototype generator clustering models. *Pattern Recognition Letters, 24*, 1563–1569.

32. Rudd, S. (2003). Expressed sequence tags: alternative or complement to whole genome sequence? *Trends in Plant Science, 8*(7), 321–329.

Chapter 32
Research on Process Algebraic Analysis Tools for Electronic System Design

Ka Lok Man and Tomas Krilavičius

Abstract Rapid software/hardware development cycle increased demand for the advanced design and implementation methods. Recently, formal methods have been put forward as a tool for modelling and analysis of electronic systems. Usage of formal semantics and syntax allows unambiguous specifications of the systems, and in such a way provides means for rigorous analysis of correctness and performance properties. We investigate applicability of two process algebra based tools for the mixed software/hardware modelling and analysis: Process Analysis Toolkit (PAT) and Software/Hardware Engineering (SHE). PAT toolkit is based on CSP-like process algebra extended with mechanisms customary for software developers and engineers. It supports reachability and deadlock analysis, complete Linear Temporal Logic (LTL) model checking and refinement. SHE methodology provides means for correctness and performance analysis by applying model-driven design methodology at the system level, i.e., high abstraction level design stage of the embedded and mixed hardware/software systems. It combines techniques for development of formal models for analysis and refinement to the actual implementation of the system. SHE toolset provides tools for modelling, simulation and real-time control code generation. Transaction Level Modelling (TLM) approach has been put forward as a tool for elaborate System-on-Chip (SoC) design. It is quite extensively applied in industry to solve a number of practical problems, occurring at the design, development and deployment stages. We apply PAT and SHE methodology for functional and performance analysis of a hardware model and a TLM model, and illustrate this by means of examples: a simple pipeline process and a process-memory communication model, respectively.

Keywords Process algebra · Verification · Simulation · Transaction level modelling · Interoperability

K.L. Man (✉) and T. Krilavičius
Solari, Hong Kong and Vytautas Magnus University, K. Donelaicio g. 58,
LT-44248 Kaunas, Lithuania
e-mail: pafesd@gmail.com; t.krilavicius@if.vdu.lt

S.-I. Ao et al. (eds.), *Intelligent Automation and Computer Engineering*,
Lecture Notes in Electrical Engineering 52, DOI 10.1007/978-90-481-3517-2_32,
© Springer Science+Business Media B.V. 2010

1 Introduction

Increase in demand for advanced techniques for design and development of mixed software and hardware systems generated a set of new approaches. It initiated a flash of activities in formal methods research and application for electronic system design. Several reasons generated so much interest in formal techniques.

Unambiguous Models Formal modelling languages allow defining systems unambiguously, because syntax and semantics are defined formally, and that includes means to define non-deterministic and stochastic behaviour precisely, too. Moreover, for the same reasons, unambiguous refinement and code generation techniques can be applied.

Strict Analysis Techniques Because models are defined using languages with strict semantics, rigorous reasoning about models is possible. Model checking, theorem proving and specifically designed algorithms, e.g., for stability analysis [18], can be used.

Over the years, through the novel language constructs and well-defined formal semantics, several timed and hybrid process algebras (e.g., [2,3,5,9,10,39,40,49, 50]) have been developed. They can be effectively used to formally specify diverse systems. Several successful case studies (e.g., [21–29,33–38]) have shown that process algebraic formalisms and their toolsets can be efficiently and effectively applied for the formal modelling and analysis of the behaviour of *Transaction Level Modelling* (TLM) [8] and *Electronic System Design*.

However, process algebraic analysis tools have not been widely used yet by the architects, engineers and researchers from the electronic community. We believe, that the following reasons are to blame:

- Usually, the syntax of the process algebraic languages and tools is not intuitive.
- The tools lack advanced programming features.
- The tools do not offer sufficient flexibility for the analysis of the models.
- The interoperability of performance and functional analysis components is missing in most of the tools.

We discuss two noteworthy process algebraic tools that are well-suited for the modelling and analysis of electronic systems: *Process Analysis Toolkit* (PAT) [42] and *Software/Hardware Engineering* (SHE) [44]. Both tools allow users to interactively and visually simulate system behaviours and verify various forms of system properties/assertions.

PAT is a CSP [11] based process algebra extended with mechanisms customary for software developers and engineers. It provides means for reachability and deadlock analysis, supports complete Linear Temporal Logic (LTL [12]) model checking and refinement. It is a generic framework that provides an user friendly editing environment, powerful simulator and model checker.

SHE [44] is a model-driven system-level design technique for modelling, correctness and performance analysis. We combine it with TLM, one of the successful techniques for *System-on-a-Chip* (SoC) development, adopted by industry and academia. Usually, it is applied as

- An early platform for software development
- Software/hardware integration tool
- Software performance analysis and system level design analysis technique
- Tool for an early estimation of the power consumption (by real software running on TLM models before *Register Transfer Level* (RTL) models are ready).

In TLM, systems are modelled as a collection of parallel components that communicate through the abstract channels. Different levels of abstraction are defined, starting from the abstract functional *specification models*, and culminating with the *implementation models*. Also, TLM models allow *"very fast simulation"*, around 1,000 times faster than pure RTL models.

Simulation is used for the performance analysis as well as a way to gain more insight on the behavioural properties of the system. *Model checking* [4] is usually applied for *functional analysis*. Different techniques are applied for the performance [53] and functional [45, 48] analysis of the TLM models, however less effort is put to combine these two approaches.

We propose to merge performance and functional analysis of TLM models to obtain a full-blown performance as well as functional analysis by combining TLM and SHE. The idea is to define TLM models using *Parallel Object-Oriented Specification Language* (POOSL [44]), the modelling language of SHE, and then translate it into other formal languages supported by the performance and/or correctness analysis tools. Such approach would allow not only to investigate models, but also examine tools interoperability [16], and compare used formal approaches.

In this chapter we investigate the following problems.

- Applicability of PAT to hardware modelling and analysis
- Compatibility of TLM and SHE, and its applicability for a simple, but practical system analysis
- Interoperability of modelling, performance and verification tools.

We illustrate both approaches by taking simple, but meaningful examples. We model a simple pipeline process using PAT. This example captures current trend of the microprocessors design, i.e., a techniques that allows considerably improving performance by starting the execution of an operation before completing the previous one. SHE and TLM benefits are exemplified by a simple processor-memory communication model that nicely describes pure master (process) and slave (memory) interaction via processor and memory interfaces supporting a specified set of operations.

This book chapter is organised as follows. Section 2 provides an overview of PAT. A sample (modelling and verification of some properties of a pipeline process) of the application of PAT is shown at the same section. In Section 3, we first present the SHE methodology including the input language and toolset. Then, by means of a TLM model (a CPU and memory system), we present the application of the SHE methodology to combine performance and functional analysis of such a TLM model. Finally, a summary of the chapter is given in Section 4.

2 PAT: Process Analysis Toolkit

Process Analysis Toolkit (PAT) is designed for applying state of art model checking techniques for analysis of various systems. PAT supports reachability analysis, deadlock-freeness analysis, complete Linear Temporal Logic (LTL) model checking and refinement checking.

Informally, the PAT modelling language (i.e., the input language of PAT) is *Communicating Sequential Processes* (CSP)-style process algebra which combines high-level modelling operators like (conditional or non-deterministic) choices, interrupt, (alphabetised) parallel composition, interleaving, hiding, asynchronous message passing channel, etc., with programmer favoured low-level constructs like variables, arrays, if-then-else, while, etc. Due to these, PAT offers a great flexibility for modelling systems.

PAT includes an user friendly editing environment introducing CSP-style input models, powerful simulator and model checker. PAT is intended to be developed as a generic framework, which can be easily extended to support new modelling languages or new assertion languages (as well as dedicated verification algorithms).

2.1 A Pipeline Process

The concept of *pipelining* is widely used in the design of microprocessors. It greatly improves the performance of microprocessors by overlapping the execution of operations. A pipeline process typically starts the execution of an operation before the completion of the previous operation.

Figure 1 depicts a classical pipeline process which consists of the three processes that form individual stages of a pipeline. Each stage of the pipeline process performs certain operation and consumes a number of time units, e.g.,

The first stage of the pipeline accepts two inputs and computes their sum and difference consuming one time unit.

The second stage accepts the results of the first stage and computes their product and quotient consuming two time units.

The third stage accepts the outputs from second stage and computes the first input raised to the power of the second consuming three time units.

Fig. 1 Three staged pipeline

2.2 PAT Model of the Pipeline Process

```
#define Max 10;
#define STAGE1 1;
#define STAGE2 2;
#define STAGE3 3;

var stage1=1;
var stage2=0;
var stage3=0;
var time;

S1()=[time <= Max && stage1==1 && stage2==0 && stage3==0]
      S1_go_S2 {stage2=1; time=time+STAGE1;} -> S2();

S2()=[time <= Max && stage1==1 && stage2==1 && stage3==0]
      S2_go_S3 {Stage3=1; time=time+STAGE2;} -> S3();

S3()=[time <= Max && stage1==1 && stage2==1 && stage3==1]
      S3_go_Stop {time=time+STAGE3;} -> Stop;

Pipeline()= S1(); S1_go_S2-> S2(); S2_go_S3-> S3(); Stop;

#define goal (time<=Max && stage1==1 &&
              stage2==1 && stage3==1);

#assert S1() reaches goal;
#assert Pipeline() reaches goal;
#assert S2()|=[]<> S1_go_S2;
#assert S3()|=[]<> S2_go_S3;
```

The above PAT model of the pipeline process mainly implements the sequential execution behaviour of the pipeline process. The processes *S1*(), *S2*() and *S3*() describe the sequential execution behaviour of the *stage1*, *stage3* and *stage3* in the pipeline process, respectively. Additionally, several constants (specified by the keyword "*#define*") and variables (specified by the keyword "*var*") are defined according to the intuitive syntax of the PAT modelling language:

- *Max* – represents the maximum number of time units.
- *STAGE(i)* – stands for the number of time units needed in the *stage(i)* of the pipeline process to complete the operation performed in such a stage, where $i \in \{1..3\}$.
- *stage(i)* – is used to state whether the pipeline process is being executed at the *stage(i)*, where $i \in \{1..3\}$; it is of value 1 if the pipeline process is executing at the *stage(i)*, otherwise 0.
- *time* – records the number of time units spent so far, with default value 0.

In the PAT model, the language construct

```
[condition]event{code} -> P();
```

is used to express that if the *condition* holds, then the *event* can occur (i.e., the *code* attached with the event can be executed) and leads to the process $P()$.

Intuitively, the process

```
S1()=[time <= Max && stage1==1 && stage2==0 && stage3==0]
        S1_go_S2 {stage2=1; time=time+STAGE1;} -> S2();
```

describes that if the control/execution of the pipeline process is at the *stage1* and the total execution time by far is within the *Max* (maximum number of time units), then the event *S1_go_S2* can occur (it reads as *S1()* terminates and switches the control to *S2()*), which in terms means that the variable *stage2* is assigned to be 1 and the *time* is incremented by the number of time units (*STAGE1*) needed by the completion of the operation at the *stage1*; and eventually the process *S2()* takes over.

In analogy with process *S1()*, processes *S2()* and *S3()* are defined. After the execution of the process *S3()*, the whole pipeline process terminates (through a *Stop* process).

At a high level description, the process:

```
Pipeline()= S1(); S1_go_S2-> S2(); S2_go_S3-> S3(); Stop;
```

captures the same sequential execution behaviour of the whole pipeline process.

2.3 Verification

The verification of the pipeline process needs to show that such a process must behave as its sequential execution model does (i.e., the execution order is *S1()*, *S2()*, *S3()* and then terminates via the process *Stop*) within the maximum number of time units scheduled.

Such a verification aim for the pipeline process is expressed as a proposition named *goal* in the PAT model:

```
#define goal (time<=Max && stage1==1 && stage2==1 && stage3==1);
```

By means of the assertions below:

```
#assert S1() reaches goal;
#assert Pipeline() reaches goal;
```

the verification aim of the pipeline process was verified successfully in a few seconds using a modern PC by PAT. Loosely speaking, two assertions are the same. The assertion *S1()/Pipeline() reaches goal* states that starting from the process *S1()/Pipeline()*, the pipeline process will reach a state in which the proposition "goal" is true.

Furthermore, PAT also easily proved the following assertions:

```
#assert S2()|=[]<> S1_go_S2;
#assert S3()|=[]<> S2_go_S3;
```

hold for the pipeline process. Informally speaking, such assertions require that *S2()/S3()* has to terminate before the event *S1_go_S2/S2_go_S3* occurs (i.e., trivial sequential execution property).

2.3.1 Observation and Simulation

In the PAT model of the pipeline process, *Max*, *STAGE1*, *STAGE2* and *STAGE3* are set to be 10, 1, 2 and 3, respectively. Due to all these, it is not hard to see that the sequential execution property is preserved in the PAT model; as the total execution time from the first stage to the third stage is only 6 time units (*STAGE1* + *STAGE2* + *STAGE3* = 6) which is smaller than *Max* (10 time units).

If we set the *Max* to 3, then the verification will fail. This also means that the sequential execution property can not be preserved in the PAT model.

In addition to the verification, the PAT model of the pipeline process was simulated using the simulator in PAT. Figure 2 shows the simulation traces of the pipeline process described as a transition graph. It is easy to see that the sequential execution property is also preserved in the simulation traces (where *State 4* in the graph denotes the termination state). This is a sort of interoperability analysis result which will be explained later in details in this book chapter.

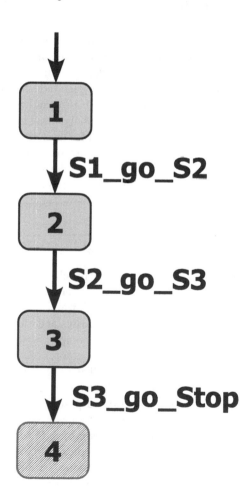

Fig. 2 Simulation
of the pipeline process

3   SHE Methodology

Software/Hardware Engineering (SHE) is a model-driven system-level design methodology that allows analysing correctness and performance properties of design alternatives based on models. The actual evaluation is based on the application of several techniques for performance analysis and a formal verification of correctness properties (i.e., functional analysis). The designer is assisted in constructing models and applying the analysis techniques with various guidelines and modelling patterns. The main key feature of the SHE methodology is its foundation on formal methods, which ensures that the obtained analysis results are unambiguous. Moreover, SHE is accompanied with a set of user-friendly computer tools (most of them are also based on formal methods).

3.1   POOSL

The SHE methodology mainly relies on the modelling language called *Parallel Object-Oriented Specification Language* (POOSL). It is a language for system-level design, which intends to deal with the complexity of hardware and software systems by bridging the gap between industrial practice and formal methods. POOSL is also a formal language with a semantics that is defined in a similar mathematical way as is common for process algebras [1]. Due to space restrictions we refer to the publications available at [44] for the formal syntax and semantics of POOSL.

The SHE methodology relies on POOSL for formalising a model into an executable model, which then allows for validation by interactive simulation and formal verification of correctness properties for such a model.

3.2   Tools for SHE

A number of tools have been developed for building and analysing a POOSL model. The most commonly used tools are *SHESim* and *Rotalumis* [44].

SHESim is a graphical tool, which allows designers to incrementally specify and modify classes of data, processes and clusters and easily express hierarchical and topological structure of the complex systems. Furthermore, it can create log files for performance evaluation purposes. An interaction diagram tool helps designers to inspect the history of all messages exchanged between different processes by generating interaction diagrams automatically during simulation. These diagrams can be used for validation purposes of a model.

Rotalumis is a textual tool which is initially for high-speed simulation after completing the validation of a POOSL model. Later, it is enhanced for the correctness-preserving code generation from a POOSL model. It has been demonstrated in [13] that Rotalumis can automatically generate code for time-critical systems and preserve the timing properties proven in the model.

Next to these two tools, several formal verification tools (e.g., SPIN [12, 47] and UPPAAL [19]) can be used as back-end verification tools by translating POOSL models to the corresponding input models/specifications for such verification tools. Currently, SHE includes a few guidelines for constructing various input models/specifications of verification tools from a POOSL model [7].

3.3   TLM Model: A CPU and Memory System

A simulation is performed on a system modelling one memory and one CPU in which the memory is a pure slave and the CPU is a pure master. Such a system is taken from http://mij.oltrelinux.com/devel/systemc/ and its implementation in SystemC [14] is also available at same link.

A memory interface dictates what services complying memories must feature; and the CPU implementation demonstrates all the supported operations with the memory.

More precisely, the CPU first gets to know if it is possible to write to a random memory address. Eventually, the CPU issues a *write* and *read* request operation with a data unit and repeats the same with more operations provided by the memory.

Two module interfaces are introduced in the system for describing the actual communication between the CPU and memory: *CPU adapter* and *memory adapter* for the CPU module and memory interface, respectively. The CPU and memory system is shown in Fig. 3.

The hardware implementation of such an communication is specified using the usual signals paradigm. More specifically, CPU adaptor (CA) and memory adaptor (MA) communicate via a set of signals wrapped to realize a full-featured bus.

A protocol must be specified for the communication to be accomplished across the bus. This is entirely enclosed into adaptor implementations and can be changed in the future without impact against the rest of the system relying on them (this is the key idea of TLM).

3.4   Simulation

The (CPU and memory) system was first described in POOSL and then simulated using SHESim. A simulation run is a list of service-issue messages from the CPU, interleaved with messages from the memory adaptor responding to the bus protocol.

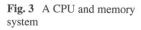
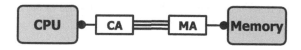

Fig. 3 A CPU and memory system

As an example of a simulation run:

```
MemAdaptor: WRITE request income,
            processing... done.
MemAdaptor: waiting for requests...
CPU: getting the data...
MemAdaptor: waiting for requests...
MemAdaptor: READ request income,
            processing... done.
MemAdaptor: waiting for requests...
```

3.5 Formal Verification

According to the guidelines given in [7], the (CPU and memory system) POOSL model was easily constructed into the equivalent PROMELA model [12] which is the input model for SPIN.

The SPIN model checker is a software package that allows the simulation of a specification written in the PROMELA. It accepts correctness claims specified in the syntax of standard LTL.

Applying the SPIN model checker to the PROMELA model, the following properties (expressed in LTL) were verified successfully in few seconds using a modern PC:

1. *Deadlock Free*   There is deadlock free in the state space generated for the (CPU and memory system) POOSL model system.
2. *Liveness Property*   CPU eventually gets the data after each write request operation is executed.
3. *Safety Property*   No read request operation is executed before a write request operation is called.

3.6 Interoperability

It is not hard to see that the simulation and formal verification results of the (CPU and memory system) POOSL model are match. More specifically, the properties (deadlock free, liveness and safety) are also shown to be held in the simulation runs (by inspection of the simulation traces) as presented in the previous section.

This combined approach of the simulation and formal verification (i.e., interoperability approach) aids the designer to perform simulation-based analysis as well as functional analysis of the POOSL model in the SHE Methodology.

4 Summary

Performed experiments show that application of process algebras for modelling and analysis of electronic systems seems promising. We were able to model and verify several different systems and gain a considerable insight into their behaviour. Application of PAT shows that it can be used to analyse application of fundamental design techniques in microprocessors. Combination of SHE and TLM shows that the same model can be used for both simulation and model checking, in such a way demonstrating that designer can use the same model to analyse the same properties.

Potentially interesting future work directions are following:

Analysis of Industry Relevant Systems   Analysis of the industry relevant systems using the above mentioned tools could be beneficial for further development of tool theories as well as directly to industry.

Application of (Hybrid) Process Algebras in Electronics   Hybrid process algebras, e.g., BHPC [3, 15, 17], Hybrid Chi [50] or HyPA [5], are well-known in Computer Science. It could be interesting to apply them for modelling and analysis of mixed software/hardware systems with a specific interest in power consumption modelling. Early experiments with BHPC [33], ACP_{hs}^{srt} [30, 31], HyPA [29] and Timed Chi [20, 32] have shown promising results. New developments of the BHPC simulation toolset Bhave (see also [52]) look quite interesting, therefore we are planning to apply Bhave simulator for electronic system simulation, and try to integrate msp-svg [15,43] (available at http://msp-svg. sourceforge.net/) with it or other hybrid simulation tools (e.g., [6,41,46,51]).

References

1. Baeten, J.C.M., & Weijland, W.P. (1990). *Process algebra*, volume 18 of *Cambridge tracts in theoretical computer science*. Cambridge, United Kingdom; Cambridge University Press.
2. Bergstra, J.A., & Middelburg, C.A. (2005). Process algebra for hybrid systems. *Theoretical Computer Science*, *335*(2/3), 215–280.
3. Brinksma, E., Krilavičius, T., & Usenko, Y.S. (2005). Process algebraic approach to hybrid systems. *Proceedings of 16th IFAC world congress* (pp. 1–6). Prague: Czech Republic.
4. Clarke, E.M., Grumberg, O., & Peled, D.A. (2000). *Model checking*. Cambridge: MIT.
5. Cuijpers, P.J.L., & Reniers, M.A. (2005). Hybrid process algebra. *Journal of Logic and Algebraic Programming*, *62*(2), 191–245.
6. Fritzson, P., & Engelson, V. (1998). Modelica – a unified object-oriented language for system modelling and simulation. In Proceedings of European Conference on Object-Oriented Programming (ECOOP'98). Brussels, Belgium, July 20–24, 1998.
7. Geilen, M.C.W. (2002). *Formal verification of complex real-time systems*. PhD thesis, Eindhoven University of Technology.
8. Ghenassia, F. (Ed.). (2005). *Transaction-level modeling*. Dordrecht, The Netherlands: Springer.
9. Groote, J.F., Mathijssen, A. H.J., van Weerdenburg, M.J., & Usenko, Y.S. (2006). From μCRL to mCRL2: motivation and outline. *Electronic Notes on Theoretical Computer Science*, *162*, 191–196.
10. Groote, J.F., Mathijssen, A. H.J., van Weerdenburg, M.J., & Usenko, Y.S. (2007). The formal specification language mCRL2. *Methods for modelling software systems*. *Dagstuhl seminar*. Dagstuhl.

11. Hoare, C.A.R. (1985). *Communicating sequential processes*. Englewood-Cliffs: Prentice-Hall.
12. Holzmann, G.J. (2003). *The SPIN model checker: primer and reference manual*. Boston: Addison Wesley Professional.
13. Huang, J., Voeten, J., & Corporaal, H. (2007). Predictable real-time software synthesis. *International Journal on real-time systems, 36*, 159–198.
14. IEEE (2005). *IEEE standard for SystemC language reference manual (IEEE STD 1666TM-2005)*. IEEE.
15. Krilavičius, T. (2006). *Hybrid techniques for hybrid systems*. PhD thesis, University of Twente.
16. Krilavičius, T. (2007). Study of tools interoperability. Technical Report TR-CTIT-07-01, University of Twente, The Netherlands.
17. Krilavičius, T. (2008). Simulation of mechatronic systems using Behavioural Hybrid Process Calculus. *Electronics and Electrical Engineering, 1*(81), 45–48.
18. Langerak, R., Polderman, J.W., & Krilavičius (2003). Stability analysis for hybrid automata using conservative gains. In S. Engell, H. Guéguen, & J. Zayton (Eds.), *Proceedings of conference on analysis and design of hybrid systems (ADHS 03)* (pp. 377–382).
19. Larsen, K.G., Pettersson, P., & Yi, W. (1997). UPPAAL in a nutshell. *Int. Journal on Software Tools for Technology Transfer, 1*(1–2), 134–152.
20. Man, K.L. (2007a). Formal specification and analysis of hardware systems in Timed Chi. *Nordic Journal of Computing, 14*(1–2), 65–86.
21. Man, K.L. (2007b). PAFESD: process algebras for electronic system designs. *The 7th IEEE international conference on ASIC ASICON*. Guilin, P.R. China: IEEE.
22. Man, K.L. (2008a). μCRL: A computer science based approach for specification and verification of hardware circuits. *The 5th IEEE international soc design conference*. Busan, South Korea: IEEE.
23. Man, K.L. (2008b). PAFSV: a process algebraic framework for systemverilog. *International multiconference on computer science and information technology*. Wisla, Poland: IEEE.
24. Man, K.L. (2008c). Specification and analysis of hardware systems using timed process algebras. *Journal of the World Scientific and Engineering Academy and Society Transactions on Electronics, 4*(4), 71–80.
25. Man, K.L. (2009a). Specification and verification of hardware circuits using μCRL. *International Journal of Computer Sciences and Engineering Systems, 2*(3).
26. Man, K.L., Fedeli, A., Mercaldi, M., Boubekeur, M., & Schellekens, M.P. (2007). SC2SCFL: automated SystemC to $SystemC^{\mathbb{FL}}$ translation. *The 7th international symposium on systems, architectures, modeling and simulation*, Lecture Notes in Computer Science 4599 (pp. 34–45). Springer, Germany.
27. Man, K.L., Fedeli, A., Mercaldi, M., & Schellekens, M.P. (2006). $SystemC^{\mathbb{FL}}$: an infrastructure for a TLM formal verification proposal (with an overview on a tool set for practical formal verification of SystemC descriptions). *The 4th east-west design & test workshop EWDTS*. Sochi, Russia: IEEE.
28. Man, K.L., Mercaldi, M., Garberoglio, F., Trischitta, A., Lai, H.Y., & Ho, C.M. (2008a). $SystemC^{\mathbb{FL}}_{tlm}$: motivation and development. In *The proceedings of the international multiconference of engineers and computer scientists 2008*. Hong Kong.
29. Man, K.L., Reniers, M.A., & Cuijpers, P.J.L. (2005). Case studies in the hybrid process algebra HyPA. *International Journal of Software Engineering and Knowledge Engineering, 15*(2), 299–305.
30. Man, K.L., & Schellekens, M.P. (2007a). Analysis of a mixed-signal circuit in hybrid process algebra ACP^{srt}_{hs}. In S.I. Ao, O. Castillo, & C. Douglas (Eds.), *The proceedings of the international multiconference of engineers and computer scientists 2007 volume I, IMECS '07, March 21–23, 2007, Hong Kong*, LNECS (pp. 568–573). International Association of Engineers, Newswood Limited.
31. Man, K.L., & Schellekens, M.P. (2007b). Analysis of a mixed-signal circuit in hybrid process algebra ACP^{srt}_{hs} (extended and revised version). *International Journal of Engineering Letters, 15*(2), 317–326.

32. Man, K.L., & Schellekens, M.P. (2007c). Mathematical modelling of digital hardware systems in Timed Chi. *The 26th IASTED international conference on modelling, identification and control.* Innsbruck, Austria.

33. Man, K.L., & Schellekens, M.P. (2008). *Current trends in intelligent systems and computer engineering*, Chapter Interoperability of Performance and Functional Analysis for Electronic System Designs in Behavioural Hybrid Process Calculus (BHPC). Springer, Germany.

34. Man, K.L., & vander Wulp, J. (2008). Specification and analysis of hardware designs using mCRL2. *The 21st IEEE Canadian conference on electrical and computer engineering.* Niagara Falls, Canada: IEEE.

35. Man, K.L. (2009b). PAFSV: formalisation of SystemVerilog. *International Journal of Computers, Systems and Signals, 10*(1), 3–26.

36. Man, K.L., & Mercaldi, M. (2009). $SystemC_{\mathrm{tlm}}^{\mathbb{FL}}$: the successor of $SystemC^{\mathbb{FL}}$. *International Journal of Computer and Electrical Engineering, 1*(4).

37. Man, K.L., Mercaldi, M., Leung, H.L., & Huang, J. (2008b). Performance and functional analysis of tlm models in the she methodology. *The proceedings of IEEE international conference on computer science and software engineering.* Wuhan, China: IEEE.

38. Man, K.L., Mercaldi, M., Leung, H.L., & Huang, J. (2009). She methodology: combining performance and functional analysis of tlm models. *International Journal of Computer and Electrical Engineering, 1*(4).

39. μCRL (2009). http://homepages.cwi.nl/~mcrl/.

40. Nicollin, X., & Sifakis, J. (1994). The algebra of timed processes, ATP: theory and application. *Information and Computation, 114,* 131–178.

41. OpenModelica System website (2009). Openmodelica system.

42. PAT (2009). PAT: process analysis toolkit. http://www.comp.nus.edu.sg/pat.

43. Schonenberg, M.H. (2006). Discrete simulation of behavioural hybrid process algebra. Master's thesis, University of Twente.

44. SHE (2009). SHE: hardware/software systems. http://www.es.ele.tue.nl/poosl.

45. Shyamasundar, R.K., Doucet, F., Gupta, R., & Kruger, I.H. (2007). Compositional reactive semantics of SystemC and verification in rulebase. *The proceedings of the workshop on next generation design and verification methodologies for distributed embedded control systems.*

46. Silva, B.I., Richeson, K., Krogh, B.H., & Chutinan, A. (2000). Modeling and verification of hybrid dynamical system using CheckMate. In S. Engell, S. Kowalewski, & J. Zaytoon (Eds.), *Hybrid dynamical systems – proceedings of 4th international conference on automation of mixed processes* (pp. 323–328). Dortmund.

47. SPIN (2009). On-the-fly, LTL model checking with SPIN. www.spinroot.com.

48. Traulsen, Claus, Cornet, Jerome, Moy, Matthieu., & Maraninchi, Florence (2007). A SystemC/TLM semantics in Promela and its possible applications. *The 14th workshop on model checking software SPIN.*

49. van Beek, D.A., Man, K.L., Reniers, M.A., Rooda, J.E., & Schiffelers, R. R.H. (2005). Syntax and semantics of timed Chi. Technical report CS-report 05-09, Eindhoven University of Technology, Department of Computer Science, The Netherlands.

50. van Beek, D.A., Man, K.L., Reniers, M.A., Rooda, J.E., & Schiffelers, R. R.H. (2006a). Syntax and consistent equation semantics of hybrid Chi. *Journal of Logic and Algebraic Programming, 68*(1–2), 129–210.

51. van Beek, D.A., Man, K.L., Reniers, M.A., Rooda, J.E., & Schiffelers, R.R.H. (2006b). Deriving simulators for hybrid chi models. *The proceedings of IEEE symposium on computer-aided control systems design*, Munich, Germany: IEEE.

52. van Putten, A. (2006). Behavioural hybrid process calculus parser and translator to modelica. Master's thesis, University of Twente.

53. Viaud, E., Pecheux, F., & Greiner, A. (2006). An efficient TLM/T modeling and simulation environment based on conservative parallel discrete event principles. In *The proceedings of design, automation and test in Europe.*

Chapter 33
Behavioural Hybrid Process Calculus for Modelling and Analysis of Hybrid and Electronic Systems

Tomas Krilavičius and Ka Lok Man

Abstract Progress in electronics requires novel techniques for modelling, design and production. Formal modelling is a well-known tool for modelling and analysis of diverse systems. Recent studies of the hybrid process algebras show that their application in electronics could improve design quality and reliability of the electronic systems. We present Behavioural Hybrid Process Calculus (BHPC), a formalism for modelling and analysis of hybrid systems combining process algebraic techniques and the behavioural approach in such a way providing means for modelling of instantaneous changes as well as continuous evolution. It is supported by BHAVE TOOLSET, a tool collection for modelling, simulation and visualisation of hybrid systems. The toolset contains MSP-SVG, a tool that provides a novel way for visualisation of hybrid systems – Message Sequence Plots and related tool. Moreover, we illustrate application of BHPC, and in particular, show how it can be effectively used for modelling and analysis of electronic systems in combination with Open-Modelica System.

Keywords Hybrid systems · Process algebras · Visualisation · Electronics

1 Introduction

Process algebras/calculi [1, 7, 13] are formal languages in *Computer Science* that have formal syntax and semantics for specifying and reasoning about different systems. They are also useful tools for verification of various systems. Generally speaking, process algebras describe the behaviour of processes and provide operations that allow to compose systems in order to obtain more complex systems. Moreover, the analysis and verification of systems described using process algebras can be partially or completely carried out by mathematical proofs using equational theory.

T. Krilavičius (✉) and K.L. Man
Vytautas Magnus University, K. Donelaicio g. 58, LT-44248 Kaunas, Lithuania
and Solari, Hong Kong
e-mail: t.krilavicius@if.vdu.lt; pafesd@gmail.com

S.-I. Ao et al. (eds.), *Intelligent Automation and Computer Engineering*,
Lecture Notes in Electrical Engineering 52, DOI 10.1007/978-90-481-3517-2_33,
© Springer Science+Business Media B.V. 2010

In simple words, process algebras are theoretical frameworks for the formal specification and analysis of the behaviour of various systems. Serious efforts [2,4,9,19] have been made in the past to deal with various systems (e.g., *discrete event systems*, *real-time systems* and *hybrid systems*) in a process algebraic way. Over the years, process algebras have been successfully used in a wide range of problems and in practical applications in both academia and industry for analysis of many different systems.

Hybrid systems are systems that exhibit discrete and continuous behaviour. Such systems have proved fruitful in a great diversity of engineering application areas including air-traffic control, automated manufacturing, and chemical process control.

Recently, through novel language constructs and well-defined formal semantics in a standard *Structured Operational Semantics* (SOS) style [15], several process algebras/calculi (HyPA [4], Hybrid Chi [19], ACP_{hs}^{srt} [2] and Behavioural Hybrid Process Calculus-BHPC [9]) have been developed for hybrid systems.

Computer simulation is a powerful tool for analysing and optimising real-world systems with a wide range of successful applications. It provides an appealing approach for the analysis of dynamic behaviour of processes and helps decision makers identify different possible options by analysing enormous amounts of data.

Amongst the above-mentioned process algebras/calculi, BHPC was specifically designed for the description of the dynamic behaviour of hybrid systems along with a powerful simulator. Currently, simulation results obtained by means of the BHPC simulator can also be visualised and analysed via *Message Sequence Charts* (MSC) [17].

On the other hand, mathematically, the behaviour of electronic system design (e.g., digital, analog and mixed-signal design) can be described by discrete variables, continuous variables and a set of differential equations, whereas switching-modes can be used for modelling mixed models (i.e., mixed-signal design). In simple words, digital, analog and mixed-signal design can be mathematically described as hybrid systems (with various level of abstraction) by nature.

Several attempts (e.g., [12]) have been made over the last few years to apply process algebraic formalisms in the context of the formal specification and analysis of electronic system design; and analysis results are shown to be very promising.

In this book chapter, we first give an overview of BHPC and its toolset; and then present the latest development of BHPC including the toolset which is well-suited for modelling and analysis of hybrid systems as well as electronic system design. This book chapter is organised as follows. Section 2 provides the theoretical foundation of BHPC and an overview of BHPC toolset is given in Section 3. The latest development of BHPC and its toolset (including several applications) is presented in Section 4. Section 5 illustrates the application of BHPC for electronic system design. Finally, a summary of this book chapter is drawn in Section 6.

2 Behavioural Hybrid Process Algebra

One of the useful techniques for simulation of combined functional analysis that includes continuous evolution and discrete changes, is Behavioural Hybrid Process Calculus (BHPC) [3,9], an extension of classical process algebra that is suitable for the modelling and analysis of continuous and hybrid dynamical systems and can be seen as a generalisation of the behavioural approach [16] in a hybrid setting. The main strengths of the BHPC are the following.

Sound Mathematical Foundations BHPC has sound mathematical foundations. It means that rigorous reasoning can be applied to investigate diverse properties of models.

Behavioural Approach Continuous evolution in BHPC is defined in the behavioural setting [16], i.e., in contrast to other hybrid process algebras (Hybrid χ [19], HyPA [4], ACP_{hs}^{srt} [2]) it is define using trajectories (solutions of differential equations is only one case of defining trajectories), not solutions of differential equations, in such a way making it more general.

Separation of Concerns Continuous and discrete behaviours are specified orthogonally, therefore they can be changed and analysed separately as well as in hybrid setting.

Bisimulation Is Congruence in BHPC It means, that bisimilar (processes, that exhibit the same observable behaviour up-to branching structure) processes can be substituted, and it does not change systems behaviour.

Tools Support BHPC is supported by Bhave toolset, see Section 3.
    We present main ideas of the BHPC in this section, see [9] for the details.

Trajectories We define trajectories over bounded time intervals $(0, t]$, and map to a *signal space* $\mathbb{W} = (W_1 \times \cdots \times W_n, (q_1, \ldots, q_n))$. Components of the signal space $W \in \mathcal{W}$ correspond to the different aspects of the continuous-time behaviour, such as temperature or pressure, and are associated with *trajectory qualifiers* $q_i \in \mathcal{T}$ that identify them. A *trajectory* in signal space $\mathbb{W}$ is a function $\varphi : (0, t] \to W_1 \times \cdots \times W_n$, where $t \in \mathbb{R}_+$ is the duration of the trajectory. We define conditions on the *end-points* of trajectories or the *exit conditions*. $\Downarrow$ denotes such conditions, as the restrictions on the set of trajectories: $\Phi \Downarrow Pred_{exit} = \{\varphi : (0, u] \to W_1, \ldots, W_n \in \Phi \mid Pred_{exit}(\varphi(u))\}$, where u is a time parameter, Φ is a set of trajectories and $Pred_{exit}(\varphi(u))$ is a predicate that defines restrictions. The set of trajectories Φ can be defined in different ways, e.g., by ODE/DAE. See [9] for the formal treatment.

Hybrid Transition System All behaviours of BHPC specification are defined by a *hybrid transition system* $HTS = \langle S, \mathcal{A}, \to, \mathbb{W}, \Phi, \to_c \rangle$, where

- S is a state space.
- $\mathcal{A}$ is a *finite set of (discrete) actions names.*

- $\to \subseteq S \times \mathcal{A} \times S$ is a *discrete transition relations*, where $a \in \mathcal{A}$, we will denote it $s \xrightarrow{a} s'$.
- $\mathbb{W}$ is a *signal space*.
- Φ is a *set of trajectories*.
- $\to_c \subseteq S \times \Phi \times S$ is a *continuous transition relation*, where $\varphi \in \Phi$ are trajectories, we will denote continuous transitions $s \xrightarrow{\varphi} s'$ for the convenience.

Language A core language is used for defining evolution and interaction of systems

$$B ::= \mathbf{0} \;\Big|\; a \cdot B \;\Big|\; [f \mid \Phi] \cdot B \;\Big|\; \sum_{i \in I} B_i \;\Big|\; B \parallel_A^H B \;\Big|\; P$$

We will require a *consistent signal flow*, i.e., only the parallel composition is allowed to change the set of trajectory qualifiers in the process.

Only a subset of complete language is introduced, see [9] for auxiliary operators, such as renaming or hiding. Moreover, other operators can be defined on top of the core language for convenience. We demonstrate it by introducing *parametrised action prefix* and *guard*.

Stop 0 is the process that does not exhibit any behaviour.

Action Prefix $a \cdot B$ first performs a and then engages in B. A special *silent action* τ defines directly unobservable behaviour, and is usually used to specify a non-determinism (e.g., as *internal actions* in [13], p. 37–43).

We will use parametrisation of action prefix like in [13], p. 53–58$a(v : V) \cdot B(v) = \sum_{v \in V} a(v) \cdot B(v)$. Parametrisation is frequently used for *value passing*.

Trajectory Prefix $[f \mid \Phi] \cdot B\,(f)$, where f is a trajectory variable, starts with a trajectory or a prefix of a trajectory from the set of trajectories Φ. If a trajectory or a part of it was taken and there exists a continuation of the trajectory, then the system can continue with a trajectory from the trajectory continuations set. If a whole trajectory was taken, then the system may continue with B.

Choice $\sum \{B(v) \mid v \in I\}$ is a generalised *nondeterministic* choice of processes (I is an arbitrary index set). It chooses before taking an action prefix or trajectory prefix. Binary version of choice is denoted $B_1 + B_2$.

Parallel Composition $B_1 \parallel_A^H B_2$, where A and H are sets of synchronising action names and trajectory qualifiers, respectively, models the behaviour of two parallel processes. Synchronisation on actions has an interleaving semantics. Trajectory prefixes can evolve only in parallel, and only if the evolution of the coinciding trajectory qualifiers is equal.

Recursions allows defining processes in terms of each other, as in the equation $P = B$, where P is the process identifier and B is a process expression that may only contain actions and signal types of B.

Guard $\langle \mathcal{P}red \rangle$ operator checks some conditions explicitly, and and they are not satisfied, stops the progress of process.

$$\langle \mathcal{P}red(\overline{x}) \rangle \cdot B(\overline{x}) = \sum_{\overline{w} \models \mathcal{P}red(\overline{w})} B\,(\overline{w})$$

Here $\overline{x}$ are process parameters variables. Behaviour is very simple, i.e., if a transition can be taken, then it is taken, if and only if the guard is satisfied.

Strong Bisimulation for hybrid transition systems requires both systems to be able to execute the same trajectories and actions and to have the same branching structure.

The *hybrid strong bisimulation relation* (equivalence) defined for the HTS is a congruence relation w.r.t. all operations defined above [9]. Hence, bisimilar components can be interchanged without changing systems behaviour, and that can be effectively employed while building and improving systems (models).

3 Bhave Toolset

BHPC is supported by Bhave toolset [8]. The toolset allows modelling, simulation and visualisation of the hybrid models [10]. It consists of several tools.

BHPCC is a BHPC language parser that translates BHPC specification into the internal specification language.

Discrete Bhave allows discrete simulation of the BHPC specifications. Semantics of the discrete version and the tool are defined in [11].

Bhave Prototype allows hybrid simulation of BHPC specification. Currently, a new version of Bhave simulator is under development. It directly parses BHPC language, and supports discrete simulation. It is currently being augmented with the facilities for continuous and hybrid simulation. A snapshot of the system is available at bhpc-simulator.sourceforge.net.

BHPC2Mod allows to translate a restricted set of BHPC models to Modelica [6] language, and then simulate them using Dymola [5] or OpenModelica [14]. However, because Modelica does not have formal semantics, translation does not necessary preserves all the properties. Moreover, parallel composition is not translated [20].

MSP-SVG is a visualisation tool that implements novel Message Sequence Plots (MSP) [9, 18]. It is available at http://msp-svg.sourceforge.net/. See Section 4 for more details about MSP-SVG.

The current versions of both tools (BHAVE and MSP-SVG) are built not just as a prototypes, but also as a hybrid "sand-box", a place to experiment with BHPC and related developments. Consequently, the architecture and implementation of the tools are being designed in such a way that it is easy to accommodate the changes and to test the algorithms developed for diverse hybrid process algebras or MSP-based visualisation techniques.

Our plans include further development of the process algebra and BHAVE TOOLSET. We are planning to augment BHAVE with continuous-time evolution, as it was done with BHAVE prototype, and then integrate it with MSP-SVG. For MSP-SVG several augmentations are in the pipeline: **folding/unfolding** would allow to hide (fold) or expose (unfold) parts of MSP, e.g., certain processes or participants of parallel composition; **forking** of processes can be employed to visualise parallel composition; action and trajectory prefixes can be used to **decorate** respective MSP elements and in such a way increase descriptivity of visualisation; **recursive calls** can be depicted as boxes with a new process identifier; an (additional) legend can be used to describe *renaming*.

3.1 Application of BHPC

Advanced Thermostat Simple thermostat controls temperature in the room by switching a heater on and off at the preset temperature intervals. When the heater is off, the temperature decreases according to the exponential function $l(t) = \theta e^{Kt}$, where t is time, l is the temperature in the room, θ is the initial temperature, and K is a constant determined by the room. When the heater is on, the temperature increases according to the function $l(t) = \theta e^{-Kt} + h(1 - e^{-Kt})$, where h is a constant that depends on the power of the heater. The temperature should be maintained between $temp_{min}$ and $temp_{max}$. Temperatures $temp_{on}$ and $temp_{off}$ are the minimal and maximal thresholds, when the heater can be turned on and off, respectively.

$$\mathbf{Thermostat}(l_0) \triangleq \mathbf{ThOff}(l_0)$$

$$\mathbf{ThOff}(l_0) \quad \triangleq [l \mid \Phi_{\mathrm{Off}}(l_0) \Downarrow tempOn \geq l \geq tempMin] \cdot \mathsf{on} \cdot \mathbf{ThOn}(l)$$

$$\mathbf{ThOn}(l_0) \quad \triangleq [l \mid \Phi_{\mathrm{On}}(l_0) \Downarrow tempOff \leq l \leq tempMax] \cdot \mathsf{off} \cdot \mathbf{ThOff}(l)$$

$$\Phi_{\mathrm{Off}}(l_0) \quad = \{l : (0, t] \to \mathbb{R} \mid l(0) = l_0, \dot{l} = -Kl\}$$

$$\Phi_{\mathrm{On}}(l_0) \quad = \{l : (0, t] \to \mathbb{R} \mid l(0) = l_0, \dot{l} = K(h - l)\}$$

Thermostat consists of two processes. In process **ThOff** the heater is off and the trajectory prefix defines the temperature fall. When the temperature reaches the interval $[tempOn, tempMin]$, the process can perform action on and switch to **ThOn**, which defines heating.

It is easy to upgrade such thermostat without changing the specification itself. Let us add a controller that observes temperature and forces the thermostat to switch on and off at exactly tOn and $tOff$, correspondingly, and compose it with thermostat.

$$\mathbf{Ctrl}(l_0) \triangleq [l \mid \mathsf{any}(l_0) \Downarrow l = tOn] \cdot \mathsf{on} \cdot [l \mid \mathsf{any}(l_0) \Downarrow l = tOff] \cdot \mathsf{off} \cdot \mathbf{Ctrl}(l)$$

$$\mathbf{UpgradedThermostat}(l_0) \triangleq \mathbf{Thermostat}(l_0) \parallel^l_{\mathsf{on,off}} \mathbf{Ctrl}(l_0)$$

$\mathsf{any}(l)$ is a special function that models an observer, i.e., it accepts any behaviour for l. It works only in parallel composition. Such trajectory prefix just adds exit conditions to the parallel composition of trajectory prefixes.

The results of simulation of the simple and controlled thermostat are depicted in Fig. 1. Dashed line depicts the evolution of the simple thermostat and solid line depicts the evolution of the coupled version.

Fluid Level Control Consider the two tanks model, where both tanks have one common fluid source that provides fluid at the rate of l_{in} units per second. Through

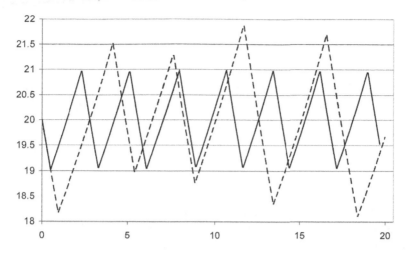

Fig. 1 Simulation of the advanced thermostat

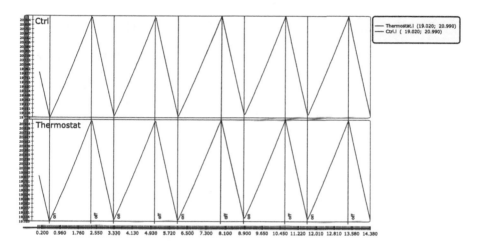

Fig. 2 Visualisation of the simulation of the advanced thermostat using MSP-SVG

a pipe, the fluid source can be directed either to the left tank or to the right tank. Both tanks have openings at the bottom, and from the tanks water drains at the rates of d_{left} and d_{right} units per second, respectively. Initially, the tanks contain l^0_{left} and l^0_{right} units of fluid, respectively. The pipe can switch between the tanks instantaneously. The objective is to keep the fluid volumes in the interval $(l_{\text{min}}, l_{\text{max}})$.

Let l_{left} and l_{right} are volumes in the left and right tanks, correspondingly.

$$\textbf{TwoTanks}(l^0_{\text{left}}, l^0_{\text{right}}) \triangleq$$

$$\left[l_{\text{left}}, l_{\text{right}} \mid \Phi_{\text{left}}(l^0_{\text{left}}, l^0_{\text{right}}) \Downarrow l_{\text{left}} = l_{\text{max}} \vee l_{\text{right}} = l_{\text{min}} \right] . \textsf{FillRight}.$$

$$\left[l_{\text{left}}, l_{\text{right}} \mid \Phi_{\text{right}}(l_{\text{left}}, l_{\text{right}}) \Downarrow l_{\text{left}} = l_{\text{min}} \vee l_{\text{right}} = l_{\text{max}} \right] \cdot \textsf{FillLeft}.$$

TwoTanks$(l_{\text{left}}, l_{\text{right}})$

$$\Phi_{\text{left}}(l_{\text{left}}^0, l_{\text{right}}^0) \triangleq \{l_{\text{left}}, l_{\text{right}} : (0, t] \to \mathbb{R} \mid$$
$$l_{\text{left}}(0) = l_{\text{left}}^0, l_{\text{right}}(0) = l_{\text{right}}^0, \dot{l}_{\text{left}} = d_{\text{left}} + l_{\text{in}}, \dot{l}_{\text{right}} = d_{\text{right}}\}$$
$$\Phi_{\text{right}}(l_{\text{left}}^0, l_{\text{right}}^0) \triangleq \{l_{\text{left}}, l_{\text{right}} : (0, t] \to \mathbb{R} \mid$$
$$l_{\text{left}}(0) = l_{\text{left}}^0, l_{\text{right}}(0) = l_{\text{right}}^0, \dot{l}_{\text{left}} = d_{\text{left}}, \dot{l}_{\text{right}} = d_{\text{right}} + l_{\text{in}}\}$$

This system is of interest, because by ignoring physical reality one can devise a controller that keeps both water levels within the required bounds: whenever l_{left} falls to l_{min}, direct the pipe to the left tank, and whenever l_{right} falls to l_{min}, direct the pipe to the right tank (or, corresponding switching at the l_{max}). However, such a controller cannot be realised physically, because it would cause the pipe to switch back and forth infinitely often within a finite amount of time, if $d_{\text{left}} + d_{\text{right}} \neq l_{\text{in}}$.

To demonstrate compositional modelling advantages we propose a slightly different version of the same system. In this specification l is water level in the tank, d_{out} is a drain rate and l_{in} is an inflow rate.

TankIn$(l_0, d_{\text{out}}, l_{\text{in}}) \triangleq \left[l \mid l(0) = l_0, \dot{l} = d_{\text{out}} \Downarrow \text{true}\right] \cdot \text{off} \cdot \textbf{TankOut}(l, d_{\text{out}}, l_{\text{in}})$

TankOut$(l_0, d_{\text{out}}, l_{\text{in}}) \triangleq \left[l \mid l(0) = l_0, \dot{l} = d_{\text{out}} + l_{\text{in}} \Downarrow \text{true}\right] \cdot \text{off} \cdot \textbf{TankIn}(l, d_{\text{out}}, l_{\text{in}})$

Controller$(l_{\text{left}}^0, l_{\text{right}}^0) \triangleq \left[l_{\text{left}}, l_{\text{right}} \mid \text{any}(t) \Downarrow l_{\text{left}} = l_{\text{max}} \vee l_{\text{right}} = l_{\text{min}}\right] \cdot \text{fill}_{\text{right}} \cdot$
$\qquad \left[l_{\text{left}}, l_{\text{right}} \mid \text{any}(t) \Downarrow l_{\text{left}} = l_{\text{min}} \vee l_{\text{right}} = l_{\text{max}}\right] \cdot \text{fill}_{\text{left}} \cdot$
$\qquad \textbf{Controller}(l_{\text{left}}, l_{\text{right}})$

System$(l_{\text{left}}^0, l_{\text{right}}^0, dl_{\text{out}}, dr_{\text{out}}, ll_{\text{in}}, lr_{\text{in}}) \triangleq$
$\qquad (\textbf{TankOn}(l_{\text{left}}^0, dl_{\text{out}}, ll_{\text{in}}) \left[l_{\text{left}}/l, dl_{\text{out}}/d_{\text{out}}, ll_{\text{in}}/l_{\text{in}}, \text{fill}_{\text{left}}/\text{on}, \text{fill}_{\text{right}}/\text{off}\right] \parallel$
$\qquad \textbf{TankOn}(l_{\text{right}}^0, dr_{\text{out}}, lr_{\text{in}}) \left[l_{\text{right}}/l, dr_{\text{out}}/d_{\text{out}}, lr_{\text{in}}/l_{\text{in}}, \text{fill}_{\text{right}}/\text{on}, \text{fill}_{\text{left}}/\text{off}\right])$
$\qquad \parallel_{\text{fill}_{\text{left}}, \text{fill}_{\text{right}}}^{l_{\text{left}}, l_{\text{right}}} \textbf{Controller}(l_{\text{left}}, l_{\text{right}})$

Simulation results for $d_{\text{left}} + d_{\text{right}} \geq l_{\text{in}}$ are depicted in Fig. 3.

4 Visualisation of Hybrid Evolutions: Message Sequence Plots and MSP-SVG

Simulation results usually visualise the evolution of the system in time. Event traces or message sequence charts (MSC) [17] adequately represent discrete system behaviour, and graphs are convenient for the ordinary continuous systems. However, in hybrid systems we have both the evolution of system variables and events. Hence a combined view is crucial to fully analyse hybrid system behaviour. See [9], pp. 118–124 for the details.

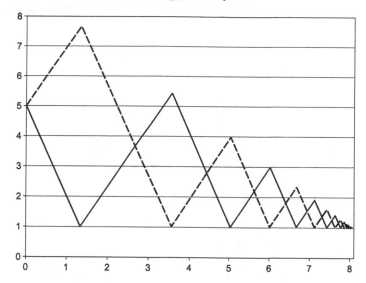

Fig. 3 Simulation of the fluid level control

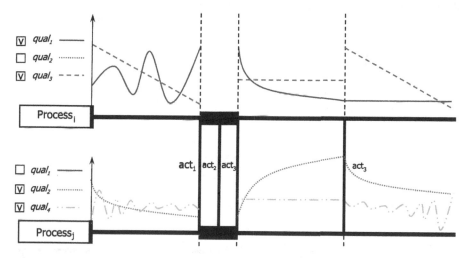

Fig. 4 MSP example

MSP has two compounds: message-sequence charts rotated 90° combined with plots. We explain MSP by an example depicted in Fig. 4. **Plots over time-lines** depict continuous-time evolution. **A legend** allows selecting qualifiers of interest, that are depicted in the plot. If several processes evolve concurrently, the synchronising qualifiers appear for both processes. In Fig. 4 qualifiers $qual_1$, $qual_2$, $qual_3$ and $qual_4$ are depicted. **Process$_i$** is related with qualifiers $qual_1$, $qual_2$ and $qual_3$, and only qualifiers $qual_1$ and $qual_3$ are selected to be visible. **Process$_j$** is related with qualifiers $qual_1$, $qual_2$ and $qual_4$, and qualifiers $qual_2$ and $qual_4$ are selected to be visible.

Single horizontal lines connected to the corresponding boxes with process identifiers, represent processes and the *time-line* (or *life-line* in MSC terminology). Time is assumed to flow to the right along each time-line at the same speed. **Process**$_i$ and **Process**$_j$ are represented by the horizontal lines and boxes with processes identifiers in the example.

Labelled vertical lines going across time-lines represent communication, i.e. (parameterised), action prefixes in BHPC. Notice that we use simple lines instead of arrows, because communication in BHPC is not directed. Communication of **Process**$_i$ and **Process**$_j$ consists of actions act_1, act_2 and act_3.

Triple horizontal time-lines depict suspension of the time-flow. Single actions are placed on the time-line at the time that relates to their moment of occurrence. A sequence of actions occurs at one moment in time, when there is no continuous behaviour between the actions. We suspend the flow of time to allow insight in the ordering of these actions. In the example, suspension of the time is depicted on the time-line as three parallel solid lines.

Figure 4 contains all information that would be available in an ordinary plot. Correspondingly, all information that is visible in message sequence charts, is also visible in MSP. Furthermore, in MSP all processes and communication between them are visualised. The proposed technique can be easily adopted to other hybrid system modelling frameworks with minimal changes, e.g., if communication is directed, arrows can be used to depict it.

Additional Notation Some additional notation could provide even more information about evolution of the system under investigation. We exemplify a potential use of some additional notation by using an upgraded thermostat as an example. The evolution with explicit recursive calls is depicted in Fig. 5.

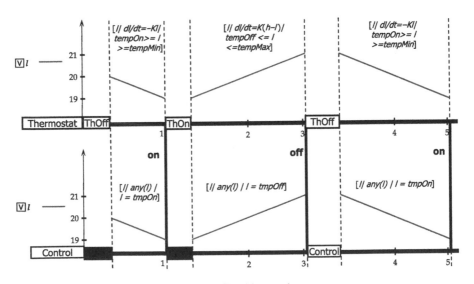

Fig. 5 Upgraded Thermostat evolution in MSP with recursion

Adding Details Details of the hybrid model can be added to the results to make their relation more apparent. For example we can decorate the results with process expressions, e.g., @, action- and trajectory-prefixes at the corresponding parts of communication lines and graphs, respectively.

Recursion Recursive calls can be depicted as boxes with a new process identifier on the time-line of the calling process. The width of the box is determined by the length of the process identifier and does not denote time-flow. We also suspend the time-flow of other processes by this width, to keep the time-lines of all processes synchronised.

Renaming Renaming of qualifiers can be depicted in the legend, or as detail along the plot. The latter gives the most distinct relation between the qualifier in the legend and its renaming for that part of the plot.

Introduction of Parallelism At the introduction of parallelism, we fork the original process into two processes. Again, we keep the time-lines of all processes in the MSP synchronised. The forked time-lines receive a process identifier when a recursive call is executed.

From Fig. 5 it is clear that the developer and user will have to choose between the amount of represented information and clarity. Therefore, the user should be allowed to choose what the user expects to see, and be able to hide (fold) or expose (unfold) parts of MSP.

For example, folding and unfolding can be introduced to control visibility of the parallel composition components. When the components are folded into a single process, the communication between these processes should be depicted in some other ways, e.g., as lines with action names that only cross this process line perpendicularly.

In Fig. 2 depicted an evolution of the advanced thermostat using a proof of concept tool MSP-SVG. It is the same evolution, as depicted in Fig. 1, but evolution of the processes is depicted separately. Moreover, synchronising actions on and off show behaviour change moments. The tool still needs some improvements to produce figures like Figs. 4 and 5.

Executable and source code of MSP-SVG are available at sourceforge.net/projects/msp-svg/.

5 Application of BHPC for Electronic Systems

A Half Wave Rectifier Circuit Model in BHPC A wide range of common electrical/electronic devices and systems can be easily and effectively modelled in BHPC. To allow the simulation of BHPC specifications in various platforms/environments, as mentioned previously, the tool BHPC2Mod translates a subset of BHPC specifications to the corresponding models described in Modelica which be further simulated using OpenModelica System (without a formal semantics defined but along

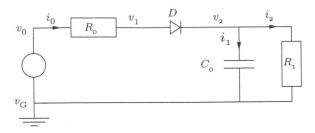

Fig. 6 Half wave rectifier circuit

with a very powerful simulator). In this section, we apply BHPC and BHPC2Mod
to specify and analyse a half wave rectifier circuit (revised from [12]).

Figure 6 depicts the half wave rectifier circuit, consisting of an ideal diode D,
two resistors with resistance R_0 and R_1, respectively, a capacitor with capacity C_0,
a voltage source with voltage v_0 and a ground voltage v_G. In the **off** mode the *ideal
diode* voltage must be ≤ 0 and the current equals to zero. In the **on** mode, the voltage
equals to zero and the current must be ≥ 0, i.e.

$$\textbf{on:}\quad v_1 = v_2 \wedge i_0 \geq 0 \qquad \textbf{off:}\quad v_2 \geq v_1 \wedge i_0 = 0$$

The state equations of other components of the half wave rectifier circuit are given
by $v_0 = F_{\text{time}}$, $v_0 - v_1 = i_0 R_0$, $C_0(\dot{v_2} - \dot{v_G}) = i_1$, $v_2 - v_G = i_2 R_1$, $v_G = 0$ and
$i_0 = i_1 + i_2$. Note that F_{time} is an arbitrary function of time, $v_0, i_0, v_1, i_1, v_2, i_2, v_G$
are continuous variables and R_0, R_1, C_0 are constants.

BHPC specification of the half wave rectifier circuit consists of the several
processes.

$$\text{IdealDiode}(i_0{}^\circ, v_1{}^\circ, v_2{}^\circ) \quad \triangleq \text{IdealDiodeOff}(i_0{}^\circ, v_1{}^\circ, v_2{}^\circ)$$

$$\text{IdealDiodeOff}(i_0{}^\circ, v_1{}^\circ, v_2{}^\circ) \triangleq [i_0, v_1, v_2 \mid \Phi_{\text{off}}(i_0{}^\circ, v_1{}^\circ, v_2{}^\circ) \Downarrow i_0 \geq 0]\cdot$$
$$\text{on} \cdot \text{IdealDiodeOn}(i_0, v_1, v_2)$$

$$\text{IdealDiodeOn}(i_0{}^\circ, v_1{}^\circ, v_2{}^\circ) \triangleq [i_0, v_1, v_2 \mid \Phi_{\text{on}}(i_0{}^\circ, v_1{}^\circ, v_2{}^\circ) \Downarrow v_2 \geq v_1]\cdot$$
$$\text{off} \cdot \text{IdealDiodeOff}(i_0, v_1, v_2)$$

$$\Phi_{\text{off}}(i_0{}^\circ, v_1{}^\circ, v_2{}^\circ) \quad = \{i_0, v_1, v_2 : (0, t] \mapsto \mathbb{R} \mid$$
$$i_0(0) = i_0{}^\circ, v_1(0) = v_1{}^\circ, v_2(0) = v_2{}^\circ, v_2 \geq v_1, i_0 = 0\}$$

$$\Phi_{\text{on}}(i_0{}^\circ, v_1{}^\circ, v_2{}^\circ) \quad = \{i_0, v_1, v_2 : (0, t] \mapsto \mathbb{R} \mid$$
$$i_0(0) = i_0{}^\circ, v_1(0) = v_1{}^\circ, v_2(0) = v_2{}^\circ, v_2 = v_1, i_0 \geq 0\}$$

$$\text{Generator}(v_0{}^\circ) \quad \triangleq [v_0 \mid \Phi_{\text{Generator}}(v_0{}^\circ) \Downarrow true] \cdot \text{generator} \cdot \text{Generator}(v_0)$$

$$\Phi_{\text{Generator}} \quad = \{v_0 : (0, t] \mapsto \mathbb{R} \mid v_0(0) = v_0{}^\circ, v_0 = F_{\text{time}}\}$$

$\text{Others}(v_0{}^\circ, v_1{}^\circ, v_2{}^\circ, v_G{}^\circ, i_0{}^\circ, i_1{}^\circ, i_2{}^\circ) \triangleq$

$\quad [v_0, v_1, v_2, v_G, i_0, i_1, i_2 \mid \Phi_{\text{others}}(v_0{}^\circ, v_1{}^\circ, v_2{}^\circ, v_G{}^\circ, i_0{}^\circ, i_1{}^\circ, i_2{}^\circ) \Downarrow \textit{true}]\cdot$

$\quad \text{others} \cdot \text{Others}(v_0, v_1, v_2, v_G, i_0, i_1, i_2)$

$\Phi_{\text{others}}(v_0{}^\circ, v_1{}^\circ, v_2{}^\circ, v_G{}^\circ, i_0{}^\circ, i_1{}^\circ, i_2{}^\circ) = \{v_0, v_1, v_2, v_G, i_0, i_1, i_2 : (0, t] \mapsto \mathbb{R} \mid$

$\quad v_0(0) = v_0{}^\circ, v_1(0) = v_1{}^\circ, v_2(0) = v_2{}^\circ, v_G(0) = v_G{}^\circ, i_0(0) = i_0{}^\circ, i_1(0) = i_1{}^\circ,$

$\quad i_2(0) = i_2{}^\circ, v_0 - v_1 = i_0 R_0, C_0(\dot{v}_2 - \dot{v}_G) = i_1, v_2 - v_G = i_2 R_1, v_G = 0, i_0 = i_1 + i_2\}$

$\text{HalfWaveRectifier}(v_0{}^\circ, v_1{}^\circ, v_2{}^\circ, v_G{}^\circ, i_0{}^\circ, i_1{}^\circ, i_2{}^\circ) \triangleq \text{IdealDiode}(i_0{}^\circ, v_1{}^\circ, v_2{}^\circ) \parallel_\emptyset^H$

$\quad (\text{Others}(v_0{}^\circ, v_1{}^\circ, v_2{}^\circ, v_G{}^\circ, i_0{}^\circ, i_1{}^\circ, i_2{}^\circ) \parallel_\emptyset^{\{v_0\}} \text{Generator}(v_0{}^\circ))$

IdealDiode models the switching-mode behaviour of the ideal diode by means of processes IdealDiodeOn and IdealDiodeOff. Initially, the ideal diode is in the **"off"** mode (described by the process IdealDiodeOff) and the trajectory prefix defines the current rise of i_0. When $i_0 \geq 0$, the process may perform action on (an unimportant action name) and switch to the process IdealDiodeOn. Analogously, IdealDiodeOn defines the period of the ideal diode being in the **"on"** mode. Notice that $i_0{}^\circ, v_1{}^\circ$ and $v_2{}^\circ$ are the initial values for i_0, v_1 and v_2 respectively; and off is an unimportant action name.

Others models the behaviour of all components of the half wave rectifier circuit excluding the ideal diode and the generator (i.e., the voltage source with voltage v_0), according to the dynamics defined by the trajectory prefix by Φ_{others}. Notice that $v_G{}^\circ, i_1{}^\circ$ and $i_2{}^\circ$ are the initial values for v_G, i_1 and i_2 respectively; $\textit{true}$ denotes the predicate "$\textit{true}$" and others is an unimportant action name.

Generator models the behaviour of the voltage source with voltage v_0, according to the dynamics defined by the trajectory prefix by $\Phi_{\text{Generator}}$. Notice that $v_0{}^\circ$ is the initial value for v_0 and Generator is an unimportant action name.

HalfWaveRectifier defines the complete systems as a parallel composition of the processes IdealDiode, Others and Generator. Notice that $H = \{v_1, v_2, i_0\}$ is the set of trajectory qualifiers for the synchronisation of trajectories and $\emptyset$ denotes an empty set.

Currently, only a reasonable subset of the BHPC language can be taken for the translation to Modelica. Due to this, applying BHPC2Mod with some manual adaptations, the above half wave rectifier circuit described in BHPC was translated to the corresponding model in Modelica and simulated using the OpenModelica System. A typical simulation result for such a half wave rectifier circuit is depicted in Fig. 7 (the output voltage of the half wave rectifier is never negative and always above some given value).

6 Summary

BHPC and its toolset can be reasonably and effectively used for the formal specification and analysis of hybrid and electronic systems as indicated in this book chapter. The latest development of BHPC including the toolset has also been presented in this book chapter.

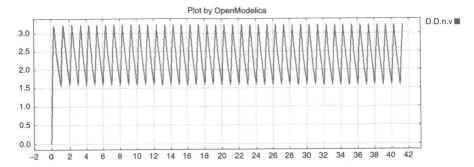

Fig. 7 Simulation of the half wave rectifier circuit using OpenModelica System

Our future work will mainly focus on adding new features and correcting bugs for the BHPC toolset as well as extending BHPC language for supporting the modelling and verification of electronic system design.

References

1. Baeten, J.C.M., & Weijland, W.P. (1990). *Process algebra*, volume 18 of *Camb. Tracts in TCS*. Cambridge, UK: Cambridge University Press.
2. Bergstra, J.A., & Middelburg, C.A. (2005). Process algebra for hybrid systems. *TCS*, 335(2/3), 215–280.
3. Brinksma, E., Krilavičius, T., & Usenko, Y.S. (2005). Process algebraic approach to hybrid systems. In *Proceedings of the 16th IFAC world congress* (pp. 1–6). Prague, Czech Republic.
4. Cuijpers, P.J.L., & Reniers, M.A. (2005). Hybrid process algebra. *JLAP*, 62(2), 191–245.
5. Dymola. (2006). Dynasim. dynasim.se
6. Fritzson, P., & Engelson, V. (1998). Modelica – a unified object-oriented language for system modelling and simulation. In Proceedings of European Conference on Object-Oriented Programming (ECOOP'98). Brussels, Belgium, July 20–24, 1998.
7. Hoare, C.A.R. (1985). *Communicating sequential processes*. Englewood-Cliffs: Prentice-Hall.
8. Krilavičius, T. (2006a). Bhave: simulation of hybrid systems. fmt.cs.utwente.nl/tools/bhave
9. Krilavičius, T. (2006b). *Hybrid techniques for hybrid systems*. PhD thesis, University of Twente.
10. Krilavičius, T. (2008). Simulation of mechatronic systems using Behavioural Hybrid Process Calculus. *Electronics and Electrical Engineering*, 1(81), 45–48.
11. Krilavičius, T., & Schonenberg, H. (2005). Discrete simulation of behavioural hybrid process calculus. In P.M.E. Bra, & J.J. van Wijk (Eds.), *IFM2005 doctoral symposium* (pp. 33–38). Eindhoven, Netherlands: Tech. Univ. of Eindhoven, Dept. of Math. and CS.
12. Man, K.L., & Schellekens, M.P. (2008). Interoperability of performance and functional analysis for electronic system designs in behavioural hybrid process calculus (BHPC). *Current trends in intelligent systems and computer engineering*, Springer, Germany.
13. Milner, R. (1989). *Communication and concurrency*. Prentice-Hall, USA.
14. OpenModelica System website (2009). OpenModelica System. www.ida.liu.se/~pelab/modelica
15. Plotkin, G.D. (1981). A structural approach to operational semantics. Technical Report DIAMI FN-19, Computer Science Department, Aarhus University.

16. Polderman, J.W., & Willems, J.C. (1998). *Introduction to mathematical systems theory: a behavioral approach*. New York: Springer.
17. Rudolph, E., Graubmann, P., & Grabowski, J. (1996). Tutorial on message sequence charts. *Computer Network ISDN System*, 28(12), 1629–1641.
18. Schonenberg, M.H. (2006). Discrete simulation of behavioural hybrid process algebra. Master's thesis, University of Twente.
19. van Beek, D.A., Man, K.L., Reniers, M.A., Rooda, J.E., & Schiffelers, R. R.H. (2006). Syntax and consistent equation semantics of Hybrid Chi. *JLAP*, 68(1–2), 129–210.
20. van Putten, A. (2006). Behavioural hybrid process calculus parser and translator to modelica. Master's thesis, University of Twente.

Chapter 34
Structured Robust Control for a Pmdc Motor Speed Controller Using Swarm Optimization and Mixed Sensitivity Approach

Somyot Kaitwanidvilai and Issarachai Ngamroo

Abstract This paper proposes a new technique for designing a robust DC motor speed controller based on the concepts of fixed-structure robust controller and a mixed sensitivity method. Performance is specified by selecting the closed-loop objective weight, and uncertainties caused by the parameter changes of motor resistance, motor inductance and load are used to formulate the multiplicative uncertainty weight. Particle Swarm Optimization (PSO) is adopted to solve the optimization problem and find the optimal structured controller. The proposed technique can solve the problem of complicated and high order controller of conventional full order H_∞ controller and also retains the robust performance of conventional H_∞ optimal control. The performance and robustness of the proposed speed controller are investigated in a Permanent Magnet DC (PMDC) motor in comparison with the controllers designed by conventional H_∞ optimal control and conventional ISE method. Results of simulations demonstrate the advantages of the proposed controller in terms of simple structure and robustness against plant perturbations and disturbances. Experiments are performed to verify the effectiveness of the proposed technique.

Keywords Particle swarm optimization · H_∞ optimal control · PMDC motor speed control

1 Introduction

PMDC motor has been widely used in many applications such as fan/pump driving in a photovoltaic system [1], robot's actuators [2], etc. Although AC motor has recently been an attractive choice for an electric drive system; however, DC motor is

S. Kaitwanidvilai (✉) and I. Ngamroo
Center of Excellence for Innovative Energy Systems, King Mongkut's Institute of Technology Ladkrabang, Bangkok 10520, Thailand
e-mail: drsomyotk@gmail.com; knissara@kmitl.ac.th

S.-I. Ao et al. (eds.), *Intelligent Automation and Computer Engineering*,
Lecture Notes in Electrical Engineering 52, DOI 10.1007/978-90-481-3517-2_34,
© Springer Science+Business Media B.V. 2010

still required for certain applications. Several techniques for controlling a DC motor have been proposed by some researchers [3–5]. In [3], a nonlinear fuzzy control was proposed to control a drive system with DC motor. Particle and Kalman filtering was developed to control a DC motor in [4]. In [5], a novel technique for designing a robust controller was presented. Although all techniques mentioned above can be successfully adopted to design an effective controller at certain load conditions; however, uncertainty specifications are not incorporated in the design. In control system, it is important that robust performance of the designed system should be guaranteed. A method, called robust control, can be used for designing a robust controller which can perform good performance under both nominal and perturbed conditions. At present, there are many kinds of robust control techniques such as H infinity loop shaping control, Mixed-sensitivity approach, mu-synthesis, etc. These techniques incorporate the system uncertainties into the design of controller; however, the order of resulting controller is much higher than that of the plant, making it difficult to implement practically.

In recent years, many researchers have tried to propose an effective technique to design a controller for general plant. A more recent control technique uses computational intelligence such as genetic algorithms (GAs) or Particle Swarm Optimization (PSO) in adaptive or learning control. Karr and Gentry [6, 7] applied GA in the tuning of fuzzy logic control which was applied to a pH control process and a cart-pole balancing system. Hwang and Thomson [8] used GA to search for optimal fuzzy control rules with prior fixed membership functions. Somyot and Manukid [9] proposed a GA based fixed structure H_∞ loop shaping control to control a pneumatic servo plant. To obtain parameters in the proposed controller, genetic algorithm is proposed to solve a specified-structure H_∞ loop shaping optimization problem. Infinity norm of transfer function from disturbances to states is subjected to be minimized via searching and evolutionary computation. The resulting optimal parameters make the system stable and also guarantee robust performance.

In DC motor speed control, many engineers attempt to design a robust controller to ensure both the stability and the performance of the system under the perturbed conditions. One of the most popular techniques is H_∞ optimal control in which the uncertainty and performance can be incorporated into the controller design. Unfortunately, order of the resulting controller from this technique is usually higher than that of the plant, making it difficult to implement the controller in practice. In this paper, we illustrate the design of a DC motor speed controller which can guarantee stability under the specified perturbed conditions and which also has a simple structure. The remainder of this paper is organized as follows. Section 2 presents the plant and the proposed design. The genetic algorithm for designing a fixed structure is also described in this section. Section 3 shows the simulation and experimental results. Finally, Section 4 concludes the paper.

Fig. 1 PMDC motor diagram

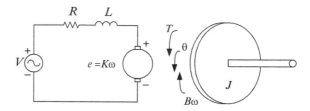

2 DC Motor Modeling and the Proposed Technique

A well known model of DC motor for a speed control system is shown in following:

$$P = \frac{\omega(s)}{V_i(s)} = \frac{K}{(Ls + R)(Js + B) + K^2} \tag{1}$$

where $J(\text{kg m}^2/\text{s}^2)$ is the moment of inertia of rotor, B is the damping ratio of mechanical system, $R(\Omega)$ is electrical resistance, L (H) is electrical inductance, and K (Nm/A) is electromotive force constant. Figure 1 shows a typical DC motor diagram.

According to the standard procedure of robust control [10], there are many techniques for designing a robust controller in a general plant; for example, mixed sensitivity function, mu-synthesis, H_∞ Loop Shaping, etc. However, controllers designed by these techniques result in a complicated structure and high order. The order of the controller depends on the order of both the nominal plant and the weighting functions. It is well known that a high order or complicated structure controller is not desired in practical work. To overcome this problem, a fixed-structure robust controller is designed.

2.1 PSO Based Fixed Structure Robust Control

PSO is used to solve the H_∞ fixed-structure control problem, which is difficult to solve analytically. The proposed technique is described as follows:

2.1.1 Controller's Structure Selection

Assume that $K(p)$ is a structure-specified controller. The structure of the controller is specified before starting the PSO optimization process. In most cases, this controller has simple structures such as PID configuration or lead-lag configuration. A set of controller parameters, p, is evaluated to maximize the objective function. In this paper, PID with a derivative first-order filter controller is selected.

$$K(p) = K_p + \frac{K_d s}{s + \tau_d} + \frac{K_d}{s} \tag{2}$$

The controller parameters set is:

$$p = [K_p, K_d, K_i, \tau_d] \tag{3}$$

2.1.2 Cost Function in the Proposed Technique

The cost function in the design is the infinity norm based on the concept of robust mixed-sensitivity control, which can be briefly described as follows [10]. In the mixed-sensitivity method, firstly, the weighting function of the plant's perturbation and/or performance must be specified. In this paper, W_2 is specified for the uncertainty weight of the plant and W_1 is specified for the performance of the system. The cost function can be written as:

$$J_{\cos t} = \left\| \begin{array}{c} W_1 S \\ W_2 T \end{array} \right\|_\infty < 1 \tag{4}$$

where T is the plant's complementary sensitivity function, and S is the plant sensitivity function.

Assume that the plant is denoted as P. The controller is denoted as K and the system is the unity negative feedback control. The sensitivity and complementary sensitivity function can be expressed as:

$$S = 1 + PK \tag{5}$$

$$T = I - S = PK(1 + PK)^{-1} \tag{6}$$

The cost function in (4) is based on frequency domain specifications. In this approach, the fitness value in the PSO is formulated by this cost function. The proposed technique can be summarized as follows:

Step 1 Specify the weighting functions in robust mixed-sensitivity function [10], and the controller's structure $K(p)$. p is the unknown controller's parameters which are referred to as '*particle*'.

Step 2 Initialize the several sets of p as particles in the first iteration of PSO. Define the PSO parameters such as population size, maximum and minimum velocities and momentum, etc.

Step 3 Generate the swarm of the first iteration randomly. Find the fitness of each particle. The inverse of the cost function in (4) is adopted as the fitness function.

Step 4 Update the inertia weight (Q), position and velocity of each particle using the following equations.

$$Q = Q_{max} - \left(\frac{Q_{max} - Q_{min}}{i_{max}}\right) i \qquad (7)$$

$$v_{i+1} = Q v_i + \alpha_1 [\gamma_{1i} (P_b - p_i)] + \alpha_2 [\gamma_{2i} (U_b - p_i)] \qquad (8)$$

$$p_{i+1} = p_i + v_{i+1} \qquad (9)$$

where α_1, α_2 are acceleration coefficients.

γ_{1i}, γ_{2i} are any random number in $(0 \rightarrow 1)$ range.

Step 5   While the current iteration is less than the maximum iteration, go to step 4. If the current iteration is the maximum iteration, then stop. The particle which has the maximum fitness is the answer of this optimization.

3   Design Example

A speed control system is used to illustrate the effectiveness of the proposed technique. In this example, the system of the speed control of the DC motor has parameters at the nominal plant as follows: $J = 0.02 \, \text{kg} \, \text{m}^2/\text{s}^2$, $B = 0.2 \, \text{N m s/rad}$, $R = 2\,\Omega$, $L = 0.5 \, \text{H}$, $K = 0.1 \, \text{Nm/A}$.

The specification of perturbation used for the design is shown in Table 1. As seen in this table, the reasonable tolerance and changes in system parameters are specified.

Performance weights can be selected properly by the well-known concept shown in [10].

$$W_1 = \frac{0.5\,s + 10}{s + 0.001} \qquad (10)$$

To specify the uncertainty weight, the plots of several multiplicative plant perturbations are shown, and then the transfer function which has higher amplitude than all of uncertainty models is specified as the uncertainty weight. Figure 2 shows the plot of set of multiplicative uncertainty models $|[G(s)/G_n(s)] - 1|$. Where $G(s)$ is the plant and $G_n(s)$ is the nominal plant. By using mathematical software, i.e. MATLAB, the uncertainty weight can be specified as:

$$W_2 = \frac{(0.2619 s^2 + 5.649 s + 19.06)}{(s^2 + 26.28 s + 106.7)} \qquad (11)$$

Table 1 Parameters changing in the design

Parameter	Nominal value	Uncertainty
J	$0.02 \, \text{kg} \, \text{m}^2/\text{s}^2$	$\pm 30\%$
R	$2\,\Omega$	$\pm 30\%$
L	$0.5 \, \text{H}$	$\pm 30\%$

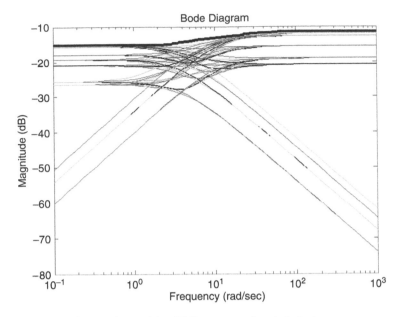

Fig. 2 The design of uncertainty weight of DC motor speed control plant

The structure of the controller is selected as PID with a derivative first order filter which has the same structure as (2). The PSO parameters are selected as: population size = 50, minimum and maximum velocities are 0 and 2, acceleration coefficient = 2.1, minimum and maximum inertia weights are 0.6 and 0.9. When running the PSO for 17 iterations, an optimal solution is obtained as shown in Fig. 3. By the proposed technique, the optimal PID with derivative first order filter controller is evaluated as follow:

$$K(p) = 191.73 + \frac{452.19}{s} + \frac{16.50s}{0.0001s + 1} \tag{12}$$

The infinity norm obtained by the evaluated controller is 0.500 which is less than 1. Consequently, since this norm is less than 1, then the system is robust according to the concept of mixed sensitivity robust control. A conventional mixed sensitivity controller is also designed for comparison. In the conventional technique, the order of the final controller is 4. The controller obtained by this method is

$$K(s) = \frac{135897685781.0162(s + 21.27)(s + 9.78)(s + 4.085)(s + 5.018)}{(s + 5.145)(s + 21.83)(s + 0.001)(s^2 + 3.494 \times 10^5 s + 6.031 \times 10^{10})} \tag{13}$$

Cleary, the order of the conventional robust controller is high and its structure is complicated. Thus, the advantage of simple structure can be obtained by the proposed technique.

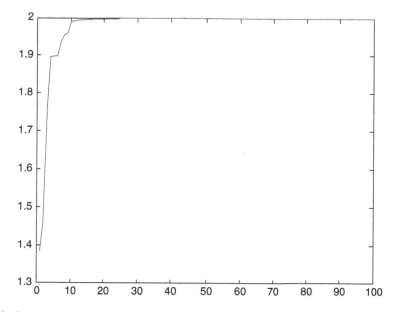

Fig. 3 Convergence of solution of the proposed technique

In addition, in this paper, we also design the conventional PID controller based on the ISE method. In this method, the controller parameter is tuned in such a way that the integral of square error between output and desired response is minimized. However, to prevent the oscillations in the response, ISE with the model reference can be applied [11]. In this paper, we adopted the ISE with model reference to design a PID controller to make the appropriate settling time (about 0.3 s). By this model reference, response of ISE controller is close to the response from mixed-sensitivity controller. By ISE method, the following controller can be evaluated.

$$K_{ISE} = 33.58 + \frac{99.28}{s} + \frac{2.506s}{0.005s + 1} \tag{14}$$

The step responses of both proposed and conventional techniques at nominal conditions are shown in Fig. 4. This figure shows that the settling time from the proposed controller is better than that of the conventional controllers while overshoot does not appear.

To verify the effectiveness of the proposed controller, the system with plant perturbation is examined. In this case, parameters of the system are changed to: $J = 0.014 \, \text{kg m}^2/\text{s}^2$, $B = 0.2 \, \text{N m s/rad}$, $R = 1.6 \, \Omega$, $L = 0.35 \, \text{H}$, $K = 0.1 \, \text{Nm/A}$. Step responses of all controllers at the perturbed plant are shown in Fig. 5. As seen in this figure, the settling time of the proposed controller is almost the same as in the conventional robust controller, and the responses are similar to that of the nominal plant. There is a little overshoot (2–3%) appeared in the response of our proposed

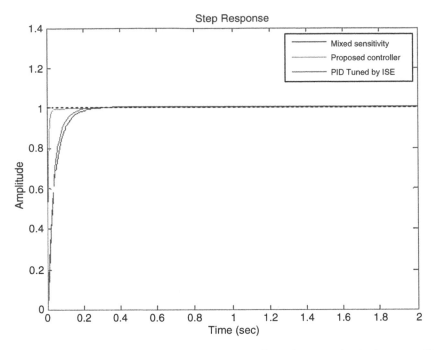

Fig. 4 Step responses of the proposed optimal PID controller and a conventional robust controller at a nominal plant

technique. In the response from ISE controller, large overshoot and slow settling time are occurred in the response. Cleary, both the proposed and mixed-sensitivity controllers are robust.

Some experiments are performed to verify the effectiveness of the proposed controller. The nominal values in Table 1 are parameters of DC motor used in our experiments. Figure 6 shows the experimental setup used in this paper. A proposed robust controller in (12) and PID controller tuned by ISE method in (14) are used to control the speed of motor. As seen in Figs. 4 and 7, the response of experimental result is almost the same as that of the simulation result.

To verify the robust performance of the system, an experiment is performed. The shaft load of motor is reduced to a certain value. This change is equivalent to the changing of parameter in the motor's dynamic. The performance is verified by step response. As shown in Fig. 8, the step response of the proposed controller is almost the same as the response in nominal conditions while the step response of the PID controller tuned by ISE method has a large overshoot and slow settling time. This can be verified that the robust performance of the proposed technique is better than that of PID controller tuned by ISE method.

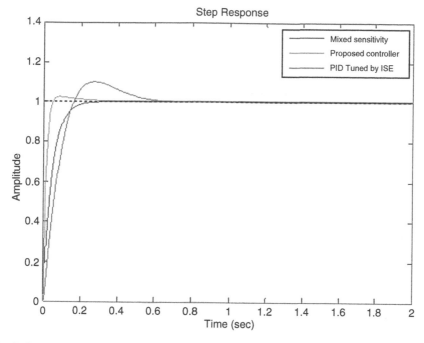

Fig. 5 Step responses of the proposed optimal PID controller and a conventional robust controller at a perturbed plant

Fig. 6 Experimental setup

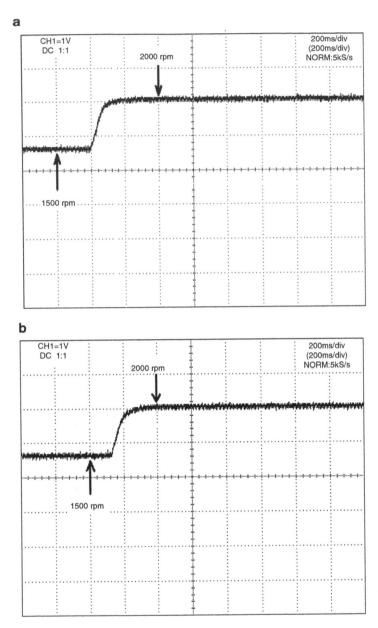

Fig. 7 Experimental results of DC motor speed control at nominal conditions. Step responses of speed of motor from (**a**) proposed controller (**b**) PID controller tuned by ISE method

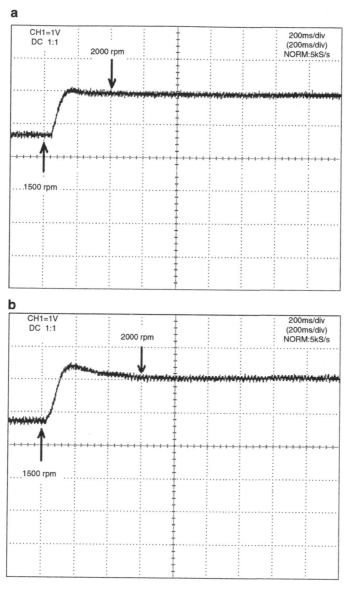

Fig. 8 Experimental results of DC motor speed control at perturbed conditions. Step responses of motor speed from (**a**) proposed controller (**b**) PID controller tuned by ISE method

4   Conclusions

The proposed technique can be applied to control the speed of a DC motor. Based on the incorporation of robust control and the PSO, the proposed technique can achieve robustness and good performance while the structure of the controller is simple.

Robustness of the controlled system can be guaranteed via the theory of mixed sensitivity robust control. In conclusion, by combining of these two approaches, particle swarm optimization and mixed-sensitivity approach; fixed-structure controller design can be achieved. Implementation in a PMDC motor speed controller assures that the proposed technique is valid and flexible.

Acknowledgements This research work was funded by King Mongkut's Institute of Technology Ladkrabang Research fund. This work was also supported by DSTAR, KMITL and NECTEC, NSTDA.

References

1. Altas, I.H., & Sharaf, A.M. (2008). A novel maximum power fuzzy logic controller for photovoltaic solar energy systems. *Renewable Energy, 33*(3), 388–399
2. Shugen, Ma (2002). Time-optimal control of robotic manipulators with limit heat characteristics of the actuator, *Advanced Robotics, 16*(4), 309–324
3. Ahmad M. Harb, & Issam A. Smadi (2009). Tracking control of DC motors via mimo nonlinear fuzzy control. *Chaos, Solitons & Fractals, In Press, Corrected Proof*, Available online 13 May 2009.
4. Gerasimos G. Rigatos (Jan 2009). Particle and Kalman filtering for state estimation and control of DC motors. *ISA Transactions, 48*(1), 62–72.
5. Thirusakthimurugan, P., & Dananjayan, P. (Oct 2007). A novel robust speed controller scheme for PMBLDC motor. *ISA Transactions. 46*(4) 471–477.
6. Karr, C.L., & Gentry, E.J. (1993). Fuzzy control of pH using genetic algorithms. *IEEE Transaction On Fuzzy System, 1*, 46–53
7. Karr, C.L., Weck, B., Massart, D.L., Vankeerberghen, P. (1995). Least median squares curve fitting using a genetic algorithm. *Engineering Application Artificial Intelligence, 8*(2), 177–189.
8. Hwang, W.R., & Thompson, W.E. (1993). An intelligent controller design based on genetic algorithm. *32nd IEEE Conference on Decision Control Conference Proceedings*, pp. 1266–1267. San Antonio, USA. 15–17 Dec. 1993.
9. Kaitwanidvilai, S., & Parnichkun, S. (2004). Genetic algorithm based fixed-structure robust h infinity loop shaping control of a pneumatic servo system. *International Journal of Robotics and Mechatronics, 16*(4), 362–373.
10. Skogestad, S., & Postlethwaite, I. (1996). *Multivariable feedback control analysis and design* (2nd ed). New York: Wiley
11. Jietae, L., & Edgar, T.F. (2004) ISE tuning rule revisited. *Automatica, 40*(8), 1455–1458.

Chapter 35
Agents for User-Profiling, Information Filtering, and Information Monitoring

Kwang Mong Sim and Paul C.K. Kwok

Abstract This paper presents an enhanced holistic information retrieval (*IR*) system that aims to automate the entire process of Web-based *IR*. The system consists of four types of agents: (1) a User profiling agent (*UPA*) that filters and reorders URLs based on a user's interests, (2) ontology-enhanced Web browsing agents (*WBAs*) that are used to autonomously browse and scan multiple Websites to determine and rate the relevance of Websites, (3) Web monitoring agents (*WMAs*) that are used for tracking and reporting changes in selected Websites, and (4) price watcher agents (*PWAs*) that monitor product prices from competing suppliers' Websites. A *UPA* generates a profile of a user's interests, then filters and reorders URLs based on the interests of the user. *WBAs* perform information filtering by considering three relevance metrics: ontological relations, frequency, and nearness of keywords. The general idea of Website monitoring is that each *WMA* is programmed to download a new copy of a Website and compare it with the old copy. *WMAs* allow users to specify monitoring rules, and provide user interface for specifying patterns and data to be monitored. *PWAs* invoke the functionalities of *WBAs* and *WMAs* for browsing and monitoring multiple Websites displaying different prices of a product. Whereas empirical results show that *WBAs* are likely to rate the relevance of Website with a small degree of error, the *UPA* can generally identify URLs that a user is more likely to be interested in. Proof-of-concept examples demonstrate the major functionalities of *WMAs* and *PWAs*.

Keywords Web information retrieval · software agent

K.M. Sim (✉)
Gwangju Institute of Science and Technology, South Korea
e-mail: kmsim@gist.ac.kr

P.C.K. Kwok
School of Science and Technology, Open University of Hong Kong
e-mail: ckkwok@ouhk.edu.hk

S.-I. Ao et al. (eds.), *Intelligent Automation and Computer Engineering*,
Lecture Notes in Electrical Engineering 52, DOI 10.1007/978-90-481-3517-2_35,
© Springer Science+Business Media B.V. 2010

1   Introduction

Web users searching for information are often overwhelmed with very large numbers of URLs returned from search engines. Whereas many of the URLs are often quite relevant, it is not uncommon that irrelevant Websites containing query keywords are among the suggested URLs because words can have several meanings (senses). For example, a typical user using Google search to search for Websites about "mountain chain" may find the URL http://www.chainreactioncycles.com/, which contains words such as "chain" (as in bicycle chain). Consulting WORDNET [1], "chain" has several senses (meanings). One of the senses of chain refers to "a series of (usually metal) rings or links fitted into one another to make a flexible ligament" and another refers to "a series of hills or mountains". In such situation, one possible solution is to program a software agent to distinguish between relevant and irrelevant URLs by searching for evidence phrases by consulting an ontology [2]. Evidence phrases may include ontologically related words such as synonym, hyponym, hypernym, meronym, and holonym. For example, Websites containing words such as "Adirondack Mountains" and "Alaska Range" (a hyponym of mountain chain) are more likely to contain relevant information about mountain chain than a Website with words like "anchor chain" and "tire chain" (a hyponym of "iron chain").

Furthermore, even though users can use search engines to locate URLs and program software agents to autonomously browse selected Website(s), they still need to repeatedly and regularly visit the Websites to retrieve up-to-date information. Due to the ever-changing content of Webpages, tracking the changing contents of Websites may be tedious and time-consuming. Examples of Websites with ever-changing contents include financial Websites that display stock prices and Websites of retail companies that display prices of computer products and accessories. It is not uncommon that investors constantly visit multiple financial Websites and continuously monitor stock prices, and analyze stock trends. One way of assisting such users is to build software tools that visit selected Websites to monitor and track changes in the contents of these Websites. Additionally, software agents for bolstering price comparisons among multiple Websites selling the same product may also be useful tools for retailers and e-shoppers.

The objective of this project is to develop a holistic information retrieval (*IR*) system (Section 2) that augments the functionalities of existing search engines by supporting the following:

1. Autonomous filtering of contents in Websites
2. Regularly monitoring and reporting (selected) changes in Websites, and
3. Regularly comparing and reporting product prices from competing suppliers

This project is designed to support the information gathering activities in Ellis' model [3] that are not supported by existing search engines. Ellis' model [3] of information gathering includes (1) activities that form initial search for information by following and linking to other information sources, (2) browsing (scanning information source), (3) differentiating (filtering and selecting among the sources), (4) monitoring (regularly following a particular source), and (5) extracting (identifying materials of interest from some sources). Whereas activity (1) is supported by existing search engines, the enhanced holistic IR system in this project is designed to bolster activities (2) through (5). To this end, this project complements and augments the functionalities of existing search engines. In particular, it is reminded here that this project does not compete with existing search engines and is certainly not designed to replace existing search engines, but rather to supplement their functionalities.

2   A Holistic IR System

This section presents the prototype of an enhanced holistic IR system consisting of four types of agents: (1) User Profiling Agents (*UPAs*), (2) Web Browsing Agents (*WBAs*), (3) Web Monitoring Agents (*WMAs*), and (4) Price Watcher Agents (*PWAs*).

User Profiling Agent: A *UPA* supports a user by bolstering activity (1) of Ellis' model of information gathering (Section 1). It performs two tasks: (1) identifies and generates a profile of a user's interests, and (2) filters and reorders URLs based on the interests of a user.

Web Browsing Agent: A *WBA* supports a user by bolstering activities (2) and (3) of Ellis' model of information gathering (Section 1). That is, it performs the following tasks: (i) browsing and scanning the information contents of a Website and (ii) determining the relevance of and rating the contents in a Website. Based on Sim's previous works [4–6], details of the functionalities of a *WBA* are given in Section 3.

Web Monitoring Agent: A *WMA* supports a user by bolstering activities (4) and (5) of Ellis' model of information gathering (Section 1). It carries out the following tasks: (1) regularly monitoring a selected Website and tracking changes in the Website, and (2) identifying and reporting selected changes in the contents of the Website that it is monitoring.

Price Watcher Agent: A *PWA* supports a user by (1) invoking a search engine to search for Websites containing the prices of a product, (2) deploying multiple *WBAs* for determining and rating the relevance of a list of Websites displaying that product, (3) deploying multiple *WMAs* for monitoring changes in product prices in multiple Websites, and (4) displaying in ascending order the prices of that product from different Websites.

Stages of Information Gathering: The stages of the information gathering process are listed and described as follows:

1. *Locating Information Resources*. This is typically the first step that a user would do when searching for information through the Web – compose a query using a set of keywords, then enter the query to a search engine. When a search engine returns a set of URLs, a *UPA* filters and reorders URLs based on a user's interests.

2. *Browsing and Evaluating Selected Websites.* Multiple *WBAs* are deployed to simultaneously visit and browse the contents of the set of URLs and verify if the contents are relevant to a query. This corresponds to the second step that a user would typically do – visiting, browsing, and deciding if the contents of the URLs are relevant.
3. *Monitoring Changes in Selected Websites.* In this stage, a set of *WMAs* is activated to monitor and track changes in selected relevant URLs. This stage corresponds to a user bookmarking a set of favorite URLs and perhaps repeatedly visiting the URL to retrieve updated and ever-changing information (e.g., stock prices).

3 User Profiling Agent

The *UPA* carries out three functions: (1) capturing user profile, (2) updating user profile, and (3) filtering and reordering URLs.

3.1 Capturing User Profile

The *UPA* captures and tracks changes in users' interests by monitoring the bookmark files of users. It creates and maintains a theme file consisting of a set of keywords that describes the interest of a user. Generation of user profiles is carried out in five steps as follows:

Step 1: Extraction of Web browsing history

The *UPA* creates a user profile based on a collection of documents extracted from the bookmark file (which contains URLs that are previously visited by the user and are classified as interesting categories) of the user.

Step 2: Generation of document vectors

To represent each document from a user's bookmark in a standard format, the *UPA* adopts a vector space model [7] that represents each document by a document vector. A document vector consists of a set of high information bearing words appearing in and characterizing the document, and these words are used for grouping documents into clusters. Algorithm 1 shows the steps for generating a document vector.

Algorithm 1
1. All unique words in a document are extracted and converted to lower cases.
2. Stop-words (e.g., "a", "an", "the", etc.) are removed.
3. For each document, the frequency of occurrence of each word is recorded.

4. Each word is associated with a *TF-IDF* (*Term Frequency-Inverse Document Frequency*) [8]. Let F_{ji} be the number of occurrences of word j in document i, and m_i be the maximum absolute term frequency achieved by any term in i. The relative term frequency is $TF_{ji} = F_{ji}/m_i$. Let d_j be the number of documents containing j and N be the total number of documents. The inverse document frequency is $IDF = \log(N/d_j)$. The *TF-IDF* of j in i is given as $TF\text{-}IDF_{ji} = TF_{ji}{}^* \log(N/d_j)$.

  4.1. For all non-stop words in each document, find the top k words that have the highest *TF-IDF* in *all* documents and place it in a set S.
  4.2. For each document i, determine TF_{ji} and $TF\text{-}IDF_{ji}$ of each word in S.
  4.3. The document vector of i is ($[TF\text{-}IDF_{1i}, TF\text{-}IDF_{2i},..., TF\text{-}IDF_{ki}]$).

5. Normalize all the document vectors.

The similarity of two documents is reflected by the magnitude of the dot product of their document vectors. The magnitude of the dot product decreases when the documents differ more.

Step 3: Generation of interest clusters

To model and predict the user interests from the set of documents from bookmarks, a clustering process is carried out to group similar documents into an interest cluster. A top-down statistical approach for clustering [9] is adopted. Details are given in Algorithm 2.

Algorithm 2

1. A $d \times d$ dot-product table is created, where d refers to total number of document vectors. Each entry E_{ij} refers to the value of dot product of document vectors i and j. E_{ij} represents a measure of similarity between two documents.
2. The mean μ and standard deviation σ are calculated for all non-zero entries in the dot-product table. Typically, 60% of the entries in the table are zero.
3. A document vector is randomly chosen to start a cluster C_1. To grow C_1, a document vector not in C_1 is tested for inclusion by determining the *new* mean $\mathbf{m}$ for C_1 if the document vector is added. The process is repeated for all the document vectors not already in C_1 and the best mean $\mathbf{m}_{best}$ is recorded.
4. If $\mathbf{m}_{best} > \mu + \sigma$, the document vector corresponding to the best mean is added to the C_1.
5. When no more document vectors can be added to C_1 the clustering process is repeated from step 3 to 4 for the remaining document vectors to form other clusters.
6. At the end of the clustering process, the clusters generated are recorded. The remaining document vectors that are not clustered are also recorded in a file (unclustered.txt) for use in a later stage.

Step 4: Identification of user interests

Each cluster created consists of a set of similar documents, representing a particular interest of a user. To identify the interest represented by a cluster, a cluster vector is generated by first summing up all the information content of common words in the document vectors. Subsequently, the top 20 words with the highest sum are treated as representative of the user interests. Additionally, the size and age of a cluster are also determined. Cluster size refers to the total number of documents in the corresponding cluster, and represents the degree of interest of a user in this particular topic. Cluster age represents the duration between the earliest and latest visited documents in a cluster. A longer cluster age implies a more persistent or longer-term interest. Both cluster size and cluster age reflect the importance of each cluster.

Step 5: Extraction of theme keywords for generation of user profile

All clusters created are ranked in descending order of cluster size. For clusters with the same size, clusters with older age are ranked higher. Keywords of cluster vectors are extracted and recorded in a theme file in the order of their importance. The theme file consists of representative keywords for characterizing user interests.

3.2   Updating User Profile

Updating user profile is carried out in four steps as follows:

Step 1: Extraction of the latest Web browsing history

To track changing users' interests, users' profiles need to be updated and the modification date is recorded. In updating a user's profile, only URLs in the bookmark that are visited later than the last modified date are examined.

Step 2: Updates of existing interest clusters

Documents extracted from new URLs in the bookmark file are checked against all the existing clusters for similarity. If a new Web document shares a similar topic with an existing cluster, it will be added to that cluster and the size and age of the cluster will be updated. Details are given in Algorithm 3.

Algorithm 3

function **UpdateExistingClusters** (Cluster[] allclusters, Document[] array_of_documents)
for each cluster C_i in allclusters
    for each new document d_i in array_of_documents

> Process $\mathbf{d_i}$ by creating document vector $\mathbf{v}$ using the same set of keywords in $\mathbf{C_i}$
> Compute dot-product $\mathbf{p}$ of $\mathbf{v}$ and cluster vector of $\mathbf{C_i}$
> If $\mathbf{p} > $ **threshold**//*threshold is defined by user*
>   Add document $\mathbf{d_i}$ to cluster $\mathbf{C_i}$
>   Update cluster size and age of $\mathbf{C_i}$
>   Remove $\mathbf{d_i}$ from array_of_documents
> end for
> end for

Step 3: Generation of new interest clusters

New documents that are not similar to existing clusters are combined with the documents that are recorded in a special file called unclustered.txt. Clustering is carried out for this new set of Web documents to explore new user interests and new clusters are recorded.

Step 4: Updates of theme file of user profile

Both existing and newly created clusters are re-ranked according to their size and age. The theme file of User Profile is updated with the latest user interests.

3.3 Reordering URLs

The *UPA* re-ranks the set of URLs based on the interest of users captured in the user profile. Re-ranking by the *UPA* is carried out in three steps as follows:

Step 1: Extract keywords from theme file of the user profile

The *UPA* reads in the most updated theme file of the user profile and the set of theme keywords are stored in a set S.

Step 2: Compute the total occurrence of theme keywords in the Web document

The content of each URL is examined. The total occurrence C of all theme keywords in S that appear in the Web document is computed.

Step 3: Output a set of ranked URLs

The set of URLs is ranked in descending order according to the corresponding value of C. By reordering the URLs, a user is more likely to read Web documents that he/she is more interested in first.

4 Web Browsing Agent

A *WBA* carries out two functions: (1) information filtering and (2) information rating.

Information Filtering: In this stage, a *WBA* adopts WORDNET's ontology [1] for determining the relevance of a Website. This is achieved by constructing a set of evidence phrases for a user query by considering ontological relations from WORD-NET such as *meronym* and *holonym*. Meronym and holonym refer to the part-whole relations of words [10]. Whereas a meronym is the name of a constituent part of a concept, a holonym is the name of the whole of which the meronym is a part (i.e., P_1 is a meronym of Q_1 if P_1 is a part of Q_1, and Q_2 is a holonym of P_2 if P_2 is a part of Q_2) [1]. Furthermore, in WORDNET, some of the categories of meronym relations include:

1. Part mernonym: P_1 is a *part meronym* of Q_1 if P_1 is a *component part* of Q_1.

   Example Query Word: battery
   Part Meronym(s): electrode, pole (terminal).
2. Member meronym: P_1 is *a member meronym* of Q_1 if P_1 is a *member* of Q_1.

   Example Query Word: Forest
   Member Meronym(s): tree
3. Substance meronym: P_1 is a *substance meronym* of Q_1 if P_1 is the *stuff that* Q_1 *is made of*.

   Example Query Word: chalk
   Substance Meronym(s): calcium carbonate

When filtering relevant URLs, a *WBA* examines the content of a URL for meronyms of a query word by consulting *WORDNET*. Given that each word can have different senses, an irrelevant URL is identified by searching for meronyms of query keywords of other senses. For instance, to identify an irrelevant Webpage for the query "battery" (in the sense of "electric battery"), the *WBA* filters out Web-pages with meronyms such as "gun" and "missile launcher" which are meronyms of battery in the sense of gunnery.

Information Rating: A *WBA* rates the information contents in a Website by considering three heuristic factors: (1) ontologically related words, (2) frequency of occurrence, and (3) nearness of keywords.

1. By searching for ontologically related words in a Webpage, a *WBA* is more likely to detect information related to a query. To ensure high precision, the heuristic in [5, p. 96] is used to guide a *WBA* in identifying relevant information of different degrees.
2. The relevance metric used by a *WBA* favors Webpages with higher occurrence of keywords in a user query. For example, if the term "car" occurred reasonably frequently in a Webpage, it seems plausible to think that the Webpage contains information that deals with "car" [11, pp. 279–280].

3. When nearness [11, pp. 237] is included in the relevance metric, the probable relevance of the information retrieved is likely to be higher. For instance, consider the query "nature picture", if both "nature" and "picture" occur adjacently in a given Webpage, then it is more likely that the Webpage contains more relevant information than when both "nature" and "picture" occur within a sentence but are separated by some words".

5 Web Monitoring Agent

A *WMA* carries out two functions: (1) monitors changes in a Webpage, and (2) extracts specific information from a Webpage periodically. The general idea of information monitoring is to download a new copy of a Website and compare it with the old copy. An algorithm for monitoring changes in a table within a Website is given in Algorithm 4 [4, 6].

Algorithm 4

1. Retrieve the Webpage from the given URL at time t as a string of characters $S(t)$.
2. Extract all tables from $S(t)$ as $\{T_1, T_2, \ldots, T_n\}$. For each T_x in $S(t)$, T_x contains a set of cells $\{c_{1,1}, c_{1,2}, \ldots, c_{r,c}\}$
3. Let Value$(c_{x,y}, t)$ return the value of $c_{x,y}$ at time t and Type$(c_{x,y}, t)$ be the data type of the $c_{x,y}$ at time t that can either be a string or numeric type. If Value$(c_{x,y}, t)$ only contains digits, period(".") and comma(","), it is assumed that Value$(c_{x,y}, t)$ is numeric. All other values are considered as string.
4. Select a cell, $c_{x,y}$, from $\{c_{1,1}, c_{1,2}, c_{1,3}, \ldots\}$ to monitor changes.
5. If the Type$(c_{x,y}, t)$ is string, report changes if Value $(c_{x,y}, t_{n+1})$ is different from Value $(c_{x,y}, t_n)$. This is accomplished by comparing the strings at t_n and t_{n+1}.
6. If the Type$(c_{x,y}, t)$ is numeric, report changes if

$$\left| \frac{\text{Value}(c_{x,y}, t_{n+1}) - \text{Value}(c_{x,y}, t_n)}{\text{Value}(c_{x,y}, t_n)} \right| > n\%$$

where n is a user defined threshold

Monitoring rules: In a *WMA*, each monitoring task can be represented by a task script using monitoring rules. Some of the monitoring rules in a *WMA* are given as follows:

    R1: If (modified()) notify();
    R2: If (modified()) download();
    R3: If (new(2) > 2.0) notify();
    R4: If (new(2)-old(2) >= 0.1) notify()

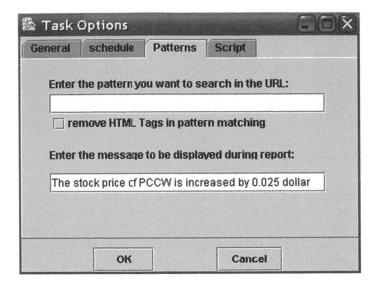

Fig. 1 *WMA* User Interface: Pattern to Monitor

R1 (respectively, *R2*) simply specifies that the *WBA* should notify a user (e.g., by sending email) (respectively, download the Webpage to local disk instead of notifying the user) if there is any changes in a Webpage. Used in conjunction with a pattern, the functions old() and new() in *R3* and *R4* refer to the old and new values of the data associated to a pattern. For instance, if a user instructs a *WMA* to monitor the value of a stock called "tom.com" (see Section VI) then in *R3* and *R4*, "2" inside the functions old() and new() is a marker that points to a data value associated to the pattern "tom.com", i.e., "2" refers to the stock price of "tom.com" displayed in the Webpage that the *WMA* is monitoring.

  Whereas regular expression is used for specifying the instructions for a *WMA* to extract specific information from a Website, a user interface is developed to allow a user to specify: (1) the pattern that a *WMA* should monitor and the message it should display (Fig. 1), (2) the interval for monitoring the Webpage (Fig. 2), and (3) the position of the data that is associated with a pattern (Fig. 3).

6 Experimentation and Evaluation

Experiments were carried out to evaluate both the *UPA* and the *WBA*. The experiment settings and empirical results for evaluating the *UPA* and the *WBA* are given in Sections 6.1 and 6.2, respectively.

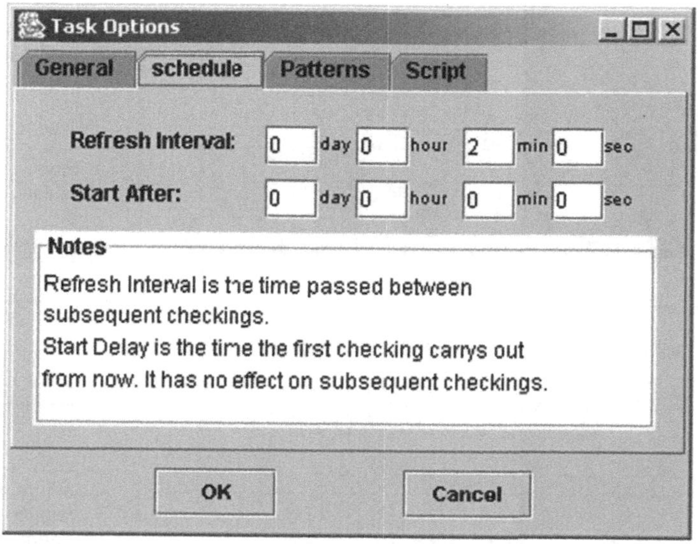

Fig. 2 *WMA* User Interface: Schedule to Monitor

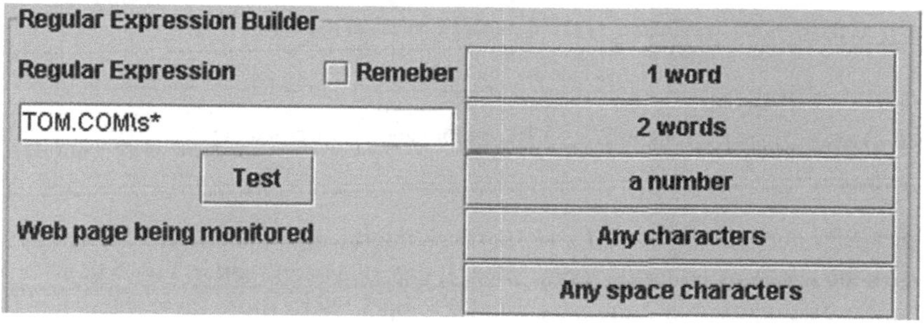

Fig. 3 *WMA* Pattern Builder Interface

6.1 Evaluating the UPA

In this experiment, a total of 50 queries were used. Two human users are asked to rate whether each of the URLs returned by the testbed is interesting to them when (1) the *UPA* is not used and (2) when the *UPA* is used. When rating the URLs, the users are given the following instructions:

(i) For each URL, assign an evaluation score between 0 and 10. A score of 10 indicates that the user is very interested in the URL, while 0 indicates that the URL is very uninteresting to the user.

(ii) For each query, compute the average scores of the top 3 and top 5 URLs.

The results for parts (1) and (2) of this experiment are shown in Figs. 4 and 5. From Fig. 4, it can be seen that among the 50 queries, there were 40 queries in which the

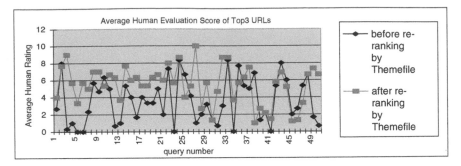

Fig. 4 Average Human Rating of Top 3 URLs before and after Re-ranking

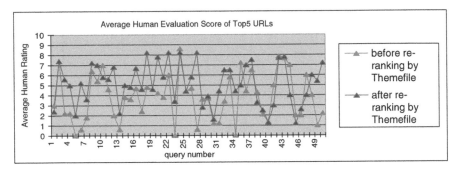

Fig. 5 Average Human Rating of Top 5 URLs before and after Re-ranking

users have rated the top 3 URLs to be more interesting when the *UPA* is used to re-rank the URLs using users' theme files. It can also been seen in Fig. 5 that for 41 out of the 50 queries, the users have rated the top 5 URLs more favorably when the *UPA* is used.

6.2 Evaluating the WBA

Evaluation of the *WBA* consisted of (1) user study, and (2) a series of experiments using the *WBA* for rating the same set of URLs rated by human users. For user study, two human users were asked to rate the relevance of the top 5 URLs returned by Google for 100 queries. Using a *WBA* to rate the relevance of contents in URLs, two series of experiments were conducted [4–6]. For the first series of experiments, the *WBA* was programmed to *incrementally* recognize exact words, synonyms, hyponyms, and hypernyms (see Table 1). The second series of experiments examined the effect of using five different combinations of weightings of the three heuristics in Section 3: (1) ontological related words (*OR*), (2) frequency of occurrence (*FO*), and (3) nearness of keywords (*NK*) (see Table 2).

Table 1 *WBA* combination of related terms

Simulation	Word relations
WBA1	{Exact words}
WBA2	{Exact words} + {synonyms}
WBA3	{Exact words, synonyms} + {hyponyms}
WBA4	{Exact words, synonyms, hyponyms} + {hypernyms}

Table 2 Weightings of the three heuristics

Weight combination	*OR*	*FO*	*NK*	Difference between users and *WBA* ratings
1	0.6	0.2	0.2	29%
2	0.34	0.33	0.33	16%
3	0.5	0.25	0.25	21%
4	0.6	0.3	0.1	29%
5	0.4	0.4	0.3	18%

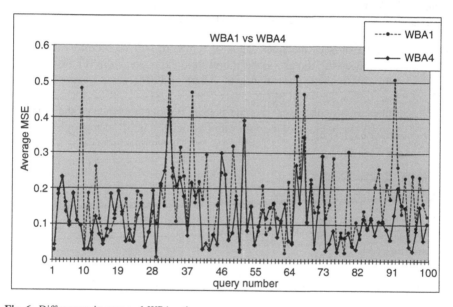

Fig. 6 Differences in user and *WBA* ratings

Empirical results: Empirical results obtained show that in the first series of experiments, a *WBA* adopting *WBA4* (i.e., scanning a Webpage for exact words, synonyms, hyponyms and hypernyms) achieved the minimum mean square error (*MSE*) relative to human users' rating when rating the relevance of Websites. Whereas space limitations preclude all results from being included here, the results showing the *MSE* between users and the *WBA* in the experiments when the *WBA* searched for related words using exact words, synonyms, hyponyms and hypernyms (i.e., *WBA4*) and when the *WBA* only searched for exact words (i.e., *WBA1*) is shown in Fig. 6. The results showing that *WBA4* attained lower *MSEs* than *WBA1* suggest that a *WBA* is more likely to reduce its error in rating URLs if it is programmed to recognize related words including exact words, synonyms, hyponyms and hypernyms.

For the five combinations of weightings shown in Table 2, empirical results in the second series of experiments show that the *WBA* achieved the minimum MSE when it adopted combination 2 in Table 2. This generally suggests that a *WBA* is more likely to reduce its error in rating URLs if it is programmed to place almost equal emphasis on all the three heuristics (*OR*, *FO* and *NK*).

7 Proof-of-Concept Examples

Two examples are provided in this section to illustrate the major functionalities of *WMAs* and *PWAs*.

Example 1. A *WMA* was deployed to monitor the changing value of the data value associated with "stock.com" in a Webpage shown in Fig. 7.

Step 1:   In this example, the stock value of tom.com will be monitored, and the user enters the pattern to be monitored (i.e., "tom.com") and the URL http://hk.finance.yahoo.com/q?m = h&s = 8001&d = v1 using the interface screen shown in Fig. 8.

Step 2:   Subsequently, the user uses the pattern builder interface of the *WMA* to specify the location of the data value associated with "tom.com". In Fig. 7,

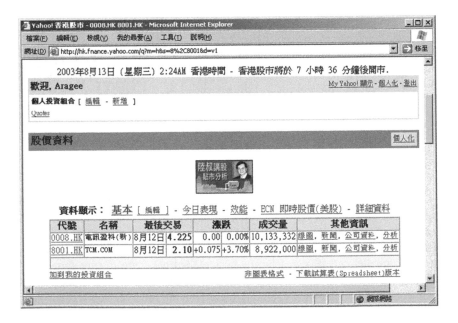

Fig. 7 A Website to be monitored

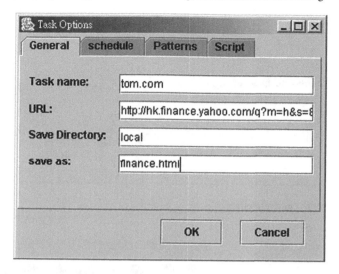

Fig. 8 *WMA* task specification interface

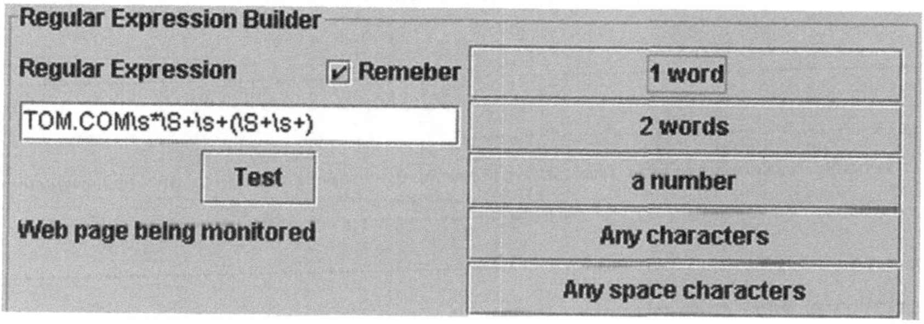

Fig. 9 *WMA* pattern builder interface

it can be seen that the data value (i.e., "2.10") of "tom.com" and the pattern "tom.com" is separated by (1) some whitespaces (this is represented in regular expression in Fig. 9 as "\s*", i.e., zero or more whitespace(s)), and (2) a sequence of characters followed by at least one space character (this is represented in regular expression in Fig. 9 as "\S + \s+"). In Fig. 9, "(\S + \s+)" represents the data value to be monitored.

Step 3: The user specifies the monitoring rule using a *WMA*'s script builder shown in Fig. 10. The instruction in Fig. 10 indicates that the *WMA* should notify the user if the stock value of tom.com is above 2.

Step 4: The user specifies the monitoring interval using a *WMA's* schedule interface shown in Fig. 2.

Step 5: When the stock value of tom.com is above 2, the *WMA* notifies the user.

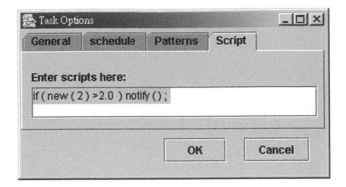

Fig. 10 *WMA* script builder interface

Fig. 11 Price watcher agent
user interface

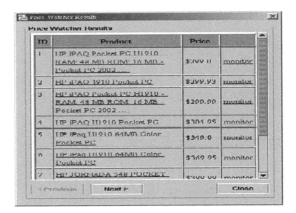

Example 2. A PWA supports a user by invoking a search engine to search for the price of a HP Pocket PC, deploying *WBAs* to browse and determine the relevance of a list of Websites displaying HP Pocket PC, deploying *WMAs* to monitor changes in the prices of HP Pocket PC in selected Websites then displaying in ascending order the prices of HP Pocket PC from different suppliers' Websites (Fig. 11).

8 Discussion and Conclusion

This paper has presented an enhanced holistic *IR* system. It serves the emphasis to mention that the system in this project is not designed to replace or compete with existing search engines. Rather it is designed to augment and complement the functionalities of existing search engines.

The novel features of this project are as follows. To assist users in identifying ULRs that they are more interested in, the *UPA* is used to: (1) capture and maintain user profile, and (2) filter and reorder URLs. Multiple *WBAs* can be deployed in parallel to simultaneously visit, browse, and scan the information contents of multiple Websites and autonomously determine and rate the relevance of the contents in multiple Websites. Multiple *WMAs* can be deployed in parallel to simultaneously monitor, track, and report changes in multiple Websites. Whereas preliminary ideas of holistic *IR* were reported in [4,6], this work extends the work in [4,6] as follows. Whereas the information filtering agents in [4,6] *only* rate the relevance of Webpages by considering exact words, hyponyms, and hypernyms, *WBAs* in this work (1) consider meronyms and holonyms when determining the relevance of Webpages and (2) consider exact words, hyponyms, and hypernyms when scanning and rating the relevance of Webpages. Whereas *only* the general algorithm of Webpage monitoring was presented in [4,6], *WMAs* in this work are built with user interfaces for specifying information monitoring rules, pattern and data value to be monitored and schedule for monitoring. Additionally, the *UPA* and the *PWAs* were not considered in [4–6]. *PWAs* in this work can be viewed as "meta-software-agents" invoking on the functionalities of *WBAs* and *WMAs* for browsing and monitoring multiple Websites containing prices of a product. Whereas both the UPA in this work and the Query processing agent (*QPA*) in [4] and [6] are designed to support activity (1) of Ellis' model of information gathering, they are different functionalities. While a *QPA* select appropriate numbers of ULRs by specializing or generalizing a search query, the *UPA* in this work filters and reorders ULRs to identify URLs that users are more likely to be interested in.

Acknowledgment This work was supported by the DASAN Project (GIST International Faculty Fund) from the Gwangju Institute of Science and Technology, South Korea. Thanks to Chan Siu Man, Cheung Ting Yee, Wong Yuk Chuen and Tsang Lai Ling for carrying out the portions of the implementations of this work.

References

1. Miller, Wordnet, G.A. (1990) An on-line lexical database. *International Journal of Lexicography, 3–4*, 235–312.
2. Fridman, N., & Hafner, C. (Fall 1997) The state of the art in ontology design. *AI Magazine*, pp. 53–74.
3. Ellis, D. (1989) A behavioral model for information retrieval system design. *Journal Documentation, 49*(4), 356–369.
4. Sim, K.M., & Wong, P.T. (2004) Towards agency and ontology for web-based information retrieval. *IEEE Transactions on Systems, Man and Cybernetics, Part C: Applications and Reviews, 34*(3), 1–13.
5. Sim, K.M. (2004) Toward an ontology-enhanced information filtering agent. *ACM SIGMOD Record, 33*(1) March, 95–100.
6. Sim, K.M. (2003) Towards holistic web-based information retrieval: an agent-based approach. In *Proceedings of the 2003 IEEE/WIC International Conference of Web Intelligence*, Oct. 13–16, Halifax, Canada, pp. 39–46.
7. Frakes, W.B., & Baeza-Yates, R. (Eds.) (1992) Information retrieval: Data structure and algorithms. Prentice-Hall, USA.

8. Salton, G., & McGill, M.J. (1983) Introduction to modern information retrieval. McGrawHill, USA.
9. Leonard N. Foner (1995) Clustering and information sharing in an ecology of cooperating agents. In *AAAI Spring Workshop on Information Gathering from Distributed, Heterogeneous Environments*.
10. Winston P., & Chaffin R. (1987) A taxonomy of part-whole relations. *Cognitive Science, 11*, 417–44.
11. Salton (1989) *Automatic text processing*. Reading, MA: Addison Wesley.

Chapter 36
Form-Based Requirement Definitions of Applications for a Sustainable Society

Takeshi Chusho, Noriyuki Yagi, and Katsuya Fujiwara

Abstract It is expected that information technology (IT) will contribute to resource saving and environmental preservation for a sustainable society. For this purpose, application software is required and then the fund is needed for its development by IT professionals. However, the preparation of the fund must be difficult. The end-user initiative development of application software is indispensable for the solution of this dilemma. This paper describes requirement definitions based on abstract forms in a method that business professionals build applications by themselves. The abstract forms are considered as interfaces of Web services based on the simple concept that "one service = one form." Therefore, the business logic can be defined as the form transformation from input forms into outputs form by business professionals.

Keywords EUC · Form-to-form transformation · Visual tool · Web service integration

1 Introduction

It is expected that information technology (IT) contributes to saving resources and environmental preservation for a sustainable society. For this purpose, application software is required and then the fund is needed for its development by IT professionals. However, the preparation of the fund is difficult unless a profit is calculated over the development cost. The end-user initiative development of application software is indispensable for the solution of this dilemma.

T. Chusho (✉) and N. Yagi
Department of Computer Science, School of Science and Technology, Meiji University, Kawasaki, 214-0033, Japan
e-mail: chsho@cs.meiji.ac.jp

K. Fujiwara
Department of Computer Science and Engineering, Akita University, Akita, 010-8502, Japan
e-mail: fujiwara@ie.akita-u.ac.jp

S.-I. Ao et al. (eds.), *Intelligent Automation and Computer Engineering*,
Lecture Notes in Electrical Engineering 52, DOI 10.1007/978-90-481-3517-2_36,
© Springer Science+Business Media B.V. 2010

For example, let's consider a thrift store which sells limited goods to limited customers in a local area. The number of goods and the number of customers will increase if business professionals develop the application for the Web site in which customers can register goods to be reused or search the list of registered goods for their own use easily.

As for another example, let's consider service counters which exist everywhere. Although some service counters already support the Internet usage, many service counters have not yet done this because of lack of funds, not lack of technologies. If business professionals at a service counter can develop the application for a Web site, they will save resources because of the paperless system and reducing the cost of electricity by not using elevators when going to the actual counter.

Furthermore, let's consider online shopping. Some customers may want to buy goods from the nearest shop to reduce carbon dioxide (CO_2) emission in transportation. It must be useful to open a Web site where you can search several online shops for the specified goods and display the information alongside the transportation distance. This application may be developed by non-professionals of IT if Web service integration of online shopping sites and an online map service site is performed by using recent mash-up technologies.

There are several approaches for the end-user initiative development. That is, the UI-driven approach makes it possible to develop applications for the UI-centered front-end subsystems easily. It is strengthened by using framework technologies. The model-driven approach makes it possible to develop applications for the workflow-centered back-end subsystems easily. It is strengthened by using a visual modeling tool. Furthermore, the form-driven approach must be easier than the aforementioned two approaches for business professionals since they are familiar with forms in daily work. It is strengthened by the form-to-form transformation and Web service integration.

Terms for end-user computing (EUC) and papers on EUC often came out in 1980s. Some papers describe definitions and classifications of EUC [7] or the management of EUC [2]. A recent paper summarizes the trends of end-user development without IT professionals' assistance [22].

There are some other works related to EUC. In the programming field, the technologies for programming by example (PBE) [16] were studied. The PBE implies that some operations are automated after a user's intention is inferred from examples of operations. The non-programming styles for various users including children and for various domains including games were proposed. In the database field, the example based database query languages [21] such as QBE (Query-By-Example) were studied. QBE implies that a DB query is executed by examples of concrete queries. User-friendly inquiry languages were proposed in comparison with SQL.

Our research target is different from these technologies and is for business professionals and business domains. The user's intention is definitely defined as requirement specifications without inference as business professionals with domain expertise develop software which executes their own jobs.

Therefore, this paper pays attention to a Web application in which the user interface is a Web browser because most users are familiar with how to use the Internet.

Furthermore, the three-tier architecture is supposed, which has been popular recently. Generally, there are three approaches corresponding to the user interface (UI), business logic and database (DB). In our studies, application frameworks, visual modeling tools based on components and form transformation tools for Web service integration were developed for EUC.

This paper presents Web application development technologies in Section 2, examples of applications in Section 3, issues on EUC in Section 4 and abstract forms and form transformation in Section 5.

2 Web Application Development

2.1 Basic Approaches

The approaches to the end-user initiative Web application development methodologies based on the three-tier architecture are classified into the three categories of UI-driven, model-driven and data-driven processes by first focusing on any one of the UI (user interface), the model (business logic) or DB. These approaches are described in this section.

2.2 A UI-Driven Approach

Recently, a UI-driven approach has emerged as Web applications are increasing. A typical example of this approach is the Struts framework [1] which is an open source framework for building Web applications in Java. The visual forms are defined first and then components for business logic and access to the DB are defined. In this approach, it seems to be easier for the end-user to define the UI in comparison with definitions of the model or the DB.

For example, there are recent success stories on end-user computing. One is a paper that the European Union's SmartGov project transforms public-sector employees into developers of the government e-services used directly by the public [15]. An intelligent e-forms development and maintenance environment and associated framework are delivered.

Another one is performed at the office of Nagasaki prefecture in Japan [11]. The staff of business professionals designed and described the user interfaces without IT professionals' assistances. Furthermore, while the staffs specified requirements on business logic and DB tables, IT professionals described documents of design specifications. Based on these documents, the system was divided into subsystems as the development cost of each subsystem was less than about 50,000$ and a small local IT company could undertake the small-scale subsystem development. As a result, the risk of ambiguous requirements was omitted, the cost was reduced to about 50%, and customer satisfaction based on usability etc. was improved. This story suggests that the UI-driven approach makes it possible for end-users to define the requirements of their own software.

We have also been studying this approach for several years [5]. The UI-driven approach is proposed for the front-end subsystem based on CBSE (Component-Based Software Engineering) [3, 8]. The systems are constructed by using UI-centered frameworks [10] and agent technologies [13]. The effectiveness of the UI-driven process is confirmed through experiences with the development of frameworks.

Business professionals define requirements for an application to be developed by using the framework. Services at the counter are defined first. Next, forms for these services are defined with navigational information. Finally these form definitions are registered into the corresponding servers.

An example of a browser defining the library system is shown in Fig. 1. The left part implies a hierarchical directory. The right part implies definitions about the service for taking out books. Intelligent navigation by agents is implemented in XML. The metadata for a window is described in an RDF (Resource Description Framework) style. While forms are defined in HTML in the conventional way, the semantics of forms are defined in RDF style also.

However, this framework for a service counter does not support the back-end subsystem with the workflow and DB. When another framework for a reservation task such as a room reservation system was developed, a visual tool for defining the DB table easily was developed simultaneously. Although end-users can use this tool, the target DB table is limited to a simple reservation table.

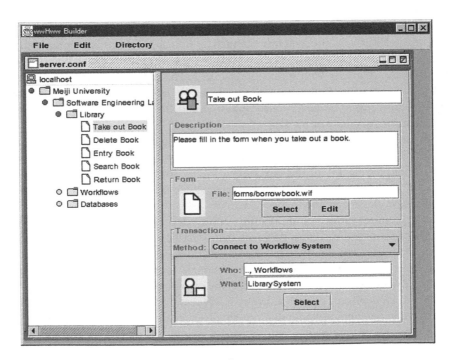

Fig. 1 The browser for system definitions by end-users

2.3 A Model-Driven Approach

Around the 1990s, object-oriented analysis and design (OOAD) technologies came out and have become the major methodologies. Some of them match the waterfall model and others match the iterative and/or incremental development process [12, 14]. In the recent OOAD methodologies, the unified modeling language (UML) [20] is used for definitions of the system model. OOAD is a model-driven approach. In addition, UML2.0 requires more rigorous definitions of models for automatic generation of program codes based on the model-driven architecture (MDA) [19].

We have also been studying this approach for several years [4]. The model-driven approach based on CBSE is proposed for the back-end subsystem which the main part is a workflow. Our solution is given as a formula of "a domain model = a computation model." This formula implies that one task in a domain model of cooperative work corresponds to one object in object-oriented model. Therefore, it is not necessary for end-users to convert a domain model into a computation model with application architecture. The domain model is considered as the requirement specifications. This process requires necessarily the fixed architecture and ready-made components such as business objects.

Our approach is different from most conventional object-oriented analysis and/or design methods which need defining an object model on static structure of objects prior to a dynamic model on interactive behavior among objects. At the first stage, the system behavior is expressed as a message-driven model by using a visual modeling tool while focusing on message flow and components. At the second stage, a user interface is generated automatically and may be customized if necessary. Then the transition diagram of user interfaces is generated automatically and used for confirmation of external specifications of the application. Finally, the system behavior is verified by using a simulation tool.

This component-based development process was confirmed by feasibility study on a given problem of the IPSJ (the Information Processing Society of Japan) sigRE group. The problem is how to define requirement specifications for a program chair's job of an academic conference. A dynamic model was constructed while introducing eleven kinds of objects. These objects are defined by drag-and-drop operations from the palette of icons. A message between objects is defined by drawing an arrow from the source object to the drain object.

In addition, branch conditions are described in rule expressions. For example, in the "produce" method of the CFP Production object, the following rules are described for branch conditions:

- if Printer = yes then print;
- if CFP Distribution = yes then distribute;

Furthermore, simulation is executed for validation of the requirement definitions as shown in Fig. 2, both on the domain model and on the sequence diagram, while displaying traces of the message flows.

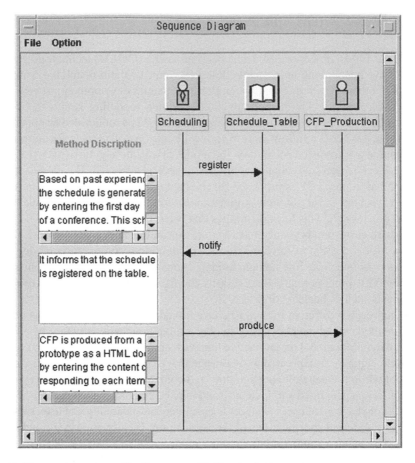

Fig. 2 The requirement specifications are verified by simulation

2.4 A Data-Driven Approach

As for a data-driven approach, a data-centered or data-oriented approach was introduced in the 1980s. In this method, the data flow diagrams (DFD) are sometimes used for a definition of the workflow. The data model is defined with entity relationship diagrams (ERD). In many mission-critical applications, the DB design is the most important. Since data structures are more stable than business logic, the data model is defined prior to business logic.

However, the design of a large-scale DB is difficult for end-users. In our UI-driven approach and/or the model-driven approach, it is supposed that DB components are used. As for a small-scale DB, visual tools are introduced for defining the DB tables.

3 Typical Applications

These technologies for end-user computing will be applied to various applications for a sustainable society. As for a Web application for a thrift store, the UI-driven approach for a front-end subsystem and the model-driven approach for a back-end subsystem are applied. If a Web site for a thrift store can be opened easily by business professionals, natural resources can be saved and the opportunity to reuse them great.

As for a Web application for a service counter, framework technologies are suitable because the domain expertise is embedded into a framework which is prepared in advance. As for the latest example, the framework for a reservation system was developed and applied to a meeting room reservation system for our department. This application is in practical use. As a result, use of natural resources is reduced by a paperless system, and the energy for elevators, trains, cars, etc. is saved since it is not necessary to visit the actual service counter.

Furthermore, if an Internet shopping site and a map site are combined using Web service integration, clients can buy goods from the nearest shop and there will be a reduction of CO_2 emission usually in transportation.

4 Issues on End-User Computing

In our experiences, sometimes end-users needed IT professionals' assistance. It is difficult for end-users to develop new components, to modify ready-made components for complicated business logic and to implement user interfaces in JSP. However, if end-users can describe requirement specifications, some IT professionals may continue application development as volunteers or some IT companies may undertake application development at a low cost. Recently, the ratio of the requirement definitions cost to the total cost on software development has been increasing since the productivity of design, implementation and testing has been improved by using various tools and object-oriented platforms.

Furthermore, in the business world, the external specifications of application software are recently considered as services as shown in keywords such as ASP (Application Service Provider), Web service, SOA (Service-Oriented Architecture) and SaaS (Software as a Service) [9, 17, 18]. Our new approach to end-user computing is that end-users develop Web applications by service integration for both the front-end subsystem and the back-end subsystem because end-users consider their applications as a level of service, not as a level of software.

That is, the service counter is considered as a metaphor to describe the interface between a service provider and a service requester for Web services. Such a service counter is not limited to the actual service counter in the real world. For example, in a supply chain management system, data exchange among related applications can be considered as data exchange at the virtual service counter.

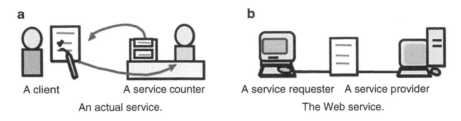

a A client   A service counter   **b** A service requester   A service provider

An actual service.   The Web service.

Fig. 3 A service counter as a metaphor for Web service

Generally, the service counter receives service requests from clients as shown in Fig. 3. Forms are considered as the interface between them. That is, the following concept is essential for our approach:

$$\text{``One service = One form.''}$$

The integration of some individual Web services is considered as transformation from some input forms into some output forms. Although most of these forms are not visual forms, end-users can consider this form as a visual form for the requirement specification. Such a form is called an abstract form in this paper. Since end-users can consider such Web service integration as the workflow with visual forms which they are familiar with, IT skills are not required of end-users. Furthermore, our previous two approaches are unified by these concepts. The UI-driven approach with frameworks for front-end subsystems is considered as the special case that a part of abstract forms are actually visual forms for interaction between the system and the external world. The model-driven approach with visual modeling tools for back-end subsystems is considered as the special case that the message flow is used instead of the form flow as the workflow. That is, cooperative work at an office is expressed by using a form flow model with the abstract forms.

5 Abstract Forms and Transformation

5.1 Form Transformation in XSLT

The best solution is that end-users can get application software by form definitions and form-to-form transformation definitions. An application which generates individual examination schedules for each student has been selected for applying our solutions to practical Web service integration. Actually, the university supports the individual portal sites for each student. The student gets the examination schedule in PDF and the individual timetable for classes in HTML. In our experiment, an actual examination schedule in PDF can be transformed into an XML document manually.

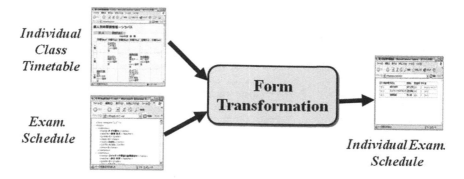

Fig. 4 Form transformation for Web service integration

The target application generates an individual examination schedule for each student from the individual timetable for classes and the examination schedule. The form transformation is shown in Fig. 4.

One input is the individual timetable for classes in HTML which is extracted from the individual portal site for each student. This document includes information about subjects for each student, that is, subject names, instructor names and class numbers. This HTML document is transformed into an XML document by using the wrapping technology. The other input is the examination schedule in XML, which includes information about subject names, instructor names, dates and periods, and room numbers. These two XML documents are merged into the individual examination schedule in XML format for each student.

This individual examination schedule in XML is transformed into an HTML document which can be displayed on the Web browser of each student. There are some conventional tools used for this transformation. The XSLT stylesheet for this application is generated by using one of the conventional tools.

The key technology of this system is the form-to-form transformation from two XML documents into an XML document. The system administrator of this application is not an IT professional but a clerk in the university office. Such an end-user does not have the ability to perform programming, but needs to modify the system when the inputs change.

For the solution of this problem, basically, the procedure of this application is described in a script language. Furthermore, a visual tool supports the end-user. The system generates the XML document by extracting classes which are included in the both input files. The early opinions on this approach are described in detail in [6].

5.2   Form Transformation by Mapping

One solution to the problem of the form transformation in XSLT is the form transformation by mapping from input forms to output forms. The end-users do not need

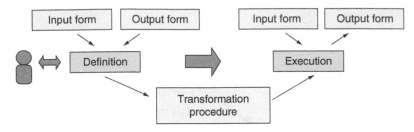

Fig. 5 Form transformation by mapping

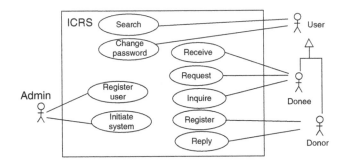

Fig. 6 Use cases of the system for the reuse site

to learn XML and XSLT technologies since they can define the form transformation procedure by only mouse manipulations to relate items in input forms to items in output forms. After the definition of this procedure, the form transformation from input forms into output forms is executed as shown in Fig. 5.

For this study, a Web application for the reuse of laboratory equipment was selected. In the School of Science and Technology which we belong to, a lot of secondhand equipment such as PCs are thrown away although many of them can still be used. If the reuse site is open, the available but unnecessary equipment is registered there and someone can find and receive the reusable equipment easily. Therefore, our end-user initiative requirement definitions method is applied to such a system, ICRS (the Ikuta Campus Reuse System).

Main rules for this system are as follows. Users are limited to members who have mail addresses which are managed by the university, that is, teachers, officers, students etc. for security check. The equipment to be given should be free for avoiding illegal dealing of the university property. The site administrator takes no responsibility for any troubles since this site is supported by volunteers.

Main functions are described in the usecase diagram of UML as shown in Fig. 6. The user actor is the superclass of both the donor actor and the donee actor. The donor registers the unnecessary equipment, or replies to the donee. The donee receives the equipment for reuse, requests the necessary equipment, or inquires about the registered equipment. The user searches a list of the registered equipment, or changes his/her password. The administrator registers a user, or initiates the system.

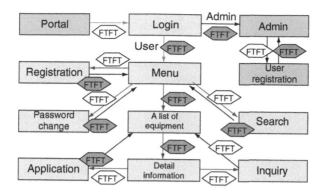

Fig. 7 UI transition and two types of FTFT

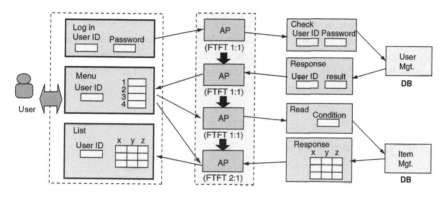

Fig. 8 A part of form-to-form transformations and database accesses

The user interfaces (UI) and the UI transition diagram are designed as shown in Fig. 7. The business logic is specified by the form-to-form transformation (FTFT) with abstract forms. In this figure, two types of FTFT are discriminated. FTFT with DB accesses via abstract forms is marked with the FTFT by a gray hexagon and implies that a transformation between abstract forms or between abstract forms and visual forms are performed before moving onto the next user interface with a visual form. On the other hand, FTFT between visual forms is marked with the FTFT of a white hexagon and implies a transformation between visual forms.

Figure 8 shows a part of form flows and form-to-form transformations. The three forms of left-hand side are visual forms for actual user interfaces corresponding to Login, Menu and List windows. The four forms of the right-hand side are abstract forms for end-users support, which are not displayed visually at the application execution time. Such an abstract form is used under construction of an application by end-users. M:N of FTFT implies the transformation from M input forms to N output forms. First three transformations are 1:1 and the last transformation is 2:1.

The Login form is transformed into the abstract form of 'Check' and it is sent to the user management DB. Next, the abstract form of 'Response' is transformed

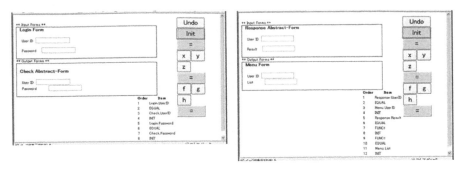

Fig. 9 Examples of FTFT by using a visual tool

into the Menu form for selection of the next operation from a display of a list of the registered equipment, registration of unnecessary equipment, search of registered equipment or password change. If the user selects a display of a list of the registered equipment, the menu form is transformed into the abstract form of 'Read' and it is sent to the Item management DB. Then two inputs of the Menu form and the Response form are merged and transformed into the List form.

5.3   A Visual Tool for FTFT

A tool for defining the form-to-form transformation was developed. The user interface was implemented in HTML and JavaScript. The generated procedure in XML is sent to the server and stored there. The interpreter of this procedure in XML was implemented in Java.

Figure 9 shows examples of the form-to-form transformation. The input form and the output form are displayed on the left-hand side. The palette with buttons for operation items is displayed on the right-hand side. Whenever a column of forms or an operation item of the palette is clicked, the order and the name of the clicked item are displayed below for confirmation.

The transformation from the Login form into the Check abstract form in the left side of Fig. 9 is defined as a sequence of operations: {1 Login.UserID; 2 EQUAL; 3 Check.UserID; 4 INIT; 5 Login.Password; 6 EQUAL; 7 Check.Password; 8 INIT}

The first four operations define that the value of the User ID column in the input form is copied into the User ID column in the output form while clicking the mouse button in order of {Login.UserID, =, Check.UserID, INIT}. The INIT operation implies the initialization as the previous execution result is not used. The following four operations define that the value of the Password column in the input form is copied into the Password column in the output form. This example is very simple.

The transformation from the Response abstract form into the Menu form in the right side of Fig. 9 is defined as a sequence of operations: (1 – Response.UserID;

2 – EQUAL; 3 – Menu.UserID; 4 – INIT; 5 – Response.Result; 6 – EQUAL; 7 – FUNCf; 8 – INIT; 9 – FUNCf; 10 – EQUAL; 11 – Menu.List; 12 – INIT).

The first four operations define that the value of the User ID column in the input form is copied into the User ID column in the output form. Next, one of the functions of f, g and h is used for complex business logic. That is, the following four operations define that the value of the Result column in the input form is the input of the f function while clicking the mouse button in order of {Response.Result, =, f, INIT}. The last four operations define that the output of the f function is assigned to the List column in the output form likewise. The body of the function, f, will be implemented in a scripting language later. The variables of x, y and z are used for temporary stores of execution results.

6 Conclusion

The end-user initiative requirement definitions are necessary for developing application software for a sustainable society. This paper described the requirement definition method based on abstract forms since business professionals are familiar with visual forms. The business logic can be defined as the form transformation from input forms into output forms. Our experiments confirmed the effectiveness of this approach.

References

1. The Apache Software Foundation, Struts, http://struts.apache.org/.
2. Brancheau, J.C., & Brown, C.V. (1993). The management of end-user computing: status and directions. *ACM Computing Surveys, 25*(4), 437–482.
3. Brown A.W. (Ed.) (1996). *Component-based software engineering*. Los Alamitos, CA: IEEE Computer Science Press.
4. Chusho, T., Ishigure, H., Konda, N., & Iwata, T. (2000). Component-based application development on architecture of a model, UI and components. *Proceedings of the APSEC2000*, pp. 349–353. IEEE Computer Society.
5. Chusho, T., Tsukui, H., & Fujiwara, K. (2004). A Form-base and UI-driven approach for enduser-initiative development of web applications. *Proceedings of the Applied Computing 2004*, pp. II/11–II/16. IADIS.
6. Chusho, T., Yuasa, R., Nishida, S., & Fujiwara, K. (2007). Web service integration based on abstract forms in xml for end-user initiative development. *Proceedings of the 2007 IAENG International Conference on Internet Computing and Web Services (ICICWS'07)*, pp. 950–957. Hong Kong.
7. Cotterman, W.W., & Kumar, K. (1989). User cube: a taxonomy of end users. *Communications of the ACM, 32*(11), 1313–1320.
8. Crnkovic, I., et al. (2002). Specification, implementation, and deployment of components. *Communications of the ACM, 45*(10), 35–40.
9. Elfatatry, A. (2007). Dealing with change: components versus services. *Communications of the ACM, 50*(8), 35–39.

10. Fayad, M., & Schmidt, D.C. (Ed.) (1997). Object-oriented application frameworks. *Communications of the ACM, 39*(10), 32–87.
11. Hirooka, N. (2005). Nagasaki prefecture (in Japanese). *Nikkei Computer*, No.2007.7.25.
12. Jacobson, J., et al. (1999). *The unified software development process*. Reading, MA: Addison-Wesley.
13. Jennings, N.R. (2001). An agent-based approach for building complex software systems. *Communications of the ACM, 44*(4), 35–41.
14. Larman, C. (2002). Introduction to object-oriented analysis and design and the unified process. Prentice-Hall, USA.
15. Lepouras, G., Vassilakis, C., Halatsis, C., & Georgiadis, P. (2007) Domain expert user development: the smartgov approach. *Communications of the ACM, 50*(9), 79–83.
16. Lieberman, H. (Ed.) (2000). Special issue on Programming by example. *Communications of the ACM, 43*(3), 72–114.
17. Margaria, T. (Ed.) (2007). Guest editors' introduction:service is in the eyes of the beholder. *IEEE Computer, 40*(11), 33–37.
18. Nano, O., & Zisman, A. (Ed.) (2007). Guest editors' introduction: realizing service-centric software systems. *IEEE Software, 24*(6), 28–30.
19. OMG, OMG Model Driven Architecture, http://www.omg.org/mda/.
20. OMG, Unified Modeling Language, http://www.uml.org/.
21. Ozsoyoglu, G., & Wang, H. (1993). Example-based graphical database query languages. *IEEE Computer, 26*(5), 25–38.
22. Sutcliffe, A., & Mehandjiev, N. (Guest Ed.) (2004). End-user development. *Communications of the ACM, 47*(9), 31–32.

Chapter 37
A Dynamic Nursing Workflow Management System: A Thailand Hospital Scenario

Nantika Prinyapol, Sim Kim Lau, and Joshua Poh-Onn Fan

Abstract In this chapter, we propose the use of dynamic compilation of web services to support workflow management using a hospital scenario in Thailand. Web service based on work practices of nursing system in Thailand is discussed. A dynamic platform for workflow management (DPWFM), which integrates decision making process of workflow management by allowing nurse supervisors to customize workflow requirement is proposed.

Keywords Workflow management · Web services · Work practices

1 Introduction

Recent advancements in information technologies have resulted in business organizations investing in innovative ways of delivering products and services. However, business organizations still face with problems of developing working schedule and managing resources effectively although huge investment has been made in expensive tools and sophisticated software [1]. Very often enterprise software dealing with workflow management does not meet specific requirements of business organizations, as a result end users encounter difficulty in using the systems [3]. In addition, attempting to modify business rules to cater for changes in business environment often result in high maintenance cost [2].

In this chapter, we propose a Dynamic Platform for Workflow Management (DPWFM) system. The proposed system employs dynamic compilation of web services to support workflow management. It integrates decision making process with

N. Prinyapol (✉) and S.K. Lau
School of Information System & Technology, Faculty of Informatics, University of Wollongong, NSW 2522, Australia
e-mail: np926@uow.edu.au; simlau@uow.edu.au

J.P.-O. Fan
Sydney Business School, University of Wollongong, NSW, Australia 2522
e-mail: joshua@uow.edu.au

S.-I. Ao et al. (eds.), *Intelligent Automation and Computer Engineering*,
Lecture Notes in Electrical Engineering 52, DOI 10.1007/978-90-481-3517-2_37,
© Springer Science+Business Media B.V. 2010

scheduling to reduce complexity of managing workflow. The main feature of the DPWFM is the flexibility it offers to users to allow customization of services that fit into their workflow requirements. We illustrate the working model of the DPWFM using a hospital scenario in Thailand. The workflow patterns were analyzed based on several reputable hospitals in Bangkok, Thailand through interviews conducted with heads of nurses and nurse supervisors from public as well as private hospitals. The interviews provide valuable insights into workflow practices of nursing in Thailand.

The rest of the chapter is organized as follows. In Section 2, we discuss the DPWFM platform architecture. Section 3 describes work practices of nurses in Thailand. This is followed by discussion of the proposed DPWFM as applied to the Thailand's nursing system in Section 4. Finally, conclusions and further study are presented in Section 5.

2   The DPWFM Architecture

The main feature of the DPWFM system is the capability of customizing workflow to meet organizational requirements. The DPWFM integrates decision making process of workflow management with scheduling. The workflow web applications are designed to describe business processes, tasks and business functions. Selecting appropriate web services related to work specifications of employees are based on criteria set by supervisor, job descriptions of the task, process enactments, or any other relevant requirements that may be related to real-time processes.

Table 1 shows four web service repositories of the DPWFM architecture: work profile service (WPS), function service (FS), function allocation service (FAS) and scheduler service (SS). In the table, we have used the nursing case study to illustrate instances of each repository. Firstly, the WPS and FS are organized: the WPS stores information related to work profile such as job description, work qualifications, job duties and responsibilities; the FS stores task descriptions associated with each task, business process or function. Information stored on the WPS and FS are often preset prior to the deployment of the DPWFM system. FAS, the third component, deals with work assignment for a particular instance of a time period, such as a shift, a day or an appropriate time period of work. The work assignment is entered to the FAS by the supervisor. Then work schedule and workflow assignment are produced by the SS which compiles the workflow and work schedule by matching work assignments stored in the FAS and information gathered from the WPS and FS.

Templates can be designed and used for routine assignments. This allows standard sets of web services and work specifications to be used. The supervisors of each functional unit will take full control in customizing the workflow schedules and deliverables require for each of their subordinates. In addition, the FAS can also allow changes and modifications upon requests from the supervisor. This will provide more flexibility and better control in organizing work schedules in the functional unit.

Table 1 Four web service repositories of DPWFM

Service repository	Characteristics	Example of instances
Work profile service (WPS)	Describes and stores career positions, job description, main responsibilities, work qualifications, work experiences, routine tasks, minor tasks, extra tasks, and ad hoc tasks	'Doctor', 'nurse', 'medical nurse', 'in-charge nurse', and 'pharmacist'
Function service (FS)	Manages and analyzes web services from business processes	'Writing report' main function has sub-functions such as 'recording vital signs', 'fill-in dose of drug', 'fill-in frequency of drug provided', and 'special requirements'
	Provides web service descriptions and communicates information Collects main functions and their sub-functions.	'Recording vital signs' main function has sub-functions such as 'record weight', 'record blood pressure', 'record respiration rate', and 'record temperature'
Function allocation service (FAS)	Gathers assigned functions based on supervisor's decision	Services that are assigned to each subordinate
Scheduler service (SS)	Schedules and customizes appropriate tasks	Unique web pages containing selected scheduler services for each user

In the case of nursing workflow, these web services were designed to comply with organizational work practice requirements during the design-build-store process. Finally, the FAS module contains the desired job functions that allocate to every individual medical nurse in the unit. At run-time, nurse supervisors in each ward or department can assign new tasks or modify previous assignments when it is necessary. The system, as a web service application, will recompile the changes into the best schedule based on working functions required and provide a newly-arranged customized task to individual staff.

Figure 1 shows the entities designed for the DPWFM. The WPS entity provides the staff profiles that will be assigned to the SS entity. The SS entity has links to the FS and FAS entities. The FS entity has relationship with the Function Task entity which contains the Function Sub-tasks entity.

3   Background of Thailand Hospital

According to the Thai beliefs and cultures, Thai people held the government hospitals in high esteem compared to the private hospital in term of reputations. Doctors and nurses who graduated from the top government universities or health collages

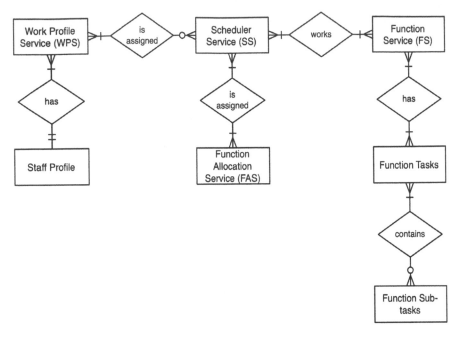

Fig. 1 ERD of DPWFM

have more opportunities to work in the government hospitals. Consequently, the doctors in the government hospitals, generally, have higher reputations in terms of quality in their specialist fields and experiences, compared to the doctors who work in the private hospitals. However, the technologies and the administration computer systems of the private hospitals are more advanced than the government hospitals.

The management of the private hospitals controls their organizations as a business company. They tend to invest more on high medical technologies and equipments as well as hiring more senior and professional doctors and nurses from the government hospitals to work in their hospitals on a part-time, or even, full-time basis. Therefore, there are significant differences between the private and government hospitals in Thailand in term of hospital management system, fees and sources of their financial support. The government hospitals obtain their funding mainly from the government, while the private hospitals need to generate their own incomes. As a result, the computer systems in the private hospitals for nurses are more advanced compared to those in the government hospitals, where the computers are mainly used in for financial applications, such as for billing. Nevertheless, the use of the computer systems is also restricted within the Thai health regulations because doctors and nurses are still required by law to sign-off and identify themselves using their own signatures and hand-writings. This results in limitation of using electronic or computer system in patient care.

3.1   General Workflow in Thailand Hospital

In general, the Thai nursing care system consists of head nurse of each ward or in-charge nurse of each shift is assigned to be a nurse supervisor who directly communicates with the doctors. Once the doctors order the medical receipts and set the treatment plan for each patient, the nurse supervisor will distribute the doctor's order to each nursing staff.

Generally there are three shifts of nurses working in a day, both in the government as well as private hospitals. Normally a shift consists of 8 h. The first shift is called the 'morning shift', which starts from 7 am in the morning until 3 pm in the afternoon. Then the 'afternoon shift' starts from 3 pm until 11 pm. This is followed by the 'night shift' which starts from 11 pm until 7 am the following morning. The nurse supervisor usually set a time, called 'transferring time', to hand-over patients' information (the patient is referred to as 'case' in nursing term) and treatment notes between shifts. This means the medical nursing staffs in the following shift have to commence their work earlier, about 15–30 min, to meet with the nurses in the current shift. For example, the nurses in the 'afternoon shift' should come to the ward at 2.45 pm and receive the case information from the owners of the cases (owner here refers to the nurse who is in-charge of the case).

Usually a case is assigned to one nurse, who is called 'primary nurse', who acts as the owner of the case. If the patient gets well and discharged from the hospital within the primary nurse's shift, then the information related to the discharged patient will not be transferred to the nurse in the following shift, although it is still kept in the case's profile. In which case, the nurse will be assigned another new case. On the other hand, if the patient is not discharged, then the primary nurse will pass information to the nurse in the following shift and the assigned primary nurse should also follow the treatment plan as noted in the treatment note. This way of transferring information during the *transferred time* between shifts is advantageous as face-to-face communication is performed and notes are transferred accordingly.

Figure 2 shows a general workflow between doctors and nurses. First, the primary doctor diagnoses a patient and then records the treatment procedures, drugs prescribed and correct doses of medicine in a paper worksheet called 'chart' or 'order' in nursing term, which is attached with an aluminum clipboard as shown in Fig. 2 step (1). The doctor then passes his orders to the in-charge nurse. The in-charge nurse then organizes the new case to a medical nurse who will be assigned as the primary nurse for this new case or patient.

In Fig. 2 step (2), the in-charge nurse rewrites the original orders of the doctor into the *nurse notes*. Once the medical nurses receive the *nurse notes* from the in-charge nurse, they will treat and care for their respective patients/cases as shown in steps (3) and (4) in Fig. 2. The medical nurses will record all vital signs such as weight, blood pressure, respiratory rate, and body temperature, monitor and make a note of every treatment outcome of the cases in their *nurse notes*. These *nurse notes* will be handed over to the nurse in the following shift.

During each shift, the pharmacists also have to dispense drugs to the medical nurses for each case based on the in-charge nurse order chart. Finally, the medical

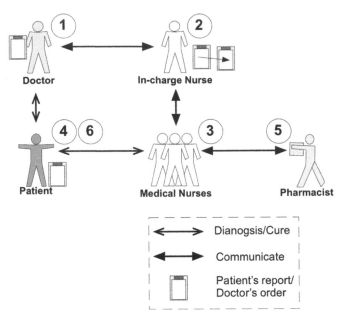

Fig. 2 Workflow in nursing care

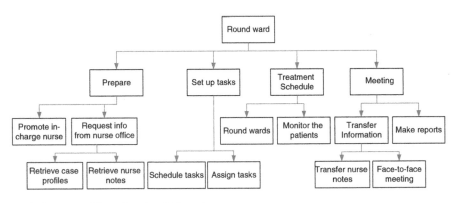

Fig. 3 Nurse workflow: functional hierarchy

nurses give the medicines or drugs to the patient in the frequency rearranged by the in-charge nurse as shown in steps (5) and (6) in Fig. 2. The nurse supervisor usually works only in the morning shift therefore, in the afternoon and night shifts, the most senior medical nurse in the ward during that shift will be assigned as an in-charge nurse (Fig. 4).

We will apply the DPWFM platform to the above nursing care process. Figure 3 shows the workflow of the nurse supervisors, which include organizing everyday works, assign and schedule tasks for nursing staff, design treatment plan and conduct meeting. When the morning shift finishes, the nurses working in the afternoon shift arrive to exchange work in the ward, information of patients and cases are

Fig. 4 Promote in-charge
nurse

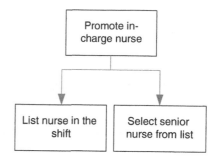

transferred to the medical nurses in the afternoon shift via meeting that takes place during the *transferred time* (Fig. 4). The process is repeated between the afternoon and night shifts, as shown in Fig. 5. This procedure is repeated every day in a similar manner.

In general the major shift is the morning shift, which often consists of more staffs to handle the enormous amount of patients in a day. The night shift often requires fewer medical nurses because the patients are generally asleep in the night. Thus the nursing care in each shift can be different; however, the workflow of the nurses is basically the same. Figure 5 shows the nurse supervisor has to rearrange and organize the staff tasks manually in each shift. It is worthwhile pointing out that patients' reports are handed over to the primary nurse in the next shift in the form of paper nurse notes and through face-to-face verbal communication during the transfer meeting.

3.2 Case Study

We have conducted personal interviews with six nurse supervisors who work in different In-Patient Department (IPD) of three famous hospitals (one public hospital, one private hospital, and one semi-government hospital) and three medical nurses from other IPD wards of two government hospitals in Bangkok, Thailand. The interview was conducted during the month of February 2009.

Based on the interviewing results, we have found that the workflow in these hospitals is similar to the one that we described in Section 3.1. However, due to funding sources available to these hospitals, there exists gap in managing human resource and computer technologies. Due to the nature of fees charged in the public hospitals, which is generally less expensive, there are many patients in the government hospitals. Thus each medical nurse in each shift generally is assigned a higher number of patients in one ward. For example, one medical nurse has to take care of eight to nine patients. Thus, a ward that consists of 32 beds will only have 4 medical nurses and 4 assistant nurses in the morning shift, 3 medical nurses and 3 assistant nurses in the afternoon and night shifts. In the government hospital, computer systems are used to record patients' details (such as name, insurance or concession) with the purpose of billing. The nursing care system is still being done manually.

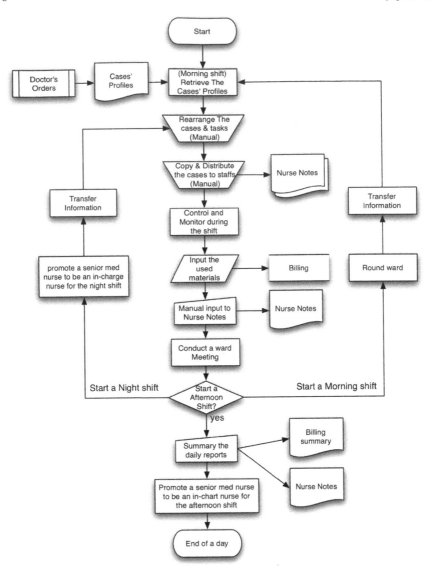

Fig. 5 Workflow of a nurse supervisor

Private hospitals often focus on superior service provided to their patients. The patients are charged for more expensive overheads compared to the government hospitals. It is common that nurses from the public hospitals also work in the private hospitals on a part-time basis. Generally, the nurses in the private hospital do not have to take care of as many patients as their government hospital counterparts. On the other hand, they are expected to provide a higher quality of care to their patients. The semi-government hospitals refer to hospitals that are funded by the government but managed as a private sector. Although this kind of hospital is managed as a private hospital which results in more efficiency, it is constrained by the funding source which is from the government.

The hospitals are equipped with computer information system. However the computer systems are mainly used for patients' profiles, human resource, inventory control, and financial management. They are not used for patients' treatment care. The level of competencies in using new information technologies is considered to be 'well'. This is because most of the younger nurses have gone through either college or university educations while the senior nurses have undertaken training of using information technology as a result of continuing career development. As a result, the nurses feel 'comfortable' in learning or retraining for new technology if that can help them in their daily work. Nurses in Thailand often prefer 'face-to-face' communication between staffs and supervisors, followed by 'written instructions' on the nurse notes or reports. However, the nurses also think that e-mail can be effective if a message or news need to be broadcasted to the entire ward.

During the interview conducted with the nurse supervisors and medical nurses, it is perceived that the proposed dynamic platform is useful if it reduces their workload. The nursing staffs, in particular from the semi-government hospitals, are more open-minded with the new technologies. Generally nurses 'feel comfortable' in using new technologies at work if they are required to do so. For instance a computer-based nursing system called 'Trend Care System' has been installed in the private hospital. The 'Trend Care System' is a nurse management system from Australia. One module of the Trend Care System is to allocate the number of nursing staffs in each day. As the software was originally being designed for the private and public hospitals in Australia and New Zealand, the nurses in Thailand found that they require more training in using the system.

The nurses also pointed out that human error is the main problem in their daily work, follow by equipment and communication errors. A majority of nurses think that new technologies should be able to help them to reduce errors and improve efficiency in carrying out their daily work. However, the nurses interviewed also indicated that the introduction of new technology can increase the stress levels in their work due to changes in work practices.

4 Proposed DPWFM

The proposed DPWFM platform is developed using Apache as the web server, MySQL as the database server, PHP, and AJAX as web design languages. It will be implemented as web services to facilitate nursing care functions using dynamic re-compilation of workflow web application. Information on the application will be displayed in XML/HTML format. The purpose of developing the DPWFM using web services technology is to provide a standard platform for interoperability among different software and applications. In addition, web services are best suited to work across multiple platforms that can help to reduce investment costs of chang-

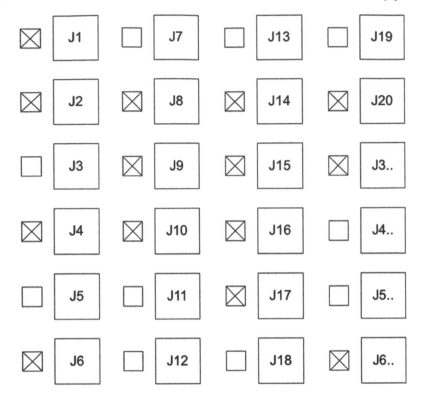

Fig. 6 The alternatives function services (FS)

ing infrastructure [4]. Generally, the government hospitals in Thailand have limited financial support for computer system implementation. Therefore, the tools and techniques of the proposed DPWFM platform are available as open source which can be used freely for development. In this case, it will suit sector such as the government hospitals in Thailand as minimum investment outlays are required (Fig. 6).

Web services stored in the WPS repository are assigned by nurse supervisors to their subordinate nurses. Each WPS has the set of FS, which has inter-connection to each other. The priority of each FS was set according to the nursing workflows. Once the nurse supervisors assign the jobs to their nursing staffs, the platform will re-compile all FS to provide optimal scheduling tasks (in form of arranged SS) to an individual staff.

Figure 7 shows an example of preparing the proper sequences of WPS list into individual worksheets, which are unique web pages of each nurse staff. In the beginning, a nurse staff has to complete the group of high priority tasks (or pre-requisite tasks), and then, allowed to do the rest of the tasks, which are of lower priority until all jobs are completed. The priorities are determined according to the business process of a particular nurse work practice. The sequence of each FS has setup and link to WPS during the analysis process. The workflow recompiles the routine schedule

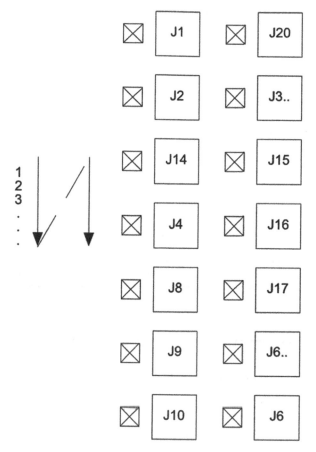

Fig. 7 The scheduled services (SS) allocate a new set of function service (FS)

tasks for the nurse staffs. It not only helps individual staff to complete the daily tasks, but also provides simple work flow transactions in the ward.

The DPWFM aims to design the friendly user interface (UI) to cater for all level of user skills. The nurse supervisor can allocate the function services (tasks) to each medical nurse by simply selecting the graphically display system report that provides all function services and work profile services. The UI designs have incorporated common widgets, such as checkboxes, textboxes, drop-down lists and buttons into the nurse supervisor's web pages as shown in the Fig. 8.

The individual web pages of the nursing staffs are generated by recompiling the function services with suitable sets of work sequences in the worksheet styles. The pre-requisite tasks have been set by the nurse supervisors. This way the medical nurses simply have to follow the sequence of jobs as determined by supervisor's decision.

In Fig. 8, there are three sections of the tasks/projects: urgent, ad hoc and common tasks. Each section demonstrates the importance and priority of the tasks. If a

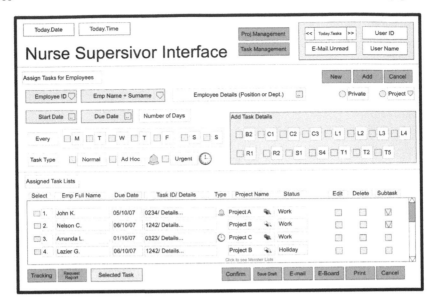

Fig. 8 The interface design for a nurse supervisor

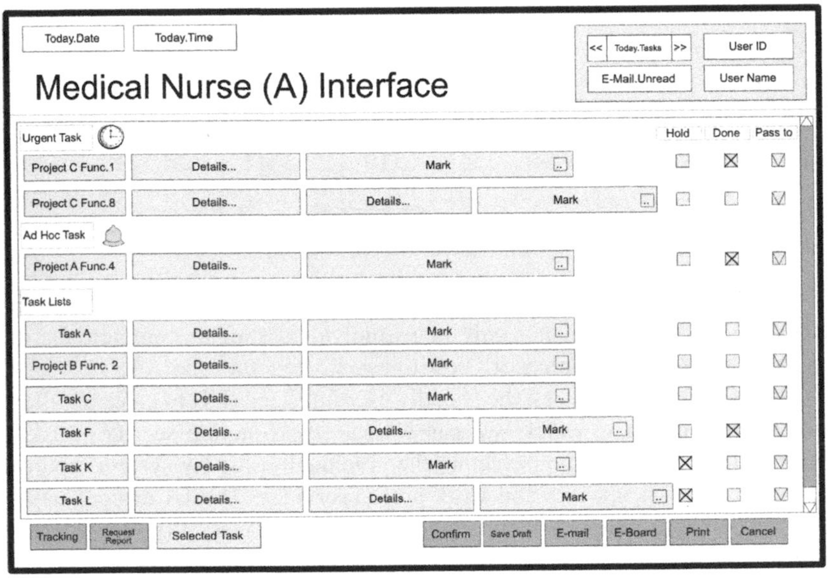

Fig. 9 The interface design for a medical nurse

task has some form of pre-requisite task, then the system will send the alert messages to let the nurses know which tasks should be completed first. Finally when the medial nurse (A) logs in to the system, she will see the overall tasks for the day. As soon as she completes the task, she can click the 'submit' button and her supervisor will be able to view the medical nurse (A)'s report (see Fig. 9).

5 Conclusions and Further Study

We have described the DPWFM, a platform for supporting workflow practices that provides the flexibility, expressivity necessary for the nurses in Thailand. The DPWFM provides dynamic compilation of web services that were customized by supervisors. In general the DPWFM aims to satisfy the requirements of customization of workflow through the process of recompilation via a friendly user interface design. We have illustrated four main web service repositories using a scenario of ward management. The mundane task of organizing workflow for individual medical nurses by the nursing supervisors is achieved through the DPWFM. Through the discussion with the nurses in Thailand, it shows that the Thai nurses are open-minded with new technologies. This is good news from the deployment of the DPWFM system, as we expect less resistance from the nurses. The next phase of our research will investigate issues related to implementation and deployment of the DPWFM to actual work setting in Thai hospital.

References

1. Chen, J., He, D., & Anquan, W. (2001). *e-Commerce and Innovation Business Process Reengineering in the Portland International Conference on Management of Engineering and Technology (PICMET '01)*. Hangzhou, China: Coll. of Manage, Zhejiang University.
2. Chukmol, U. (2008). A Framework for Web Service Discovery: Service's Reuse, Quality, Evolution and User's Data Handling in the 2nd SIGMOD. PhD Workshop on Innovative Database Research. Vancouver, Canada.
3. Grimes, B., & Jackson, J. (2009, March 2). Google: Enterprise Software too Difficult to Use [Online]. Available: http://gcn.com/Blogs/Tech-Blog/2006/05/Google-Enterprise-software-too-difficult-to-use.aspx.
4. W3C. (2007, November 12). Web Services Architecture [Online]. Available: http://www.w3.org/TR/2004/NOTE-ws-arch-20040211/.

Author Index